Sanjan Reports Volume 1

Mariners and Merchants: A Study of the Ceramics from Sanjan (Gujarat)

Rukshana J. Nanji

BAR International Series 2231
2011

Published in 2016 by
BAR Publishing, Oxford

BAR International Series 2231

Sanjan Reports Volume 1
Series Editors: Dr S. P. Gupta (deceased), Dr Homi Dhalla and Mr K. N. Dikshit

Mariners and Merchants: A Study of the Ceramics from Sanjan (Gujarat)

ISBN 978 1 4073 0793 0

BAR Publishing is the trading name of British Archaeological Reports (Oxford) Ltd. British Archaeological Reports was first incorporated in 1974 to publish the BAR Series, International and British. In 1992 Hadrian Books Ltd became part of the BAR group. This volume was originally published by Archaeopress in conjunction with British Archaeological Reports (Oxford) Ltd / Hadrian Books Ltd, the Series principal publisher, in 2011. This present volume is published by BAR Publishing, 2016.

Printed in England

BAR
PUBLISHING

BAR titles are available from:

BAR Publishing
122 Banbury Rd, Oxford, OX2 7BP, UK
EMAIL info@barpublishing.com
PHONE +44 (0)1865 310431
FAX +44 (0)1865 316916
www.barpublishing.com

FOREWORD

The World Zarathushti Cultural Foundation (WZCF), Mumbai and the Indian Archaeological Society (IAS), New Delhi are pleased to present the final report of the Sajan excavations. This report on the ceramic assemblage is the first in the series. Other volumes, similarly authored by specialists dealing with other aspects of the excavations, are now ready for publication and will follow this volume in the near future.

WZCF has been actively involved with the research and conservation of Parsi Zoroastrian heritage in India. It was Dr. Homi Dhalla, the President of the WZCF who initiated the Sanjan excavations. The Parsis of India are descendents of Zoroastrian Persians who are known to have migrated to the shores of India from Iran in the Early Medieval period. Rich oral traditions and literary sources describe this historical event, mentioning Sanjan as the earliest such settlement on Indian soil. The *Kisse-i-Sanjan* or Story of Sanjan, a Persian poem written in 1599 A.D., is a prime source for the history of this settlement. However, the veracity of this document and the oral traditions of this community had never been archaeologically established. It was in 2000 that the Indian Council of Historical Research granted the project "Historical and Archaeological Study of Parsi / Zoroastrian Religious Places of Importance in Western India, with special reference to Bahrot Caves" to Dr. Homi Dhalla. The excavations at Sanjan (2002-2004) were undertaken as part of this endeavour and were financed by the Indian Council of Historical Research, the Archaeological Survey of India and the Sir Dorabji Tata Trust, through WZCF.

The Sanjan excavations were undertaken by the WZCF in collaboration with the IAS, (Late) Dr. S. P. Gupta, Chairman of IAS, was the Director of Excavations, providing guidance and direction to the project with his rich experience and expertise, until his unfortunate dennse in 2007. Dr. Gupta recognized the immense importance and relevance of the Sanjan project, not only for understanding Parsi history, but also in the study of Medieval Archaeology in India. It is with some pride that WZCF and IAS can claim to have broken new ground and to have conducted the first Early Medieval excavations in India. We also wish to express our gratitude to (Late) Dr. Mani Kamerkar for her contribution to this project.

This is a period of history that has been unfortunately ignored in Indian Archaeology. The Sanjan excavations have generated interest and opened up a whole new approach in the discipline, so that several projects focusing on Indian Ocean trade and the Early Medieval period are now being undertaken by archaeologists and historians in India. The Sanjan excavations have not only succeeded in providing a fresh perspective to Parsi / Zoroastrian history, but have also provided evidence for a prominent, multi-cultural, commercially active port settlement which formed an important component of the Indian Ocean trade network of the Early Medieval period. The excavations brought to light a rich body of ceramics, artefacts, human remains, structures, botanical and faunal remains, all of which have been subjected to detailed study by experts and specialists.

The ceramic assemblage at Sanjan is one of the most important collections of the Early Medieval period to be found in India to date. This body of material, comprising of West Asian, Chinese and Indigenous wares, was completely unknown in the Indian context although scholars abroad had long been working on similar collections. The lack of corroborative material from India was regarded as an impediment in the understanding of Indian Ocean maritime trade and cultural interactions. No parallel collection or published reference material was available in India for study. Although preliminary analysis was undertaken in the first two years of excavation, it was found to be completely invalid.

Dr. Rukshana Nanji who worked as Ceramic Specialist and Field Archaeologist on the Sanjan project undertook the difficult task of reanalysing, cataloguing and interpreting the entire collection towards her doctoral thesis, which is also the final ceramic report of the Sanjan excavations. Her meticulous and detailed data collection and analysis is the first of its kind in India and has been welcomed by scholars working in the Indian Ocean littoral who have long rued the lack of information from the sub-continent. Originally titled *"The Study of Early Medieval Ceramics in India, with Special Reference to Sanjan (Gujarat)"*, her thesis was submitted to the Deccan College (Deemed University), Pune (India) towards her doctoral degree. Dr. Nanji has produced the first fully documented, researched and analysed work on Early Medieval ceramics in India, despite several limitations. In many ways, this report is path-breaking. No other Early Medieval ceramic collection from India has been catalogued and studied in such detail, including in its methodology, fabric analysis, petrographic analysis and statistical data.

WZCF and IAS are happy to bring this report to the attention of the academic world. Other volumes in the series are to follow. The publication of the final reports will certainly put Sanjan on the archaeological and historical map of the world.

Dr. Homi Dhalla
World Zarathushti Cultural Foundation,
Mumbai (India)

Mr. K. N. Dikshit
Indian Archaeological Society,
New Delhi (India)

AUTHOR'S NOTE

Sanjan was excavated between 2002 and 2004 when the Early Medieval period was still a neglected area in Indian archaeology. This situation is now marginally improved. But a lot of work is still required in the study of this period in India. The study of the Sanjan ceramics was especially difficult as there was neither an archaeological precedent to follow nor any available published data for reference in India. Preliminary analysis, done along with Dr. A Dandekar in the initial years of excavation was, unfortunately, found to be almost completely invalid, both in methodology as well as classification. My doctoral thesis "The Study of Early Medieval Ceramics in India, with Special reference to Sanjan (Gujarat)" was thus conceived and developed as the final ceramic report for the WZCF Sanjan excavations based on my own research. I sincerely thank all the institutions and individuals who made this study possible by offering me their valuable inputs. It is for the first time that a detailed analysis of Early Medieval ceramics has been undertaken in India. In many ways, this may be seen as a starting point. The conclusions of this study are entirely my own. I am aware of the many mistakes and flaws in this endeavour, all of which I take responsibility for. Despite it's limitations, this study needs to be published so that the long-felt need for data on Early Medieval ceramics in India may be addressed. As the first volume of the final Sanjan Excavation reports, it is hoped this aim is achieved.

A paucity of space does not permit the inclusion of colour photographs of the entire Sanjan collection. These may be viewed at http://www.wzcf.org/sanjan-excavations/pottery.html.

ACKNOWLEDGEMENTS

The Sanjan excavations were carried out between 2002 and 2004 by the ***World Zarathushti Cultural Foundation*** *with permission granted and financial assistance offered by the Archaeological survey of India under the overall direction of (Late) Dr. S. P. Gupta, Chairman, Indian Archaeological Society, New Delhi. The excavations were part of the Indian Council of Historical Research's project entitled "Historical and Archaeological Study of Parsi / Zoroastrian Religious Places of Importance in Western India with special reference to Bahrot Caves" given to Dr. Homi Dhalla, Founder-President WZCF and (Late) Dr. Mani Kamerkar. Excavation of the Sanjan Dokhma in the year 2004 was financed by Sir Dorab Tata Trust through the WZCF. The excavation team included co-director Dr. Kurush Dalal, Dr. Rukshana Nanji, Rhea Mitra, Dr. Abhijit Dandekar, Rohini Pandey, P. Aravazhi, Suresh Bomble, Shivendra Kadgaonkar, Reza Abbas, Pranab Sharma and Gency Chaudhury. Subject experts Prof. Shobana Gokhale, Dr. Walimbe, Dr. Veena Mushrif-Tripathy, Dr. Pramod Joglekar, Prof. Rajaguru, Dr. Sushma Deo, Dr. Kajale and Dr. Arti Mukherjee made important contributions to this project with their expertise. Illustrations have been done by Mr. Mani, Mr. Laxminarayan, Mr.Jassuram, Dr. Shrikant Pradhan and Mr Karanjkar. Mr. Yazdi Tantra and Mr. K. Ramanathan set up the website. Our thanks and appreciation to all the contributors of this project.*

This ceramic analysis would not have been possible without the help and guidance provided by several institutions and individuals to the author. The WZCF, IAS and the author would like to put on record their gratitude and appreciation to all of them:

Prof. V. D Gogte, (retd.) Deccan College Pune - Research Guide for the author's doctoral thesis.

The Director and staff of Deccan College Post-Graduate and Research Institute (Deemed University) Pune

Prof. K. Krishnan, Head of Department, and Prof. Kuldeep Bhan, Department of Ancient Indian Culture, History and Archaeology, M. S. University, Baroda.

Dr. Derek Kennet, University of Durham

The British Museum – Dr. St. John Simpson, Dr. Sheila Canby and Dr. Roberta Tomber

The Ashmolean Museum, Oxford – Dr. Oliver Watson

The Victoria and Albert Museum – Susan Stronge and Tim Stanley

Dr. Elizabeth Lambourne and Seth Priestman

Bombay Parsi Panchayat and the World Zoroastrian Organisation
This project would not have been possible without the help and support of the Davierwalla family, the workers and the people of Sanjan. Sincere thanks to all of them

The author would personally like to thank several close friends and colleagues who provided both assistance and support. This endeavour would not have been possible without the encouragement and support provided by her family, Veera and Freddy Nanji.

* * *

CONTENTS

* * *

CHAPTER I : INTRODUCTION

SANJAN

"A place in the desert was fixed upon
It was a pleasant spot, and there they fixed their abode.
They found that spot acceptable,
And they raised a city where there had been a desert."

- *Kisse-i-Sanjan (Bahman Kaikobad) 1599A.D*

Sanja'n, north latitude 20° 12′ and east longitude 72° 52′ (*Gazetteer* vol. XIV 1882 (reprint 2000): 301) is located on the western seaboard of the Indian sub-continent. Once considered as part of Thana District, it presently falls under the administrative jurisdiction of Umbargaon Taluka, Valsad District in South Gujarat. Its strategic location on the north bank of the Varoli River with the Arabian Sea a mere 4 to 5 km. downstream, and its simultaneous accessibility to the hinterland, offers it a natural advantage which, by all accounts, it has enjoyed in the past. This part of the north Konkan is a long, flat alluvial strip of land following the profile of the coastline and flanked on the interior by hills. The area, however, is rich in deep and navigable creeks with secure berthing facilities for sea-going, coasting and riverine vessels. Sanjan provides one such workable facility. With the thriving commercial ports of Khambat, Bharuch and Surat to the north and Sopara, Thana, Kalliana and Chaul to the south on the same seaboard, Sanjan has enjoyed a prominent position in the geography of the coast through history. As V.K Jain points out, "the greater importance of the region lies in the navigational creeks on the coast which served as the nucleus of exchange for various types of commodities. The openings through the Sahyadris, i.e., the Western Ghats, which flank Thana in the east by the Tal, the Nana, and the Bor passes have from the beginning of local history caused trade to concentrate in the Thana ports" (Jain 1990: 18). The geo-morphological survey and study carried out by Prof. Rajaguru and Dr. Sushma Deo in 2004 (Rajaguru and Deo 2005: 93 -98) has provided a profile for the site and the surrounding area. They observe that the Varoli River "is a rocky stream originating approximately 20 km southeast of its mouth in the hill ranges (150 – 200m) in district Thane of Maharashtra. It flows through a rocky gorge (10 – 15m deep) in its middle course. A river course, initially east-west abruptly turns south-north and becomes straight for 3.5 km and is probably controlled by north-south lineament, and it changes again to east-west trend in the estuarine portion (i.e., from its mouth near Umbargaon to Railway crossing). The river is tidal 6 to 7 km upstream and drains a flat alluvial fill surface of 10 – 15m thickness." The geological explorations covered the areas of Umbargaon, Nargol, Asan Devi, Dara, and Sanjan itself. An outcrop of Karal or a beach dune complex, dateable to within the Holocene, was observed at Umbargaon. There is a well developed modern sandbar at the mouth of the river at Nargol. A relatively older beach-dune-karal complex was observed at Nargol At Asan Devi, they observe "an isolated outcrop of the alluvial fill (5-6 m) above the modern mud flat occurs as an island in Asan Devi village" and note that the calcretised alluvial fill is a relict fluvial landform of the late Pleistocene age. The alluvial fill near Dara on the left bank of the Tukkad nala, a small tributary of the Varoli contains a sandy silt with potsherds, charcoal and burnt clay. This deposit may be a result of "monsoonal floods strong enough to disturb the medieval site existing in the area" (Rajaguru and Deo 2005: 96). The site of Sanjan is itself situated on an undulating landscape developed on a mud flat formation younger than the alluvial fills or the beach-dune-karal formations. The geologists also note that the peripheral areas of the site have been affected by flood erosion. One of the interesting observations made while studying the topographic map (46 D / 16 – 1884-85) pertinent to this area is that a series of ponds or small lakes are marked along the coast in a north-south orientation between Agar in the south and Saronda in the north. These ponds are located approximately 0.5 to 1 km inland from the coast. This indicates recession of the sea from the lake-line to the present coastline. "It is suggested that the estuarine zone around Sanjan has been affected by moderate sea level fluctuations and by coastal configurational changes. Such changes are seen in the form of strong monsoonal floods, development of the sand bar at the mouth of the creek and sudden variation in the transporting of silt / clay load due to complex relationship between fluvial and marine processes operating in a humid monsoonal estuary." (Rajaguru and Deo 2005: 97). One of the significant observations is that Late Holocene floods may have disturbed the peripheral part of the habitation around the 13th to 14th century - this can be seen in the excavated trenches as a sandy-silty deposit. This may have been caused by a slight increase in aridity at the beginning of the Little Ice Age – a global climatic change. The geological explorations are of great significance in the light of the archaeological explorations undertaken for the purpose of this study and help in providing the context for the data so found.

The hill ranges in this area are rich in woods and timbers, bamboos and teaks. The area in and around Sanjan is dotted with baobab trees some of which have an extremely large girth (*Gazetteer* vol.XIV 1882 (reprint 2000): 302). Date palms, coconut and other tropical trees common to the South Gujarat region can be seen. Local populations speak of timber as an item traded from the Sanjan *Bandar* or port until some decades ago. Rice, inferior cereals, etc are cultivated. The forests provide for other products such as gum, honey and herbs (Jain 1990: 21). A recent introduction to the area is floriculture and the commercial cultivation of roses but this is in a limited area. The average annual rainfall in the area is approximately 200 cm (Jain 1990: 20)

HISTORY OF SANJAN :

The history of Sanjan can be pieced together from the literary, epigraphic and epistolary sources available to us and on the basis of the accounts and records left by various travellers, sailors and geographers.

Literary Sources :

The historical evidence for Sanjan comes from diverse sources such as literary records, epigraphic sources, travel accounts etc. all of which bear testimony to its antiquity and its relevance. However the most important and detailed description of the settlement is found in a Persian narrative poem called *Kisse-i-Sanjan* or the story of Sanjan. Considering that Sanjan is known primarily in legend and history as the first Persian Zoroastrian settlement on Indian soil, this text becomes extremely important as source material for the history of the site. This exhaustive work was written in 1599 AD by a Parsi priest, Bahman Kaikobad Sanjana, in Navsari. He claims that the narrative is based on an oral tradition which has been passed down since generations and that he is only retelling what he has heard from his teacher. Although the poem is titled the Story of Sanjan, in actual fact it covers a much longer time-span and deals with events and issues other than the settlement itself.

The poem begins with the conventional doxological verses wherein the poet calls upon divine help with all humility and asks for guidance in his endeavor to tell the tale. He briefly describes the three great disasters to befall the Zoroastrian faith in ancient Persia – first, the invasion by Alexander and the resurgence of the faith under Ardeshir, the founder of the Sasanian dynasty; second, the decline and chaos that followed this event for a brief period and resurgence again under the rule of Shapur, and; third, the fall of Yazdegard III which brought to an end the Sasanian dynasty and Islamised the Persian empire. This last event triggered the migration of a group of Persian Zoroastrians to the mountains of Kohistan where they resided for about one hundred years. Led by a wise priest, they moved to Hormuz where they spent another fifteen years, after which they sailed for the shores of India. The first place they settled at was the island of Diu, where they resided for nineteen years. After this short stay at Diu, they once again boarded their ships and sailed southwards to the mainland. A great storm befell them and the migrants prayed for safe deliverance, promising to build an *Atash Behram* or a fire temple as thanksgiving. They made landfall at Sanjan where the local ruler, Jadi Rana, granted them permission to reside upon the fulfillment of certain conditions. The Persians thus settled at Sanjan and built a city in the wilderness, naming it after a town in Iran. Some time later the Atash Behram was built and the holy fire, the Iranshah, was consecrated. The settlement thrived and the population grew. After three hundred years, the Zoroastrians spread to other places such as Surat, Bharuch, Variav, Khambat, Navsari, etc. This peaceful sojourn came to an end seven hundred years after the arrival of the migrants, when the armies of Sultan Mahmud attacked Sanjan under the command of his general, Alf Khan. The Zoroastrians fought on the side of their ruler and a fierce battle is described in which fourteen hundred able-bodied Parsi men took part. They were led by a hero called Ardeshir. Unfortunately they were defeated and Sanjan was abandoned. The population moved away, taking with them the holy fire. They sought refuge in the caves of the Bahrot mountains about 20 Km. away. Here they remained for twelve years, after which they moved to Bansda where they resided for another fourteen years. People from Navsari and other places would visit the Iranshah each year as a pilgrimage. A philanthropist called Changa Asa made it possible to shift the Iranshah to Navsari where the holy icon was installed and where it remained to date of the writing of the *Kisse-i-Sanjan*. The text then ends with another conventional doxology and mentions the author's geneology and the date of composition of this poem.

Bahman Kaikobad makes no claim that he is writing a historically accurate account, only that he is repeating what has been orally passed down to him as tradition. As such, to expect historical exactitude is unrealistic. One may also keep in mind that the events he is narrating are of the distant past from his own time hence some distortions and interpolations are to be expected. The dates are more or less round figures and the historical identities of individuals remain a matter of great speculation and debate amongst scholars such as Jivanjee Modi, S.K.Hodiwala, S.H.Hodiwala, Eduljee, etc (Eduljee 1996; Paymaster 1915 : Menant 1898 (English ed.M.M.Marzban,1917); Kamerkar and Dhunjisha 2002). The issue is compounded by the fact that no individuals by the names of Jadi Rana or Sultan Mahmud occur in

Indian records. The author appears convinced that Sultan Mahmud is Mohammed Begada of Champaner but this neither bears out in the chronology of the site nor in the epistolary evidence provided by sthe *Rivayats* (see below). The chronology that scholars try to derive on the basis of the text is also open to interpretation since the periods mentioned are in approximations. The only figures that appear to be specific are the numbers of years spent in Hormuz, Diu and, to some extent, Bahrot and Bansda. Some other discrepancies are noticeable as well. For e.g. the establishment of Sanjan is attributed to the newly arrived migrants, as is the naming of the settlement. This is belied by the epigraphic records and inscriptional evidence, some of which predate the migration as mentioned in the *Kisse-i-Sanjan* (see below). Despite these drawbacks, the information provided by the *Kisse-i-Sanjan* remains invaluable and provides a documented record for a pact or a treaty between a local ruler and a foreign migrant community. It is also worth noting that the poem mentions the various places to which the Persian Zoroastrians spread. Almost all the places mentioned are coastal settlements along the Gujarat shoreline. It appears that there was no attempt to move into the hinterland. This may well be due to the mercantile nature of the community and the important role of the sea in their lives. Another pertinent observation that can be made at this point is that the poem never speaks of agriculture or the agrarian nature of the settlement.

Epistolary Evidence :

The Persian Zoroastrians or Parsis, as they came to be called, never returned to the settlement of Sanjan once they had left. The relocation of the population in other towns and settlements such as Khambat, Bharuch, Surat, Randher, Variav, Navsari, etc and the installation of the holy fire at Navsari meant that Sanjan lost all significance as the center of religious authority, a position it had previously enjoyed. This is obvious from the epistolary evidence provided by the *Rivayats*. The *Rivayats* are a series of letters exchanged between the Zoroastrians in India with the Zoroastrians of Iran. There are twenty-six such letters which contain extremely important religious information and are historically significant since they contain dates, genealogies, place names, routes used by the carriers, topical issues, etc. The earliest *Rivayat* is that of Nariman Hoshang, dated to 1478 AD. It is significant that the letters mention other places settled by the Zoroastrians in Gujarat but Sanjan finds no mention at all. It is by this omission that the *Rivayats* provide evidence for the terminal date for the settlement at Sanjan. Had Sanjan still been a settlement of Zoroastrians at the time of writing in 1478, it would surely have found mention since the presence of the holy fire, the Iranshah, and the prominent enclave of high priests would have made it the most important religious center of the community. It appears from the *Rivayats* that Sanjan had not only lost significance by this time but that it was not even a memory from the recent past (Dhabar, 1932). Hence if Mohammed Begada was indeed the Sultan Mahmud who attacked Sanjan, his dates and the event itself would have been more or less contemporaneous to the early *Rivayats*. Such a momentous incident as the sack of Sanjan and the movement of the Iranshah to the caves of Bahrot would surely have found mention in the correspondence between the co-religionists. Instead, an important point to note is that the early *Rivayets* mention that the Zoroastrians in Iran were unaware of any Zoroastrians in India until the arrival of the messenger with the letter (Dhabar 1999)

Epigraphic and Inscriptional Sources :

One of the earliest inscriptions to mention Sanjan – and one that has so far received little attention - is the Nagarjunakonda inscription of Abhira Vasushena, Year 30 which has been dated by D. C. Sircar to circa 278 AD (EI vol.XXXIV 1961-62: 197 – 204). This is a six line inscription which records the construction of a wooden image of Ashtabhujasvamin and its installation on the Seta-giri by the following persons – i) Mahagramika Mahatalavara Mahadandanayaka Sivasepa of the Peribideha family or clan; ii) the Yorajis of Sanjayapura; iii) Saka Rudradaman of the city or country of Avanti; and iv) Vishnurudra-sivalananda Satakarni of the city of Vanavasa. While conceding that the word 'Yoraji' is uncertain in its meaning and that the word 'Yorabhi' could well be an error, Sircar derives the word 'Yoraji' from "Yavanarajah" and hypothesizes that certain Yavana or Indo-Greek chiefs of Sanjayapura are being referred to. He goes on to state :

"As regards the location of Sanjayapura, it should be pointed out that Sanjaya is stated to have been another name of Sanjan in Thana district of Bombay state. The place is often identified with Sanjayantinagari mentioned in the Mahabharata in connection with Sahadeva's conquests in the south. If the above interpretation of the passage in question is acceptable, the inscription under study offers the only evidence regarding Indo-Greek rule in the Sanjan area about the close of the third century A.D. These Indo-Greeks, if they really ruled at Sanjan, appear to have been originally subordinates of the Sakas of Western India." (Sircar 1957 - 58: 199).

Interestingly enough, another inscription, dated earlier,

of the Kshaharat king Nahapana (32 AD – 77 AD) in cave no.10 in the Pandulena caves of Nasik mentions 'Nanagola' or Nargol, which is located at the mouth of the Varoli river, about 5 Km. downstream from Sanjan (Gokhale 2004: 108). The king had ordered the plantation of thirty-two thousand coconut trees at this site. The close vicinity of the two sites makes it likely that the entire estuarine area from Nargol to the Sanjan Bandar which was the navigable stretch of the river may have functioned as one unit or that the anchorage may have shifted at a later date from the Nargol area to a location upstream due to geomorphological changes.

The Sanjan copper plates of Buddhavarsa were found at Umbargaon, a short distance from Sanjan at the mouth of the Varoli River,on thebank opposite Nargol. Buddhavarsa claims to be the younger brother of Pulakeshin II. This grant establishes the control of the Western Chalukyas in this region in the seventh century. The date of this grant is 671 AD. (Konow 1917-18: 144 – 152). This is relevant in the light of the cordial and close relations maintained by theSasanian king Naoshirwan and Pulakeshin II,asis reported in historical sources (Kamerkar and Dhanjisha 2003:13;alsoWink 2002:105).

The copper plate grant of Amoghavarsa I is datable to 871 AD and records the grant of the village Jharivallika located in the twenty-four village group adjoining Sanjana to four Brahmans of Karahada (Karad). This is an important grant since it not only mentions the rule of the Rashtrakutas over this region, but it also gives an idea as to the area covered by the administrative unit of Sanjan. The village Jharivallika has been identified as the village Zaroli. It is bound in the east by the river Kalluvi (Kalu); in the south by Uppalahataka (Uplat); in the west by Nandagrama (Nandgaon); and in the north by Dhannavallika (Dhanoli). Sanjan is also spelt varyingly as 'Sanjana' and 'Sanjjana'. (Bhandarkar, EI vol.XVIII)

The Chinchani copper plates number nine in total and consist of three separate grants made by the Silahara kings between 926 AD and 1053 AD. Chinchani is located about twenty-two miles south of Sanjan and is the find-spot of these copper plates. The first of these grants is of the time of king Indra III and is dated 926 AD. Sanjan is referred to as 'Samyanamandala'. This territorial division was given by Krishnaraja II, sometime between 878 AD and 915 AD, to a Tajik general, Mohammed Sugatipa, a.k.a Madhumati, who was appointed Governor in appreciation for his success in bringing all the neighbouring harbours ('vetakula') under the control of his king. Madhumati Sugatipa patronized a monastery founded by Aunaiya and two temples, and gave the villages of Kanaduka and Devihara as a grant towards the upkeep and repairs of this monastery and for the offering of *naivedya* to the goddess Dasami. Amongst the local communities mentioned as residing in Sanjan are Panchagauda Brahmanas, Modha Baniyas who had brought their deity Bhillamaladeva or Madhusudana from Rajasthan, Tajikas or Muslims and Hamjamanpaura or the Parsis (the word 'Hamjamana' deriving from the Persian 'Anjuman' or community). The largest population appears to be Koli and Mahar and in fact one administrative unit was called the Koli-Mahar Vishaya. The relevance of this inscription lies not only in the mention of a Tajik or Arab Governor of the region but also in the information it provides regarding the demographic composition of the settlement, the cosmopolitan nature of the town and the prominent trading communities (Sircar1957 - 58; also see Sankalia 1983: 209 - 213). While the actual identity of Mohammed Sugatipa remains nebulous, it is not improbable that he may well have been either a powerful merchant prince with considerable control over the ports of the region or a mercenary who could consolidate some hold over the harbors of the area. This is not unusual since it is known that Persian and Arab traders on the west coast were sometimes powerful enough to even challenge the local rulers. The merchants who imported horses from Arabia and Persia for the Indian kings were frequently given authority over certain territory in lieu of payments. Often, these merchants were also the trainers and caretakers of these expensive animals, maintaining a kind of cavalry for the king (Lambourn 2001: 117-49). It could be possible that Sugatipa was one such merchant. It is indicative, as well, of the relations between the Hindu kings and the non-Hindu traders and mercenaries in their territories.

The second grant in the Chinchani set belongs to Chamundaraja, the subordinate of the Silahara king Chhinturaja. It is dated to 1034 AD and pertains to an oil mill which was granted to his subordinates, officers and other members of the local community. This last group also consists of the mukhyas or elders of the 'Hamyamana' or Parsis. (Sircar 1957 -58)

The third grant is of the time of Vijala, dated to 1053 AD. 'Samyanapattana' is mentioned as a district consisting of 4000 drangas and extending up to Akasika, which has been identified as Agashi near Virar. The reference to Sanjan 700 refers to the seven hundred villages of the region. But D.C. Sircar is of the opinion that the interpretation of 4000 drangas as towns or cities would be an error and that the reference is in fact to the annual revenue or tax of 4000 drammas. (Sircar 1957-58:

45 -76). The term 'pattana' is used for the settlement of Sanjan. The relevance of this is underlined by the fact that the term was used to indicate a market centre as well as a riverine settlement or river-port (H.P.Ray 2003: 20).

Accounts of Geographers, Mariners and Travellers :

There are no known records or accounts left by travelers, sailors or geographers which mention Sanjan in the pre-Islamic period. It is the Arab geographers and travellers who write about Sanjan, as indeed they write about the other ports of the Indian Ocean and India in particular. The Persian records that mention Sanjan also do so after the 9th century and not before. While there are numerous records to establish the contact of the Parthians and Sasanians with western India and with Indian kings and dynasties, there are no records which mention Sanjan as either a town, port or trade centre. Sanjan finds mention as 'Sindan' in these post-Sasanian Arab and Persian sources, and appears to have been a prominent port-of-call for Persian Gulf vessels on the India or the China run. Most of these accounts are datable from the 8th-9th centuries to the 12th century or so. After this period, the information becomes more scant and repetitive with little that is new.

There appears to be some confusion about the name 'Sindan', since it is often confused by historians with another town with a similar name, 'Sandahan', in Kutch (Janaki 1969: 56-57). In fact, some authors suggest that there were as many as four towns with the same name – one in Sind, one in Kutch, one in Konkan and another as far south as Goa (Janaki 1969: 56-57). However the descriptions and distances given by the geographers, sailors and travelers appear to indicate clearly that the identification of 'Sindan' with Sanjan is correct.

Al Biladuri mentions that he was informed by Mansur, son of Hatim, that Fazl, son of Mahan, a slave of the Samma house, subdued Sindan in c.820 AD and sent to Khalifa Al Mamun (813 – 833 AD) the gift of an elephant. He built a Jama Mosque. He was succeeded by his son Muhammad who was overthrown briefly by his brother while he was fighting the Medhs in Saurashtra. However he retrieved his throne and sent to the Khalifa the gift of the longest teak ever seen. The people of Sindan overthrew Muslim rule but spared the Mosque where the Muslim residents continued to worship (Elliott and Dawson 1952: 129). If this account is factual, then the dynamics of governance in the area during the 8th-9th centuries need to be looked at afresh. While there is some debate about the location of 'Sindan' being in Kutch or Konkan in this account, it is pertinent to point out that other writers also attest to the Jama Mosque in Sanjan and, as V.K.Jain points out, the mention of teak indicates that it was the port town of the Konkan which was famous for its teak and bamboo exports (Jain 1990: 135). Al Biladuri also locates Sindan between Bharuch and Sopara, about 200 miles from Debal.

Ibn Haukal describes Sindan as a strong and great city with a Jama Mosque and a place where Muslims are respected. He notes that mangoes, coconuts, lemons and rice grow there (Janaki 1969: 58). In his work *Hudad al-Alam* or Regions of the World written in 982 AD, he gives details of the different countries and regions of the ancient world in startling detail and notes that the region under discussion in this study was under the rule of the king "Ballaharay" or "Ballahara" who was the most powerful king in India, clearly an indication of the Rashtrakuta rulers. He further mentions:

"Samur, Sindan, Subara, Kanbaya, four towns on the coast, in which live Muslims and Hindus (*Hinduvan*). In that locality is a Friday-Mosque (*mazgit-I adhina*) and an idol temple. The people of these towns have long hair and at all seasons wear only an izar…The climate (of these parts) is hot. The government (*padshahi*) there belongs to Ballah-ray. Near these (towns) there is a mountain on which grows much bamboo (*khaizuran*), rotang (*nayza*, spelt: *nira*), pepper, and coconut. In Kanbaya shoes (*na'lain*: "sandals"?) are produced which are exported to all the countries of the world". (Tr. Minorsky 1937: 88)

Al Ishtakhri says Sindan is five days journey from Surbaya and Saymur while Al Idrisi gives the location as one and a half miles from the sea and five days journey from Saimur. He says it is rich and populous with people noted for industry and intelligence. They are rich and warlike in temper and the town has extensive commerce in imports and exports. Al Biruni places it 200 miles from Debal, between Broach and Supara. (Elliott and Dawson vol.I 1866-77: 27, 66; also, Sachau 1983: 208-209)

Buzurg Ibn Shahiyar al-Ram-Hurmuzi, a Persian sailor who collected stories of voyages from mariners and sailors on the waterfronts of Siraf, Basra and other ports in the 10th century describes the adventures of sailors making the journey to China and India from the Persian Gulf. His *Kitab Ajaib al-Hind* or the Wonders of India narrates the incident of a voyage undertaken by three ships from Siraf to Saymur (Chaul) in the year 919 AD. The ships carried 1200 men – merchants, shipmasters, sailors and others of diverse nationalities. The cargo was of great value. The route taken by this fleet of ships was via Sindan, Tanah and Saymur. This journey took eleven days but was met by a terrible storm just as they prepared to harbour at Saymur. The ships were pushed out to open seas by the winds and could not make safe landfall. Since the decision to jettison the cargo was taken too late, the ships sank, taking with them most of the men and the entire cargo. According to the author, this loss of skilled manpower, cargo and merchants was responsible for the decline of both ports, Saymur and Siraf which could not sustain such a huge blow (Hourani 1951: 119-20). This may not be strictly true since the ports of Siraf and Saymur did function as commercial centres well past the date mentioned in the account. However the mention of the route, the mountains of Sindan and Tanah, the frequency of storms along the western coast of India and the details of the voyage are of relevance.

An indirect but very important reference to 'Sindan' is found in an unusual source - the Geniza documents. Documents pertaining to the Jewish business house of IBN Awkal (980 - 1030 AD) mention the items of trade between Egypt and India. Three varieties of indigo are mentioned, the most sought after and expensive being Sindani indigo or indigo from Sindan / Sanjan (Stillman 1973 : 15-88).

Exploration and observations in the nineteenth and twentieth centuries :

Although no scientific archaeological excavations were undertaken at Sanjan prior to 2002, the site was well known and had been visited by archaeologists, historians and amateur archaeologists since the 19th century. Their observations and notes during exploration leave some record of the site and its changing morphology as habitation, human activity and environmental factors have taken their toll over the years.

The Gazetteer of the Bombay Presidency vol. XIV (1882) gives a remarkable description of the town and the ancient remains of the old settlement. The writer observes the hills to the east of the present town and the river, and the undulating landscape on the west, with the remains of brick structures, sculptures and architectural members. Most of the descriptions match observations made by archaeologists during the excavations of 2002-04. However, most of the stone slabs with fragmentary inscriptions and carved motifs are no longer traceable. One inscription in Devanagari script is supposed to have given the date 1432 AD and the other inscription is said to have been in Kufic or Arabic with few letters decipherable. This may have been a grave stone. Some of the features seen by the author of the Gazetteer, such as the blind well at the base of the Dokhma mound are also no longer in evidence. The presence of silted up lakes which are being used for the cultivation of rice is pertinent, as is the observation that the entire area is dotted with superior quality, large size bricks which are reused in modern constructions and which can be bought cheap. These bricks measure 15" x 9" x 3". One important observation is about the structure of the Tower of Silence or the Dokhma itself. The author tells us that the foundations of this structure can be seen and that it is constructed of brick and cement. The date of this structure is approximated as between 1300 AD and 1500 AD, on the basis of the construction material. A foot-note on page 305 mentions that sixty years prior to the writing of the Gazetteer, most of the structure could still be seen and that the main wall had either crumbled or vanished due to the theft of bricks over the decades until only the steps could be seen. These too had gone by the time the Gazetteer came to be written. After mentioning the large circular well – today known as the *Peer no Kuvo*, and the sculptural fragments outside a village house, the author describes the pond and the remnants of bricks. Close by to the pond, on the bank, is the landing platform where vessels of 80 tons or 250 *Khandis* can make the landing in one tide from Umbargaon. It is important to note that at the end of the 19th century boats could still make a landing on the Sanjan Bandar, considering that the vessels can only approach the South bank of the Varoli River today. Siltation and the formation of mangroves make it impossible for vessels to approach Sanjan Bandar. Small riverine crafts may have been the only vessels which could make it to this anchorage once the siltation of the river started. The Sanjan fort is dated to 1613 and is said to be Portuguese. The Portuguese called the town St. John de Vacas and the Bahrot hills as St. John's Peak.

M.M.Marzban edited the English translation of the French book, *The Parsis of India vol. I*, written by Delphine Menant in 1898 and published it in 1917 with additional notes and commentary. In this volume he mentions the remains of the ancient settlement at Sanjan and in an important footnote on page 63 he gives an overview of the observations of various visitors to Sanjan over a period of time. The first person mentioned is Dr. J. Wilson who visited the site in 1839 and found very few Parsis living there. But he did visit the ancient Dokhma and noted the remains of the structure which was "constructed before 1400 AD". B. B. Patel's account of the visits he made to various sites was published in the newspaper *Jam-e-Jamshed* (4th April, 1881). His description and notes on Sanjan have been reproduced by Marzban to record the degradation of the structure.

Like the author of the Gazetteer, he too notes the large quantity of bricks which can be seen on the ancient habitation areas. The Dokhma at this time had the foundation and the central well or Bhandar in evidence, despite the thick vegetation. He too notes a blind well in front of the structure. Local informants, he notes, are of the opinion that Sanjan once had nine Dokhmas – a fact also noted by the Gazetteer. A Nargol resident further informed him that the structure had been standing in about 1831 and this information was confirmed by another visitor to the site in 1841. Yet another source mentions the steps being extant in 1846. As in the Gazetteer, this account also mentions that Sanjan was large enough to merit description as a *Navteri Nagari* or a settlement which measured nine miles by thirteen miles.

In 1982-83, Dr. H. D. Sankalia sent Dr. Ashok Marathe and Shri Dundappa to explore Sanjan with a view to ascertaining the veracity of the Chinchani copper plate which mentions the Tajik governor Mohammed Sugatipa and the temples dedicated to the goddess Dasami and to Bhillamaladeva (see above). The explorations are reported by Dr. Sankalia who mentions that other than noting the brick structures and ruins of the old settlement, the explorers also collected potsherds and Indo-Sasanian or Gadhiya coins. An important note made by Dr. Sankalia is that "the pottery from Sanjan gives some definite idea as to the age of the ruins, still as yet there is no shred of the early Christian era. Such as the Red Polished Ware or amphora, indicating trade with the Roman world." (Sankalia 1983: 211) The pottery is unfortunately not described by Dr. Sankalia. Nevertheless, he makes an important point when he states that "a naturally rich, fertile coastal belt as Sanjan could not have remained unoccupied or settled from the earliest historical time, that is third century B.C. or at least during the time of the Andhras and Satavahanas and Indo-Greeks. Hence both from point of view of the Parsi chronicle *Kisseh-I Sanjan*, and the history of the west coast in fact Bombay – a small but full stratigraphical excavation on any of the ancient mounds at Sanjan is necessary" (Sankalia 1983: 211-12).

Excavation at the Dokhma or Tower of Silence by Dr. Sir Jivanji Modi :

Sanjan had been visited by various scholars and had been explored as the first Zoroastrian settlement on Indian soil. However, the first attempt to do any kind of archaeological work at the site was undertaken in 1917 at the Dokhma or the Tower of Silence by Shams-ul-Ulama Dr. Sir Jivanji Jamshedji Modi. But he did not himself report this excavation or his findings – a surprising fact given that he wrote profusely on Parsi history and also delivered numerous lectures. The record of this brief excavation has been published by another scholar of Zoroastrian history, Dr. J. M. Unvala in the Preface to a report on his own attempts in 1949 to excavate and study the Dokhmas at another site called Tena which is close to Surat (Unvala 1951: I – IV). Unvala reports that in 1950 he found an envelope preserved in a cupboard of the K. R. Cama Oriental Institute, Bombay. This accidental discovery brought to light some "painted pottery and other antiquities from Susa in Iran" and "an envelope containing twelve whole (unbroken) glass bangles and nine fragments, four toe-rings of mixed metal, similar to those found in the Brick Dakhma and in the First Stone Dakhma of Tena, and one bracelet, probably once covered with a copper foil, together with a letter from Mr. G. V. Acharya, former Curator of the Prince of Wales Museum of Western India, Bombay..... fixing the date of these bangles somewhere between the fifteenth and eighteenth century"(Unvala 1951: I – II). These finds are mentioned in the correspondence between Modi and Acharya as belonging to an old Dokhma of Nargol near Sanjan. This correspondence of 1925 between the Curator of the museum and Dr. J. J. Modi has been traced and published by Unvala as appendices at the end of his report (Unvala 1951: 9 – 13). None of the antiquities from this excavation are traceable at the K. R. Cama Oriental Institute and Unvala's work remains the only report on this early excavation.

In an attempt to find out more about Modi's work, Unvala visited Nargol and the Sanjan Dokhma in 1950. Here he was able to trace the documents and property papers of the land on which this ancient structure stood. He found out that in 1917 Modi had the *Bhandar* or central well of the Sanjan Dokhma "opened and thoroughly emptied of bones. The glass bangles, the toe-rings and the bracelet mentioned above were found, it is alleged, during the course of this operation. The Bhandar had been again filled up with bones and covered over with bricks." (Unvala 1951: II). This exercise can be seen more as partial clearance work rather than a full-fledged excavation. Unvala's own observations of the Dokhma in 1950 are almost the same as those of the archaeologists who worked there half a century later in 2002-04. The site was overgrown with date palms and grass but the remains of the structure were discernable. The slightly raised outer wall with a break where the entrance had once been was noted and some measurements, such as the inner diameter, were taken (36 feet). Apparently, Unvala did not find any traces of brick steps leading into the circular structure nor did he note a blind well at the base of the mound as noted by the visitors of the previous century. On the basis of the bangle types found at both Sanjan and Tena, Unvala dated the structure to between the 15th and the

18th century and concluded "Thus this old Dakhma of Sanjan is not as old as one would like to believe. It is relatively old. An older or rather the oldest Dakhma of Sanjan remains still to be discovered." (Unvala 1951: III) An interesting observation made by Mr. G. V. Acharya in his letter dated 9th April 1925, in which he approximates the chronology of the finds by Modi is that some of the bangle pieces resemble the ivory pieces which have been found in the excavations in Sind and that they have been dated between the 4th and 8th centuries. However he presumed that the more recent looking bangles and the older types had all been found together and therefore he placed them in the 15th to 18th century period (Unvala 1951: 11)

While Unvala has rendered a great service by taking the trouble to trace and publish some record of the excavations of 1917, his conclusions and chronology, like Mr. Acharya's, may require some reconsideration in light of recent archaeological data.

SANJAN EXCAVATIONS 2002 -2004 :

Sanjan was known to scholars, historians and archaeologists since a long time but no excavations had been undertaken to actually ascertain the veracity of known records or to establish the role of the site as a maritime commercial settlement during the early medieval period. **The World Zarathushti Cultural Foundation** took up this project in 2002. The site was excavated over a period of three years from 2002 to 2004. The Principals of the project were Dr. Homi Dhalla and Late Dr. Mani Kamerkar. The Director of the excavations was Dr. S. P. Gupta, Chairman of the Indian Archaeological Society, New Delhi and co-Director was Dr. Kurush Dalal, Mumbai. Preliminary excavation reports of all three seasons have been published (see Gupta *et al* 2002, 2003, 2004, 2005). It is necessary for the purpose of this study, to review the excavations in order to understand the context of the ceramic assemblage and the method by which it was retrieved.

The ancient site at Sanjan is located about a mile away from the urban center of the modern town, on an area known as the Sanjan Bandar (port). Exploration on and around the site prior to and during the commencement of excavation work established that the site was located only on the northern bank of the Varoli river and that it covered an area of approximately 1.5 x 1.5 km. The explorations on and around Sanjan Bandar resulted in a large surface collection of glazed and unglazed pottery, some glass and shell fragments. The disturbed nature of the mound with large-scale human activity and cultivation had resulted in exposed brickwork along the embankment as well as on smaller mounds slightly further away from the most dense scatters and habitation on the Bandar. These appeared to be small farmsteads, homesteads and satellite settlements of limited occupation, similar to the pattern of settlement seen today as well. Some structural features could be seen extending into the mudflats at low-tide, indicative of the fact that the Varoli had flowed further south in the past and that the northward shift of the river had in fact cut into the southern periphery of the mound. The Varoli is not navigable past the Bandar due to rock outcrops. It should also be noted that the Bandar bank cannot be approached by boats today since the entire northern bank has extensive mudflats. Small fishing vessels can be seen across the river on the southern bank, moored at Palgam. The explorations were also able to establish the location of the ancient Dokhma or Tower of Silence, constructed and used during the occupation of the site by the Persian Zoroastrians. A high small mound, separated from the Bandar by the Sanjan-Khatalwad road and located on the Koli-Khadi stream, could also be identified for excavation work. A number of sculptures, mostly broken but some under present worship could be seen during explorations. These appear to belong to the Rashtrakuta and Silahara periods (Gupta *et al* 2005: 60; also, Shivendra Kadgaonkar, personal communication). Prominent among these is the image of Kubera which was found in a field a considerable distance away from the Bandar, across the Koli-khadi stream. Although it is badly damaged, it is still identifiable. Another image of note is the goddess figure which was found under a tree outside a fisherwoman's house on the Bandar. The small image of Ganesha to the left of the figure indicates that this is a form of goddess Parvati. This image was collected as a surface find and is under study at present (Shivendra Kadgaonkar, personal communication).

The following note is based on the published preliminary reports (Gupta *et al* 2002: 182 – 198; Gupta *et al* 2003: 26 – 34; Gupta *et al* 2004: 93 – 106; Gokhale 2004: 107 – 112; Joglekar 2004: 113 -115; Gupta *et al* 2005: 55 – 61; Mitra and Dalal 2005: 62 – 68; Nanji and Dandekar 2005: 69 – 72; Mushrif and Walimbe 2005: 73 -92; Rajaguru and Deo 2005: 93 - 98) and on observations made by the present author as part of the excavation and exploration team.

Season 1 (2002) :

The first season of excavations in 2002 commenced atop the Sanjan Bandar, as close to the highest part of the mound as possible. Present human occupation made it impossible to conduct horizontal excavations. The excavations had to be restricted to a small area measuring 5m x 7.5m within the present settlement. The aims of the excavation were defined by the excavating archaeologists (Gupta *et al* 2002: 182 – 183) as under :

1. To establish the existence and location of the ancient settlement of Sanjan;

2. To establish the date of the founding of the settlement;

3. To establish the nature of the occupation;

4. To establish the date of desertion, and

5. To ascertain the role of Sanjan in the East-West trade in the Indian Ocean.

The trenches of the first season were trial trenches and were designated as such, TT1 and TT2. TT1 measured 5m x 5m and TT2 was an extension of TT1 on the northern side, measuring 2.5m x 5m. Each trench was divided into quadrants and each quadrant was dug separately. The excavators used the lot system to number each dig. Hence each quadrant had a designated series of lot numbers assigned to it. All material collected in a dig in a quadrant carried the relevant lot number, which was depth specific. The presence of structures and features made it necessary to maintain varying levels in each quadrant. Hence the depths reached in each quadrant made uniformity impossible to maintain. However, the preliminary report describes the seven stratigraphic layers identified in quadrants where the maximum depth of 5.35m was reached. To recap briefly,

Layer 1 : The Humus was highly disturbed and contained a dense concentration of modern material due to intense human activity on and around the site. The upper 5cm was removed and discarded. The succeeding deposit was slightly more compact, though intrusive material was present. The gradient of the slope ensured that the layer was not even and that it thinned towards the northwest. The average thickness of the layer was about 15cm. The base of this layer had a dense concentration of brickbats.

Layer 2 : Immediately under the brickbats at the base of layer 1, the collapsed remains of brick walls were encountered. The bricks of these walls had collapsed in an accordion manner. There appeared to be a floor-like feature of brickbats rammed into a sticky dark clayey matrix. According to the published report, "this layer resembles a burnt platform of rammed rubble" (Gupta *et al* 2002: 183).

Layer 3 : This layer was thin and consisted of a dark brown sticky clay-like material which was devoid of brickbats. It rested on a brickbat floor and was rich in artifacts and ceramics. The base of this layer in TT1 SW and NW Quadrant revealed the presence of two ring-wells at a depth of 1m from the datum. A brick wall separated the ring-wells from the floor. These features, i.e. ring-wells, wall and floor form the terminal occupational phase at the site.

Layer 4 : This layer was lighter in colour and had some potsherds and small brickbats. It was silty, sandy and grey-brown interspersed with tiny calcrete nodules. This layer was encountered only in TT2 in the Southeast Quadrant.

Layer 5 : This layer was similar in appearance to layer 4. It was encountered at a depth of 2.4m and was slightly darker and clayey compared to the previous layer.

Layer 6 : This layer was noticed only in a small patch between ring-well 1 and ring-well 2 and could well be the residual material from the breakage and collapse of ring-well 3 which is immediately underneath, or of the ring-wells between which it occurs. Bands of decomposed organic material are evident.

Layer 7 : This layer is the natural soil upon which the settlement was established. It is sandy, silty and yellowish in color. It appears to be a fluvial deposit. The upper 15cm of this layer contained some artefactual and ceramic material. The lower portion of this layer was progressively sandier and interspersed with small pieces of *murum*. At the base of this layer, at a depth of 5.35m, was encountered the weathered crust of the bedrock.

Two phases of structural remains were identified during the excavations. The first phase was represented by a large brick built structure, of which parts of the northern and western walls were seen in the trenches. The bricks measured between 35 and 40cm in length, 21 and 25cm in breadth and 6 to 7cm in thickness. The walls had fifteen extant courses of brick underlying which was a foundation of three courses of cobbles and clay. Bricks from this structure had apparently been removed or robbed by later inhabitants as is evident in the southern part of the excavation. To the west of the structure was a ring-well (Ring-well III) with nine extant rings, each measuring approximately 33 – 39cm in height. The ring-well had a depth of 3.2m, of which 1.5m had been excavated into the *murum* representing layer 7. The upper rings had collapsed upon one another but the lower rings were intact and could be exposed *in situ.* To the northwest could be seen a rammed cobbled floor. A flimsy 'S' shaped extension of five courses of brick and having a breadth of one brick was designated Structural Phase IA.

The second structural phase was identified at a depth of 1.05 -1.10m and was represented by a small wall with two extant courses running north-south, with a floor rammed with cobbles and brickbats to the east, and two large ring-wells (Ring-well I and II) to the west. The diameter of the ring-wells was approximately 80cm. Five to seven rings were extant. The height of the each ring was about 20cm. At the bottom of both ring-wells was seen a thick deposit of charcoal, clay and potsherds. Above the ring-wells fallen bricks bore evidence of the collapse of an adjacent structure. These ring-wells flanked Ring-well III and all three appeared in section to be contained within the same pit. The ring-wells and the pit within which they were sunk yielded a high amount of artefactual and ceramic material.

The artefact assemblage consisted of glass and glass objects; iron objects such as nails, arrowheads, knives, ladle, spike, axe, angles, etc; beads made of various materials such as glass, stone, terracotta, paste, copper, etc; bangles and rings; copper objects such as antimony rods, needles, bell, etc; terracotta discs and lamps; some stone objects such as hammer stones, pestle, muller etc; and coins, of which the fragmentary Sassanian / Abbasid-Sassanian issue, the coin of the Sultan of Sind, the coin with Brahmi legend and elephant motif, and the Indo-Sassanian / Gadheya coins are of relevance (Gupta *et al* 2002: 189. also, Gokhale 2004: 107 – 112)

The preliminary conclusion, drawn on the basis of the evidence, was that the site was a thriving, economically strong centre and was involved in local as well as foreign trade during the 8th and 11th centuries AC. (Gupta *et al* 2002: 189).

Season – 2 (2003) :

The excavations of the second season were located about half a kilometer away from the trenches of the first season due to circumstances beyond the control of the archaeologists. The site chosen for excavation was the mound on the bank of the Kolikhadi stream. This site appeared less disturbed. Although the original mound may have been much larger, it appeared trifurcated by rain-gullies and a mud track. The Sanjan-Nargol road separated it from the Bandar mound. The preliminary reports of the second season were published in 2004 (Gupta *et al* 2004: 93 – 106; also, Joglekar 2004 : 113 – 115).

It was possible to excavate many more trenches at this location since human activity did not interfere with the work. The top of the mound was cleared of vegetation and a grid was laid. The grid consisted of trenches measuring 5m x 5m each, which were subdivided into four quadrants each, following the methodology of season 1. These trenches were given alphabetical and numerical designations. The excavations at this location were of a horizontal nature. The methodology followed was the same as the previous year, with depth-specific lot numbers within allotted series assigned to each quadrant of the trenches.

The excavations revealed a large structural complex consisting of floors and platforms rammed with bricks, brickbats and cobbles. One such well-constructed floor had three large troughs embedded in it and was designated Structure 1. To the north of this structure were two platforms, one of well packed cobbles measuring 5m x 5m identified as Structure 2, and a second platform made of large brickbats and lined along its southern periphery with semi-dressed stone blocks which was designated Structure 3 (Gupta *et al* 2004: 95). The presence of six human skeletons in extended positions to the south of the structural complex and in association with it made interpretation of the site more complex. The skeletons were placed in a north-south position with the head to the north and faces turned to the west or the right. The arms were either flexed over the midriff, stomach or the lower part of the abdomen. Of the six exposed skeletons, five could be retrieved for study. The preliminary report on these human remains identified two skeletons as definitely male and two as female, while one specimen could possibly be male and the sex of one specimen could not be determined since it was only partially exposed in section The estimated age at time of death ranged between 30 to 45 years (Mushriff and Walimbe 2005:73 - 92). A pertinent observation made in the anthropological report on these human remains is that "Morphologically the Sanjan population appears to be delicately built, in comparison with the other specimens recovered from the Medieval or Historic levels in western India, like the ones from the sites like Padri, Navdatoli, or Kuntasi. This general observation, however, should not be taken as indicative of different 'ethnic' stock of this series" (Mushriff and Walimbe 2005: 77). There were no discernable burial pits for most of the skeletons, except for HS-4 and HS-5. They were all contained in Layer 2 and were encountered at a depth of approximately half a meter.

The stratigraphy at Kolikhadi revealed four layers. The habitation deposit, at this location, did not exceed 1m. in totality. (Gupta *et al* 2004: 97 – 98)

Layer 1 : This layer was hard and compact, interspersed with brickbats. Some modern material was found in the humus or the upper 5cm. Root activity, burrowing by small rodents and reptiles, etc had also disturbed this layer. The thickness of the layer was about 10 – 15cm in almost all trenches. The layer appeared to be lined at its base by small brickbats at a number of places.

Layer 2 : This layer represents the most intense occupational phase at this location. The thickness varies from 18cm to 41cm. It is reddish brown in color and darker than the preceding layer. It is granular and has a number of brickbats, brick and stone features and structures at its base. The three troughs and the skeletons were encountered in this layer, as were ceramics and artifacts such as coins and beads.

Layer 3 : This layer was dark brown and clayey. It was the natural soil upon which the habitation was established. The upper few centimeters of this layer yielded some habitational material such as potsherds, brickbats, etc. the thickness of the layer varied from 5 – 33cm and in one trench (A3, NW) it reached a depth of 65cm. A whitish ashy band was noticed in the first 5cm of this layer in trenches B3 NE, A3 SW and A3 NW. This could signify some pyrotechnic activity (Dr. Vishwas Gogte, 2004: personal communication).

Layer 4 : This yellowish silty, sandy layer constituted the virgin soil encountered all over the area. The layer was interspersed with kankar nodules and heavily decomposed marine shells. It appears to be a fluvial or marine deposit which rests upon the murum or disintegrated beach-rock. A stone and brickbat platform / pit base was constructed into its upper 10cm and was then plastered with the same material as layer 4.

Other than the ceramics, a large number of artifacts were also retrieved. These were noted to be of a 'poorer' nature than those found at the Bandar in the previous season (Gupta *et al* 2004: 98 -99). The assemblage consisted of beads made of various materials such as glass, stone, terracotta, etc; glass bangle fragments; coins made of copper, silver and lead; terracotta objects such as discs, lamps and a zoomorphic vessel; worked garnet chunks; some stone artifacts, etc. A coin of Allauddin Khilji (1296 – 1316 AC) was found on the surface. As compared to the previous season, the second season of excavations did not yield as many glass fragments or glass artifacts.

One pertinent observation made about the distribution of faunal remains at this location was that animals such as pig, horse and domestic ass were represented in the collection at Kolikhadi but were missing from the assemblage at the Bandar during the first season (Joglekar 2004; unpublished report; also, Nanji and Dhalla 2007, in press). This is of relevance considering the variation in the distribution of pottery as well (Gupta *et al* 2004: 101).

As mentioned in the preliminary report, the monsoons of the previous year had resulted in the erosion of deposit along the embankment of the Bandar and exposed extensive structural remains all along the northern bank of the Varoli River. A section along this embankment was cleared and scraped, exposing well built brick structures. Pottery and artefacts were collected as part of this work. At the base of this section, at the level of the river, was noticed a small square feature which was either a well or a tank. It was extremely well made of burnt brick and was excavated to a depth of about 1m. This small excavation trench was numbered TT3. It yielded a very large quantity of potsherds. The walls of the well had indents at six or seven courses. It could only be excavated briefly since it was located in the inter-tidal zone and the water levels of the river were rising fast. Some artifacts and animal bones were also retrieved from this feature (Gupta *et al* 2004: 97 -98).

The preliminary conclusions drawn at the end of the excavations of the second season were that "the settlement at SJN-K appears to have been part of and more or less contemporary to SJN-B. It is possible that the difference in the distribution pattern of the ceramic assemblage and the artifacts coupled with the lack of brick walled structures could be indicative of social / economic / occupational stratification." (Gupta *et al* 2004: 106).

Season 3 (2004) :

The excavations of Season 1 and Season 2 had not been able to establish a clear stratigraphy for the site nor the ethnicity of the population that was supposed to have migrated to Sanjan from the Persian Gulf. These two issues were addressed by the excavations of the third season. The two separate locations chosen for the excavations were the ancient Tower of Silence or Dokhma, and a small trench measuring 5m x 5m just beside the embankment road on the Bandar. The first location was specifically chosen since it was the only uniquely ethnocentric structure at Sanjan that could provide irrefutable evidence for the presence of Persian Zoroastrians at the site. The second location was selected so as to establish the stratigraphic profile of the port site of Sanjan. The preliminary reports on the excavations of the third season were published in 2005 (Gupta *et al* 2005: 55 – 61; also, Nanji and Dandekar

2005: 69 – 72; Mushriff and Walimbe 2005: 73 – 92; Rajaguru and Deo 2005: 93 – 98).

The Sanjan Dokhma was located atop a naturally high mound close to the confluence of the Kolikhadi and Tukkad-nallah streams. The construction appeared to be basic in nature with bricks, brickbats and cobbles along with mud mortar and lime plaster used as the construction materials. The three sections of the Dokhma, i.e. the outer wall, the pavis or parabolic floor upon which the corpses are laid and the Bhandar or central well which acts as a receptacle for the osseous remains had been constructed by using the simple method of excavating a crater on top of the mound. The wall of the crater was riveted and enforced with brickbats and cobbles to check erosion by the elements. A facing of bricks on the internal side of this wall indicated that it originally had a veneer or that there had been at least one attempt at renovation and repair. The wall had a gap which was located due east and would have served as the entrance to the structure. The inner floor was plastered with mud mortar and some bricks. A shallow well having a diameter of 5m and a depth of 1m was excavated in the centre and walled with burnt brick and lime plaster (Nanji and Dandekar 2005: 69 – 70).

The methodology used for the excavation of this circular structure had to take into account the varying depths and gradients of the mound and the structure. The structure was first divided into four quadrants on the basis of the four cardinal directions. Two baulks running north-south and east-west dissected the circle and were maintained throughout the excavations. However it was found that this method was not feasible since it was difficult to mark the locations of artefacts and other material as it occurred. The entire mound was gridded into 5m x 5m trenches and excavation was taken up in select trenches both on the interior and exterior of the structure. Almost three-quarter of the interior of the structure was exposed and excavated (Gupta *et al* 2005: 56 – 57). A large number of bones were found in the central well. Potsherds which were mixed in the mud mortar were found during clearance. The presence of human activity in the vicinity of the structure has also caused the deposition of potsherds in the upper layer and on the surface. Most of this pottery was in secondary context and appeared rolled (Nanji and Dandekar 2005: 70). A large number of bangles, bangle fragments, rings of mixed metal and silver, a gold-foil earring and a small gold bead were also found (Gupta *et al* 2005: 57 – 58). It is estimated by the anthropologists studying the bones from the Dokhma that the Bhandar or central well contained the remains of at least 350 to 400 individuals (Mushrif and Walimbe 2005: 74).

The second excavation of Season 3 was undertaken on the embankment of the Sanjan Bandar, just beside the road, diagonally across from the section which had been scraped the previous season. This trench measured 5m x 5m and was designated TT4. Continuing the convention of the previous seasons' methodology, the trench was divided into quadrants and each quadrant was allotted a depth-specific lot number series. The stratigraphy consisted of six layers (Gupta *et al* 2005: 58 – 59). A brief review of these, on the basis of the published Preliminary report, is in order at this point, so as to put the structures, artifacts and pottery in context.

Layer 1 : This layer was brown in color and constituted the humus. The soil was fine and did not appear to contain any brickbats.

Layer 2 : This layer was almost identical to layer 2 in TT1 and TT2. It was dark red-brown in color and had a large number of brickbats and brick debris. At the base of this layer, located almost in the centre of the trench, was a platform of burnt brick, measuring approximately 2.5m x 2.5m. A burnt patch with a large amount of charcoal embedded in a flue-like depression was seen on top. The platform consisted of three courses of bricks resting on a bed of cobbles. This was designated Structure 1.

Layer 3 : This layer was thick, brown and sticky. The number of brickbats was less than the preceding layer. The base of this layer had a cobble-packed floor in the SE quadrant. To the west of this floor was a brick-lined well which extended from layer 2 down to a depth of 5m. The north face of the well made up the southern face of the platform (Structure 1). A number of pits filled with brickbats were seen in this layer.

Layer 4 : The NE quadrant of the trench was the only quadrant which could be dug up to layer 6 since structural remains in the other quadrants restricted the excavations. This layer was grayish brown sticky and clayey and was interspersed with brickbats and potsherds. An earlier structural level was encountered in this layer. Two courses of bricks which appear to be a wall were seen extending east-west and against the section of the SE quadrant. The excavators noticed a pit filled with potsherds, brickbats and lenses of sandy clay adjacent to this structure.

Layer 5 : Was the sandy silty river alluvium, natural soil on which the settlement was established. It was homogeneous in nature and yellowish brown in color.

Layer 6 : This layer was encountered at the base of the

pit in a restricted area of the quadrant. It was darker, brownish-black and was sticky alluvium. The base of the pit was lined with a large storage vessel base.

The number of artifacts recovered from this excavation was not very high. Bangle fragments, glass fragments, beads of glass, carnelian and agate, terracotta hop-scotches, nails and iron fragments, copper fragments, coins of silver and copper, fragment of a stone vessel, a small terracotta Ganesha figurine and a silver ring were amongst the artifacts found. The ring is of significance since it is identical to two rings recovered from the Dokhma excavations (Gupta *et al* 2005: 59)

Palaeobotanical remains were also recovered through wet flotation by Dr. Kajale of Deccan College Pune. The grains identified in the preliminary analysis are rice (*Oryza sativa* Linn.), wheat (*Triticum* sp., *Triticum* cf. *Aestivum*), lentils (*Lens esculenta* Moench.), Hyacinth beans (*Dalichus lablab* Linn.), gram (*Cicer arietinum* Linn.) and ber (*Zizyphus nummularia*). Animal bones and shells were also retrieved (Gupta *et al* 2005: 60; Dr Kajale, personal communication).

The preliminary conclusions presented at the end of the third season state that "The structural remains, ringwells (found only in the first season), foreign and local ceramics, artifacts, and the extensive sculptural and structural remains of stone temples clearly point to a very active and flourishing urban milieu at Sanjan from the 8th to the 13th centuries A.D. The foreign ceramics and artifacts clearly point towards Sanjan playing an active role in the Indian Ocean Trade during the above mentioned period. The evidence of a big Zoroastrian mortuary structure clearly confirms the presence of a large Zoroastrian (Parsi) population at Sanjan at this time."(Gupta *et al* 2005: 60 – 61)

Caveats and Limitations of the Excavations :

The excavations at Sanjan had certain caveats and limitations which it would be pertinent to note so that a balanced and realistic perspective can be maintained.

One of the important drawbacks faced by the archaeologists was the lack of space on the Sanjan Bandar to facilitate horizontal excavations which would have been the ideal requirement for a site such as this. The Bandar mound had the thickest habitation deposit and was probably representative of the most sustained and intense occupation at the site. A horizontal excavation would have facilitated a better understanding of the nature of the site, the structures and settlement and of the cultural material. Present day settlements and human activities on the mound have severely restricted the area available for archaeological work.

The presence of modern habitation, cultivation, human activity, vegetation, erosion, flash floods, animal activity, etc have all contributed to the disturbance of the site.

The method of excavation using the allotment of lot numbers by depth meant that horizontal digs carried certain reference numbers which would place all finds from that dig within a certain depth-specific context. During the excavations of first season, the exposure of the ring-wells and excavation of the area around them was done vertically. These vertical digs were also given lot numbers but it was found that the depth variations could not be accounted for, nor were the lot numbers indicative of the part of the quadrant they came from. Hence pottery from these lots was discarded as being totally out of context. Only diagnostic sherds could be retained.

The presence of pits in excavated trenches often resulted in the material from inside the pit and the outside getting mixed up. These pits were not always immediately identified as such, resulting in the admixture of material from the stratified context and the contents of the pits.

The well found in the third season of excavations, in TT4 SW, was capped by and belonged to layer 2. However, the entire deposit from inside the well which had a depth of 5m. was also allotted to layer 2. This implies that the well deposit which is akin to pit material, is in fact the same as the layer 2 encountered in the stratified context of the other quadrants. In actuality, the well deposit is a silty sandy fill deposit. The material from the quadrant has been taken for study of the representative ware classes and typological analysis. But the statistics for this quadrant cannot be taken at par with those of the other quadrants having a stratigraphic profile.

The intense structural activity of the past was evidenced by the structural remains, debris and brickbats in the excavations. The presence of these frequently made it difficult to maintain levels while excavating or to get a clear stratigraphic profile. Only one quadrant in the excavations on the Bandar could provide a stratigraphy for the ceramics, despite some disturbances.

The necessity to excavate trenches in varying localities during the three seasons resulted in a lack of continuity and proved to be a detriment to the interpretation of structures and features when encountered in trenches. The interpretation of artifacts and ceramics and the recognition of distribution patterns in order to establish activity areas were also severely restricted, especially on the Bandar mound.

It is with these caveats in mind that the present study has been undertaken so that a cautious approach may be taken in the study of the rich body of material brought to light by the Sanjan excavations.

(Images at :_
http://www.wzcf.org/sanjan-excavations/pottery.html.)

* * *

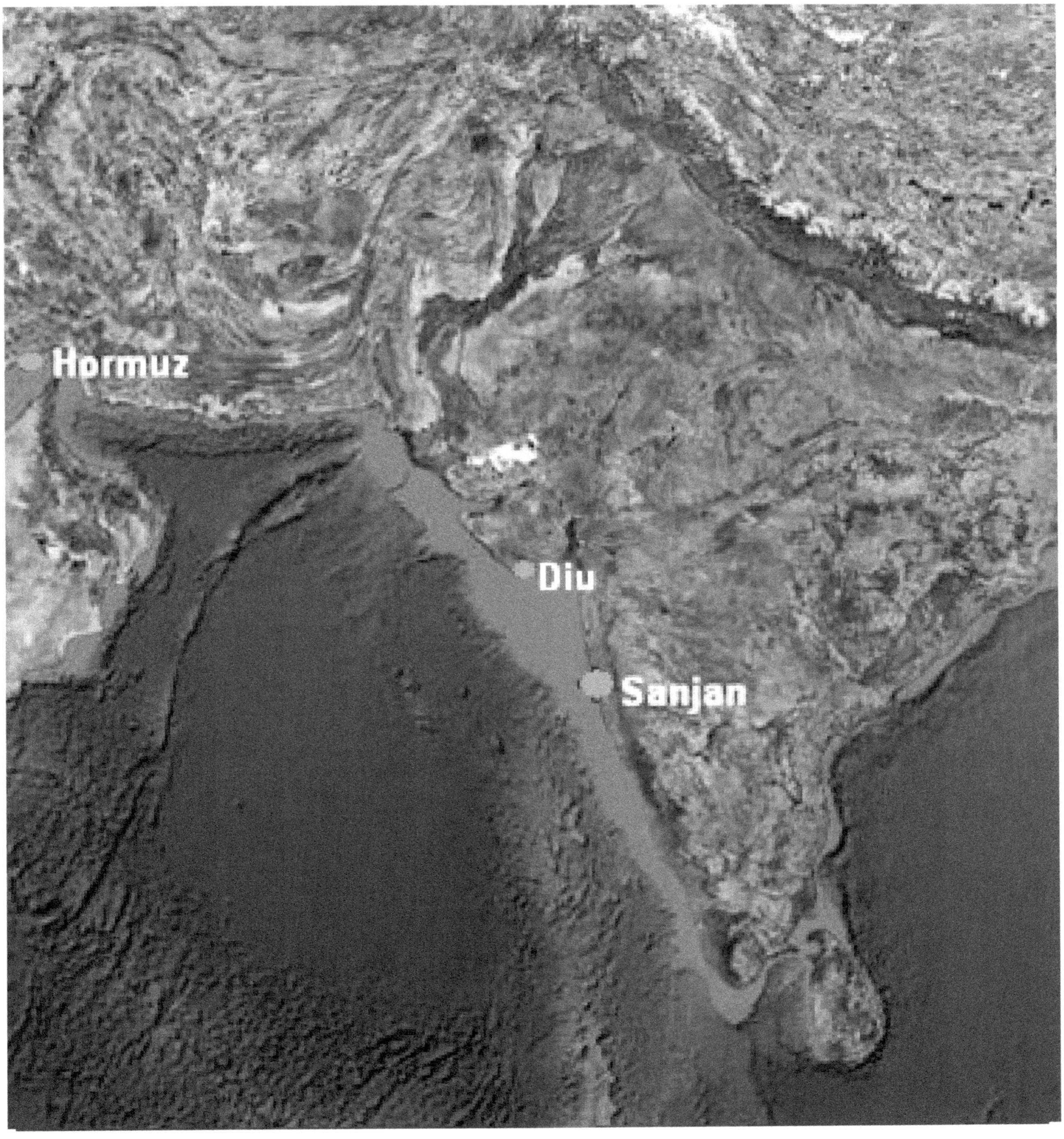

Fig. 1 - Location of Sanjan on the West Coast of India.

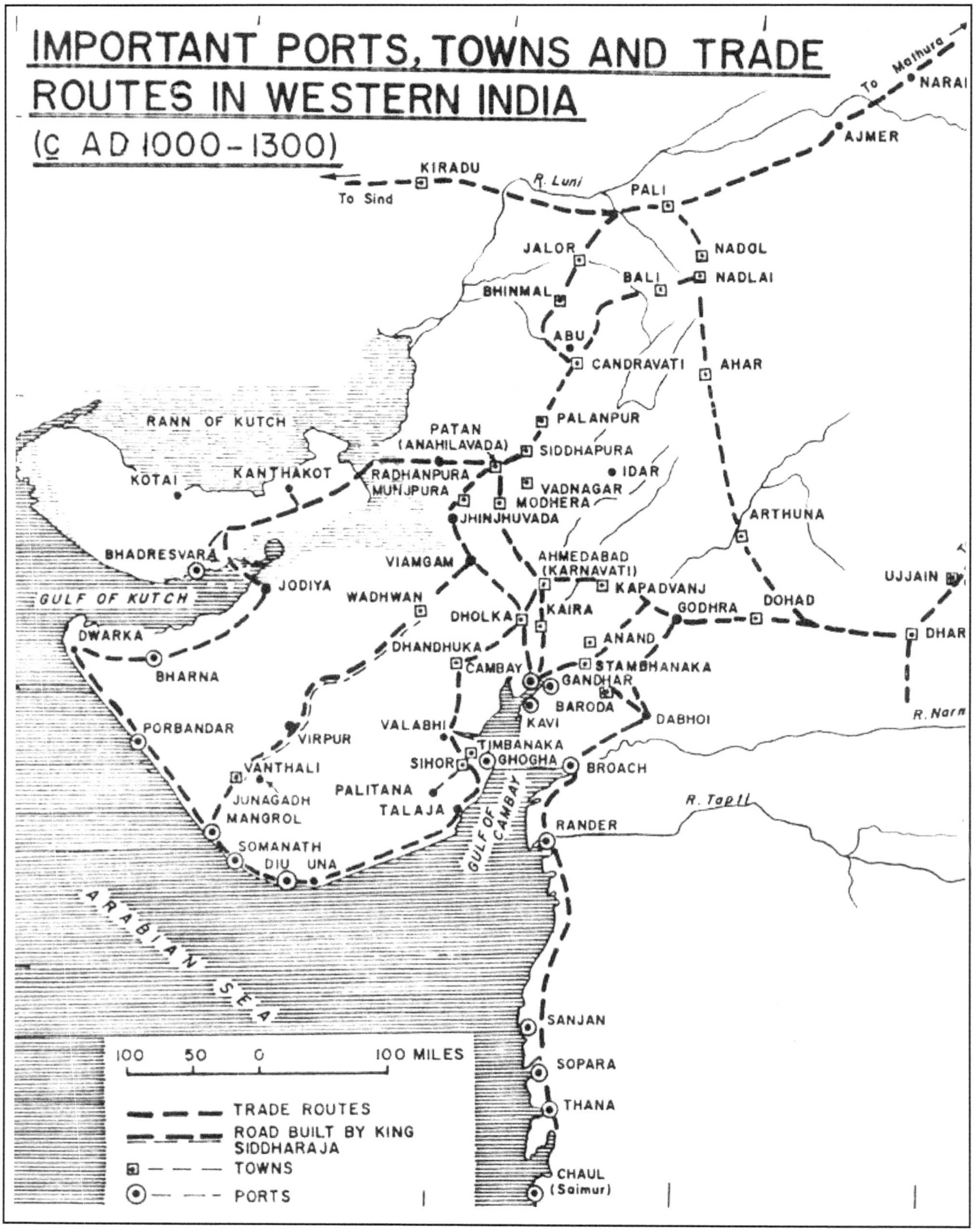

Fig. 2 - Important Trade Routes (Courtsey - Janaki, 1969)

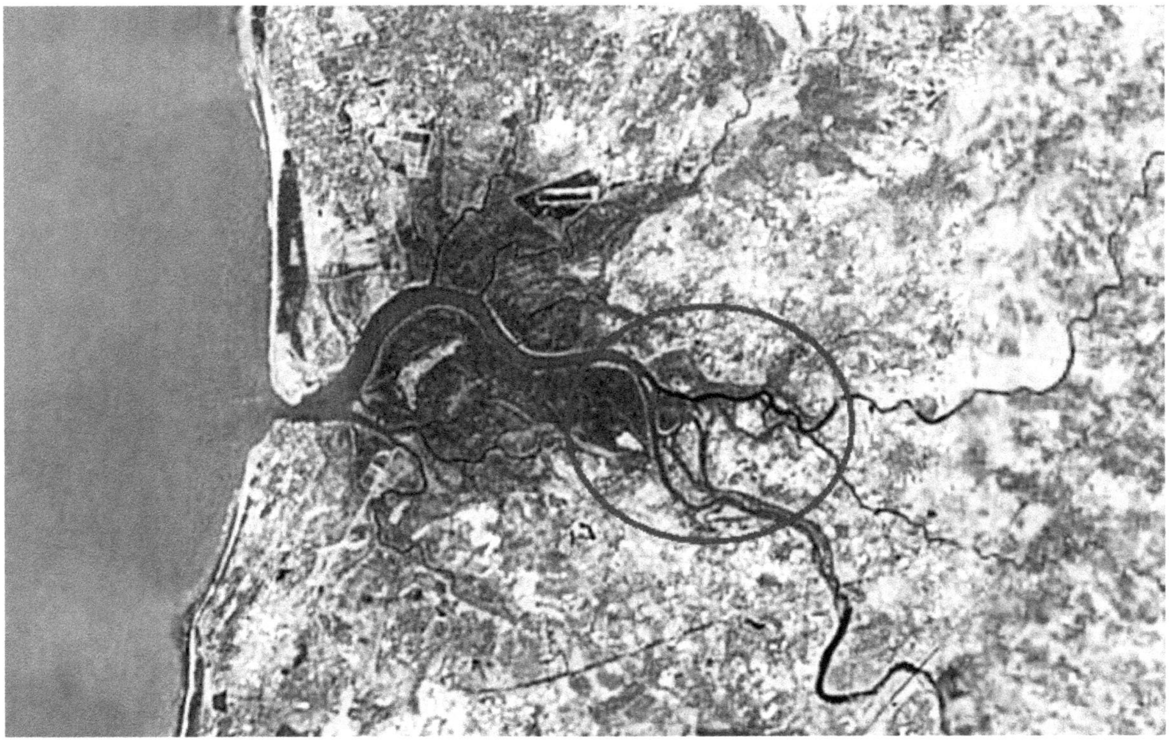

Fig. 3 & 4 - Estuary of the Varoli River (Satellite image above and drawing below)

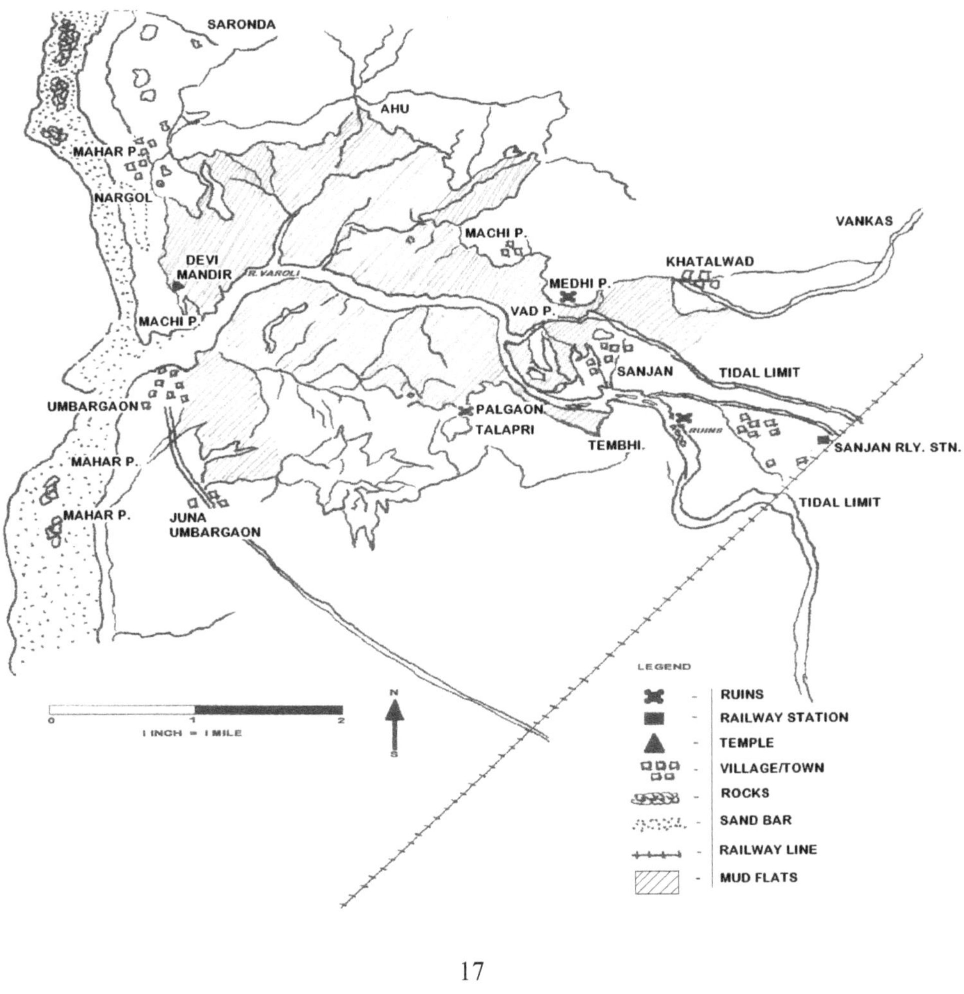

Fig. 5 - View of Sanjan Bandar from the South Bank of Varoli

Fig. 6 - Structures and Features in TT1 and TT2.

Fig. 7 - Structures and Features in TT1 and TT2.

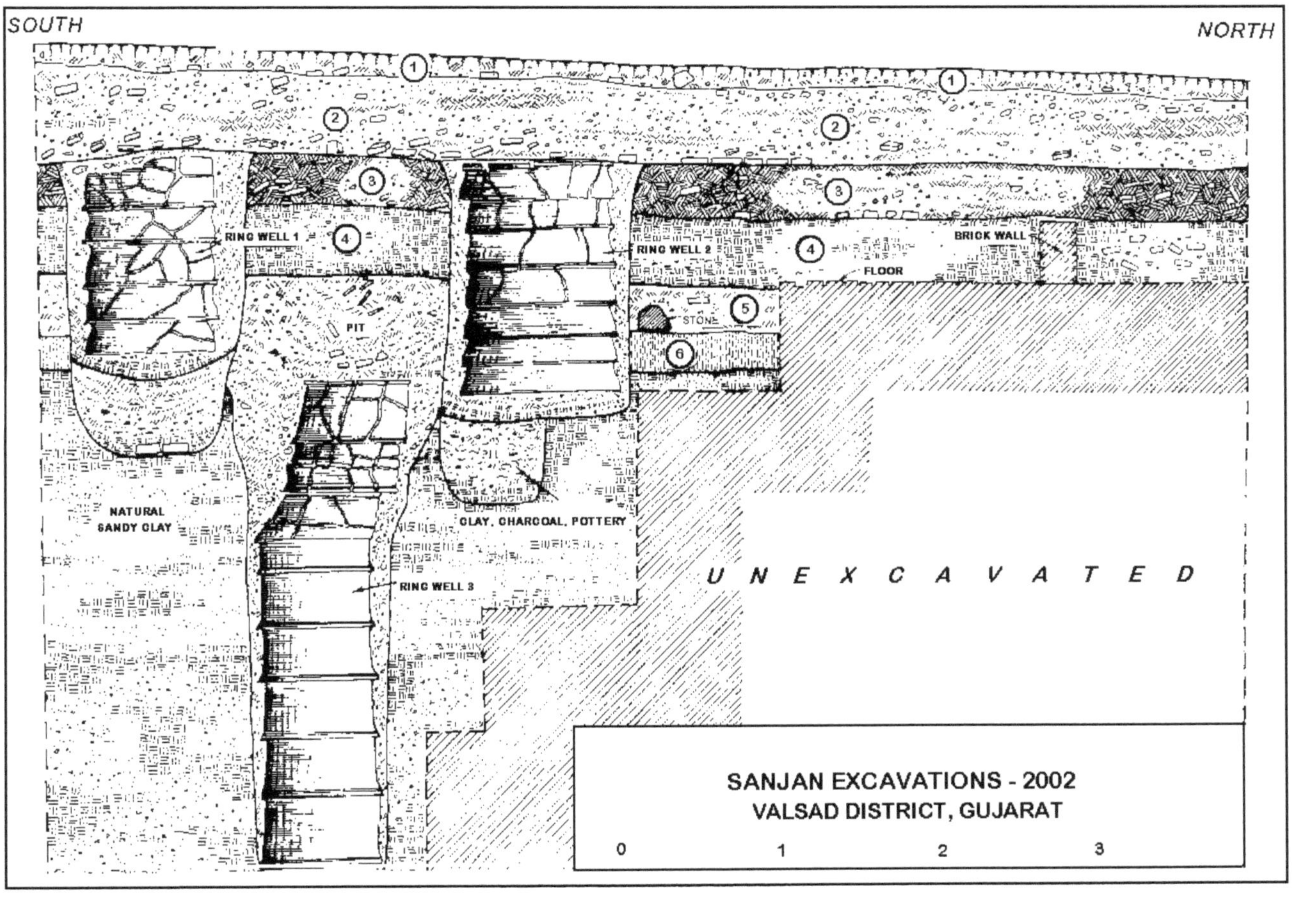

Fig. 8 - Ringwells in TT1.

Fig. 9 - Structures and Features at Koli Khadi.

Fig. 10 - Human remains at Koli Khadi.

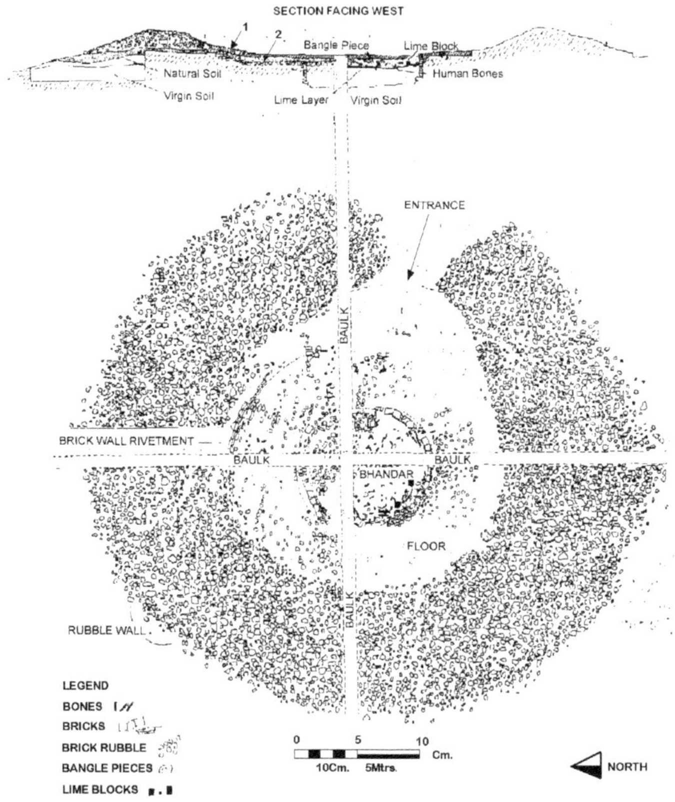

Fig. 11 - Dokhma Structure.

Fig. 12 - Central Well of the Dokhma with Lime-plastered bricks.

Fig. 13 - Brick Platform, Well and Floor in TT4.

Fig. 14 - Early Structural Phase in TT4.

CHAPTER II - SANJAN CERAMICS

"If a potter has an idea, she makes it into a pot and it exists beyond her, in its own separate life. She uses a physical substance to display her thoughts."

- Jeanette Winterson

Aims of the Present Study :

This report aims at filling the lacuna that exists in Early Medieval archaeology in India. No comprehensive study of pottery belonging to this period has been undertaken to date nor is there a referential collection available as a precedent model. West Asian and Far Eastern wares have been subject to study on sites outside India and have been the focus of work by archaeologists as well as art historians. Hence they can provide a chronological model for the study of indigenous pottery which has been found along with these traded ceramics at Sanjan. This has relevance for the study of similar material from sites both in India and abroad since Indian pottery has been reported from numerous sites outside India but its chronology and provenance remains speculative. The role of Sanjan in the maritime trade of the Indian Ocean and the nature of the contacts between the distant ports in the Indian Ocean littoral can be assessed and defined through the study of these trade ceramics. Information gleaned from historical and literary sources about Sanjan and its migrant, mercantile and local populations can be evaluated and either verified or refuted on the basis of the archaeological evidence, a prominent part of which is the pottery assemblage.

This study aims at -

1 the identification of meaningful categories of pottery which may have remained unknown in previous Indian archaeological research.

2. the co-relation of known and identified categories of pottery with previously unrecorded wares
3. the morphological, statistical and petrographical analysis of ware categories in order to establish a data-base.

4. the establishment of a chronology for the wares and, by extension, for the site.

5. the interpretation of the site on the basis of the ceramic assemblage.

6. the assessment of trade networks, cultural contacts and inter-regional dynamics during the Early Medieval period in the Indian Ocean on the basis of the ceramic evidence, and,

7. the establishment of a precedent for future studies of Early Medieval ceramics in India.

Field Methods and Preliminary Analysis of Pottery at Sanjan :

The pottery collected at Sanjan during the three seasons of excavations and the pottery collected during explorations forms the corpus of the material under present study. Some of the unique and diagnostic sherds found on the surface have been included in the morphological study of the pottery and for illustration.

As explained in the Preliminary Report of the first season of excavations at Sanjan (Gupta et al 2002: 184), explorations on the Sanjan Bandar mound and in the surrounding areas prior to commencement of excavations brought to light a wide range of glazed and unglazed pottery. Most of this pottery was unknown and unfamiliar to the archaeologists. A methodology was therefore adopted which it was thought would serve the requirements of this unfamiliar body of material. The preliminary identification and nomenclature used for the wares was meant to be both descriptive as well as distinctive, indicating both the fabric colour and the surface treatment. Texture and composition of the fabric was not taken as a criterion and when terms such as 'Coarse' were used at all (such as Coarse Red Ware or Coarse Grey Ware), it was in a general and loose sense implying that there was no surface treatment (Gupta et al 2002: 184). The field methodology established in the first season continued to be used for all the seasons of excavation. This preliminary work was done by the author along with Dr. Abhijit Dandekar in the first two seasons and was continued by the author singularly in the third season and after.

Pottery yards were organized close to the excavation trenches and the material collected during the digs was subjected to preliminary study in them. The usual conventions of washing, sun-drying, sorting, counting, weighing and registering the pottery were followed. Since the lot system of numbering the finds was the methodology followed by the excavators, the same system was used for the pottery (A note should be made here – the lot number specifically mentions a particular depth. However, this does not imply that the material collected came only from that particular depth. It refers to the material collected from the deposit between the previous lot and the present one. For example, lot no.101 records a depth of 0.35m, indicating that the material collected does not occur only at a depth of 0.35m, but in actuality occurs from the datum level to a depth of 0.35m. Similarly, lot no.102 records a depth of 0.40m, indicating that the material so labelled comes from the deposit between 0.35m and 0.40m.). Pottery collected in each dig was therefore labelled with the depth-specific lot number, trench and quadrant designation and other excavation details. All diagnostic sherds were retained. These were rims, bases, body features such as handles, spouts, finials, etc. and sherds with decoration or ornament. Some sherds were set aside for verification and illustration.

The pottery collected during the first season of excavations was categorized into general classes which were meant to be flexible and to allow for modifications in nomenclature as and when confirmed. Two wares were identified in the second season – Turquoise Glazed Ware and Celadon (Gupta et al 2004: 101) – and were therefore classified as such, distinctly separate from Glazed Buff Ware and Glazed Grey / Stone Ware. Classes such as Lustre Painted Ware and Hatched sgraffiato Ware were identified later but were not used in the categorization (Gupta et al 2004: 106).

The ware categories identified in the preliminary analysis and published in the preliminary reports (Gupta et al 2002, 2003; also, Nanji and Dandekar 2005: 69 - 72) are mentioned below:

Porcelain, Glazed Grey / Stone Ware, Celadon, Glazed Pink Ware, Pink Ware, Red Slipped Pink Ware, Glazed Buff Ware, Turquoise Glazed Ware, Mica Washed Red Ware, Slipped Grey Ware, Coarse Grey Ware, Slipped Red Ware, Coarse Red Ware, Red Polished Ware and, in the case of pottery from the Sanjan Dokhma in 2004, Hatched sgraffiato Ware.

The pottery of the third season was initially sorted on the basis of the same methodology as the previous years. Only a brief preliminary report on the pottery found at the Dokhma was been published (Nanji and Dandekar 2005: 69 – 72). The pottery from the main habitation area, trench TT4 was not included.

Limitations of the Preliminary Methodology : The methodology adopted in the preliminary analysis, both in the field and during the preparation of preliminary season reports, was a general system adopted to facilitate the organization and understanding of a body of material which was unknown and had no precedent model or study as a reference point in India. This system had several limitations which made it totally unsuitable for the final detailed analysis of the pottery and for the interpretation of the assemblage. These are listed below in brief :

1. The nomenclature used for the ware classes was totally different from the international nomenclature used for the same wares found on sites outside India. The categories formulated at Sanjan were general, broad classes which, at most times, encompassed many distinctly separate wares in one group when in reality they had little in common. For example, Glazed Buff Ware included Turquoise Glazed Ware, Lustre Painted Ware, White Glazed wares, and sometimes even unglazed wares, with the presumption that the glaze had worn off or eroded. Similarly other classes too encompassed categories which should, by rights, have been distinct and separate.

2. The nomenclature identified the colour of the fabric as a distinction – buff, pink, red, grey, etc. However, the system used to identify the fabric colour was not scientific and was highly subjective. The method of streaking the sherd on a stone and checking the tone of the colour in order to identify it was found to be totally unacceptable and arbitrary to scientists working on pottery analysis (Prof. V. D. Gogte, personal communication; also, Prof. Krishnan, personal communication) as well as to the international scholars who are more familiar with this body of material (Dr. Derek Kennet, personal communication). This method was not used in conjunction with the usual conventions of pottery analysis – i.e. the study of fabric by lens along with observations of inclusions, texture, production technology, tempering material, firing, etc. The Munsell colour chart which is used as basic reference for colour identification was not employed in this analysis either. Hence a certain amount of subjectivity and error was bound to be present.

3. Erroneous identification was done in the case of most wares. This was due to a lack of referential material and access to other collections. For example, in the second season all stonewares having a green glaze were identified as Celadon.

4. No sub-categories were made to distinguish within broader classes.

5. An attempt was made to keep the fabric as a constant and surface treatment as a variable. This resulted in infusing a certain amount of rigidity in the system so that the most distinctive features of

some wares were not taken into consideration. For example, decorative techniques were the distinguishing features of certain classes such as Lustre Painted wares, Hatched sgraffiato Wares etc. However they had been categorized as Glazed Buff Ware and Glazed Pink Ware in the first two seasons. It should also be noted that the fabric was not identified on any scientific basis or examination.

6. Some of the nomenclature was misleading and could create misconceptions. For example the ware categorized as Mica Washed Red Ware leads one to suppose that the fabric is the same as the slipped and plain red wares. In reality the fabric is quite different and that the mica is not present only as surface treatment but is present in the clay as well.

7. No statistical analysis was done in the preliminary analysis.

8. No petrographic studies were undertaken to scientifically establish the profiles for the various wares and fabrics.

9. No parallel collections could be accessed to provide a reference point for identifying the wares.

10. Very little published data and reference material was available in India to help in the analysis of this assemblage.

The preliminary methodology had to be completely abandoned and a new methodology had to be adopted by the author to re-sort and study the pottery of all three seasons in detail when it was found that the earlier system was unsuitable. A system of analysis, nomenclature and classification has been used to study the entire corpus of retained material based on the conventions followed by other scholars working in the Persian Gulf region.

Methodology devised and used for Final Analysis :

The system adopted for the final analysis has tried to address the issues which could not be done earlier. However, it was not possible to reassess the entire collection since much of the material had been discarded in the field either due to its non-diagnostic nature or due to non-contextual collection during excavation. All the diagnostic sherds had been retained and these have been subjected to analysis on the basis of a methodology that attempts to accommodate the established categorization of West Asian and Far Eastern Wares, formulate categories for unreported / unidentified wares, define fabric by code and note its variations, scientifically verify such classification and present a typological model for each class. Flexibility has been retained in this methodology so that distinctive traits become the defining hallmarks of each class.

i. The pottery has been sorted and registered on the basis of Locality, Trench and Quadrant, Lot number and depth, Layer and any specific context such as well, pit, section, etc. which was mentioned on the excavation labels.

ii. Each sherd has been noted as a separate entry with all the above mentioned contextual details.

iii. An x 10 hand lens has been used for observation of all details which have then been entered along with the contextual data. The details noted are as follows:

a. Ware Class and sub-class

b. Fabric class and sub-class

c. Categorization in the preliminary analysis

d. Surface treatment

e. Shape, design and diagnostic features

f. Measurements such as diameter, thickness, length and breadth,

g. External and Internal colour

h. Description and observations – tempering, texture of the fabric, firing, production technology, inclusions, etc.

iv. The terminology adopted for most of the West Asian and some Far Eastern classes of wares is the accepted nomenclature which was verified both through first-hand archival reference work and through personal consultation with subject experts. Wherever nomenclature had to be formulated, it has been done taking into account the most distinguishing and distinctive feature as definition. This was found to be more specific than would be possible within a system that rigidly aims at maintaining one constant criterion and others as variables.

v. Sub-classes have been formed on the basis of distinctions noticed within a class.

vi. Some sherds which are distinctly different from known classes and which are too few in number have been classified as Miscellaneous. These have been described and numbered so that they may provide a reference point. Some of these are not diagnostic sherds and so not much can be said about them. Similarly a number of glazed wares which were represented by only one or two sherds and whose identities were not established - all of which belonged to fabric group PINK - have been respectively described in the section Unique Glazed Pink Ware.

vii. Fabric identification has been done on the basis of observations made by use of hand lens. This identification tries to take into account the inclusions, temper, texture, colour, firing, etc.. Sub-groups within the fabric have been distinguished wherever variation has been noted.

viii. The fabric identification done visually by use of lens has been verified through petrographic analysis of thin-sections wherever possible, given the paucity of time.

ix. Measurement of rim diameters has been done by taking the diameter of the orifice or mouth of the vessel into consideration. Diameters of ring bases have been measured taking the outer rim into consideration. Thickness of body sherds can vary depending upon the parts of the vessel they belong to. Hence average thickness has been taken. Features such as spouts have been measured for length and for the diameter of the perforation.

x. Illustrated Sherds have been numbered using codes. All rim shapes and body features such as handles, spouts, finials, etc have been allotted numbers with code 'T', for example, T.100. Sherds with designs and decoration have been assigned 'D', for example D.15. All bases are numbered 'Bs' for example Bs.10. These can occur in combinations if the rim or base has a decoration on it. For example, rim T.47.1 can have a design on it which is D.14. In which case it is numbered T.47.1 D.14. Similarly a base sherd can have a design number combined with it.

xi. Statistical analysis of the assemblage has been undertaken.

xii. Cross-referencing of the Sanjan assemblage with similar material from other sites both in India and abroad has been done. Archival material from museums as well as referential published material has been accessed and subject experts have been consulted.

xiii. Trench TT4 has been considered the index trench for the construction of a ceramic sequence since the trenches TT1 and TT2 were greatly disturbed. Maximum depth was reached in the North-East (NE) quadrant of trench TT4 which was dug down to virgin soil. This quadrant, measuring 2.5m x 2.5m, has been taken as an index. TT4, NW and TT4, SE have also been considered for co-relation where necessary as they were relatively undisturbed. The presence of structures in these quadrants prevented complete excavation. Trenches TT1 and TT2 have been mentioned as context for certain wares but the stratigraphy remains somewhat uncertain.

Classification : Several classes of pottery wares have been identified at Sanjan. These have been allotted a code which is written in upper case throughout this report. The term 'ware class' is used in this classification for a body of material within the assemblage which displays "a consistently similar range of fabric, decoration, form and other variables" (Kennet 2004: 86). A subclass consists of a body of material from within the ware class which displays a variation from the parent group, while still maintaining the defining characteristics. This variation could be in decoration, shape, surface treatment, manufacturing technique, etc.

Fabric classes have been similarly defined on the basis of common characteristics. Sub-classes within these have been noted on the basis of inclusions, texture or some variation in the material. In the following table, fabric groups identified in the petrographic analysis are mentioned alongside the archaeological categories to facilitate co-relation. It has been found that sometimes visually diverse archaeological fabric classes have similar or identical mineralogical profiles or vice versa.

The term 'ceramics' has been used in the course of this report in a general sense, sometimes interchangeably with the term 'pottery', and does not indicate any technologically specific group.

Petrographic Analysis : The analysis of the thin-sections and the preparation of the report has been done by Prof. K. Krishnan, Department of Ancient Indian History and Archaeology, Maharaja Sayajirao University, Baroda. The petrographic analysis of the pottery from Sanjan was undertaken as a pilot study since constraints of time did not permit a more elaborate study. It is possible that a more detailed and fuller analysis may change some of these preliminary findings. However, this is an extremely important and innovative step since no attempt has been previously made to chemically analyze or mineralogically define Early Medieval pottery in India. The full report and graphs are given in the Appendix. However it would be prudent to include the methodology and the aims of the study at this point.

The thin-sections of forty-five samples were subject to this analysis. They represented the ware classes from West Asia and indigenous wares which were prominent in the assemblage. Far Eastern Wares and Miscellaneous classes were not included in the study. Some fabric samples of less prominently represented wares could not be analyzed due to a paucity of time.

The objectives of the study and the methodology used have been defined by Prof. Krishnan in his report as follows:

(i) "to characterize the ceramics from Sanjan into different petrographic groups based on their composition,

(ii) to see how far these petrographic groups are comparable to the archaeological categories,

(iii) to understand the geological provenance of the raw-material,

(iv) to understand the texture of the clay paste,

(v) to understand the technique of manufacturing, and

(vi) to understand the microstructural features of different wares.

The thin-sections were studied using a polarizing microscope [Leitz Laborlux (12 Pol D)]. The minerals were viewed in plane polarized light (PPL) and crossed polarized light (XPL) and identified with the help of their optical properties. Grain size measurements of the representative area of each fabric were done with the help of point counting (James Swift Automatic Point Counter with a stepping stage attachment to it). The point counting was done by setting the stepping stage movement to single jump.

Based on the mineralogy the thin-section samples were grouped into ten main groups, namely A, B, C, D, E, F, G, H, I and J. The criteria chosen for defining the group was the mineralogical composition of the grains."

The fabric groups were then co-related to the archaeological groups so that a mineralogical profile could be derived for the different fabrics. Sometimes apparently diverse wares and fabrics produced similar mineralogical profiles either indicative of different traditions and / or techniques within the same regional context or a common source of raw material for the production of diverse wares. Conversely, sometimes similar fabrics and wares have provided differing profiles. The changes and variations also reflect a stylistic, typological or technological shift which could be chronologically relevant. These aspects have also been considered in the study of these fabrics.

Four geological sources for the fabric clays are indicated in the study. These are

a. clay from the vicinity of a basaltic region

b. clay with inclusions of basaltic minerals

c. clay which is rich in quartz, and

d. clay which contains mica

Temper and texture of the fabrics has also been taken into account.

The thin-section analysis forms a part of the following ceramic report and the interpretation of the assemblage. It is discussed in the following sections and is presented in the appendix.

Morphology of the Sanjan Ceramic Assemblage :

Glazed Earthenwares :

These do not form a very large percentage of the Sanjan collection as compared to the unglazed wares. But they are of great relevance both in establishing the identity and nature of the site as well as in building a chronology. These wares are mostly West Asian in origin and their presence on the Indian west coast is significant in the light of Sanjan as a known settlement of Persian migrants as well as a centre for trans-oceanic trade contacts in the early medieval period. The range of wares is vast and diverse.

Turquoise Glazed Ware (Ware code: TGW)

Definition : This is one of the most quantitatively significant glazed earthenware at Sanjan. It is also one of the most widely distributed West Asian ware in the Indian

Ocean. It has been reported from sites as far flung as Japan in the east (Glover 2002: 165-177) and the East African coast in the west. Almost all historical and early medieval port sites in the Indian Ocean littoral report the presence of this ware. It is distinctive and easily recognizable. The tradition of this pottery is reportedly quite early and spans from the Parthian times to the tenth century or so. Arthur Lane suggests that the same blue-green 'Parthian' glaze was used through the Sasanian times (226 – 641 AD) well into the Islamic period (Lane 1947: 9). This long history of the ware has prompted many scholars to refer to it as Sasanian-Islamic Turquoise Glazed ware. The more developed and elaborate forms are seen in the late Sasanian and early Islamic periods, leading to the term Islamic Blue-Glazed ware or Alkaline Glazed Ware. Mason refers to these jars as 'Hib' (Mason 1991: 52). However the term Turquoise Glazed Ware appears to be the most appropriate as it is descriptive but neither denotes a dynastic nor cultural bias. Due to the long tradition of this particular ware, it is not always easy to use this category as fossil index for the chronology of the site, though some scholars do attempt to differentiate between the earlier and the later types.

David Whitehouse describes it well while discussing the Siraf assemblage - "The so-called "Sasanian-Islamic" ware, which has a blue-green alkaline glaze and often bears applied, incised and stamped ornament, is the commonest type of glazed pottery at Siraf. It has a soft, sandy buff fabric and is thickly covered with a glaze which varies in colour from blue to pale green and is usually darker on the outside of the pot than the interior. At Siraf the glaze is often corroded and the original colour cannot be seen; many sherds simply have a matt greyish surface. The decoration may be applied, incised, "chip-carved" or stamped and all four techniques may appear on a single pot. However, most of the ornament is applied. Sasanian-Islamic pottery was widely used in Mesopotamia and the coastal regions of Fars and Makran. Vessels of Mesopotamian or Persian origin were exported from the entrepots of the Persian Gulf and sherds have been recorded at Manda and Unguja Ukuu in Tanzania and at Bambhore in West Pakistan; examples even reached south-east Asia. At Siraf pottery with blue-green glaze was already present in Period I and it occurs throughout the later deposits." (Whitehouse 1968: 14)

Shapes and forms : The most common shapes at Sanjan seem to be straight-necked jars, neck-less or hole-mouthed jars with sloping shoulders, globular jars, jars with handles, bowls, and a few saucer lamps. None of the vessels have residue or soot deposit. The shapes, decoration and potting are clearly indicative of storage vessels, table wares and utility objects. The sturdy, large jars and pots could well have been used for the transportation of goods and could themselves have been items of trade for their aesthetic value. The blue glazed vessels belong to bowl, dish, jar, pot and saucer lamp forms. The bowls follow Chinese shapes (Tampoe 1989: 31-33), though the straight-sided bowl shapes of West Asia are also present. Strap handles, small faux handles and loop handles are represented. The neck-less jars with sloping shoulders and barbotine decoration are identical to the ones from Siraf, Susa (Lane 1994: Pl.3; also Whitehouse 1968: Pl.VI) and other sites in Iran and Iraq. High necked jars with or without handles are prominent in the Sanjan collection. One such jar has three large strap handles extending from rim to shoulder.

Glaze and surface treatment : The most distinguishing feature of this ware is the monochrome alkaline turquoise glaze which overlies a yellow to cream soft paste body. The shades of the glaze can vary, sometimes dramatically. Sub-divisions of the class can be effected by taking into account the colours / tints of the glaze, fabric composition and decoration. The glaze on the exteriors of the jars, vases, pots and other closed forms is usually quite evenly applied, with some exceptions where the glazing is noticeably uneven. The glaze is applied on both the interior as well as the exterior of the vessels and varies both in colour as well as thickness. The grooves, rills, incised wavy lines or meanders, incised designs and ribbed surfaces are darkened by the pooling of glaze. This pooling of glaze itself becomes a decorative feature of the vessel since the glaze on the raised parts of the vessel appears paler in comparison to the depressed parts of the surface which the glaze has filled. The interior glaze can be the same colour as the exterior or it can be completely different ranging from jet black to grey, pale green, pale blue, cream and sometimes combinations of these. This has been observed in the collection at Tell Abu Sarifa by Adams as well. He notes "dense, uneven glaze on a soft yellow, somewhat flaky, fabric, most characteristically applied to large strap-handled jars of type N but also found occasionally with bowl forms subsumed under type D. Particularly in the case of the jars, there are low horizontal corrugations in the underlying fabric that were formed during the construction process. On the outer surface, these lead to differences in glaze thickness that may appear as alternating bands of slightly lighter and darker hue. Probably because of firing differences, glaze on interior surfaces often is much lighter in colour, even approaching an uneven, dirty white or grey. Bowls being more open and hence uniformly fired, this interior-exterior contrast does not occur s... Although probably beginning in the late Sasanian period, blue-green glaze predominates over all others in all of the Islamic levels" (Adams 1970: 107-108). Frequently trails or swags of thick glaze can be seen on the interior, indicative of an upright firing position in the kiln. Spur marks or stacking marks on the ring bases and sometimes on the rims of jars also indicate upright stacking, perhaps one on top of the other, during firing. The few bowls found have similar marks on the interior surface. The bowls, saucer lamps and open form vessels usually have the same colour glaze on both the surfaces. Some sherds show distinct evidence of vitrification. The glaze is severely bubbled, crackled and reddened. The glaze is quite well preserved on most sherds and flaked on others. But not many sherds in this category have evidence of exfoliation at Sanjan. The glaze is generally turquoise blue in colour in the Iranian examples, with an occasional sea green sherd (Tampoe 1989: 31-33).

Sub-categories : TGW-1 has a dark green glaze on the exterior and most times on the interior as well. Sometimes the interior glaze can be grey or black. The fabric is BUFF(A). It is yellow or creamy yellow with fine quartz particles and some air-holes. It has a chalky texture. The glaze can border on a black tint, especially where it has pooled. Frequently the glaze is extremely glossy with a mirror finish. Most of the sherds appear to be medium thick to thick in potting and belong mostly to jars and heavy-set, large vessel forms. This sub-category constitutes a large quantum of the TGW corpus at Sanjan. The shapes seen in this class are T.35.2, T.44, T.53.1, T.65, T.66, T.67 D.35.1, T.72.1, T.72.2, T.318, T.373 D.168, T.373.1, T.374 and T.375 D.169. The base shapes are Bs.17, Bs.22, Bs. 24, Bs.26, Bs.27, Bs.32, Bs.33, Bs.35, Bs.67, Bs. 68, Bs.69 and Bs.70. A large number of decorations are seen in this sub-class – D.28, D.28.1, D.29, D.30.1, D.31.1, D.31.2, D.32, D.34, D.35, D.35.1, D.35.2, D.35.3, D.35.4 and D.110.

TGW-2 has a paler green glaze with a grey tinge to it. This sub-category consists mainly of bowls. The potting is quite fine as compared to the others. The glaze has a matte finish or may have weathered in this manner. The fabric, BUFF(B) is sandier and hard-fired. It ranges from a creamy white to a very pale pink. The glaze on both surfaces is the same and adheres to the body quite well in most cases. The bowl shape is usually straight-sided with a slightly inverted rim. The shapes present in this sub-class are T.14.1, T.49, T.53.1 and T.376. The bases are Bs.16 and Bs.19. Only one sherd with design is seen – D.174.

TGW-3 has a bright green glaze which resembles the colour of the monochrome wares of the thirteenth and fourteenth centuries. The fabric of this subcategory is BUFF(A), same as TGW-1. The shapes include T.30, T.42.2, T.67, t.72, T.73.2 and T.377. The only base found in this category is Bs.29. Three designs are seen – D.28.2, D.30.1 and D.35.1.

TGW-4 has a blue tone which can range from dark blue-black to turquoise blue-green. More interestingly, some sherds have the true colour of the turquoise stone. The glaze can be extremely glossy. The interior glaze is mostly black, grey, cream or pale green. The fabric is the same as TGW-1. Many shapes are the same as in TGW-1 as well. The shapes seen in this sub-class are T.65.1, T.65 D.26, T.67, T.76, T.73.2, T.376 and T.283. The base shapes are Bs.33 and Bs.34. The designs are D.29, D.30, D.31, D.32, D.33, D.35.2, D.36 and D.171. This sub-category is quantitatively well represented.

TGW-5 has a leaf green glaze thinly and unevenly applied on the exterior while the interior glaze is black and cream or grey, thickly applied with pooling in depressed parts of the surface. Occasionally the interior may have the same glaze as the exterior. At least two samples from the lower levels at Sanjan have crude potting with lumps of clay and glaze adhering to the base and to the ring foot. No decoration is seen on any of these samples and it is likely that they are chronologically earlier than the other sub-categories since they do not occur in later levels. Deep grooves created by coiling can be seen on the interiors and the internal glaze is as thick as 5mm in the well of the vessel. The shapes in this sub-category are few – T.42.1, T.64 and T.376. The base is represented by Bs.68.

TGW-6 is represented by only one small body sherd. It has a bright green and black glaze. The black appears to underlie the green and may be underglaze painted. The fragment is too small to facilitate a proper identification. The fabric, BUFF(C), is much softer and chalkier with no visible inclusions and few airholes.

Kennet divides this category at Kush and Ras al-Khaimah into six sub-categories on the basis of glaze tint and fabric (Kennet 2004: 28-31). The sub-divisions are different from the classification done in this study since the glazes, fabrics and even the post-depositional weathering of the sherds from Ras al-Khaimah and Kush appear to follow different patterns from those at Sanjan. The mustard yellow and white glazes reported by him are totally absent at Sanjan. Horton also notes these differences in glaze tones at Shanga and relates them to the colours of the interior glazes. He suggests that the sherds with black internal glaze may belong to the earlier periods (Horton 1996: 217). The shapes and decorations described by him closely match those at Sanjan.

Decoration : The decoration of these vessels can range from the plain to the elaborate. A variety of techniques and traditions are evident. While some of the vessels depend upon the vessel profile and the depth of colour to act as decoration, many vessels have bosses, stamps, appliqué designs, barbotine ornament and piped lines, spirals, dots, and bands applied to them. Bands with finger impressed chain designs, either round or square, are seen on the shoulders of some of the larger vessels. Faux handles are also commonly seen attached horizontally to the shoulder. The straight necked jars often have an incised wavy line or meander just under the rim, with / without rilling or ribbing. Some vessels with strap handles may not have been strictly utilitarian but may also have been decorative – for e.g. the three handled jar from season 1. The more decorative forms are taken to belong to the later period, i.e. the early centuries of the Islamic period. Turquoise Glazed Ware with appliqué decoration is normally dated to the 8th / 9th century (Kennet, 2004: 30; also, Whitehouse 1979b:881; and Mason & Keall 1991:52). The green glazed jar bases found from the lower levels of SJN-B TT4 closely resemble those reported by Oliver Watson from the Al-Sabah collection which he dates to the 7th or 8th century A.D. (Watson 2004: 158-159). Watson classifies Islamic wares into "families". This category he calls Early "Green Glazed" Wares (Watson 2004: 156-160). The heavy ring-foot and the somewhat careless glazing with clay still adhering to the base and the rounded globular profile of the body indicate an early date in the Sanjan samples. The prominent coils on the inside of the vessel and the smoothened exterior are indicative of a combined technique of production. The barbotine decoration seen on some of the Sanjan sherds has close parallels and, in some cases, is identical to the decoration on sherds found at sites such as Unguja Ukuu on Zanzibar (Juma 2004: 110), 'Ana (Northedge et al 1988: 93), Suhar (Kevran 2004: 307, 324-25, 361), etc. In fact, one large storage jar found at Susa in Iran and catalogued by Arthur Lane finds an exact replica in the barbotine sherds found at Sanjan. The Susa jar is dated by Lane to c.7th – 8th century (Lane 1947 : 9, and plate 3). The sherds illustrated by Chittick in the Manda excavation report (Chittick 1984: plate 26, a, b, c, d and e) are identical to those at Sanjan as well.

Fabric : The fabric of this ware is usually a soft yellow to cream or buff paste with some white quartz particles, a few air-holes and a sand temper. Most often the clay is well levigated and in some samples, a little grainy. The visual examination of the sherds indicated that these wares have three fabrics which are similar and the differences may be minor, as in firing or a variation in temper. The fabric sub-classes are BUFF(A), BUFF(B) and in one case, BUFF(C). However, two petrographic fabric groups have been identified in the chemical analysis of these wares.

Petrographic analysis : Thin section analysis of these subcategories brought forth some more information regarding the petro-fabrics. TGW-1 was found to belong to fabric groups D and E. TGW-2 occurred only in group D. TGW-4 and TGW-5 have fabric group E.

The thin-sections falling under the Fabric Group D had a fine grained smooth matrix. The samples were non-pleochroic in PPL and dark grey in XPL. The grains were sub-angular to sub-rounded in shape. The largest grains observed in the samples had sizes of 160 μ and 700 μ. The non-plastic inclusions were perfectly sorted to well sorted and had a frequency of 5 to 10%. The grain size distribution character was unimodal (Fig 5). Most of the grains showed more or less a parallel orientation. Voids were few, the majority being grain fallout. The detritals included quartz, mica, plagioclase feldspar and a few patches of hematite.

The thin-sections falling under the category of fabric E had a fine grained greyish matrix. These samples are non-pleochroic in PPL and dark grey in XPL. The grains were angular to sub-angular to sub-rounded in shape. The highest grain size in each sample was 80 μ, 100 μ, 200 μ and 600 μ. The grains were perfectly sorted to well sorted and had a frequency of < 5-10%. The grain-size distribution pattern was mostly unimodal. Some of the sections had grains showing parallel orientation, with some having no definite orientation. Voids were few and most of them were grain fallouts. The non-plastic inclusions included quartz, plagioclase feldspar, microcline plagioclase, augite and mica. Patches of hematite was also observed.

The mineralogy of the non-plastic inclusions allows one to compositionally group them into different geological terrains.TGW-1 and TGW-2 which belong to petro-fabric D have high quartz content, followed by feldspar in the clay. These wares may be products of a geological terrain containing clays of this composition. For the purpose of this study, this terrain has been designated no.V. TGW-1 appears to have also been produced in an area having basaltic clays but lacking in bioclast – an area designated as Ia in this study.

TGW-4 and TGW-5 belong to Petrofabric Group E and have a mineralogical profile which indicates a basaltic composition of the clays, the same as some of the TGW-1 samples. The geological terrain Ia is indicative of a basalt source or the vicinity of a basalt source for the clay since it lacks bioclast.

Robert Mason and Edward J.Keall did the petrographic analysis of the Abbasid glazed wares of Siraf and Basra and their analysis of the Turquoise Glazed Wares from the sites was found to belong to seven petrofabrics (Mason and Keall 1991: 51-70). Abbasid Glazed wares could have been produced at a number of sites. According to the medieval historian al-Yaqubi, potters from al-kufa and al-Basra were brought to Samarra. Hence these three sites could have produced these wares. Baghdad and Susa are the other possible production centers. The Turquoise Glazed group is defined by Mason and Keall in the manner of O.S. Rye as having "a common technology of being glazed with a thick alkaline glaze, coloured turquoise by copper under oxidizing conditions." There are many variations within this group. A study of the Basra kiln furniture as well as the sherds shows in thin-sections a well-levigated clay having a very fine featureless ground mass, and inclusions in a bimodal grain-size distribution. The fine population consists of angular quartz with trace of feldspar and amphibole, all under 50 microns. The kiln furniture shows poor mixing of clay and fine quartz, indicative that the inclusions were added as temper. The coarse population (0.25 to 1mm) consists of rounded grains of quartz, cloudy untwined feldspar, clear plagioclase and occasionally felsite. The Turquoise Glazed Ware groups are similar since they all probably have a Mesopotamian origin. Petrofabrics 1, 2 and 3 are similar to each other and are tentatively attributed to Basra. Petrofabric 2 has a higher abundance of ferromagnesian minerals such as biotite and a higher relative abundance of fine quartz. Petrofabric 3 has an even higher abundance of ferromagnesian grains. Petrofabric 4 has a high abundance of inclusions such as biotite schist. Petrofabric 5 is similar to the Basra type. Petrofabric 6 was found only in one sherd and had a high abundance of all inclusions sparticularly fragments of felsic volcanic rock. Petrofabric 7 has a high abundance of inclusions particularly quartz. But this fabric is finer than fabric 6. The pottery characteristics of the last two petrofabrics are distinctly Sasanian and so it is thought to belong to a kiln in Mesopotamia active in Sasanian or early Islamic times. (Mason and Keall 1991: 51-70)

Occurrence at Sanjan : Turquoise Glazed Ware is one of the most commonly found glazed wares at Sanjan. The surface of the Bandar or ancient port site is littered with these sherds and they can easily be collected from the fields, embankment along the river, open lands and from the undulating surfaces of the peripheral landscape which also have remnants of brick structures which may have been farmsteads or homesteads in the past as they are today. The density of the scatter is highest by the riverside, on the uppermost part of the mound and on the northern part of the settlement. This scatter thins as one moves towards the Dokhma – in fact, all surface pottery decreases in quantity as one approaches this mortuary structure. Similarly a distribution difference is noticeable at the Koli Khadi where the percentage of most glazed wares is meagre and mostly belongs to the later period (13th- 14th century). Turquoise Glazed Ware does not seem to extend past the 10th – early 11th century at most sites where it has been found. At Sanjan the ware appears from the lowest levels but the sherds that occur in the early layers are TGW-5 and few in number. They are plain and some are quite crudely made. The percentage of the wares TGW-1 to TGW-4 increases in the intervening / middle levels, i.e. layers 2 to 5 indicating a more intensified contact with the Persian Gulf region. Associated wares indicate an Abbasid date, i.e. 9th – 10th century for these levels. The later TGW sherds are not only more profuse but also more decorative and the barbotine designs become more common. The numbers again seem to decrease and eventually disappear in the later levels – upper part of layer 2 and layer 1 - associated with the sgraffiato wares and other 12th -13th century classes.

Discussion : This ware has been reported from almost all the sites in the Indian Ocean littoral. Glover did extensive exploration in South East Asia, South China and Japan and reported this ware from almost all the coastal sites he visited (Glover, 2002: 165-177). Carswell is supposed to have found these wares during his explorations on the coast of India, although the sites are not reported in print (Glover 2002: 169). At Mantai in Sri Lanka, Turquoise Glazed Wares is the highest number in glazed wares -180, followed by Dusun stoneware which number 139 (Carswell and Pritchett 1984: 64-65). Banbhore and Brahminabad in Pakistan also report these from the excavated trenches, dateable to the Abbasid period (Khan 1960, 1963). Sites such as Manda (Chittick 1984), Kilwa (Chittick 1974), Shanga (Horton 1996), Unguja Uuku (Juma 2004), etc on the East African coast also report this ware in context of trade with the Persian Gulf region. Kennet reports the class from Ras al-Khaimah and Kush (Kennet 2004: 29-31). In the Persian Gulf itself the pottery has been well documented at sites such as Siraf (Whitehouse 1968: 1-22), Suhar (Kervran 2004: Pl.25), Samarra (Northedge 1996: 229 – 258), 'Ana (Northedge et al 1988: 91), etc. The fact that the distribution of this ware is restricted to coastal regions and port sites in the Indian Ocean indicates strongly that it was transported mainly through maritime trade and may have served the dual purpose of being both an item of trade as well as container for trade goods. The ware does occur on inland sites such as Susa, Nishapur, Merv, etc in Iran and Iraq, indicating that it was a maritime trade ceramic outside of the Persian territories. Glover suggests that the import of this ware could have been for the consumption of Persian and Arab trading communities based within these foreign ports and that it may explain why these sherds are not reported from hinterland sites in Southeast Asia (Glover 2002: 174). While there is no doubt that many centers of production existed, it is generally accepted that a large bulk of these wares could have been produced at or near Basra in Iraq and perhaps at Siraf as suggested by

Tampoe. Kennet forwards a more flexible provenance – Central and / or Southern Iraq but also mentions Basra as a probable area of production.

Sarre classified these basins, jars and lamps at Samarra in the excavated houses and in the so-called 'Parthischer Friedhof' in Group I A of the ceramic assemblage. These were datable to the Abbasid period. Similar wares with cut decorations under blue-green glaze and Abbasid Sasano-Islamic barbotine jars were also found and these were datable to the Middle Islamic period as is mentioned in Northedge's review titled Sarre's Keramik von Samarra in Perspective (Northedge 1996: 229-258). Northedge and Kennet discuss the Samarra Horizon wares and mention the presence of the Turquoise wares and their long tradition in this context. "Monochrome glazing was introduced from about the middle of the second millennium, and continued to be used in the Parthian and Sasanian periods. The glazed wares which occur in Islamic strata earlier than the Samarra Horizon, or on sites abandoned before its introduction, follow this tradition, and are generally covered in an alkaline glaze which ranges in colour from a deep turquoise blue to sea green. The glaze is applied not only to large jars with barbotine decoration and storage vessels, but also to small carinated bowls. Vessels with these glazes have often been dubbed 'post-Sasanian' in an Islamic context, but it is clear that monochrome blue-green glazes did not die out with the introduction of the Samarra horizon but continued in use until the Ottoman period" (Northedge and Kennet 1994: 21-23).

The excavations at Banbhore in Pakistan in 1964 also report this ware. The large thick-textured jars with deep blue-green glaze showing relief patterns on the exterior are attributed to a Sasanian origin but persist in the later period as well (Pakistan Archaeology 1; 1964).

The 'Arja mines in Oman which were in use at the time Suhar was occupied have also yielded this ware but have not been used as index fossils in the dating of the site (Costa and Wilkinson 1987:186)). Failaka Island (Kuwait) reports Turquoise Glazed Ware of an early Abbasid date which was imported from south Mesopotamia (Kennet 1991: 102 -103). Excavations at 'Ana, Qal'a Island (Iraq) have brought this ware to light, in association with the Samarra horizon wares. Ceramic groups 4 and 5 within which this ware is classified belong to deposits which "should be at least close in date to the period of the Caliphateat Samarra (221/836–279/892), and are very probably contemporary with it."(Northedge,et al 1988: 83, 91).

At Manda, the ware is found in Period I and extends up to the start of Period II when the earliest sgraffiato wares appear. This indicates a date from mid-ninth to mid eleventh centuries AD (Chittick 1984: 11, 71 - 76). This chronology matches that at Kilwa where this ware occurs in levels belonging to Period Ia which covers the bracket ninth century to c.1000. The dates proposed by Horton for this ware at Shanga are c.750 AD, when the occupation at Shanga begins, up to c.1000 AD. (Horton 1996).

Whitehouse reports the presence of "green-glazed sherds of "Sasanian-Islamic" type" from Period I deposits at Siraf and infers that the period belongs to a phase when Siraf had begun to handle international trade but had not yet become the thriving commercial port that it did later on. (Whitehouse 1968: 7). The suggested date for period I is c.803-4 AD (Whitehouse 1971: 4).

The occurrence of Turquoise Glazed Ware at Sanjan in such considerable quantity is indicative of a sustained and intensive contact with the Persian Gulf region.

The Samarra Horizon Classes of Pottery :

Some of the glazed wares found at Sanjan fall within the bracket of the so-called Samarra Horizon classes of pottery. This issue needs some elucidation in order to put these wares at Sanjan into the right perspective.

Background : The site of Samarra has long been regarded as the benchmark site for the establishment of a chronological and typological model for Abbasid pottery. Samarra, located in present day Iraq, was the seat of the Abbasid Caliphate for a brief period in the ninth century and was presumably abandoned thereafter. The site was occupied between 221 / 836 and 269 / 892 AD. The German Samarra-Expedition in 1911-1913 brought into focus the relevance of Samarra in the history of Abbasid pottery. In 1925, Sarre's report on the ceramics at Samarra defined the characteristic typology of these wares and presumed that the date of occupation at Samarra provided a fixed dating point for the pottery of the Abbasid period (Northedge, 1996: 229). The report titled 'Keramik von Samarra' became a benchmark work and the chronology Sarre suggested was accepted as a standard viz. that the ceramics recovered at Samarra belong to the 56 years of the Abbasid Caliphal occupation of the site. Later excavation reports, the Hafriyyat Samarra 1936-1939 (Excavations at Samarra 1936-1939) by the Iraq Directorate-General of Antiquities accentuated Sarre's periodisation of the Abbasid wares. In the 1960s and 1970s this position was challenged by archaeologists such as John Hansman, David Whitehouse, Monique Kevran and others. Sites such as Siraf, Susa, etc yielded the same wares in stratigraphic and sequential contexts, thereby refuting the oversimplified chronology of Sarre at Samarra.

These excavations also proved that the introduction of the various wares stretched beyond the brief occupation bracket at Samarra, as presumed by Sarre. It was also evident that occupation at Samarra may have continued for a longer period after the Caliphate shifted than was previously thought. The Samarra Periodisation model has subsequently been reassessed by various scholars like Northedge, Kennet, Whitehouse, Tampoe, Watson, Kervran and others. The range of ware classes falling within this chronological bracket has also been reassessed. Other sites have yielded evidence for pre-Samarran wares and their development. This helps to put the Samarra debate in the right context.

The early ninth century saw a revolutionary change in the range of forms and decorative techniques in the ceramic traditions of West Asia. The Abbasid Caliphate moved to Samarra in 836 AD and this power centre proved to be the focus of the Islamic potters for the duration of the Caliphal occupation. Earlier traditions and techniques, especially in glazing and surface treatments, underwent a complete change. Innovations in forms, decorations and glazes took inspiration sometimes from the T'ang wares which were being imported into the Caliphate lands from the Far East. According to Arthur Lane, three waves of Chinese influence can be seen in the pottery of West Asia – the first is the influence of T'ang porcelain and stoneware at the beginning of the ninth century, the second in the twelfth century and a third wave in the fifteenth century (Lane 1947: 3). Of the new techniques, the most important developments were the introduction of polychrome glazing, lustre painting, cobalt painting and sgraffiato. The development of opaque glazes and the use of in-glaze colours such as cobalt blue, green, yellow, manganese purple, etc were some of the more important innovations of the period. Samarra remains an important reference point for the introduction of these features to West Asian pottery.

Northedge notes that the archaeological record at Samarra begins as far back as the Chalcolithic period, and extends, at least, up to the thirteenth or fourteenth century (Northedge 1996: 229-258). The spatial distribution of occupation during varying chronological periods has to be considered when reviewing the site in order to place the Abbasid occupation in perspective. For the relevant time period that this study is concerned with, it is sufficient to note that Samarra was occupied in the pre-Abbasid period, i.e. prior to 836 AD. Northedge also notes the presence of Late Sasanian and Ummayad structures. In 836 AD, Surra Man Ra'a was founded by al-Mu'tasim. The city was expanded and witnessed an extraordinary growth during the reign of al-Mutawakkil (847 – 861 AD). This period of economic growth is reflected in the structural activity at the site, in the construction of palaces, gardens, extensions to existing structures, new areas of the city and even in the foundation of the new city of al-Mutawakkiliya. The contraction of Samarra that followed the assassination of al-Mutawakkil took place in three phases – one, the abandonment of the palaces; two, the departure of the Caliphs and the abandonment of the Abbasid city, and: three, the establishment of a Middle Islamic pattern of settlement. According to Northedge and Kennet this abandonment of the Abbasid city would have already taken place by 895 and certainly before 903 AD (Northedge 1996 : 229-258; also Kennet 2004).

Pre-Samarran Pottery : To say that the Abbasid occupation at Samarra saw a sudden efflorescence of new ceramic typology is to deny the presence of pre-Samarran traditions of pottery. It is true that innovations and revolutionary changes in the ceramic traditions were introduced at this time. But they did not emerge abruptly and fully developed. The seeds of their development can be seen in the traditions that already existed. Monochrome glazing was already an established practice during the Parthian and Sasanian periods. The alkaline glazed turquoise blue and green wares are commonly found at sites having pre-Abbasid occupations. Late Sasanian and post-Sasanian levels at various sites have the sherds of these large jars with barbotine decorations, storage vessels and even bowls, all of which continue during the Abbasid period and well into the tenth and eleventh century. Northedge and Kennet (1994: 21-35) describe the late Sasanian and Early Islamic storage vessels made of "a creamy fabric decorated with stamps, often with animal or geometric designs". Honeycomb Ware is yet another well-known pottery type known from excavations all along the Persian Gulf coast and in Iraq. One of the more diagnostic Early Islamic introductions is Thinware or Eggshell ware which consists of fine unglazed buff pitchers, jugs or jars having a single handle, sometimes with a turban or a thumb-rest. It is believed to have been introduced during the Ummayad period and later forms of the ware have decoration by way of incised designs, circle and dot decorations etc. Similar vessels in glass are known from Raqqa (Northedge and Kennet 1994: 23). Moulded vessels with relief decoration are also known from Syria and Damascus, belonging to the period of Harun al-Rashid (786 – 809AD).

The Samarra Horizon Pottery : The glazed wares associated with the Samarra Horizon consist of the white glazed vessels with cobalt blue, green and brown decoration by way of painting, polychrome and monochrome lustre painting, plain opaque white glaze, plain grey glaze, splash decoration in green and white or green yellow and white or white and blue, scratch decoration under splash glaze (sgraffiato) and relief moulded vessels with green or golden glaze (Northedge and Kennet 1994: 21-34). Turquoise Glazed Ware continues through the Caliphal period alongside the Samarra Horizon Wares. At Sanjan the relief moulded wares are absent though most of the other categories are seen in the assemblage. These have been described below. Unglazed wares of the Samarra Horizon are treated separately in the section on unglazed wares.

White Glazed Ware (Ware code: WGW)

Definition : This is one of the early wares which belong to the Samarra Horizon group of glazed ceramics. Cream coloured T'ang porcelain had already reached the Persian Gulf by the early ninth century. The lobed bowls and dishes were sought after for the purity of their form as well as their superior fabric. While the fabric was impossible for the Mesopotamian potters to imitate, scholars such as Arthur Lane are of the opinion that they attempted to recreate the same glazed effect by using tin-oxide and a modified lead-glaze (Lane 1947:13). Analysis of the glaze in recent times has revealed that sometimes the opaque white glaze was not a product of tin or lead at all (Mason and Keall 1991: 52). The ware has been variously defined as Abbasid Opaque White-glazed Ware, Tin-glazed Ware, Opaque White Glazed ware, etc. Kennet calls it YBTIN or Plain Opaque White Glazed Ware (Kennet 2004: 32). James Allen describes these wares as earthenware copies of Far Eastern imports and says that "the lead glazes with which they are covered contain suspended particles of tin-oxide which make the glaze slightly more viscous and, more important, colour it opaque white. Pottery in this style is thus generally called "tin-glazed" ware. By the use of tin oxide the Iraqi potter was able to make pots which, though softer in their body material, were very similar in appearance to their Chinese counterparts, a resemblance enhanced by the adoption of some Chinese style bowl shapes" (Allen 1971: 10). Tampoe, like Whitehouse and others, classifies all the white glazed categories at Siraf under one umbrella term - White Glazed Wares and includes in these the splashed, painted and decorated varieties as well as the plain white glazed vessels (Tampoe 1989: 32 -36). Morgan categorizes the Abbasid Opaque White-glazed Wares in the Khalili collection into three types – Monochrome White Wares, Lustre Wares and Overglaze painted Wares (Morgan 1994: 36 – 37). In the present study, the nomenclature indicates a ware having white glaze and no decoration, the splashed, painted and decorated types being treated as separate categories. The fabric is always pale yellow buff in colour.

Shapes and forms : The White Glazed Ware vessels at Sanjan are almost all exclusively open forms of tableware – bowls and dishes, with a few saucer lamps. The class closely follows the Chinese shapes and profiles. The West Asian potters may not have replicated the body of Chinese porcelain but they tried to imitate the white glaze and the graceful profiles of the delicate Chinese bowls with their fine tipped rims and ring bases. The bowls and dishes have shapes reminiscent of the Chinese forms, sometimes with the circular central ridge and radiating ribs extending to the rim. In the cases where this ridge and ribbing are absent, the bowls are smaller in size and have the everted tapering rims of the Chinese vessels and are more finely potted than the former type. It appears that the technique employed to manufacture these vessels was moulding, with finishing done on the wheel. The bases have broad footrings with square section and with some traces of scraping on the underside. In most cases the glaze covers all surfaces of the interior and exterior. A bowl from the Nasser D. Khalili collection of Islamic Art (Grube 1994: cat.18, 26) appears identical to the bowl from Sanjan (T.382) and has been dated by Grube to the tenth century. The saucer lamps have disc bases and handles - sometimes a little misshapen due to hand-moulding - attached to the rim on the side opposite the wick channel. Similar saucer lamps are seen in the Siraf collection (Tampoe 1989: fig. 58). A variety of shapes are seen in this sub-class – T.14.1, T.41, T.42, T.42.1, T.43, T.43.1, T.44, T.45, T.46, T.47, T.47.1, T.52, T.53.1, T.56, T.57.1, T.58, T.61, T.62.1, T.63, T.63.1, T.64, T.64.1, T.75, T.381, T.382, T.383, T.384 D.177 and T.384.1 D.177. The bases are Bs.8, Bs.17, Bs.22, Bs.23, Bs.28, Bs.29, Bs.30, Bs.31, Bs.72 and Bs.72.1.

Glaze and surface treatment : The white glaze overlying the clay paste body is evenly applied and is milky white. It adheres quite well to the body, though a lot of sherds do appear flaked. It is pertinent to note that only in some cases does one observe a degradation of the glaze to a grey-brown colour or exfoliation as is the case with some of the Cobalt Painted Ware which also have a white glaze background. The flaking of the glaze is also not as severe as in the case of Splashed White Glazed Ware, except in the case of dishes from this class which follow the same profile i.e. with the ribs and ridges on the vessel interior. The smaller and finer potted bowls and cups appears to have similar glazing to Lustre Painted Ware – 1 (LPW-1), in which case too, one observes that the white glaze retains its milky opaque tone and appears to fit the body of the vessel better than some of the other glazed ware categories. Very fine black flecking can be observed on the glaze under lens. Occasionally the white of the glaze has a very pale blue tinge. This is particularly true of the finely potted smaller vessel forms such as cups or small bowls, where the thickness of the vessel is only 2 - 3mm.

Fabric : The White Glazed Wares at Sanjan all have the same pale yellow to buff soft paste body as Turquoise Glazed Ware, Lustre Painted Ware, Cobalt Painted Ware and Splashed White Glazed Ware. They belong to fabric class BUFF(A). The clay is fine and well levigated, particularly in the smaller vessel forms.

Petrographic Analysis : The thin-section analysis places this ware in petrographic Fabric group D (see above, in the section on Turquoise Glazed Ware).

Mason and Keall studied some of the samples from Siraf and say that although the traditional view is that these wares are 'tin-glazed', "chemical analysis and also petrographic examination have shown that some are actually opacified by grains of quartz and feldspar and also by bubbles present in the glaze. This effect can also be noted in opaque white glazes on Parthian and Sasanian pottery" (Mason and Keall 1991: 52). Basra is one of the pottery production centers known to have produced a large quantity and range of wares both for domestic use as well as export during this period. Siraf was another production centre; however other production centers within Iraq were also producing these wares as the samples from Samarra, Nippur and al-Hira denote (Mason and Keall 1991: 61). An important observation made in this regard by Mason and Keall is that these other production centers were making the plain White Glazed Wares but not the Lustre or Blue-Painted variety. They also propose a 'working scheme' for the chronology of these ceramics, suggesting a chronological model which consists of nine periods, each covering a century. White Glazed Ware, Blue-Painted (in this study, Cobalt Painted Ware) and Lustre Painted Ware belong to Period 4 (c.850 – 950).

Occurrence at Sanjan : This is a prominently represented ware at Sanjan. The well in TT4, SW yielded a large number of these sherds. But in the stratified context in TT4, the frequency of these wares is seen to be the highest in the lower part of layer 2, with one sherd in layer 6 (TT4 NE), which could have been intrusive. The wares found in association with White Glazed Ware are TGW, SWGW, CPW, LPW, ESW(W), BW, STONE-2, CHANG, PORC-1, PORC-2, CEL-1, etc. In Trench TT1, the ware occurs in layers 1, 2, 3 and, in one case, 4.

Discussion : This ware has been reported from sites such as Unguja Ukuu (Juma 2004: 111), Manda (Chittick 1984: 77), Shanga (Horton 1996: 277-78) and Kilwa (Chittick 1974: 303) in East Africa and Mantai in Sri Lanka (Carswell and Pritchett 1984: 64). Juma places the ware in Period Ib (9th – 10th century). Like Whitehouse, Tampoe and others, Chittick also categorizes the painted and splashed varieties under the common class of White Glazed Ware at Manda. He dates them to the later part of Period I which corresponds to the mid ninth century, a date which agrees with Whitehouse's Phase A at Siraf which is dated to c.850 AD. Chittick also reports this ware from Kilwa where it is used as one of the chronological markers for Period Ia (ninth century (?) to c.1000). A distinction has been made by him between what he calls "White tin-glaze" and "Splashed tin-glaze". Horton sub-classifies this class into four variations at Shanga – Plain, Blue splashed, Green-brown splashed and Yellow-splashed. He surmises that the first two subclasses are contemporary and precede the latter. The four sub-classes are together placed in the chronological bracket 800 – 1000 AD. At Mantai, Carswell and Pritchett place it in Phase 5 with an approximate 9th – 10th century date. Amongst the West Asian wares at Mantai, this class records the second highest sherd count, after Turquoise Glazed Ware which is the most prolific.

Basra is accepted by most scholars as one of the most important centers of manufacture for pottery particularly White Glazed Wares. The fine yellow fabric is taken as one of the distinguishing marks of the Basra kilns (Watson 2004: 36 – 38).

Splashed White Glazed Ware (Ware code: SWGW)

Definition : In earlier works by several scholars, the terms Opaque White Glazed Ware, Opaque Tin-glazed Ware, etc were used to include all white glazed wares, with or without decoration. In this study, as mentioned earlier, the distinctions have been noted. This class consists of vessels having a pale yellow to buff earthenware body with a white glaze which has green, turquoise, blue, manganese black or purple-black splashes, usually placed radially on the interiors of open form vessels. Though closely related to the previous class, this category of ware has some distinct characteristics of its own, the most noticeable being the decoration.

Shapes and forms : All the sherds in this class belong to open-forms such as dishes and bowls. The shapes appear to be imitations of the Chinese forms with both rounded or tapering everted rims and low footrings. A circular ridge in the well of the vessel and vertical ribs rising to the rim along the wall of the vessel can be frequently seen, especially in the case of dishes. The vessels appear to be moulded and wheel-finished, like the White Glazed Ware. They are medium thick (approximately 5 to 7mm) and the smaller forms of bowls and cups seen in the previous category are missing. The footrings are low, square in section and quite broad. A number of shapes are similar to WGW. The shapes seen in this sub-category are T.4.3, T.12, T.41, T.42, T.42.1, T.43.1, T.48, T.52, T.53, T.59, T.61, T.62, T.62.1, T.63, T.63.1, T.64 T.64.1 and T.386 D.180. The bases are Bs.1, Bs.5, Bs.28, Bs.29 and Bs.72.

Glaze and Decoration : Opaque white glaze covers the whole body of the vessel and can vary in thickness. The tone of the glaze can range from milky white to grey and light blue. The flaking and exfoliation of the glaze on these vessels is very heavy and severe crazing can be seen in most cases. The turquoise, green, deep sky blue, manganese black and manganese purple splashes are restricted to the interior of the vessel, sometimes continuing over the rim onto the upper part of the exterior. The most common placement of the splashes is in a radial pattern, following the vertical ribs from the floor of the vessel to the rim. In some cases, the radial splashes are only on the walls of the vessel while irregular shaped dots or linear splashes are placed on the floor of the vessel. The white glaze absorbs the colour splashes and there is no sharp demarcation of outlines – in fact, the splashes are blurred and blend with the background. Frequently the white background itself loses colour and takes on a paler tone of the splashed colour. This is particularly noted in the case of vessels splashed with deep sky-blue. Turquoise, sky-blue and green are the most common colours seen in this ware. The manganese black and purple-black are rarer and in these instances one notices the streaky, waxy nature of the glaze, while the background takes on a grey tone. Although many sites in the Indian Ocean littoral report the green, blue or turquoise splashed wares and mention the inscriptions sometimes found on them, no such inscribed sherds have been recovered at Sanjan. Only in a rare instant has a small sherd been found having more than one colour splashed on it.

Fabric : The body of these bowls and dishes is made of similar yellow clay as the White Glazed Ware, Lustre Painted Ware and some of the Turquoise Glazed Wares. It belongs to Fabric BUFF(A). Visual examination of the sections shows a slightly sandy texture with small white quartz particles and some airholes. It is chalky and can have very fine black or grey inclusions.

Petrographic Analysis : Thin-section analysis places this ware in petrographic Fabric Group D, along with WGW, TGW, LPW and CPW.

Occurrence at Sanjan : This is one of the better represented of the glazed wares at Sanjan, quantitatively next to Turquoise Glazed Ware and sometimes exceeding it in number. The sherds occur almost throughout the sequence but one must keep in mind the disturbed nature of the site and the presence of pits in the excavated trenches. The possibility of these sherds intruding to a lower level is high just as it is possible that some sherds may have been displaced by the upward movement of material during pit formation processes. However, this class is best represented in the lower part of layer 2, along with White Glazed Ware in TT4, NE. It is associated with TGW, CPW, WGW, LPW, BW, ESW(W), STONE-2, PORC-1 and CEL-1.

Discussion : A large number of scholars club all wares with opaque white glaze under one category and treat the decoration alone as a variation. They make no distinction between the painted and splashed decorations. Tampoe distinguishes the Siraf White Glazed Wares into two fabrics – Coarse buff body and Saljuq frit body and further sub-divides these fabrics on the basis of the decorations (Tampoe 1989: 35 – 36). SWGW at Sanjan corresponds to the Siraf Coarse buff body fabric, type c and, in rare cases, type d. Mason and Keall who did the petrographic analysis of the Siraf samples note that the oxide pigments are splashed informally onto the glaze. They place this ware with turquoise / brown splashes in Period 5 of their chronological scheme, later than the plain White Glazed wares. Their approximate dating for this ware is c.950 – 1050 (Mason and Keall 1991: 57). This would imply that the plain opaque white glaze was introduced first along with lustre and cobalt painting, and that the splash decoration was introduced at a slightly later stage, predating the sgraffiato wares. However Kennet argues for an earlier date for what he classifies as COBALT since, according to him, this ware occurs just before the introduction of White Glazed Ware or YBTIN – i.e. after 803-4 and before 835-6 AD. It must be noted that Kennet did not make a distinction between the splashed decoration and the cobalt painted decoration in his report on the ceramics from Kush and Ras al-Khaimah (Kennet 2004: 31 – 33). He is now in agreement that such a distinction is necessary and that the turquoise, green and blue splashes noticed on these sherds may not contain cobalt at all (Kennet, personal communication). Priestman has made this distinction in his recent thesis (Priestman 2005). Monique Kervran dates this ware with turquoise and / or brown splashes at Suhar to around 830 AD, later than the Cobalt Painted Ware which she places earlier at c.750 AD, based on the evidence from Susa (Kervran 2004: 315). Chittick calls this ware Splashed tin-glaze at Kilwa, where only two sherds have been found, and dates it to Period Ib – 9th and 10th century (Chittick 1974: 303). At Manda, he groups it with WGW and CPW into a single category and places it in Period I (Chittick 1984: 77). Horton reports the ware from Shanga and places it within the broad time bracket 800 – 1000 AD (Horton 1996: 277-78), as does Juma at Unguja Ukuu (Juma 2004: 111).

Cobalt Painted Ware (Ware code: CPW)

Definition : Cobalt Painted Ware is one of the early Samarra Horizon wares, contemporary to the plain White Glazed Ware according to some scholars. The most distinctive feature of this ware is its fine opaque white glaze and the painting of either motifs or calligraphy in deep cobalt blue on the white background, achieving an effect poetically described by Arthur Lane as "ink on snow" (Lane 1947: 13). Frequently this ware has been reported as having both, the cobalt blue calligraphy as well as splash decoration. However, at Sanjan no such sherds have been found. This ware has a glaze that adheres to the pale yellow body better than SWGW, although flaking is common.

Shapes and forms : Most of the vessels in this class are open forms such as bowls and cups. However closed forms also occur as is evidenced by the jar fragments found in the Sanjan excavations in 2002. The bowls and cups are small tableware forms with close affinity to Chinese shapes and profiles. The rims are either plain or tapering everted types and the bases have footrings. The small finely potted cups and bowls are thin in section, sometimes only 3mm in thickness. Sturdier versions with lower footrings are also seen in the more shallow bowls and dishes. The vessels appear moulded and wheel finished. The jar fragments (T.71 D.27) indicate a vessel with an everted rounded rim, sloping rounded shoulders to which two or four small faux handles are attached and a floral decoration stenciled or painted by hand onto the shoulder. A reconstructed flat base with vertical walls may well belong to the same vessel since it came from the same context. Unfortunately it was not possible to assemble the whole vessel. The shape is reminiscent of Chinese jars or vases. The same motif appears again on another sherd, the floor of a small bowl. This painted motif closely parallels that from Siraf (Tampoe 1989: Fig.52 nos.271, 275 and 277). A small bowl with an inscription in Kufic script is typical of the class, with the traditional blessing placed vertically from rim to the floor of the vessel and presumably, across. The shapes that occur in this class are T.43, T.44, T.46, T.47, T.52, T.71 D.27, T.385 D.178 and T.385.1. The bases are represented by shapes Bs.11, Bs.28 and Bs.73 D.179.

Glaze and Decoration : Two types of glazes have been noted in this ware – one type retains its milky white opaque colour and fits the body of the vessel better with less flaking, while the other type of glaze degrades to a grey-brown colour and is subject to exfoliation, flaking and degradation. The first type of glaze has been noticed on the small bowls and cups while the second has been noticed on the jar sherds and on the bases of the shallow bowls and dishes. The floral decoration of clover-leaf and club-leaf motifs, lines, dots and rosettes appear on the second type of glaze. The inscribed sample belongs to the former type of glazed vessel, deep blue on milky white (T.385 D.178). The inscribed decoration invokes blessings ('barakah') and is placed radially on the vessel interior. The floral or more abstract motifs appear to be placed on the interior floors and perhaps walls of bowls and dishes with no evidence of decoration on the rim. The jar sherds have similar motifs of clover-leaf, half rosettes with horizontal lines slashed across as connectors between them, the whole design producing a stencilled effect (D.27). These motifs resemble the ones reported from Siraf (see above).The placement of the painting on the sloping shoulder of the vessel between the faux handles and the distinctly Chinese shape of the jar make this sample unique. No parallel piece could be identified in any of the collections, catalogues or excavation reports accessed for this study. The base of a dish or bowl with a ring-foot and having motifs and stylistic features very similar to the jar could be indicative of a common kiln or source. The use of cobalt for painting was supposedly an innovation of the West Asian potters. The absorption of the pigment by the white glaze leads to a blurred or smudged outline. The flaking, degradation and exfoliation on the second type of glaze are more severe.

Fabric : The body of this ware is the same pale yellow earthenware as the previous categories – BUFF(A). Lens observation shows that the clay is better levigated but has the same fine quartz inclusions and small airholes. The whitish buff colour body, noticeable in the case of some White Glazed Ware sherds, is missing.

Petrographic Analysis : The thin section studies of this ware place it in the Fabric Group E, along with TGW-1, TGW-4, TGW-5 and WSPW-3. The description of this fabric group has already been given in the section on Turquoise Glazed Ware and so does not require repetition.

Mason and Keall's petrographic analysis of the samples from Siraf show that the "Blue-Painted" types are almost all, without exception, of the Basra fabric (Mason and Keall 1991: 61). Watson also makes a note of this exclusivity in his description of the pieces in the Al-Sabah Collection when he says "The Basra product is technically the most refined, and it is the most distinctive: its fine yellow clay fabric, its careful potting, and especially when decorated in lustre or blue... It is noteworthy that painting in blue occurs only on pieces attributable to Basra; no doubt reflecting the cost and difficulty of sourcing the cobalt pigment. The Basran style of painting, in particular the radial inscription, was most influential, being produced elsewhere in a manganese brown or green."(Watson 2004: 37 - 38).

Occurrence at Sanjan : Although this ware is not found in very large numbers at Sanjan, it is of relevance in establishing the chronology of the site and fixing a tentative date for the levels within which it occurs. The unusual jar, flat base, the ring-footed bowls and cups as well as the dishes represent the demand for these wares during the late eighth-ninth century. That these were expensive tableware produced perhaps in limited numbers in the kilns of Basra and distributed through the trading network of the Indian Ocean is borne out by their restrictive numbers at almost all sites. As with other glazed wares, the Sanjan Bandar area has yielded the maximum number of sherds in the first and third seasons of excavations. Layer 2 in TT4 had the highest number of sherds of this ware with an occasional sherd in layer 3, associated with the WGW and SWGW, as also with TGW, LPW and ESW(W). BW(SV), BW(TJ), BW and SWPW are the other wares associated with it. PORC-1, PORC-2, STONE-2, CHANGSHA and STONE-3 were some of the Far Eastern Wares found in the same layer. Sherds of this ware were neither found in the lower levels nor in the upper levels which are associated with sgraffiato ware. In TT1 and TT2, the different classes appear more mixed in almost all the layers making it difficult to identify a ceramic horizon or to establish with certainty the proper context of the wares. TT4, however, provides a clear context for these wares –WGW, CPW and SWGW all occurring nearly exclusively in the lower part of layer 2 as a distinct horizon.

Discussion : It is generally agreed by most scholars that Cobalt Painted Ware is one of the earliest wares of the Samarra Horizon, contemporary with WGW. Kervran dates the appearance of this ware to the mid-eighth century at Susa (Kervran 1977: 89) and Tampoe places the Siraf collection of this ware to a slightly later date, the end of the eighth century (Tampoe 1989: 88). However there is some debate about the acceptability of this date since a large part of Tampoe's research has been based on the pottery data cards and notes of the excavators which initially claimed the presence of these wares in the fill of the Mosque platform. This claim was later corrected in subsequent interim reports by Whitehouse. Mason and Keall also refute Tampoe's claim of such an early date for the occurrence of this ware at Siraf, while they place the appearance of it to Period 4 (850 – 950 AD) in their chronological scheme (Mason and Keall 1991: 57). Whitehouse dates the appearance of this ware at Siraf to after 825 AD. Northedge and Kennet are of the opinion that the ware may have gone out of production soon after the founding of Samarra in 836 AD (Northedge and Kennet 1994: 25), a date supported by the evidence from Siraf. Kervran reports the ware from Suhar and mentions in this context that "At Susa the white opaque glaze with cobalt blue may appear around AD 750 and the other colours, like turquoise and brown, around AD 830; since the former ends around AD 880, the latter lasts some more decades"(Kervran 2004: 315). On the basis of evidence from Siraf, Costa and Wilkinson argue for a ninth century date for the few sherds of this ware at Arja (Costa and Wilkinson 1987: 185). Despite the finer points on which the debate continues, the fact remains that CPW is one of the earliest innovations of the Abbasid potters who may have imitated Chinese forms but who introduced the use of cobalt as a decorative pigment. The most prominent centre for the production of this ware has been identified at Basra though some other centers could also have produced similar wares. The introduction and short production life of the ware make it a useful chronological marker to date sites. The middle-late 8th century date proposed by Kervran for its introduction and the continued presence of sherds only up to the middle-late 9th century contexts makes it an archetype for the Samarra Horizon. While this ware is reported from numerous sites, the quantity is very low as compared to WGW and SWGW sherds. For example, at Mantai only two sherds are reported from the excavations by Carswell and Prickett (Carswell and Prickett 1984: 64). At Sanjan, the representation is considerably more prominent.

Lustre Painted Ware (Ware code: LPW)

Definition : The technique of lustre painting on an opaque background originated in Egypt and was used to decorate glass. However it was soon adopted by the Mesopotamian potters for the decoration of pottery, particularly opaque white glazed vessels on which the designs and ornamentation were especially pleasing. Mason calls it a technological innovation, defined by the application of lustre pigment after the first firing of the vessel. The pigment consisted of an amalgam of silver, copper, ochre and a fixative and required a second firing in an anaerobic low-temperature kiln (Mason 1997:105-107). This ware is seen by many scholars as one of the most important innovations of the West Asian potters. According to Grube, the discovery by potters that the application of a thin film of metal oxides over the glaze of an already fired object would, in a second firing produce a golden lustrous sheen, was one of the innovations not only relevant to a region or period, but to the history of ceramic-making (Grube 1994: 11). Three groups of lustre decoration have been identified in the Samarra Horizon – polychrome lustre, ruby lustre and monochrome lustre. The first group employs two or three colours of ochre and brown. The second group uses a bright copper-red. The third group uses a single brownish colour which can range from yellow to greenish-red (Northedge and Kennet 1994: 29 – 33; also, Morgan 1994: 36 - 37).

Shapes and forms : This elite ware consists mostly of open forms such as bowls and dishes. Jars, pitchers, ewers and other closed forms have been reported from various sites and in museum collections. However, at Sanjan, only the open forms are represented. The bowls are small to medium sized, having a conosegmental or a hemispherical profile, with either plain or tapering everted rims (T.42.1, T.43, T.43.1, T.44, T.47, T.50D.22, T.51D.23, T.53.1D.22, T.57, T.57.1, T.58D.24, T.60, T.378D.175, T.379D.175, T.379.1 and T.380D.176). The bases are ring-footed (Bs.28, Bs.29 and Bs.30) and the shapes closely follow those of WGW and CPW categories, apparently inspired by Chinese models. The carination of the conosegmental bowls is fairly pronounced while the vessels themselves are quite shallow. One conosegmental bowl has a small circular groove with a dot placed in the centre of its floor. The vessels appear to have been moulded and then finished on the wheel.

Glaze and surface treatment : The ware is distinguished from other categories by the lustre effect produced by the use of copper, silver, and other metallic oxides as pigments for painting on the opaque white glaze of the vessel. Lane explains the technique with some lucidity when he says that the metallic oxides compounded with some form of sulphur produce both the colour as well as the lustre. The method of application is also elucidated by him – the pigment is combined with an earthy vehicle such as red or yellow ochre and painted on to the surface of an already fired glazed vessel, using wine lees or vinegar as a medium. The second firing of the vessel is done in a reducing kiln. The firing of the vessel has to be well controlled as the pigments can be volatile, frequently leaving behind only a stain but no lustre effect. The silver oxide provides a yellow colour while the copper oxide provides the lustrous element. If the application is thick, the colour produced is a solid copper or greenish-gold and a thinner application of pigments results in the rainbow colours of nacre or mother-of-pearl (Lane 1947: 14). Usually the white glaze adheres well to the body, particularly in the case of LPW-1. Some sherds have evidence of flaking. LPW-3 sherds show severe exfoliation and degradation.

Decoration : Of the three types of Lustre Wares identified by Lane, Grube, Northedge, Kennet, Kervran, and others, certain motifs and decorative techniques have been noted as distinctive features. Polychrome Lustre Ware, which is considered to be chronologically earlier than the Bichrome and Ruby Lustre Wares and the Monochrome Lustre Wares, is decorated in varying colours on a single vessel. The colours provide a contrast to the background and the potters appear to concentrate more on this aspect of the decoration and less on the motifs themselves. Half palmettes, rosettes, wing motifs, dots, dashes, swirls, foliage are common while human and animal motifs are absent. Varying tones of ochre, yellow and brown are used, sometimes with black or red. The Ruby Lustre Ware has a deep ruby-red glaze with gold painting on it. At Sanjan, no such decoration is in evidence but there are a few sherds with a white background and dark red painting executed on them. These are very few in number and badly exfoliated making the motif almost unrecognizable. The Bichrome Lustre Wares use brown and yellow. The Monochrome Lustre Wares employ a single colour palette of a greenish yellow or brownish yellow colour. The motifs are abstract, geometric, human, faunal, etc. The motifs are usually placed on the interior of the open-form vessels with spirals, medallions, dots and dashes, swirls, etc making up the decoration on the exterior. The placement of a large central figure, either human or animal, in the well of the bowl and 'contour panelling' by way of dotted backgrounds to fit the motif (Lane 1947: 16) is noted in the later samples. The figure may have a calligraphic or pseudo-calligraphic legend placed around it or in its vicinity. This is the most common type of Lustre Ware found at Sanjan (LPW-1). No human figures are represented on the sherds at Sanjan but four shallow bowls having the so-called running hare motif on the interior have been collected during the three years of excavations. This motif is also seen on various jars and bowls in other collections such as the Nasser D. Khalili collection of Islamic art (Grube 1994: 26). The rim has a band running along it and the walls of the vessel are decorated with pseudo-calligraphic symbols while the large animal figure takes up the centre of the vessel. The animal could, in actuality, be any other quadruped such as a deer, since it is quite a stiff and childish rendering. The effort of the painter to cram the figure into the well of the vessel is evident, resulting in considerable distortion. According to Arthur Lane, they are "childish incunabula, the mental symbols rather than the recorded observations of life" (Lane 1947: 16). The contour panels of dots and fillers are missing in the Sanjan samples, with the exception of one small rim fragment which has traces of a pale yellow leaf-like motif and small dots filling the background. While the interiors of the vessels have a greenish-yellow, yellow or ochre palette, the exteriors of the vessels have brown, red, reddish-green or yellow-green tones. The exterior motifs usually consist of spirals, concentric circles or roundels, medallions and squiggles. The spaces in-between the medallions or roundels are filled with small motifs such as dot and line, 'V' motifs, etc. The medallions contain dots, small hook like designs, squiggles, etc. In the case of one sample belonging to sub-class LPW-1 at Sanjan (D.40), the sherd is thickset and appears to be the shoulder of a carinated dish or platter with the decoration on the exterior consisting of a broad band in mustard yellow with horizontally placed tear-drop motifs filling the white field underneath. In some instances at Sanjan, the colours seem to have faded away leaving a yellow stain. However, most samples appear well preserved, despite the flaking.

Sub-categories : LPW at Sanjan has been sub-divided into three sub-classes. They do not reflect a chronological sequence. The variations in colour and decoration have been used as a criterion. The fabric remains the same.

LPW-1 is the best represented at Sanjan. It is monochromatic in colour-scheme, though the colour on the interior could vary from the exterior. This may be due to the firing or uneven application. The colours are yellow, greenish-yellow, mustard, reddish-green and ochre. The vessels are small shallow bowls with conosegmental or gently rounded profiles. The rims are tapered or rounded and everted. The base has a ring foot. One body fragment (D.40), mentioned above, appears different in shape from the rest – it is a thickset vessel with a sharply angled carination and the decoration only on the exterior in mustard yellow. It could be the shoulder of a carinated dish or a platter with a flanged rim. In most cases in this sub-class, the decoration on the interior consists of a band of solid colour, usually deep yellow or greenish-yellow, running along the internal and sometimes the external lip of the rim. The internal walls of the bowl commonly have pseudo-calligraphic symbols. The floor of the bowl has a large central motif. The four bowls with the 'running hare' or rabbit motif depict a stylized quadruped with long 'ears', a small circle representing the head with a dot in the centre for the eye. The legs of the animal are crammed into the available space somewhat awkwardly. On the exterior, the vessels have medallions and circles with dot and line fillers in-between them, or wide spirals. The colours of the painting appear unstable on the exterior and are seen in shades of red, reddish-green and reddish-yellow. The paintings on the vessel interiors are, in contrast, well defined and sharp against the white background. The glaze covers all surfaces and in one sample there is a trefoil motif, somewhat blurred, contained inside the ring-foot which extends on to the ring itself. The vessels appear to be moulded and then wheel finished. In the case of two other rim fragments, a border of dot and line design running along the exterior of the lip is seen. The exteriors of the bowls from the Al-Sabah Collection, catalogued and described by Watson, closely parallel the bowls at Sanjan both stylistically as well as in colour and form (Watson 2004: Cat.E11, Cat. E.12, Cat.E.16: 191, 192, 194).

LPW-2 is represented by only a few sherds. The distinguishing feature of the group is the use of a deep brown colour for the painting. The vessel shape represented by the rim is similar to the above mentioned classes – a small shallow bowl with a slightly everted rounded rim and ring-foot base. The painting on the exterior has been executed with a broad brush and appears bold. The medallion motif with three oblong dots placed within it and 'V' shaped fillers in-between can be seen. The interior of these sherds has flaked and so not much can be said about the treatment or decoration on the interior, except that the colour used for the decoration appears to have been yellow. The technique of manufacture remains the same as above.

LPW-3 is distinguished from the other two sub-classes by the use of a dark ruby red colour for the painting. The background remains opaque white, now degraded to a brownish-grey. This sub-class is very rare and only two or three sherds have been found. These are difficult to illustrate or elucidate due to heavy exfoliation and degradation. The rim sherd in this sub-class represents a small shallow bowl with a gently rounded profile.

Fabric : All three sub-classes at Sanjan have the same fabric – BUFF(A). The colour is pale yellow and the clay is fairly well levigated with small white quartz particles noticeable in section. The fabric appears to be the same as that of SWGW , WGW and TGW -1 and TGW-2. The potting is fine in most cases, with the thickness ranging from 3 to 5mm. Stonepaste fabric is absent at Sanjan, reinforcing the view that the Sanjan samples belong to the Mesopotamian type, current during the period of the Samarra Horizon.

Petrographic Analysis : The thin-section analysis of LPW-1 sherds shows that the ware belongs to Petrofabric Group D, which has been previously discussed in the section on Turquoise Glazed Ware. Mason is of the opinion that Basra was the manufacturing centre for Mesopotamian Lustre Ware of the Abbasid period, holding a monopoly on the production and distribution of this ware (Mason 1997: 103 – 107).

Occurrence at Sanjan : This ware was included in the category of Glazed Buff Ware during the preliminary analysis of the first two seasons of excavations at Sanjan. In the present study, this ware is treated as a distinctly separate class of ware. Layer 2 in TT4 has yielded the maximum number of these sherds. Some have been retrieved from the well deposit in TT4 SW All can be identified as belonging to the Abbasid period. They occur in association with WGW, CPW, SWGW, TGW, ESW(W), BW, PORC-1, CEL-1, STONE-2, etc.

Discussion : This ware has a long and eventful history, originating in Egypt, developing into an established ceramic technology in Mesopotamia, transplanting itself to Egypt again, and then spreading to Syria and Iran. Changing traditions, technological variations and regional and chronological aesthetics are reflected in the different types of Lustre Wares found. The differences between Abbasid, Fatimid, Saljuq and Safavid Lustre Wares have been drawn by various art historians, archaeologists and scholars. One of the most important differences is in the use of soft-paste fabric in Abbasid Lustre Wares and stone-paste bodies in the later wares. Variations in colours, motifs, vessel forms, painting styles, etc. distinguish each 'school' from the other. The Abbasid Lustre Ware industry to which most, if not all of the Sanjan collection belongs, is said to have been centred at Basra. Its close association with CPW may be attributed to common production centers or kilns.

LPW has a wide distribution in the Indian Ocean although the quantities reported from sites are not in large numbers. However, as Watson points out, the production of the ware appears to have been restricted to a few centers at any one time in the course of its complex history (Watson 2004: 38). The ware was reported by Sarre in the excavations at Samarra. Polychrome and monochrome lustre tiles were found in the palace of Al-Mu'tasim and Polychrome and Monochrome Lustre Ware vessel sherds were found at Qasr al'-Ashiq. On the basis of the material found during Sarre's excavations and the Iraqi excavations, Northedge and Kennet place Polychrome and Ruby Lustre Wares to the second half of the ninth century, "a dating compatible with the conventional date of AD 862-3 attributed to the lustre tiles in Qairawan". They date monochrome lustre to after the major depopulation of Samarra between 885 and 895 AD, suggesting that the production was limited to the first half of the tenth century. An important point they note is that Mesopotamian lustre ware bowls differed in form from the eastern slip-painted wares which were current after the second half of the tenth century. They never adopted the flat base and straight flaring rim of the eastern slip-painted wares (Northedge and Kennet 1994: 32-33).

Kennet discusses and evaluates the Samarra Horizon issue in his analysis of the pottery from Ras al-Khaimah and presents a well-drawn up table to summarise the dating of different wares at Siraf - by Whitehouse, Susa - by Kervran, and at Samarra - by Northedge and Kennet (Kennet 2004: 31-32). Whitehouse is of the opinion that there were three stages of the introduction of the Samarra Horizon – stage 1 had WGW, CPW and SWGW; stage 2 consisted of Splashed wares, and; stage 3 saw the introduction of Lustre Wares and early sgraffiato. Kervran suggests a four stage introduction at Susa – stage 1 which has CPW and SWGW; stage 2 with WGW; stage 3 introduces Splashed wares and Lustre Wares, and; stage 4 when early sgraffiato is introduced. Northedge and Kennet suggest a three stage introduction of wares at Samarra – stage 1 has CPW and SWGW; stage 2 introduces WGW and; stage 3 has Lustre Wares and early sgraffiato Wares. Whitehouse and Northedge and Kennet are in agreement that Lustre Wares and early sgraffiato were introduced after 885-895 AD, whereas Kervran argues for a slightly earlier date for Lustre Wares at Susa, around 861 AD. Most scholars agree, however that these wares continued until the middle of the tenth century.

The fabric of the Abbasid Lustre Wares, as far as is known, is almost entirely restricted to the Basra clays.

Mason is of the opinion that certain factors define the hierarchy of wares and systems of pottery production. He places Lustre Painted Wares of the Abbasid period in the first quality group of elite ceramics as they are defined by their "technically innovative techniques of manufacture combined with decoration executed in an artistically sophisticated manner, and production in a very restricted number of centres with a very wide distribution". Basra, according to him, was the sole producer of lustre-wares between c.800 and 975 AD, although the distribution of the ware ranged from Spain to China and southwards to the tip of Africa (Mason 1997: 103-107). This would have made it a particularly sought after luxury item. Its presence at Sanjan, then, gains relevance and reflects upon both the economy of the settlement as well as on the commercial contacts it had in the Persian Gulf. Mason's study of Iranian Lustre-Painted wares is of significance for another reason as well – he emphasizes the fact that almost all known Iranian Lustre-Painted Wares can be attributed to the kilns at Kashan, they are almost exclusively of stone-paste fabric and that they are datable to the Safavid period (Mason 1997: 111). No such Iranian samples are present at Sanjan.

Lustre Painted Ware of the Samarra Horizon types have been reported from various sites. Surprisingly very little of this ware has been found at Siraf. Mason attributes this paucity of finer wares to various factors such as a lack of arable land, dependence on trade and disruptions caused by the Qarmatians and Suhar Omanis. There could, therefore, have been times during the 9th and 10th centuries when such local economic fluctuations could have taken place (Mason 1991: 56 - 57). Tampoe distinguishes between the earlier Mesopotamian lustre and the later Saljuq lustre at Siraf on the basis of fabric and style (Tampoe 1989: 36-37). As Kennet points out, the problem with dating of these early wares at Siraf has been compounded by the fact that the pottery cards upon which Tampoe has based much of her work and which Whitehouse had drawn upon for the first interim report had been erroneously filled in the first place (Kennet 2004: 31). Kush and Ras al-Khaimah do not report Lustre Ware but Hulaylah which is close to Kush yielded two sherds of monochrome lustre in the survey. This leads Kennet to conclude that Kush was probably not occupied during the late 9th and early 10th centuries (Kennet 2004: 33-34). Horton's description of the ware at Shanga where it occurs in considerable numbers establishes its Mesopotamian source. In the light of the rarity of the ware at Siraf, Horton surmises that the ware was probably imported into East Africa directly from Basra and not from Siraf (Horton 1996: 279). If the same hypothesis is applied to Sanjan, then one may similarly deduce that LPW was a direct import from Basra and that the site had commercial links with the head of the Persian Gulf.

At Manda, Chittick notes two types of Lustre Ware – those with opaque white glaze and buff body and those with dark blue glaze and white body. The first group he dates to Period I (mid ninth to early eleventh century) (Chittick 1984: 11, 77-78). Adams calls the ware Gold or Olive Lustre Ware at Tell Abu Sarifa and assigns it to level IV (650 – 800 AD) (.Adams 1970: Table 2, 110-111). Only one sherd of Lustre Ware was found during the survey and excavations at 'Arja mines though several were found in surveys of the Lasail mines. Costa and Wilkinson ascribe to this gold lustre sherd a date concomitant with the date from Siraf which was within the immediate trading sphere of the site (Costa and Wilkinson 1987: 186). They do not account for the fact, however, that Lustre Ware at Siraf itself is a rarity. Carswell and Prickett report seventeen sherds of this ware at Mantai and attributed it to the 9th – 10th century AD (Carswell and Prickett 1984: 59, 64). The occurrence of this ware at Sanjan is relevant in the stratified context and can help in establishing the chronology for the levels at which it occurs.

Bichrome Glazed Pink Ware (Ware code: BiGP)

Splashed Wares, supposedly inspired by T'ang three-color splashed Wares, are one of the most prominent of the Samarra Horizon classes having generated much debate about their actual antecedents. Most scholars refute the theory that Chinese three-color Sancai wares were models for the West Asian imitations since it was found that the Sancai wares were mainly used as funerary furniture and were not an export ceramic. In fact, the ware has been reported from sites like Susa (Grube 1994: 12 -13), Samarra and Fustat (Northedge and Kennet 1994: 34). The ware described by most scholars as Splashed Ware of the Abbasid period is usually on a buff body, belonging to the WGW group. These wares have a mottled and splashed effect. SWGW at Sanjan does not include these splashed effects on the opaque white glazes. However, splashed decoration on a lead-glazed ware with a pale reddish pink body is present in the Sanjan collection. These are completely different from the preceding classes and could either be precursors to or contemporary with the early sgraffiatos which make an appearance towards the later part of the Samarra sequence. Watson describes and illustrates pieces from the Al-Sabah collection which closely resemble the Sanjan sherds (Watson 2004: 40-41, 199-201).

Definition : This class is defined by a white slip and a translucent glaze overlying a pink fine earthenware body. Green, yellow and brown or ochre splashes of colour can be seen on the surface. The colours are usually cloudy and no other attempt at decoration is evident. The vessel forms include closed as well as open forms. They have been variously called Splashed Wares, sgraffiato (sometimes even when lacking incised decoration), or, as Lane puts it imaginatively, "egg and spinach" ware (Lane 1956: 12-13). Tampoe defines this class as an "Islamic imitation of the mottled Tang ware". She classifies all these wares with and without sgraffiato in one category and then differentiates between them on the basis of fabric and decoration. This class she places in the Fine red fabric group (e) (Tampoe 1989: 37-39). The fine reddish pink or orange-pink body has a white slip applied to it with the colours splashed on at random with no attempt to define a design or pattern. The most common combination of colour at Sanjan is white and green or white and yellow, though one sample also has a mustard-brown colour all over with dark manganese black patches and a similar dark lip to the rim.

Shapes and forms : Both closed as well as open vessel forms are seen in this class. The closed forms are beakers and small jugs or jars. These are straight sided with short triangular rims. One small tea-cup shaped strap handle has also been found. This handle is slightly twisted and has a channel running along its length. The open forms are mostly bowls. No dishes have been found in this class. The bowls are rounded, with fine-tipped slightly everted rims, similar to the sgraffiato wares. The bowl with the mustard-brown colour is not exactly circular, perhaps due to a manufacturing defect. The bases are flat discoid types, sometimes with a groove on the interior to create a ring-foot. They are somewhat heavier-set and not as finely shaped as the sgraffiato bases. The vessels appear to be moulded and then wheel finished. Stacking marks on the underside of a base indicate the method of firing. The vessels are finely potted with an average thickness of about 4mm. The shapes seen in this class are T.10, T.12, T.14.1, T.17, T.18, T.346, T.347, T.348, T.349 and T.350. The base shapes are Bs.8, Bs.61, Bs.62, Bs.63 and Bs.64.

Glaze and surface treatment : The pink fabric of the vessel is covered with a white slip and the colours are splashed on to the surface in a cloudy, mottled or runny effect. A transparent glaze is then thinly applied over it. The treatment of the mustard-brown bowl is somewhat different, having the colour applied all over the white slip and with manganese black streaks on the interior. The rim has a dark lip and two dark crescent-shaped patches placed at a short distance from each other which could well have been caused by stacking. The bowls do not appear to have a glaze on the exterior, the slip and the glaze both stopping just under the rim, sometimes in an uneven swag. However some of the bases have traces of glaze on both surfaces and some body sherds show signs of erosion or flaking, especially on the exterior. So it may well be that some of the wares did have an external glaze as well. The closed shapes also have glaze on both surfaces, with the colour splashes restricted to the exterior. The glaze has eroded on most of the bases and some of the other sherds. White and green, white and yellow or ochre yellow, pale yellow and green, mustard-brown and manganese black are the colour combinations seen in this class.

Fabric : The fabric of this class is mostly PINK(B). It is fine well levigated clay having a few small white inclusions of quartzite and minute red specks. The texture is sometimes a little chalky. Usually the firing is even and in the few cases where it is not, a pale grey tinge can be seen in the pink fabric. PINK(B) is the same fabric as the sgraffiato wares. One or two exceptions have PINK(C) fabric which is also fine, well levigated and pale pink in colour. But this fabric is sandier and has fine black and white inclusions which can be seen under lens. The fabric has turned grey in firing in some sections. Petrographic analysis was not done for this ware class. However, the thin-section analysis for PINK(B) fabric was done for the sgraffiato wares. The mineralogical profile for the same is given below in the respective sections.

Occurrence at Sanjan : The sherds from this class were found at almost all localities excavated, though not in profuse numbers. The frequency is higher in TT1 than in TT4 but it is significant that the ware only occurs in Layer 1 and the upper part of Layer 2, along with the sgraffiato wares, TGW, PORC-1, CEL-1 and STONE-1 and STONE-2. In TT4, the upper part of Layer 2 also contains a few sherds of SWGW.

Discussion : This class closely resembles the sgraffiato ware classes. The fabric is the same and the glazing tradition appears to be common. It is possible to infer that the wares with and without sgraffiato came from the same kilns. According to Watson, "they are an indigenous development that takes advantage of the natural behaviour of the simple materials" (Watson 2004: 40-41). Northedge is of the opinion that the wares with only splashed decoration are earlier and that the ones with the sgraffiato incised decoration are a later development – the former belonging to the ninth century and the latter to the tenth (Northedge 1994: 33-34). The ware has been reported first from Samarra by Sarre (Northedge 1996: 243-44) but has subsequently been found at sites such as Siraf (Tampoe 1989: 37-39). Kennet reports a similar ware from Hulayla (Kennet 2004: 33). He calls this ware SPLASH. The white or buff fabric of this ware seems to be different from that at Sanjan. At Siraf both, the buff as well as the fine reddish pink fabrics is reported. Adams' report on the excavations at Tell Abu Sarifa includes this class. He assigns it to Level V, dateable to the period 800 – 950 A.D. (Adams 1970: 108, 119), a sufficiently large time bracket to span the Samarra Horizon period itself. Watson believes the splashed but unincised wares belong to the ninth century whereas the wares with both splash and incised design belong to the tenth century and after (Watson 2004: 47 – 48). Northedge and Kennet are of the opinion that the Splash wares were introduced in the early to middle ninth century, "certainly before the occupation of the city of al-Mutawakkiliyya in AD 861" (Northedge and Kennet 1994: 33).

Splashed Ware with sgraffiato

This ware is represented by very few sherds. The shapes are open form consisting of small bowls. These sherds have darker and more vibrant colours splashed on the interior. The dark green, brown and ochre are so densely splashed on that the underlying white slip is completely covered and can hardly be seen at all. One of the bases has a dark incised squiggle, in the form of an open '8'. The colours have run into and pooled in the incised line so that it appears black. The bowl illustrated by Watson in the catalogue of the Al-Sabah collection closely parallels this base in colour and glaze, but lacks the incised design (Watson 2004: 201). He mentions Nishapur as the probable provenance. The fabric of the Sanjan sample is PINK(B), the same as Bichrome Glazed Pink Ware, the sgraffiato wares and White Glazed Pink Ware. The difference between the splashing and incisions seen on this ware and on the other sgraffiatos is evident. Clearly this ware is an early type. Two other sherds with ochre and green splashes and incised designs appear somewhat similar to the above mentioned base. The incised designs appear simple and are located within the spots of colour. However, this design shows signs of hatching and may therefore be of a later date. It has therefore been described later in the relevant section on Hatched sgraffiato Ware.

Sgraffiato Wares

The introduction of sgraffiato is a complex issue and scholars have been divided on the inclusion of this ware in the Samarra Horizon. Sarre had included early sgraffiato in his typology at Samarra but later studies have shown that sgraffiato appears only within a certain area of the city which was occupied in the post-Caliphal period. However the occurrence of the ware with other Samarra Horizon classes at other sites suggests that it was introduced not long after the abandonment of the city, in the early tenth century (Northedge and Kennet 2004: 33-34).

The ware takes its name from the decorative technique which is its hallmark. It involves the application of a white slip over the body of the vessel and the incision of designs through it so that the fabric can be seen. Green, yellow and brown are frequently splashed on, sometimes with little relation to the incised design. A translucent lead glaze is applied over the surface. A number of production centers are known from Iran and Afghanistan by the 11th century, though the technique originated in southern Iraq in the early tenth century (Kennet 2004: 34). Sgraffiato wares are some of the most prolific West Asian wares found on Indian Ocean sites.

The differences in style, fabric, technique, etc have been used by various scholars to determine the regional and chronological variations within this class. Whitehouse defines three stylistic groups at Siraf – Style 1 which is early sgraffiato, Style 2 and 3 which are late hatched type (Whitehouse 1979: 49 – 61). Kennet classifies sgraffiato into early, hatched, yellow, monochrome green, monochrome mustard, two-tone, polychrome and degraded classes (Kennet 2004: 34 – 37).

In this study, all sgraffiatos have been classified on the basis of the splashed colours on them – polychrome, bichrome, monochrome, white, etc. Some of these are early sgraffiato types without hatching. Most have hatching and may be considered later. Champleve, which was represented by only three sherds, is also described in this group although it may have a slightly later date. The description of fabric, occurrence at Sanjan and general discussion of these classes is given in the later part of this section, in order to avoid repetition, since all the classes have a common fabric and occurrence pattern.

Hatched Sgraffiato Ware (Polychrome) (Ware code: HsG(P))

Definition : This class of sgraffiato has multiple colours splashed on its surface, usually in shades of green, brown and yellow or ochre on white. As with the other sgraffiato wares, the forms are almost without exception, open shapes such as bowls. They are mostly hemispherical and have a ring-foot or a disc base. In some cases, one sees an attempt to try and make the incised design coordinate with the splashes. The hatched designs are executed on the vessel interiors. The exterior is usually unglazed. The white slip, some remnants of colour and glaze can be seen stopping a centimetre or two under the rim on the exterior. Tampoe classifies this as Type g with a fine red fabric, Style 3 or late sgraffiato at Siraf (Tampoe 1989: 38-39). Some of the sherds in this collection also match her description of Style 1 or early sgraffiato.

Shapes and forms : As mentioned earlier, these vessels all tableware forms. The hemispherical bowl form takes its inspiration from the splashed wares which were themselves copies of the Chinese forms. The rims are simple in most cases with either a tapering lip or a rounded one. The occasional rim with a slightly inverted lip is also present. One unique rim has been cut in a zigzag manner (T.9.1D.3, T.10D.100, T.10.1, T.10D.3, T.13, T.19D.6, T.20D.7, T.42.2, T.48, T.52, T.56, T.271 and T.352D.146). The bases are disc-like and flat, or disc-like with a groove cut into the inner side to form a ring-foot (Bs.8, Bs.11, Bs.12, and Bs.66D.145). Some of the body and rim sherds have mending marks along old fracture lines, indicating repair and retention of the vessel despite breakage. Only one sherd shows a different profile. It is a body sherd with a sharply angled carination having a very deeply incised floral design but no hatching. The treatment and the shape are both unique and no such samples have occurred in the trenches, this one being a surface find.

Glaze and Surface Treatment : The white slip which covers the interior of the vessel also covers the upper part of the exterior surface along the rim in an uneven manner. The incised designs are cut into the slip and the underlying dark colour of the body brings the incisions into relief. The application of the slip appears thin and the colours, usually green, brown, yellow and ochre are splashed with no apparent relation to the design. In some rim sherds a concerted effort has been made to use the green along the rim and ochre, yellow or brown on the interior of the vessel. Some of the colour may splash or overrun on to the external surface just along the rim. A translucent glaze overlies the whole surface. The exterior walls and the base are usually left untreated to either slip or glaze. In the case of one sherd (D.103), the white slip has been covered with a yellow colour and an ochre splash has been placed in the centre, and green spots have been splashed on to correspond to the hatched design. The whole surface appears somewhat patchy. The glaze has eroded on a number of sherds though the colours and designs can still be seen quite clearly.

Decoration : The decoration of this ware depends upon the incised designs and upon the colours that have been splashed on to the surface. The sgraffiato technique of incising a design through the slip is further highlighted by the fine oblique lines used for hatching the background or field so that the motif stands out in relief. The incised designs are restricted to the bowl interiors. The rims all show the standard convention of having a double line border incised just inside the lip under which all other decoration is placed. The styles vary considerably and may be broadly divided into those having hatching and those that have incised designs but no hatching. The sherds with polychrome colouring and non-hatched decoration are represented by two types – a base with an incised cross and quadrants filled with squiggly lines radiating outwards, the whole splashed with green, brown and traces of ochre yellow; and a body sherd with a yellow and green surface having very sharply and deeply incised design of concentric circles and floral motifs. This latter sherd was a surface find and no such sherd was found in the excavated trenches. Sherds with hatched designs form the bulk of this class. The designs follow four styles – in the first and more commonly found type, the rim has a double lined border under which a row of repetitive motifs are placed to give a pseudo-calligraphic effect. These may be finely incised or more widely spaced but they follow the same style of execution. The hatched lines on one rim sherd are oriented in two different directions. The second type has the double lined border along the rim but the design is vertically placed on the walls of the vessel. It is widely spaced with blank columns interspersed with palmettes which are placed in panels. There is a second border of double lines at the base of this vertical design and a blank space, followed by a centred motif on the floor of the vessel. The third type has an ochre splash in the centre and four green spots around. Within these colour spots are small incised swirls filled with hatching. The fourth type has centred motifs of rosettes or palmettes on the floor of the bowl. Geometric and abstract motifs are used for the centers of bowls. These can be intricately drawn and hatched. A fifth type should be mentioned though it is represented by only one sample. It is a rim sherd cut in a zig-zag manner to give a serrated effect. The inside of the rim has a border of a twin-leafed vine or a creeper and a hatched background. Green and brown splashes colour the surface.

Hatched Sgraffiato Ware (Bichrome) (Ware code: HsG(B))

Definition : This ware class closely follows the previous class, differing only in the use of colour. The interiors of these vessels have the hatched incised designs cutting through the white slip but only one colour is seen in combination with the white background. The most common colour is green, although sherds with yellow, ochre or brown are also seen, though rarely. Tampoe classifies these as Type f with a fine red body, Style 2, at Siraf (Tampoe 1989: 38-39)

Shapes and forms : The shapes in this class are also open forms, mainly bowls and small to medium sized dishes simple rims which are either rounded or tapered and sometimes slightly everted (T.9, T.10, T.11D.4, T.13, T.18, T.19, T.42, T.42.1, T.43, T.353D.147, T.354D.148, T.355D.151 and T.356D.158). The bases are the same as in the previous class – flat, disc-like with or without grooves scraped on the underside to create a ring-foot effect (Bs.8, Bs.11). As with the other sgraffiato classes, many of these sherds show evidence of repair and mending by way of fine perforations.

Glaze and Treatment : The technique of these classes remains the same – the application of a white slip, incised design with hatching of the background and splashes of colour prior to the glazing. The colour is usually splashed on to the floor of the vessel and along the rim. It is blurred and has no definite outline. The exterior of the vessel is unglazed with the slip and sometimes the translucent glaze stopping just under the rim. The glaze may sometimes have dripped down the exterior surface. The white slip frequently appears to take on a cream or pale yellow tinge.

Decoration : The designs in this class are somewhat similar to the preceding class. The styles noticed in this ware are of four types. In the first style, the inside of the rim has a double line border running along it. The repetitive motif creating a pseudo-calligraphic effect is contained in this border. One such rim has a motif which resembles a stylized horse. The hatching is fine and has been done in vertical strokes instead of the usual oblique slant. The space under this border does not appear to have any design and may be a field left plain by the potter so that the central motif may be highlighted. In the second style, a similarly ornate border runs along the rim but the space under it is covered with an intricate geometric or abstract design. One such rim has the traces of broad circular bands or arches under the border of the rim. A similar patterned sherd (no.1211) is reported by Tampoe from Siraf (Tampoe 1989: 40, and Fig.66, 193). The third style has a single or double line running along the inner lip, very close to the edge, and an abstract or stylized floral design which covers the wall of the vessel and stops at this line. No border or horizontal band of design can be seen. The fourth style has a hatched curving design within which a small hatched triangle can be seen. A fifth style may actually be part of the first two since it is seen on the bases of the bowls. In this case the floor of the vessel has a double or single lined circle within which complex central motif is placed. This could be a rosette, palmettes, abstract design or geometric pattern. In one instance the circle around the motif has been dotted with manganese black colour to create a beaded border. D.157 is a body sherd with a complex design and three such black dots under it. The hatching does not usually follow a single orientation but may vary in different parts of the same design. The incised designs without hatching are missing in this class. The central motifs such as the rosette are seen on sherds from other sites as well indicating either that the motif may have been a traditional, standard design commonly used by many potters or that the sherds with identical designs may have come from a common kiln and may share a common provenance. For example, the rosette motif on the Sanjan sample is the same as that reported by Kennet from Ras al-Khaimah (Kennet 2004: CP 3, 145). One unusual rim sherd (D.149) has small circles filled with hatching floating on a blank field under the double lines which run along the rim.

Hatched Sgraffiato Ware (Monochrome) (Ware code: HsG(M))

Definition : Very few sherds of this ware have been found at Sanjan. They have a single colour which covers the entire interior surface. The most common colour is green either in a dark or a bright tone. They only occur in the upper levels of the trenches.

Shapes and forms : Bowls are the predominant vessel form in this ware, following the same profile as the other sgraffiato classes. The rim shapes are also the same (T.9.1, T.11, T.13, T.42, T.63). The bowls can sometimes be quite small. The base sherds found in the excavations of 2004 in TT4 have flat disc bases with a groove cut on the underside to form a ring foot with a very prominent disc left in the centre. Another fine groove is seen on the exterior at the point where the vessel wall and the base are joined (Bs.65 D.144).

Glaze and Surface Treatment : A white slip underlies the colour which covers the entire surface of the interior. The most common colours are dark green, bright green and in a rare instance a blackish brown tone. The exterior has the white slip and the translucent glaze but no colorant. The glaze and the colour are applied in thin coats.

Decoration : The motifs and designs are slightly different from the preceding sgraffiato wares, especially on the bases. Bs.65.D.144 is the base of a small bowl with some curved hatched design with an open space in the centre. On this open field is a small hatched triangle containing a circle and dot. The colour of this sherd is blackish brown. D.143 is also a base sherd with a curved hatched design contained within a roundel. The colour is bright green. The rim sherds in this class have the double line border along the rim and design underneath. The horizontal or vertical placement of decoration is not seen but the random incision of abstract design on the vessel walls is in evidence.

Hatched Sgraffiato Ware (White) (Ware code: HsG(W))

This class of sgraffiato wares is represented by only a few badly eroded fragmented body sherds. They have the hatched decoration of the previous classes but lack the splashes of colour. The white slip has a translucent glaze applied over it. The fragments are too small and rolled to identify any motifs, styles or patterns. What is discernable, however, is the fineness of the incisions and the closed spacing of the oblique lines. The incisions are shallow and the erosion of the sherds makes it difficult to trace the designs. Grube, like Lane, suggests that these plain white sgraffiatos are inspired by engraved silverware (Grube 1994: 115).

Hatched Sgraffiato Ware (Yellow) (Ware code: HsG(Y))

Only a few sherds of this ware have been found at Sanjan. The fragments are extremely small and very badly eroded. The fabric, PINK(F) is slightly softer than the other classes and is pale orange in colour. The glaze and decoration have totally eroded leaving a yellow 'skin' on the surface. Not much can be said about the decoration or treatment. Some incised lines are faintly visible. The known shapes are T.11, T.42 and Bs.11.

Yellow Sgraffiato

Only two body sherds of this ware have been found. They are different from the above in that there is no hatched decoration. The glaze is bright yellow and covers both surfaces. The design is restricted to the interior and consists of two incised lines. On one sherd the glaze and design are both effaced due to erosion and can be faintly seen. The fabric of this ware is PINK(B). the vessels appear to be bowls or dishes, with a thickness of about 4 – 5mm. Both sherds were found in TT1 in layers 2 and 3.

Champleve (Ware code: CHAMP)

Definition : This class refers to a ware with a pale red-pink fabric, the same as the Hatched Sgraffiato wares.

A creamy-white slip is applied to the body. The design is incised and cut into the slip in broad strips so that the darker colour of the body shows through. No hatching is seen on these sherds. The ware has been identified by art historians as "Garrus" ware (Lane 1947: 31-35. Grube 1994: 121). This name comes from the region in West Iran where it was supposed to originate. The date for this ware has been a topic of some debate since scholars like Grube place it as early as the eleventh century (Grube 1994: 115-116) and others like Lane think it is as late as the 12th or 13th century (Lane 1947: 26). Lane is of the opinion that these are rustic contemporaries of lustre-painted wares from Rayy. These wares have also been referred to as Gabri wares under the mistaken notion that they were associated with the pre-Islamic traditions of the Gabri or Zoroastrian community of Iran.

Shapes and forms : Only four sherds of this ware have been found at Sanjan. Two of the body sherds suggest that they belong to bowls or dishes but it is not possible to infer the shapes or profiles. One base, Bs.7 is flat discoid. It appears to belong to a bowl.

Glaze and Surface Treatment : The creamy-white slip and the glaze give the interior surface a pale yellow colour. The strips and bands which have been cut through the slip show the colour of the body which, seen through the translucent glaze, appears beige or pale brown. The glaze and slip has flaked in places. The exterior is left untreated.

Decoration : It is not possible to say much about the motifs or stylistic features of these designs since no sherds large enough to facilitate identification have been found. The design on the base consists of some incised curved lines which are part of some design and broad strips or bands that have been cut through the slip (D.17).

Fabric of Sgraffiato wares and Champleve : The fabric of all the above mentioned wares is PINK(B). This is also the same fabric as BiGP, CSGW, etc. The clay is well levigated, fine, reddish pink to pale orange-pink. The texture is slightly chalky and some very fine quartz and some red particles seen under lens. The firing is usually even.

The thin-section analysis for this fabric places it in the Petrofabric Group A. The thin-sections falling in Fabric A have a fairly micaceous matrix. It is made of fine grained clay. The matrix is non-pleochroic in Plain Polarised Light (PPL) and fairly birefringent in Crossed Polarised Light (XPL). The grains were sub- angular to sub-rounded in shape. Their sizes varied from 80 μ to 120 μ. Its frequency fell between 5 to 10%. The grains show more or less a parallel orientation. Voids were very few. Most of them appeared to be areas of grain fallout. The mineralogy of the detritals included quartz, plagioclase feldspar (dominant), mica and sparse occurrence of patches of hematite. The non-plastic inclusions in these sections were well sorted (Fig 1). Their grain-size distribution character was nearly unimodal (Fig). It may be added here that no other sample from the forty-five analyzed fell within this group other than the sgraffiato wares. The geological terrain from which the clay could originate has a high feldspatic component, with quartz and mica. It has been identified tentatively as terrain VI in the report by Dr.Krishnan.

The petrographic analysis of these wares from Siraf was done by Mason and Keall who identified nine petrofabrics. The date they assign to the Hatched Sgraffiato is Period 6 (1050 – 1150 AD), and they suggest that a number of production centers may have existed for these wares. Petrofabric 8, to which some of the Siraf samples belong, matches the Basra clay profiles (Mason and Keall 1991: 62 – 63).

Occurrence of Sgraffiato and Champleve at Sanjan : The sgraffiato wares at Sanjan all occur in the upper levels in trench TT4. These ware classes are totally absent from the lower part of layer 2 down to layer 6. They occur in association with PORC-1, PORC-4, STONE-2, and glazed wares with a pink fabric, such as BiGP, MGP, MGB, IGPW, WGPW, etc. They were found in larger numbers in TT1 and TT2. However the stratigraphic context was not well defined. But it may be noted that in these trenches as well they were restricted to the first three layers. Polychrome splashing is less prolific than bichrome splashing at Sanjan. Monochrome and white sgraffiatos are even rarer. CHAMP is represented by only four sherds, all of which occur in TT1 in layer 1.

Discussion on Sgraffiato wares and Champleve : Sgraffiato wares form some of the most widely distributed of West Asian wares in the Indian Ocean. The introduction of the sgraffiato tradition has generated much discussion and debate. Scholars like Northedge, Kennet and others are of the opinion that the early sgraffiato wares were introduced at the end of the Caliphal occupation or just after at Samarra and that they form part of the later Samarran assemblage, dateable to the early tenth century (Northedge and Kennet 1994: 33 -34). Grube traces the origin of the sgraffiato technique to the Coptic wares of Egypt in the fifth to seventh centuries and thinks it was introduced to Iraq and Iran later. He notes that a transition appears to take place from the splash-glazed wares to splash-glazed wares with sgraffiato and that this shift takes place in the early 10th century (Grube 1994: 119). The early sgraffiato decoration is supposedly quite simple as compared to the later more complex version of hatched design. Whitehouse designates it as Style I at Siraf (Whitehouse 1979: 50). Tampoe dates it to a slightly earlier period as compared to Northedge and Kennet (Tampoe 1989: 90 - 94. Northedge and Kennet 1994: 33 - 34). However, almost all scholars agree that sgraffiato wares occur late in the Samarra sequence, though the date remains undecided. The hatched wares with more complex and intricately executed decoration are a later development. The presence of mending marks on these sherds at Sanjan is indicative of the value placed on these vessels. This has been noted by Horton at Shanga as well (Horton 1996: 281). They are reported from almost all sites in the Indian Ocean and it is seldom that any two sherds are exactly the same even though some standard motifs such as the rosettes and palmettes or vines and scrolls are commonly used. That each piece incorporated some element of personal creativity is evident. Whitehouse refers to this hatched variety as Style III (Whitehouse 1979: 49 -61) or Late Sgraffiato. Grube suggests that it went out of production in the Gulf by 1150 AD or so (Grube 1994: 121). This is not very different from the date suggested by Mason and Keall – 1050 to 1150 AD (Mason and Keall 1991: 62 – 63). Production centers for the sgraffiato wares which have been suggested by various scholars are Basra, Nishapur, Samarqand, Lashkari Bazar (Afghanistan), Sirjan, etc. The demand for these wares was obviously great, judging by their prominent occurrence at sites spread from the East African coast to the Far East. Horton presents a detailed report on the occurrence of the types, shapes, fabrics, decorations and techniques represented in the Shanga assemblage. The closed forms reported by him are totally absent at Sanjan. A large number of hatched designs illustrated by him are seen in the Sanjan assemblage – for example, the rosette motif, palmettes, scrolls, placement of small triangles as centers in swirling abstract motifs, pseudo-calligraphic designs, etc. The rim with shaped or notched lip which is seen in a singular sample at Sanjan is also reported from Shanga. Horton presents the stratigraphic distribution of the different sgraffiatos in the excavated trenches and correlates them to the other sites in the Gulf which report this ware. He dates the hatched or late sgraffiatos to between 1000 and 1300 (Horton 1996: 279 – 89). At Manda, the ware is assigned to Period IIa, mid-11th to late-12th century. It occurs in large quantities. Some sgraffiato also occurs in Period IIb - late 12th to 13th century (Chittick 1984: 11-12, 79-81). At Kilwa they are assigned to the 11th and 12th centuries, Period Ib. Three classes are identified at Kilwa, as at Manda – Hatched Sgraffiato, Simple Sgraffiato and Champleve (Chittick 1974: 303-304). This ware is poorly represented at Mantai with only one sherd (Carswell and Prickett: 1984: 64). It is more prominently seen at Banbhore, Pakistan where it is dated to the 11th century and after (Pakistan Archaeology 1, 1964: 54 and Pl.XXIII). At Suhar, Kervran reports them from Levels V and VI, dateable from 860 – 1100 AD (Kervran 2004: 315). At Abu Tell Sarifa the wares occur in Level VI which is dated 950 – 1100 (Adams 1970: 8 -10). At 'Ana the ware is reported by Northedge, Bamber and Roaf in Group 6 along with Monochrome Lustre Ware and Splashed Ware. They suggest a late 10th and early 11th century date for it. (Northedge et al 1988: 94).

CHAMP has been reported in association with the sgraffiato wares from sites like Shanga, Manda, Kilwa, Siraf, etc. They are almost all dated to the 11th and 12th century.

Monochrome Wares

A number of scholars use the term 'monochrome' to describe wares that can be quite diverse. In this study the term has been used to define wares with a single coloured glaze, no decoration. The monochrome wares at Sanjan belong to two different fabric groups but they share the same features. Almost all occur in the uppermost layers or in the disturbed context of pits and the well. They form a very small percentage of the assemblage. They are described below on the basis of their fabrics.

Monochrome Glazed Pink Ware (Ware code: MGP)

Definition : This ware class consists of bowls and dishes made of pale pink to red-pink earthenware, the interior of which is glazed in a single colour. The most common colours at Sanjan are green and blue. The glaze is splotchy and sometimes reddened in firing. The vessels can be heavy-set or of medium thickness.

Shapes and forms : Almost all the vessels in this class are open forms, although some closed forms such as small jars are also represented by body sherds. These are very few in number. The bowls and dishes are hemispherical (T.13, T.16, T.43, T.44, T.272). Two types of bowls are seen – those which are thickset (5 – 8mm.) with rounded profile simple rims and ring-bases, and those which have medium thickness (4mm), flanged rims with a high ridge and shiny glaze. No bases of the latter type have been found so far at Sanjan. The flanged rim with the pronounced ridge is typical of the fourteenth century monochrome wares. The thickset vessels have stacking marks on the vessel interiors. The spurs or spacers used for stacking appear to have been tripods. One unusual sherd which has been included in this class is a stud finial with a dark brown-black glaze and perforations on its body, similar to modern teapot lids (T.344). This piece appears to be quite late since it was found at the Dokhma and no similar sherd was found in the excavations on the Bandar in the three years of work. Some flat bases are reminiscent of jars or basins. Base types seen are Bs.6, Bs.17 and Bs.22.

Glaze and Surface Treatment : The glaze on these vessels is monochromatic. The most common colours are green and blue. The colours can range from dark to bright. The splotchy and uneven tone of the glaze could be due to the underlying fabric and uneven application. The speckled effect may be due to bubbles. A white slip can be seen on the body under the glaze. This slip is sometimes seen on the exterior just under the rim. The rest of the exterior surface is left unglazed and unslipped. Flaking is common. None of these vessels show the careful and superior quality glazing of the previous classes. The finial is glazed all over and traces of the same brown-black glaze are also seen inside the perforations. An olive green or mossy green glaze is seen on a few sherds with severe flaking. The white slip under this glaze is very pronounced.

Fabric : The fabric of this class is a pale red-pink to pink. There are three sub-fabrics which can be seen. PINK(B) is the same fabric as the sgraffiato classes and has the same characteristics. PINK(A) is a more sandy and coarse red-pink clay with many small white inclusions. The finial is made of this fabric. A softer fabric, PINK(F) is also found. It is pale buff pink colour with a fine sand temper and resembles fabric BUFF(F) in texture. The thickset bowls and dishes usually have PINK(F) fabric. The vessels with flanged rims have PINK(B) fabric. No petrographic analysis of these sherds was done.

Occurrence at Sanjan : These wares occur along with MGB, sgraffiato wares and Chinese wares such as PORC-1, stonewares and CEL-1. They have been retrieved mainly from the disturbed quadrants which contained pits and features such as the well in TT4. In the stratigraphic context they occur in layers 1, 2 and 3 in TT1, from the upper layers at the Dokhma or Tower of Silence and a few sherds from the Koli-khadi. The presence of this ware is relevant in that it is one of the few classes in the assemblage that can be assigned a late date. Occurrence of this ware class in layer 3 in TT1 may be intrusive since the density of brickbat debris is great. Modern pottery sherds were also encountered up to a meter or more in depth in the same excavated trenches. This is indicative of the possibility that some material of a later date may have percolated to the lower levels. The presence of the ware at the site, in the uppermost level, is relevant in providing evidence for human activity at the site at a date later than that suggested by a bulk of the assemblage. The few sherds found in trench TT4 are also from the uppermost part of the trench and are restricted to the first 0.65m.

Discussion : The monochrome wares at Sanjan are important indicators of some occupation at the site or its vicinity in the thirteenth and fourteenth centuries. A few sherds of this ware and similar wares were found by the author during explorations at Palgam, which is located on the south bank of the Varoli diagonally across the river from Sanjan, and Medhi Falia, which is located at Khatalwada, across the creek from Sanjan in the North-west direction. This bears out the hypothesis that some occupation in or near Sanjan continued even after it was supposedly abandoned as the Kisse-i-Sanjan claims. This is supported by the continued use of the Dokhma or Tower of Silence in the first half of the fifteenth century, a date suggested by AMS analysis of the bones from inside the structure (Nanji and Dhalla 2007: 42, 54). Also, Dhalla and Nanji 2006: 75 – 78). The ware has been reported from various sites on the East African coast. Ungujaa Ukuu , Shanga and Kilwa are some of the sites where monochromes have been reported. Juma describes these sherds from Ungujaa Ukuu and makes a distinction between those with a red fabric and those with a buff fabric. He dates them to the mid-14th century (Juma 2004: 112 - 113). Chittick distinguishes between Early Monochrome, Standard Monochrome and Late Monochrome at Kilwa. His descriptions closely match those at Sanjan. He dates the wares to between the later part of the 14th century and the 17th century. The thickset hemispherical bowls at Sanjan could be dated to the 14th century on the basis of Chittick's periodisation. The bowls with flanged rims and ridges, which Chittick describes as 'ledge rim', are included in the Standard Monochrome group and are dated from the mid-15th to 16th century (Chittick 1974: 304). Similar wares are reported from Manda as well. The dates correspond to those at Kilwa (Chittick 1984: 11 - 12). At Shanga, Horton reports Monochromes and classifies them according to the fabric and glaze but says that they have a common range of forms, the hemispherical bowl being the most common. The descriptions match those at Sanjan. He suggests a South Arabia or Red Sea origin for these wares. The suggested date for the ware is mid-14th century (Horton 1996: 293 – 96).

Monochrome Glazed Buff Ware (Ware code: MGB)

Definition : This ware is very similar to MGP in form as well as in treatment. The fabric is pale yellow to buff in colour and the vessel forms are mainly hemispherical bowls. A thin slip and green or blue glaze covers the interior surface and stops just under the rim on the exterior. The closed forms and the flanged rims with ridges are missing in this class.

Shapes and forms : Only hemispherical bowls are seen in this ware class. They are of medium or small size.

The medium sized vessels have a thicker potting (5mm.); whereas the smaller bowls have a slightly finer potting (3 – 4mm). The rims are plain and the bases are ring-footed. The shapes in this class are T.9, T.41, T.42, T.47, T.63.1 and T.272. The base forms are Bs.9, Bs.14, Bs.15 and Bs.71.

Glaze and Surface Treatment : The most common colours are green and blue. The tones of colour can be bright and the glaze has a high gloss. The glazing is of a better quality on these sherds than on the Monochrome Glazed Pink Ware. The stacking marks are identical to the previous class.

Fabric : The pale yellow to buff fabric of this ware class can range from the fine and chalky type to the more grainy and sandy sort. The fabric sub-class in this ware are BUFF(A), BUFF(E) and BUFF(F). The first of these is the same fabric as the TGW, SWGW, WGW, LPW, CPW and BW. The second fabric is very similar to the first but is more sandy and coarse in texture with many inclusions and fine red particles which could well be grog. It is fired hard and has a more white tone than yellow. This fabric is the same as BW(SV) and BW(TJ). The third fabric is a softer, finer fabric with a fine sand temper and better levigation of clay. Fine quartz particles can be seen in section under lens. Some BW sherds have this fabric. It can be fired to a pale pink, in which case it resembles PINK(F) and may in actuality be a variation of the same.

Petrographic analysis : Chemical analysis places BUFF(F) of this class in the Petrofabric group H, the same as BW(SV). The thin-sections falling under the category of fabric H had a fine grained fairly micaceous matrix. These samples are non-pleochroic in PPL and grey in XPL. The grains were angular to sub-angular in shape. The highest grain sizes in the samples were 160 μ and 200 μ. The grains were well sorted and had a frequency of 20 to 30%. The grain-size distribution pattern was unimodal. Grains did not show any orientation. Voids were very few and most of them were grain fallouts. The non-plastic inclusions included quartz, plagioclase feldspar, cryptocrystalline silica, augite and mica. There are a few hematite patches also. The geological terrain assigned to this fabric group is Ia, i.e. a terrain having basaltic components and no bioclasts.

Occurrence at Sanjan : This ware occurs in the same contexts as the previous class and is associated with it. It is present in the assemblage from the Sanjan Dokhma or Tower of Silence where it occurs in the uppermost levels. In the trenches on Sanjan Bandar, the ware is restricted to the upper two layers in TT4 and TT1. Its occurrence in layer 3 in TT1 appears to be intrusive.

Discussion : This ware class is very similar to the previous one and occurs at sites on the East African coast, as at Sanjan, in association with the former. Chittick reports it from levels dateable to Period III (late 13th to 14th century) at Manda, and calls it 'Early Islamic Monochrome'. He places the pink fabric monochromes – what he refers to as Standard Islamic Monochromes in Period IV (15th and early 16th century) (Chittick 1984: 12). At Shanga, Horton reports the pink-buff body monochromes but the pale yellow-buff body wares seem to be absent at the site (Horton 1996: 293 – 96). At Kilwa, Chittick reports the ware as Early Islamic Monochrome and dates it to the late 14th century (Chittick 1974: 304). If this periodisation is to be accepted, it appears that the buff body monochromes are of a slightly earlier date than the pink fabric monochromes. This chronology provides continuity in the Sanjan sequence and suggests a scant but certain occupation on or near the site in the 14th and 15th centuries.

Cuerda Seca Glazed Ware (Ware code: CSGW)

Definition : This is a unique ware with no known parallels in any catalogues, reports or collections accessed for this study. The identification of this technique of manufacture was suggested by Kennet (Dr. Derek Kennet, personal communication). The ware has been tentatively named after the technique which involves the tracing of a design onto the surface of a vessel with dry manganese lines in order to contain the different colour glazes which are used for painting. The technique is supposed to have originated in Egypt where Fatimid pottery, sometimes called Fayyumi wares, display this treatment. The technique was later developed and had a long history in the west, particularly in the case of Spanish pottery. Watson traces the western tradition of cuerda seca to this technique of polychrome glaze painting "where the pattern is made up of areas of different colour glazes (brown, green, turquoise, yellow, white) applied separately and sometimes divided by a dry manganese line. These may be precursors of the cuerda seca family of ceramics." (Watson 2004: 36). It is possible that some of these sherds such as T.351 D.141 could have a North African or Tunisian provenance. But this hypothesis requires substantiation.

Shapes and forms : The shapes in this ware class at Sanjan are open forms such as bowls and dishes. The dishes have hemispherical profiles with everted rims and ring foot bases. The shapes appear Chinese in profile. The bowls are either shallow or rounded with ring-foot bases and everted rims. Another shape that occurs is that of small to medium sized, conosegmental bowls with a shallow groove running along the interior of the carination which can be seen in section. This groove is filled with glaze. The rim in this case is more everted. The vessels appear to be moulded and wheel finished. They are glazed on all surfaces. The shapes in this ware are T.9, T.14, T.14.1, T.14.2, T.43, T.43.1, T.47, T.52, T.61, T.62.1 and T.351 D.141. The base types are Bs.13, Bs.17, Bs.29 and Bs.72.

Glaze and Surface Treatment : The glazing of these vessels appears to be a complex procedure, combining many traditions on the same vessel. The splashes of colour which have run, juxtaposed with the sharpness of the designs and motifs and the sections of the design resembling in-laid tile work all combine to create a dramatic effect. The exterior of the vessels are glazed in colours of blue, ochre, white, manganese black and green. The colours are blurred and have run into each other although the effort to segregate them is seen in the faint black outlines which weakly try to establish panels. A faint black line can be seen on the exterior, at the junction of the wall and the ring-foot and traces of white and blue glaze are seen on the base itself. The glaze on the underside of the base may have been thinly applied since most of it has flaked away leaving only faint traces. The interior of the vessel has the same blurred effect of colours running into each other but some traces of the design or pattern they were meant to follow show through by way of blurred dots, lines, etc. The floor of the vessel usually has a well defined design marked out in manganese black. In one unique sample at Sanjan (T.351 D.141), the floor of a small conosegmental bowl has a geometric motif of a six-pointed star repeated and interlinked to form a complex design. In another unique vessel (T.14.2), remnants of an animal motif can be seen drawn out in black on the floor of the dish. The black outlines of these designs on the interior surfaces are well defined and appear less glossy than the rest of the surface which has a high-gloss effect. The vertical surfaces of the vessel have not been able to retain the design and the glaze has run, sometimes almost effacing the design altogether. The sherds are flaked and eroded in places but the designs and decoration can still be clearly seen, at least on the above-mentioned two examples.

Decoration : The two identifiable samples from Sanjan are different from each other, although the palette of colours appears to be the same. The first piece (T.14.2) is a rounded shallow dish with glaze covering both the surfaces. The ring-foot base is unglazed. On the exterior, blue, ochre and white are seen to form vertical panels or blocks. Blue lines define the alternate white and ochre panels. Traces of a black line running around the base at the junction of the ring-base and vessel wall can be seen. The interior of the vessel has the same blurred and splashed effect but the attempt to define broad radiating panels of black, blue, ochre and white can be seen. These extend from the circular ridge in the well of the vessel to the rim. Traces of a stylized animal motif drawn in black can be seen on the floor of the vessel. It appears to be an isolated motif. The second sample is a medium sized conosegmental bowl (T.351 D.141) with far more elaborate and complex decoration. The exterior of this bowl has similar panels of colour as the previous sample but they are better defined with blurred black lines that run vertically from rim to the foot-ring of the base. Blue, ochre and white are the dominant colours used as fillers in these blocks. A black line circles the base just above the ring-foot where a groove can be seen. The base itself has traces of white and blue glaze. The top of the everted rim has traces of black dots which have run into the blue and white glaze. The interior walls of the vessel have white, blue, ochre, black, green and brown-gold glaze, all of which have run into each other and created a splashed effect. Traces of circles or loops and dots can be seen. A groove runs along the carination and has filled with glaze. The space between the carination and the floor of the bowl has been painted as a beaded border of blue bands, white background and black dots, all contained within two black lines. The floor of the bowl is dramatically decorated with a geometric design of interconnected six-pointed stars executed in sharp black lines to form a mosaic tile work effect. The stars are elaborately decorated with blue and white dots, almost like enamel in-laid work. The whole design is laid on an ochre-gold background. The dots inside the stars and between them are in concentric circles of black, blue and white – the same combination as the beaded border which surrounds the stars. No parallel to either one of these pieces can be seen in any of the collections accessed for the purpose of this study. No such ware appears in any published excavation report. As mentioned earlier, it has been identified through consultation with Derek Kennet.

Fabric : Two fabrics have been identified in this ware class. The first is PINK(B) which is the same as the sgraffiato wares, WGPW, BiGP, etc. The second fabric may be a variation of the first. PINK(F) is a softer and finer fabric which is pale reddish pink to creamy pink in colour. It is fairly chalky and has fine white inclusions. Minute red specks can be seen under lens. It is similar to BUFF(F) in texture and the clay appears to be well levigated with fine sand temper. Some of the vessels in the class of UPW have the same fine fabric. No petrographic analysis has been done on this class as yet.

Occurrence at Sanjan : This ware is found in small numbers from the excavations of Season 1 and 3 on the Bandar area. Unfortunately almost all the sherds belonging to this class have been retrieved from disturbed contexts – the ring-well pit (TT1) and the well (TT4). Nothing can be said about its stratigraphic occurrence or associated wares. The only quadrant which was relatively undisturbed and could provide some indication of the occurrence of this ware is TT1 SE where some of these sherds have been found in layer 2 associated with SWGW, WGW, HsG, BiGP, etc. It is uncertain if this quadrant offers a completely undisturbed stratigraphy since the presence of structural debris and dense scatter of brickbats made the layers difficult to identify and mark with clarity.

Discussion : This ware is unusual and rare. The lack of available data makes it difficult to say much about it. Kennet's identification of this ware, its technique of manufacture and the tentative suggestion that it could have an Egyptian, North African or Spanish provenance, is borne out by the description given by Watson in his discussion on Polychrome Glaze-Painting and Fatimid pottery (Watson 2004: 36, 53). The technique may have spread from Egypt and Syria westwards as well as eastwards, apparently taking firm root in the Spanish and Mediterranean traditions. Watson distinguishes between the technique of polychrome glaze-painting on opaque white glaze and the cuerda seca technique of separating the different colours "by a dry line of unglazed pigment that is intended to stop the colours flowing into one another during firing". The first he defines as Fayyumi and suggests that the second is frequently mistaken for the same. The Fayyumi tradition he dates to the 9th - 10th century and the developed cuerda secca technique to the 10th – 11th century (Watson 2004: 53). The antecedents of the technique could however be earlier in date.

Unique Glazed Pink Wares

This is not a defined category or class but a group of diverse sherds, each unique in its occurrence at Sanjan. Only a single sample or two of each of these has been seen in the collection. The only thing they have in common is the fabric and the lead glaze. The decoration differs in style as well as treatment. Some of these unique sherds are early in date and may form part of the Samarra Horizon repertoire of wares. Hence they are included at this point to maintain continuity.

White Glazed Pink Ware (Ware code: WGPW)

Definition : This ware class is different from the preceding classes although the fabric is the same reddish pink earthenware. The ware resembles the slip painted wares of Nishapur and Sirjan, in that it has a white surface with manganese black painting on it. The ware is not found in prolific numbers.

Shapes and forms : The only shapes found in this class to date are small rounded bowls. No closed forms have been found. The rims are simple. The forms closely follow those of the sgraffiato wares. The vessel shapes are T.9.1, T.11, T.15, T.41, T.58D.104, T.64.1, T.269, T.270 and T.345D.139. The ring bases give the vessels a Chinese profile. The bases are Bs.11 and Bs.16. Mending marks can be seen along old fracture lines, indicating that this was a prized ware which was retained despite breakage.

Glaze and surface treatment : The interior surface of the vessel is covered with a white slip which flakes and erodes very easily. The exterior is unglazed. Manganese black pigment is used for painting on the white surface. The black appears to have a greenish tinge. The exterior has faint traces of the white glaze which appears to have been wiped off from near the rim.

Decoration : The use of white glaze on the interior of the vessels brings the black painting into relief. Three different styles of painting can be seen in the Sanjan samples. The first is a plain white surface with a black band running along the lip the walls of the vessel are left plain and traces of a black design can be seen on the floor of the vessel. The design is executed in broad bold strokes and consists of a straight line at the end of which is a short curved band forming an anchor-like motif. The painting has eroded and faded making it difficult to say more about it. In the second style, the bowl has the same black band along the rim but the walls of the vessel have a design of circles under a horizontal line. The circles contain four closely placed dots which have faded to a green in places. The design is executed in medium thick strokes, unlike the first style. The space between the circles is filled with short lines, some of which are curved. This second type has a close parallel in the Merv collection at the British Museum which is unfortunately not yet studied and published (Dr. St. John Simpson, personal communication). The third style has only a black band along the rim and no discernable design or attempt at decoration. The fabric and treatment are the same as the Sanjan samples.

Fabric : The body is of the same reddish pink clay as the sgraffiato wares, PINK(B). Some fine quartz particles can be seen. It is well levigated and a few air-holes are seen in section under lens. The firing is even. No thin-section analysis has been done on this class. But the petrographic analysis for the sgraffiato wares with the same fabric have provided a profile for PINK(B). By extension, the clay source for these ware classes appears to be common.

Occurrence at Sanjan : The ware has been found from the stratigraphic context from the excavations of Season 1 and Season 3. The sherds found in Season 2 came from the well or tank on the embankment and had no context. In the stratigraphic context, they occur in trench

TT4, SE and NW with sgraffiato wares, mainly in the upper part of layer 2.

Discussion : This ware could well be a regional product, following a local tradition. Its appearance in the Merv collection suggests that it could have an Iranian origin. The provenance of these wares is uncertain but could well be in east Iran or the vicinity of the Nishapur region where the same palette of colours was current in the more elaborate and aesthetically developed slip-painted wares.

Incised Glazed Pink Ware (Ware code: IGP)

Only one sherd of this ware is seen at Sanjan and it occurs in Layer 1 in one of the trenches at the Dokhma or Tower of Silence at Sanjan. It has no known or reported parallels from other sites. It may well be a regional product. It can be identified by its semi-coarse red-pink sandy fabric - PINK(D) - which has a thick white glaze with finely incised lines and small dots pierced into the glaze to form delicate patterns. The fine dots may be executed by some fork-like instrument. The glaze has a matte finish. The fabric is semi-coarse, dark red-pink and sandy with some quartz-like white inclusions. It almost resembles RED(C) in texture. The sherd may be intrusive since it occurs so close to the surface and comes from a structure where almost all the pottery is in secondary context. The excavated habitation areas of the Sanjan Bandar did not have any such sherds. Hence it need not be considered a representative ware. It is included in this study only in order to record its presence in the assemblage.

Black and Ochre painted Glazed Pink Ware

This ware also finds poor representation and is seen only in a few sherds. It occurred in the ringwell pit material in TT1 SW, so its stratigraphic context is suspect. The ware has the same fabric, PINK(B), as sgraffiato wares and the Splashed ware mentioned above. But the treatment is completely different. The only form seen in this case is a small bowl with rounded profile and a simple rim. The fabric is clearly seen through the thin transparent glaze which covers the interior and part of the exterior. The white slip is applied in uneven horizontal streaks, mainly towards the centre and along the rim. Some traces of ochre are evident near the rim. The lip of the rim has a black border of elongated narrow crescents which appear more dabbed than painted. The centre of the bowl has black spots dabbed to form a floral motif in a somewhat simplistic manner. Another such rim sherd has a plain black line running along the lip with a white band under it and ochre seemingly covering the rest of the vessel. This difference in treatment from the other classes of pottery could be indicative of regional variations and local traditions which may have been current in the different production centers. One similarly coloured sherd has a double line border running along the interior of the rim and traces of a hatched design. It is therefore difficult to assign any date to this ware.

Blue Glazed Pink Ware

Only one small rim sherd (T.47.1 D.14) has been found of this ware to date. It belongs to a bowl or dish. The fabric is PINK(F) pale pink, slightly sandy clay with a chalky texture. A white slip covers the body. On the exterior, this white slip can be seen near the rim with a dark band running along it. Another black band is seen along the lip. The interior has a bright blue glaze and a boldly executed, broad border painted in black along the inside of the rim. The design is contained in two horizontal lines and consists of a series of oblique lines with hooks at their upper ends. The black strokes of the painted design appear bold against the blue background. The black manganese pigment appears to have burnt in the firing and sunk into the glaze. In this, the black lines resemble the outlines present on the CSGW vessels. Another horizontal band is seen under the border on the body of the vessel.

Painted Glazed Pink Ware

One small fragment of this unusual ware has been found. The fabric is a soft, slightly sandy pale pink earthenware - PINK(F). A complex design has been painted on the surface in thick black lines. Bright green has been used as filler in between two black lines, one of which has small hook-like spikes. The extant part of the design has small squares filling in the spaces. The decoration appears to be underglaze painting. The black stands out in relief. A translucent glossy glaze covers the whole surface. The exterior is unglazed. The details recorded on the sherd have been erased and so no context is known, other than that it was retrieved in the first season of excavations.

Splashed Glazed Pink Ware (Ware code: SGPW)

Two small sherds of this ware have been found in layer 1 of trench TT1. The fabric is PINK(F). It is fine and chalky with a thin application of glaze on both surfaces. The ware closely resembles the BiGP in treatment but has many more colours used on the surfaces. The external surface has white, yellow and green while the internal surface has green, yellow and brown. The only identifiable shape is T.61.

Yemen (?)

The identification of this ware has been done on the basis of the description given by Kennet in the report on the pottery from Ras al-Khaimah (2004: 41 - 42). It appears to match the description of the Sanjan sherds. However, this identification needs to be confirmed. The three sherds found at Sanjan have fabric PINK(B). The fabric of the sherds at Ras al-Khaimah seems to be different. The thickness of the sherds is approximately 5mm. They are covered on the interior with a bright yellow glaze and black or brown-black painted design. The design appears to be vertical lines or loops on the interior of the bowl or dish and a blackish band may appear along the simple rim. The designs and the glaze are both faded and flaked. The exterior has a pale whitish wash-like appearance. All the sherds occur in TT1 from layers 1 and 2. Horton refers to it as Black-on-yellow at Shanga (Horton 1996: 291). It appears to be well known on the East African coast since it has been reported from Kilwa (Chittick 1974: 304) and Manda (Chittick 1984: 81 – 82) amongst other sites. At almost all sites the ware can be dated to the 13th – 14th century. Kennet adopts this nomenclature as it is presumed that the ware was produced in Yemen (Kennet 2004: 42).

Khambat Type Glazed Pink Ware (Ware code: KTGW)

This nomenclature has been used to distinguish between the monochrome glazed ware bowls and dishes from the Persian Gulf region and those locally produced at Khambat. The Khambat glazed ware industry which thrived during the Medieval period was unknown until it was discovered by Bhan (personal communication 2003) who has since done extensive exploratory work in the area. However it may be noted here that Carswell had explored Khambat or Cambay briefly in 1976 and has mentioned that they "found some sherds of local earthenware, crudely glazed in a fissure at a point where the town wall had collapsed. These included a triple kiln-spur and a sherd with three spur marks. On this evidence, Cambay may be considered as a centre for the production of this type of crude local ware, which may well have been exported far afield through its foreign contacts" (Carswell 1979: 29). The industry appears to have drawn inspiration from the glazed ware imports from the Middle East and most vessels appear to be imitations both in shape and in surface treatment. The fabric, however, is completely different from the Persian Gulf wares. Bhan reports the presence of thousands of wasters, setters, fragments of kiln-lining, crushers, nozzles or tuyeres, etc from the area known as Lashkarshah, located 1km southeast of the fortified town of Khambat (Bhan 2006: 90 – 95). The fact that a large glazed ware industry existed in Khambat during the medieval period is not surprising since Khambat was a thriving port with a large population of foreign merchants, mariners and migrants at the time. The presence of Arab and Persian traders is attested to by the architectural and inscriptional evidence still extant in the town as well as the accounts by foreign travellers. The local glazed ware industry probably catered to the needs of the expatriate communities. However the presence of the Khambat type glazed wares from South-east Asia as well as the East African coast indicates that it was also an export item. A variety of glaze colours are seen in the surface collections from Khambat. Various shades of blue, turquoise, green, cream, manganese purple, black, maroon, lilac, etc are present. A large number of sherds show designs in black, manganese brown, green, etc painted on a white background. As in the case of the Persian Gulf wares, most of these bowls and dishes are unglazed on the exterior. The flanged rim with high ridge which is seen on the fourteenth century monochrome vessels from the Persian Gulf is also seen in the Khambat repertoire. The Khambat glazed vessels have fabrics which are much coarser than the fine buff and the fine pink fabrics of the Persian Gulf. One fabric is almost the same as RED(C). The other is slightly better levigated and has a pale brown-red colour. The coarseness of the fabric makes the glaze appear bubbled, pitted and uneven on a large number of sherds, and speckled in many cases. Sherds of Chinese Blue and White Porcelain and frit ware with imitation blue and white decoration can be collected from the surface at Lashkarshah. This helps in dating the medieval glazed ware industry of Khambat to about the fourteenth century. The excavations at Chaul have yielded a large number of these sherds indicating the movement of these goods along the coastline.

Only one sherd of this type has been found at Sanjan. It occurs in layer 1 in trench TT1. It is relevant because few sherds of such a late date have been found. The sherd is a body fragment of a dish or a bowl (not illustrated). It has no glaze on the exterior. The internal glaze is white with two dark manganese black lines which may be part of a more elaborate design.

Glazed Red Ware (Ware code: GRW)

Definition : This ware class is represented by a few sherds found during the excavations of season 1 and 3. Most come from a disturbed context – the ringwell pits and the well in TT4. A few come from the stratified context in the trenches. The ware is defined by its semi-coarse body of red clay over which a glaze is applied. The glaze colour could be a pale white, opaque white, olive green or moss green. It usually flakes easily and most sherds hardly have any glaze left.

Shapes and forms : The most common forms are bowls with the glaze usually on the interior. They are straight-sided or hemispherical and thickly potted. Closed forms are also present, as is seen in the complex jar rims. The bases of the bowls are flat. The shapes seen are T.401, T.401.1, T.9, Bs.6, Bs.17 and Bs.79.

Glaze and surface treatment : The most common colour for the glaze is white, pale creamy white and a mossy green. The white glaze is opaque and flakes very easily. The green glazed sherds appear to have a white slip on the body. No attempt is made to decorate these sherds. The glaze on the rims is thinly applied so that the red fabric is seen through, giving it a pinkish tinge.

Fabric : The fabric class of this ware is RED(B) which is red in colour and can range from the coarse to the semi-coarse. The clay has a large number of airholes and inclusions which are evident even without the lens. It is sand tempered and also has organic material in it. The coarseness of the texture, the presence of mica in the clay and the heavy sand temper mark it as different from the other glazed ware classes. This could be a regional or indigenous product.

Discussion : The glazed earthenwares at Sanjan have either a buff or pink fabric which is, in most cases, quite fine. This ware is different both in its fabric as well as in the glazing, perhaps indicating production centers in regions other than the Persian Gulf. The clay is similar to the glazed wares reported by Bhan from the Laskarshah area in Khambhat (Bhan 2006: 90 – 95). The shapes at Lashkarshah are open forms – straight-sided and hemispherical bowls and dishes with ring-foot bases and plain or everted rims. At Sanjan, the vessels in this class include closed forms as well as the open ones. While it is possible that the local kilns at Lashkarshah could have produced such wares, it is just as likely that the production centers could be in the Sind or Makran region or in an area where the clays are similar to Gujarat. The glaze on these vessels is different from the glaze on the preceding class in colour, application and quality. The paintings are entirely missing.

Frit Ware (Ware code: FRIT)

Three sherds which were found in the second excavation season at SJN-K and which were initially classified as Chinese Blue and White Porcelain have been reclassified as Frit Ware. Lens examination of the fabric and of the glaze in the final analysis showed the body to have a crumbly, grainy frit composition. The texture is granular and the colour of the fabric is white. But it entirely lacks the fineness and the hardness of porcelain. The glaze is applied in a thin coat of white with cobalt blue painted designs. The paintings lack sharp outlines and can be quite blurred. The white glaze has a greyish tinge and is badly crazed. The designs seen in this class are D.86, D.96 and Bs.44. The base is a high ring-foot in imitation of the Chinese vessels. Five sherds of frit ware without any glaze were found in TT4, mainly from layer 1 and the upper part of layer 2. The glaze may well have eroded. The fabric of some sherds was fired pink.

The fabric designation for this class is FRIT. The petrographic analysis of one sample showed that it belongs to Petrofabric Group J. The thin-section falling under the category of fabric J had a fine grained matrix. The sample is non-pleochroic in PPL and red in XPL. The matrix looked ferruginous. The grains were angular to sub-angular in shape. The highest grain size was 200 μ. The grains were perfectly to well sorted and had a frequency of 10 to 15%. The grain-size distribution pattern was unimodal. Grains did not show any orientation. Voids were many and most of them were grain fallouts. The non-plastic inclusions included quartz, plagioclase feldspar, cryptocrystalline silica and mica.

Unglazed Earthenwares :

The unglazed wares at Sanjan form a bulk of the assemblage. A vast range of fabrics, forms, types and styles can be seen in the collection. Some of these classes have fabrics similar to or the same as the Glazed classes, indicative of common sources for the clay and possibly regional industries which may have produced both the glazed as well as the unglazed wares.

Eggshell Ware (White) (Ware code: ESG(W))

Definition : This is one of the earliest Islamic wares belonging to the pre-Samarra period and extending into the Samarra Horizon sequence. The ware is identified by the thinness of its potting, from which it takes its name. The ware is also referred to as Thinware by Sarre in his Samarra typology and by Northedge in his commentary on Sarre's classification (Northedge 1996: 229 – 258). The class consists of closed form vessels which are made of white or creamy white fine clay and are extremely thin in section, rarely exceeding 2mm.

Shapes and forms : The most common shapes in this class are small pitchers, jugs and jars. The profiles have some minor variations but the form remains more or less the same (T.54, T.74, T.74.1, T.74.2, T.393, T.393.1, T.393.2, T.393.3, T.393.4D.190, T.393.5D.190, T.394, T.394.1 and T.395). They have a globular body which is sometimes squared and a high vertical collar which ends in a simple featureless rim with a tapered lip. Teacup handles with thumb-rests extend from the shoulder of the vessel to just below the rim or sometimes to the rim itself. Some of them appear to have had two handles while some appear to have a single handle. None of the vessels have the spout channels for pouring out the contents and so it may be presumed that the two handled vessels may have been for drinking. At least two fragments of filters have been found at Sanjan, suggesting that some of the vessels served as flasks or pitchers. They have a fine moulded design through which the holes have been pierced. Similar filters are reported by Watson who illustrates the samples from the Al-Sabah collection and says they are from Fustat (Egypt), dateable to the 10th – 12th century (Watson 2004: 132 -33). The vertical collars of the vessel can sometimes have low offsets above the shoulder and below the rim. On this same space one can find, in some cases, an inscription which could either be the name of a person, blessing to the owner or just a pseudo-calligraphic motif. This is incised with a sharp instrument. The globular body is plain. The bases are sharply defined non-contiguous convex-centred with a prominent 'kick' on the interior of the vessel (Bs.23, Bs.25, Bs.76, Bs.76.1). The handles are slender, well made, of even thickness and round in section. They have a small thumb-rest which is luted on top of the curve. In most cases it appears as if a strip of clay was rolled into shape for the handle and a small bit was cut out of the same roll to be luted on as the thumb-rest. Only in one instance it appears different. The small triangular thumb-rest has a small cut on the top to give it the appearance of an animal head, similar to the one reported from Siraf by Tampoe (Tampoe 1989: 25-26, 239 fig.39 no.943). This may be by design or by accident. One rim sherd appears to belong to the upper part of a flask. It has gently sloping sides and a slightly everted rim with a sharp lip.

Surface treatment and decoration : This is a plain ware with no visible surface treatment on most sherds. On some, minute observation under lens shows the application of a thin self-wash. Most of these vessels retain their white or creamy white colour. The surfaces are smooth. The sherds from the well deposit in TT4 SW have a yellowish silty encrustation. Most sherds are undecorated. Incised calligraphic characters are seen on the collars of some sherds and incised design on some slightly thicker sherds which appear to be intermediate between Eggshell Ware and Buff Ware. These consist of wavy lines and an incised border like design on the exterior of a rim. The incised designs have been drawn with a sharp pointed instrument. The filters have a fine moulded fretwork design with holes pierced through them. The beaded bands are minutely crafted. Only one small body sherd (D.47) has an unusual decoration – traces of painting with dark blue glaze can be seen on the unglazed surface. No other sherd has been found with similar treatment.

Fabric : The fabric of these vessels is extremely well levigated and fine with no inclusions. It belongs to Fabric BUFF(G). No other ware class has this fine fabric. It is fired hard but still has a chalky texture. The colour can be milky white or a creamy white. The fractures are sharp. The sherds which have a thickness of 2.5 – 3mm have a slightly more chalky and soft texture.

Petrographic Analysis : The thin section analysis of this ware places it in Petrofabric Group G. The thin-sections falling under the category of fabric G had a fine grained smooth matrix. These samples are non-pleochroic in PPL and dark grey in XPL. The grains were angular to sub-angular in shape. The highest grain sizes in the samples were 120 μ and 160 μ. The grains were perfectly to well sorted and had a frequency of 5%. The grain-size distribution pattern was unimodal. Grains showed a parallel orientation. Voids were very few and most of them were grain fallouts. The non-plastic inclusions included quartz, plagioclase feldspar, and mica. No other ware from the analyzed samples was found to belong to this petrographic group.

Occurrence at Sanjan : A majority of these sherds have come from the ringwell pit, other pits and the well in TT4. Some sherds have occurred in the stratified context. They are best represented in TT4 NW and SE, although the well has yielded the maximum number of these sherds. The highest number of sherds in the stratified context comes from layers 2 and 3. A few sherds occur in TT4 NE in layer 4 but this may be intrusive as the sherds are very small. Two other fabric types in this same range of wares are found associated with white Eggshell Ware – one is a pink fabric and the other is a fine red. These are discussed below in greater detail.

Discussion : Thin ware or Eggshell Ware is considered as one of the important markers for the early Islamic period and Samarra Horizon. Given the long history of this ware it is necessary to differentiate between the Sasanian, early Islamic and later types. The presence of the ware is generally assigned to the early Islamic period. At Samarra the sherds are found in contexts which were abandoned in circa 895 AD, indicating that they were in circulation prior to this date. As at Sanjan, the associated wares are blue-green glazed jars (Turquoise Glazed Ware), green and white splashed glaze, cobalt blue and green on white glaze and white glazed wares. The ceramic grouping of the ware is I.A. Northedge's evaluation of Sarre's typology and chronology places this ware along with Cobalt Painted Ware, Buff Wares with barbotine designs, Turquoise glazed ware jars, basins and lamps, Splashed White Glazed Ware (green) and monochrome green glazed ware to the last fifteen years of the 9th century (Northedge 1996: 240 – 44). At 'Ana, the Thinwares, with variations in their fabrics, belong to Group 4 and are dated to the early Abbasid period, i.e. 836 – 892 AD (Northedge et al 1988: 90-91). Adams dates them along with plain wares to the early Islamic period at Tell Abu Sarifa, noting the differences between the Pre-Islamic and Early Islamic forms (Adams 1970: 100). Other sites that report the ware are Kush (Kennet 2004: 61), 'Arja (Costa and Wilkinson 1987: 187), Siraf (Whitehouse 1971: 15. Tampoe 1989: 25-26), Suhar (Kervran 2004: 301 -308) and Manda (Chittick 1984: 94). Kervran suggests as early a date as 700 – 900 AD for this ware at Susa and about 860 – 890 AD at Suhar where it occurs in layer 10 (Level V). Costa and Wilkinson suggest that bases with raised centers were current in the 9th century at Nishapur but by the 10th century they had developed the flat base form. One of the known centers for the manufacture of these wares was Siraf, though other centers could well have existed.

Eggshell Ware (Pink) (Ware code: ESG(P))

This ware is almost exactly the same as the previous class, except for the fabric which is pale pink. It is, in essence, a variation of the previous class or perhaps a regional product, indicating that the tradition of popular wares could have been located in various regional centres of production. The clay used for this class is pale pink, hard fired and thinly potted with reddish-gold mica flecks seen in the fabric. This pink fabric is designated PINK(G). The texture is grainy and sandy as compared to the white eggshell ware fabric, and the finish is not as smooth. The surface has a rough texture with some small lumps of clay still stuck to the surface, perhaps due to careless trimming. The body sherds and handles suggest the shapes of small jar, jug and pitcher; though no rim sherds or bases of this ware have been found (T.396). One body sherd has the broken part of a luted handle still evident on the globular body. This ware is relatively scarce compared to the white fabric vessels. The thickness is about 2mm. Incised designs and calligraphy does not appear on the sherds retrieved so far. A thin pale yellow coating can be seen on the surface. It is difficult to tell if it is a wash or if it is post-depositional discoloration caused by deposition in the well. The description of the sherds from Manda includes a distinction made between the white fabric and the pale pink fabric found in this class. Chittick refers to them as "gudulia (water cooler) and egg-shell ware", used for cooling and storing water (Chittick 1984: 94). He dates these to Period I, from the mid-ninth to the eleventh century. However, the plain buff wares with thicker body and larger vessel forms are also considered in the same group by Chittick, with a differentiation drawn on the basis of thickness alone. The presence of two body sherds of this ware in layer 2 in TT4 SE may indicate that it belongs to the same period as the other ware classes at this level.

Eggshell Ware (Red) (Ware code: ESW(R))

This is a completely different fabric, having nothing in common with the previous two types of Eggshell Wares. The shapes appear to be the same – small drinking vessels and pitchers with handles (T.397, T.398). But the fabric and even the potting appear to be different. The fabric, RED(A), is an extremely fine, well levigated red clay which has no inclusions or visible temper. Some very fine mica particles are present. Its texture, compactness and smooth surface are reminiscent of the superior quality Red Polished Ware. It is hard-fired and the fracture of the ware is as sharp as broken glass. The thickness of the vessels is never more than 1mm. They are wheel-made vessels with striation marks of a fast wheel evident. The exterior surface has been burnished. A pale yellowish silty coating can be seen on these sherds due to post-deposition weathering. But this is not as marked as in the case of ESW(W) or ESW(P). The tea-cup handles are very similar to ESW(W), but are more delicate (T.399 and T.399.1). These are also round in section and have thumb-rests which have been luted on top of the curve. The thumb-rests are not shaped but seem to be part of the rolled clay from which the handle was made. The rims are vertical and sharp at the lip which is ever so slightly inclined inwards.

Unfortunately the few sherds found of this ware all come from the well deposit in TT4 SW. There is no stratigraphic context to these sherds but it may be added that they occur in the well only from depth 3.67m to 4.04m. They are associated with the other eggshell ware classes and the mixed material of the well. Kennet reports Red Eggshell Ware from Kush which he dates from the ninth to the 11th century (Kennet 2004: 61). It remains to be ascertained if these are the same type as the ones found at Sanjan.

Buff Ware (Ware code: BW)

Definition : This ware class is defined by its white and creamy-white fabric which has no surface treatment and which may or may not have decoration by way of stamps, bosses, incised designs and appliqué or barbotine ornament. The range of forms and types vary and the ware can be mistaken for Eggshell Ware or Thinwares in the case of finely potted vessels. The criterion used in this study to distinguish between the two, in case of doubtful samples, is the average thickness and fabric. Sherds with an average thickness of 2mm are considered as Eggshell Ware and those exceeding 3mm are taken as Buff Ware vessels. The fabric of Eggshell Ware is almost exclusively BUFF (G) whereas the Buff Ware sherds have variations of buff fabrics. Some sherds are ambiguous despite this differentiation.

Shapes and forms : Almost all the vessels in this class at Sanjan are closed forms, with the exception of one small shallow bowl which has a thin section, an everted rim and a deeply incised design on the exterior just under the rim. The most common forms are small to medium size jars, jugs and pitchers. One pipe-like sherd was mistakenly classified as a pedestal base but, in actuality, may be the neck of a flask or bottle. The presence of slightly inverted club rims having incised oblique lines on the top and a high neck (T.387 D.181 and T.387.1) and a fairly thick body suggest large jar forms. The thickness of sherds belonging to the small vessels is between 3 to 5mm. These are probably the small pitchers and jars identical or similar to Eggshell Ware shapes. A few small stems of handles, either rolled or hand-moulded, are found. The body sherds sometimes have moulded, incised or applied ornament. Overturned beaked rims (T.388) occur, as do the sharp lipped vertical collared small jugs or pitchers like the Eggshell Ware vessels (T.389). In this case, the thickness is slightly greater and there are two prominent ridges and a deep groove at the juncture of neck and shoulder. The bases represent the larger vessels and are almost all flat with prominent coiling ridges on the interior. None of these sherds have any residue in them. The thickness of these is approximately 5 to 8mm. A large number of handles are also evidence that jugs and pitchers form a large part of this class. One large curved stem handle, luted to a small fragment of the rim, a well-moulded turban handle and smaller hand-moulded or rolled stems of handles are present in the collection. Body sherds of the smaller vessels, especially those belonging to Fabric BUFF(A) and BUFF(F) have incised or appliqué or barbotine designs, many of which are common to the Turquoise Glazed Ware class as well. One sherd has a thickness of 20mm and may belong to a flat base since it has no curvature whatsoever. More body sherds of this class have survived than rims or bases.

Surface treatment and Decoration : The surfaces of these vessels are left untreated, the porosity of the vessel being of functional importance. The white or pale buff colour of the clay may rarely have a light self-wash, but in most cases the surface is not treated to more than smoothening. The interior and exterior of the vessels have the same colour and the treatment. The coarser fabrics have a grainier and rough texture. The decorations noted on these sherds can be divided into the following types – a) moulded decoration, as is evident from the shoulder fragment having the square and circle design (D.39.1) b) "riffling" or textural modification of the surface (D.39) c) incised decoration consisting of straight and oblique lines, curved lines, etc (D.184, D.184.1, D.185) placed either on the body or on just below the rim and in one case on the rim (T.387 D.181) d) incised decoration done with a comb-like instrument (D.186, D.186.1, D.36.1, D.36.2) e) grooves with short vertical lines cutting them at intervals to form a chain like design. f) barbotine ornament (D.183) g) stamps and bosses – medallions, rosettes, etc (D.182). A large number of the decorations such as the comb incised designs, barbotine designs, stamps, etc are common to both Turquoise Glazed Ware as well as to Buff Ware classes. In some cases they are identical.

Fabric : The fabrics of these vessels can be semi-coarse to fine. Three fabrics have been identified – BUFF(A), BUFF(E) and BUFF(F). The first is common to Turquoise Glazed Ware, Lustre Painted Ware, White Glazed Ware, Cobalt Painted Ware and Splashed Glazed White Ware. BUFF(E) is creamy white and hard-fired. It is quite compact and has a grainy texture, perhaps due to the sand temper. It is usually seen in the more thickly potted vessels and in BW(SV) and BW(TJ) classes. Sometimes the fabric has a lot of airholes, making it porous. Inclusions are seen under lens. These are fine grains of black, white and occasionally red. BUFF(F) is a fine soft fabric with a chalky texture. It is well levigated and can be creamy white to milk white, sometimes turning a pale pink in firing. Only in one sherd could a visible quartz particle be seen. Inclusions are rare. The decorated sherds belong mostly to this fabric and some to BUFF(A). No petrographic analysis was done for the fabrics in this class. However the petrographic analysis for these fabrics was done when they occurred in other classes such as BW(TJ) and BW(SV). It is discussed later in the relevant context.

Occurrence at Sanjan : This ware is well represented at Sanjan. It occurs mostly from layer 2 down to layer 6 in the stratigraphic context in TT4. The North-East quadrant of the trench which provides the sequence for the site and in which all six layers were exposed has maximum number of these sherds occurring in layers 2, 3, 5 and 6. Layer 2 has a limited number of these sherds. In Layer 3 the quantity increases. No sherds of this ware occur in layer 4. In layer 5 the count goes up and the maximum sherds are retrieved from this level. Layer 6 has some sherds but not in the same numbers. This distribution is relevant in the light of the associated material. The glazed wares decrease in the lower levels, with only some TGW sherds and one or two stoneware sherds in Layer 5. This decrease in glazed wares sees a corresponding increase in Buff Ware as the depth increases. The other wares present in these layers, other than the indigenous ones, are BW(TJ), BW(SV), WSPW, UPW, SBW and some miscellaneous classes. While it is associated with the Samarra Horizon wares in layers 3 and 2, it appears to predate them in the earlier levels – the lower part of layer 5 and layer 6. This is of significance in the building of the Sanjan chronology.

Discussion : This class has a pre-Islamic history which extends into the Islamic period. Adams suggests that "the indices of change through time pertaining to thin ware, as to plain ware, involve vessel form and surface decoration rather than fabric" (Adams 1970: 95). He calls this class Plain Ware at Tell Abu Sarifa. The open forms reported by him are totally absent at Sanjan, with the exception of one small rim sherd which belongs to a bowl. The turbans found on the handles of the larger vessel forms can have a range of forms, though at Sanjan only a few pieces are found. The narrow-waisted dome shaped turban from Sanjan is not seen in the Tell Abu Sarifa sequence. The number of thumb-rests on handles of medium and small sized vessels is greater than the properly moulded turbans. Some of these resemble the illustrated sherds from Tell Abu Sarifa. According to Adams and also Costa and Wilkinson these are unambiguously Islamic, dateable from 800 - 1100 AD (Costa and Wilkinson 1987: 187). Kervran reports the ware from Suhar as it occurs at various levels. The flat base sherd reported by her as belonging to layers 18 to 20, Level III (Kervran 2004: fig.11 no.19. 278-79) is similar to the ones from Sanjan. This level at Suhar is dated to the Sasanid period – from the 3rd century to the early 7th century. Similarly, the barbotine and applied decoration sherds she assigns to Level IV (Kervran 2004: fig.13 no.24. 282-83) is dated to the same Sasanid Period. These sherds are extremely similar to those at Sanjan. However it must be also noted that some of the other decorated sherds from Suhar such as the comb-incised sherd and the rosette from Level V, illustrated in fig. 21, which Kervran dates to the early Islamic period – mid to late 7th century are also present in the Sanjan assemblage (Kervran 2004: fig.21 no.9, 12. 295). At Manda, Chittick reports this ware along with gudulia and eggshell wares as belonging to Period I. He doubts the early date (late 7th and early 8th century) proposed by Rosen-Ayalon at Susa but agrees that some of the motifs go back to the Umayyad period or even earlier (Chittick 1984: 94). Horton calls these wares Fine Creamwares at Shanga, suggesting that they were made at Siraf. He dates them to the 8th century by corroborating the dates with Siraf and Susa (Horton 1996: 297-98).

Buff Ware (Torpedo Jars) (Ware code: BW(TJ))

Definition : This ware has been treated as a class because of the distinct features observed about it. These transportation vessels have been reported from various sites and at most of them the classification includes these with the plain wares or buff wares. The distinctive shapes of these vessels, the varying fabric types and the residue on the interiors make these a separate category. These amphorae jars are hole-mouthed with thick potting and a distinctively elongated or pointed toe base. The interiors have a black coating of bitumen. The fabrics can vary but almost always have sand temper and inclusions.

Shapes and forms : The rims of these tall elongated vessels are usually simple, clubbed and rounded (T.400). Two perforations near the rim of T.400 could indicate a rope or cord handle which was passed through. The shoulders slope down and the body tapers to a point. Two types of bases have been found at Sanjan – the first is a short toe with a flaring body; the other is a long narrow toe, hollowed in the centre and filled with bitumen. A large number of body sherds having a black application on the interior surface have been found. This has been identified as asphalt or bitumen at other sites where similar sherds have been found. The vessels vary in thickness from 7mm to 20mm. the hand made sherds being thicker and less evenly crafted. A number of body sherds have the finger-press marks of smoothening or shaping on the interior surfaces. Some body sherds of more even thickness (7 – 10mm) are square or diamond shaped indicating that they may have broken along the sections which had been joined to form the vessel.

Surface treatment : Most sherds in this class have no surface treatment. Only in rare cases can one see traces of a self-slip or salt wash on the exterior. The interior surface is treated to a black application which is sometimes seen to have trailed or dripped on to the exterior surface. No decoration is seen on any of these sherds.

Fabric : The most common fabric in this ware class is BUFF(E), although BUFF(A) and BUFF(F) also occur in rare cases. BUFF(E) is a creamy white to pale buff yellow clay with some air-holes and inclusions which are seen under lens. It is hard-fired and has a compact body. Sometimes the core is pale pink in firing. There is a fairly heavy sand temper and the texture can be chalky. Some sherds have higher porosity than others. A salt-wash or self slip is seen on the occasional sherd. BUFF(A) is the same pale yellow fabric as the TGW classes and the Samarra Horizon Wares. BUFF(F) is a finer, better levigated fabric with fewer inclusions or air-holes and a more chalky texture. But this fabric occurs rarely in this class.

Petrographic analysis : The samples from this class which were subject to thin-section study belong to BUFF(E) fabric. Three samples were selected (RN/33, RN/20 and RN/16). These came from layers 3, 4 and 5. RN/33 and RN/20 were found to belong to Petrofabric Group C and RN/16 to Group F.

The matrix of the sections falling under the group fabric C is medium grained. It is ferruginous and non-pleochroic in PPL and dark brown in XPL. The grains are mostly sub-angular. The largest grain sizes observed in the thin-sections were of 200 μ, 240 μ, 260 μ, 300 μ and 340 μ. They are well sorted and the frequency of grains varies from 20-50%. The grain-size distribution pattern is uni-modal. The grains do not show any kind of orientation in most of the cases except in one case where it shows a parallel orientation. Voids are few. The dominant minerals are quartz, plagioclase feldspar, microcline feldspar, cryptocrystalline silica, argillaceous inclusions, augite, mica, very few bioclasts and few patches of hematite. The presence of basaltic components, argillaceous material and very few bioclasts indicates geological terrain Ib as its source, i.e. a basaltic source.

The thin-section falling under the category of fabric F had a coarse grained ferrugenous matrix. This sample is non-pleochroic in PPL and dark red in XPL. The grains were angular to sub-angular in shape. The highest grain size in the sample was 300 μ. The grains were well sorted and had a frequency of 40%. The grain-size distribution pattern was unimodal. Grains did not show any definite orientation. Voids were many and most of them were grain fallouts. The non-plastic inclusions included quartz, plagioclase feldspar, augite, mica, crypto crystalline silica and bioclasts. Patches of hematite were also observed. The geological source for this clay is designated Ia which is also basaltic but with no bioclasts. Of all the samples analysed, this is the only sample which is present in this petrographic group.

It is possible that human agency has played a role in the modification of a common source material. But it is equally likely that these large transport vessels were made in a number of production centers. BUFF(A) which belongs to Petrofabrics D and E in the case of TGW and the Samarra Horizon Ware samples which were analysed, and which is also the fabric of some of the torpedo jar sherds, is suggestive of production centres perhaps in the vicinity of the kilns producing the glazed wares.

Occurrence at Sanjan : These sherds occur in large numbers even though the diagnostics are few. They occur in layers 2 to 6, with the highest count in layer 5 of TT4 NE. Their occurrence in the collection of season 1 is also considerable but has no stratigraphic validity since they occur mostly in the ringwell pit and in disturbed contexts. They first appear in trench TT4 NE at a depth of 0.60m and continue up to. 3.38m. They have a noticeable but low count in Layer 6 and then suddenly increase in quantity in layer 5. The numbers again decrease in layers 4 and 3, with another surge in layer 2 when the counts go up.

Discussion : Torpedo jars are reported from various port sites which are known to have contacts with the Persian Gulf region. This is not surprising, considering that they were transportation vessels. While it may not be possible to speculate as to the contents of the vessels until residue analysis is done, it is apparent that the black application on the interiors of the vessels probably addressed a specific requirement. Scholars like Adams have tried to differentiate between the Sasanian and the early Islamic torpedo jar shapes, given their long history. At Tell Abu Sarifa, Adams classifies these as Group G of the Plainwares and Thinwares. He suggests that the Sasanian examples have more elongated and cylindrical bodies while the early Islamic types usually have a more globular profile and wider body diameter (Adams 1970: 100). The torpedo jars reported by Northedge, Bamber and Roaf from 'Ana cover the Parthian to the Islamic period. Although the profiles match those at Sanjan for the Sasanian as well as the Abbasid period torpedo jars, the fabrics appear to be completely different (Northedge et al 1988: 1 – 145). The torpedo jar reported by Kervran from Suhar layers 12-14, Level 5 is dated 670 – 890 AD (Kervran 2004: fig. 24 no.2. 307) and matches Bs.77.1 at Sanjan. They occur at Kush (Kennet 2004: 63), Manda (Chittick 1984: 83, 89) and Unguja Ukuu (Juma 2004: 113).

Buff Ware (Storage Vessels) (Ware code: BW(SV))

Definition : These vessels belong to the Buff Ware group and are similar in the fabrics seen in the above classes. However, they are distinct in their forms and surface treatment. This class is defined by large sized storage vessels which have a pale creamy white to buff fabric. Unlike the Torpedo jar sherds, these sherds do not have any residue or black application on the interior surface. The exterior surface may or may not have decoration by way of incised design, meanders or ridges. The bases are usually flat.

Shapes and forms : No rim sherds of this ware have been found, despite a large number of body sherds and some bases. The thickness can range from 5mm. in some vessels to 22mm. in others. The average thickness is about 12mm. The base (Bs.78) is contiguous flat, the walls of which are sloping. The interior of the vessel flakes as if it has been treated to a thick self wash but this is difficult to tell with certainty. The body sherds indicate that these large jars were piriform in shape with sloping shoulders. The surfaces are not treated to any wash or slip. Only smoothening has been done.

Decoration : The decorations are somewhat plain as compared to the Buff Ware vessels. The application of a ridge to the shoulder and a comb-incised meandering line under it is seen in one sample (D.36.1).

Fabric : The fabric of this class is almost exclusively BUFF(E), with one exception which was BUFF(A). The fabric has been described above since it is common to Buff Ware (Torpedo Jars).

Petrographic Analysis : Of the four samples analysed from layers 2, 4, 5 and 6, two belong to Petrofabric Group C and the other two belong to Group H. The description of petro-fabric Group C has been discussed earlier and does not require repetition. WSPW-1 and 2, BW(TJ) and SBW-1 are included in this fabric group.

The thin-sections falling under the category of fabric H had a fine grained fairly micaceous matrix. These samples are non-pleochroic in PPL and grey in XPL. The grains were angular to sub-angular in shape. The highest grain sizes in the samples were 160 μ and 200 μ. The grains were well sorted and had a frequency of 20 to 30%. The grain-size distribution pattern was unimodal. Grains did not show any orientation. Voids were very few and most of them were grain fallouts. The non-plastic inclusions included quartz, plagioclase feldspar, cryptocrystalline silica, augite and mica. There are a few hematite patches also. The geological source is Ia – basaltic with no bio-clasts. One sample of Monochrome Glazed Buff Ware belongs to this group.

Occurrence at Sanjan : This ware has maximum representation in TT4 NE with only a few sherds occurring in the other quadrants. The occurrence in the stratigraphy is also noticeably different from other wares. It occurs in highest quantity in layers 6 and 5, decreasing tremendously in layer 4, disappearing in layer 3 and making a weak appearance in layer 2. Its presence along with BW(TJ) and TGW– 5 in the lower levels is relevant since it corresponds to the earlier structural phase in the quadrant. Although it continues in the upper levels which contain the later structural phase, the numbers are less than those in the lower levels. It is not clear if the absence of rim sherds and high number of body fragments found in such a localised context is indicative of a specific activity in the area.

Discussion : The large jars that make up this class were probably both storage as well as transport vessels. The lack of residue or bitumen or asphalt on the interior indicates that they were used as containers for goods different from the contents of the torpedo jars. The ware has been included in the same category as buff ware or cream ware or plain ware at most sites and has not been treated separately by most scholars.

Buff Ware (Spot) (Ware code: BW(SPOT))

This class of Buff Ware is identified by its fabric which is coarse and has large black inclusions that appear like spots. It rarely has a surface treatment, though some sherds appear to have a grey wash which, in most cases, has worn off. Occasionally incised design by way of a meander or an incised line is seen on the surface. Few sherds of this ware have been found at Sanjan. The ware was first identified by Kennet who called it SPOT or Spotty Ware. He identifies two sub-groups of this ware – SPOT.C (coarse) and SPOT.F (fine). The Sanjan class appears to belong to the former sub-group (Kennet 2004: 64).

The fabric of this class is its most distinctive feature. It has been classified at Sanjan as fabric BUFF(H). The body of these vessels is a pale cream or greyish cream with a large number of angular inclusions, many airholes and a rough texture. The ware is best represented in layers 5 and 6 of TT4 NE, though some sherds have been found in layer 3 as well. Almost no diagnostic sherds have been found in this ware, making it difficult to identify the vessel types. But the thickness (5 – 10mm) and curvature of the ware suggests that they were jars or other closed form vessels. Not many sherds of this class have been found.

Unglazed Pink Ware (Ware code:UGP)

Definition : This class consists of an unglazed and unslipped ware which has a pale pink to reddish pink fabric. The fabric sub-classes and shapes vary. However they are similar to the range of wares having pink fabrics and belong to the same genre. They may be decorated by way of applied or impressed design, grooves or ridges. Most sherds belong to large vessels.

Shapes and forms : Most of the sherds in this class belong to large vessels such as storage jars with and without handles. The jars can be high-necked and have inverted tapering rims and luted handles (T.24), or wide-mouthed jars with inverted rounded rims with a ridge on the exterior and double corrugation below (T.22). The rim can also be inverted with an external rounded projection like an overturned or rolled rim (T.25). The jars can have broad strap handles with a wide channel on the outer side (T.28.1), or simple cylindrical handles (T.26.1). The decoration almost always appears on the shoulder or neck of the vessel. A large number of them appear to be luted and moulded vessels. Some of them have the luting lines clearly visible at the shoulder and neck. The base sherd found in this ware has a contiguous ring (Bs.19). One or two small beakers, with more delicate potting, are also seen (T.361). The thickness of these heavy-set vessels can vary from 7 to 10mm. and 15 to 22mm.

Surface Treatment : No surface treatment is evident on the surface of these vessels. Some of them have a pale creamy pink surface which looks like a light wash but is in fact a result of the firing and not of any application.

Decoration : The large jars in this class have decoration mainly on the shoulder near the neck and sometimes on the body and shoulder. It consists mainly of impressed and applied decoration. T.359 D.162 is the large, curved shoulder of a heavy-set storage jar with the remnants of a broken handle still luted to it. Three horizontal bands of impressed oblique dots can be seen. The dots are in sets of four and the dots seem more like short dashes. The impressing has been effected with some sharp tined tool. The luting of the body to the shoulder is evidence of the manner in which this vessel was made. The interior has been manually smoothened and is uneven. T.360 is the shoulder fragment of another high-necked storage jar with evidence of a high ridge which has been applied to the shoulder. This vessel is less crudely potted than the previous one. The small beaker sherd has deep grooves and ridges on the exterior giving it a corrugated effect. The same can be seen in D.163. A large shoulder fragment, D.163.1, has a band of four sharply defined grooves placed horizontally on the shoulder while a diagonally placed, broad, shallow groove cuts it and continues downwards on the body. Traces of shallow grooves can be seen on the body under the diagonal groove. The sharpness and even execution of the design has the appearance of being moulded into the fabric.

Fabric : The fabrics identified in this class are PINK(F), PINK(B) and in one case PINK(A). PINK(F) is a fine well-levigated fabric with a soft chalky texture. It has a pale pink to creamy pink colour and fine sand temper. Fine white inclusions and some red particles can be seen under lens. The fabric is very close to BUFF(F) in texture. PINK(B) is the same fabric as the sgraffiato wares and the other glazed pink classes (see above). PINK(A) is a sandy, coarse fabric with a darker colour. It is almost red and shows many white inclusions under lens. The fabric is unlike the other pink types and is seen only in two samples in the collection – the monochrome finial found at the Dokhma (T.344) and the beaker in this ware class (T.361).

Occurrence at Sanjan : The ware occurs in layers 2 and 3, with one sherd in Layer 6 in TT4. The occurrence in TT1 and TT2 is confined to layers 1, 2 and 3 but in these trenches the context is suspect.

Discussion : The fabric of this ware has some similarities with the buff fabric. This makes it difficult to identify or correlate with fabrics described by scholars at other archaeological sites. While a large number of sites report pink to buff or light red fabrics, it is not easy to ascertain that the fabric they are describing is the same as the ones found at Sanjan. At Siraf, for example, Tampoe reports a Fine Pink Earthenware (Tampoe 1989: 18). However, neither the fabric description, surface treatment nor the shapes match those at Sanjan. Chittick's report of Manda includes jars with tall necks and a red or pink fabric. But the black slip, thin potting and description does not match the Sanjan sherds. The Massive jars of the previous period have a pink or red fabric but do not conform to the Sanjan samples (Chittick 1984: 84 – 87). The only report that comes close to matching the Sanjan sherds is of Shanga, where Horton classifies this class as Group 11, Siraf Storage Jars, c. He calls them Fine Pink Earthenware. The description of fabric, shape and surface treatment seem to be similar (Horton 1996: 296 -297, fig.220).

White Slipped Pink Ware (Ware code: WSPW)

Definition : This ware class consists of those vessels which have a semi-coarse pink to red-pink fabric and which have a white slip on the exterior surface. These are mostly large vessel forms with medium thick to thick potting and a heavy sand temper. They are mostly storage jars and transportation vessels which could include amphorae shapes. Some vessels in this class have the bitumen or asphalt coating on the interior. The class has been subdivided into three at Sanjan.

WSPW-1: This subclass consists of large storage jars and basins with a white slip on the exterior and no coating or residue on the interior. There are very few rims in this sub-class. T.357 has a wide flanged, rounded rim with a thick white slip which covers the exterior of the vessel and the rim but does not extend to the interior. The thickness of this basin is 17mm. There are a number of body sherds which show evidence of slip on the exterior alone and never on the interior. Some body sherds have

(D.166). One moulded small fragment which appears to have flaked off a SBW-2 vessel consists of a double zig-zag line and dot design (D.18). Other designs consist of incised lines and dots (D.21, D.21.1). One interesting sherd has a sharp fluted edge and a curvature suggesting that it is part of a luted vessel. This may be the carinated shoulder part which joins to the body with a fluted effect (D.165). to this may be added the spout with the serrated edge which was luted on a jar. Most decoration is seen on SBW-1 and especially SBW-2.

Fabric : Two fabrics could be identified in this class. The first three sub-classes have fabric PG(A) which is pale grey to pink in colour it is semi-coarse and hard fired with a sand temper. The fabric is brittle and thin in section with evidence of grog and black inclusions which can sometimes be large enough to pit the surface. Some quartz grains can be seen as well. In SBW-3, the fabric is the same but it is fired to grey. SBW-4 and SBW-5 have a far more coarse and grainy fabric, PG(B). This fabric is similar to the previous one but one frequently notices red shiny particles which could well be red mica. The fabric itself appears more red. It is not a common fabric.

Petrographic analysis of SBW-1 shows that fabric PG(A) belongs to Petrofabric Group C, the same as WSPW-1 and WSPW-2, BW(TJ) and BW(SV).

Occurrence at Sanjan : This ware class is found mostly in layers 2 and 5 at Sanjan. In TT4, SBW-1 occurs mostly in layer 2. A large number of the sherds come from the well. But the other quadrants also have this class in layer 2. SBW-2 Occurs in layer 2 with a few sherds in layer 5. SBW-3 which is poorly represented is only found in layer 2 and the occasional sherd in layer 1. SBW-4 is found mainly in layer 2 and a stray sherd in layer 5. SBW-5 does not occur in TT4 at all. It is present in TT1 in layers 2 and 3 but this is suspect given the disturbed nature of the site.

Discussion : No parallel could be found for this ware in any reports of other sites. Some of the rim shapes are somewhat similar to those reported from Suhar. They are mostly ascribed to level III, dateable to the early Islamic or Umayyad period. Kervran identifies the wares with these rims as being of Indian origin (Kervran 2004: 263 – 381). But it is difficult to tell if the wares at Suhar are the same as those at Sanjan. The unglazed wares reported from most sites and the supposedly Indian pottery from some of them may include sherds of this ware but it is difficult to ascertain this from the brief descriptions given.

Large Incised Storage Vessels (Ware code: LISV)

Definition : The nomenclature for this ware class follows that used by Kennet in the Ras al-Khaimah pottery report (Kennet 2004: 58). The earlier classification of this ware as a sub-group of Coarse Red Ware was found to be totally wrong and misleading (Gupta et al 2002, 2003, 2004). This is a distinct category with various identifiable sub-groups. The forms are almost without exception large storage jars with both rilling and incised design or meanders as decoration, or broad bands and grooves which may have a cream slip. Only one sub-group, LISV-3, may consist of smaller jars. The jars are all piriform and may have handles. The ware is sturdy, compact and well fired with sharp fractures and a stony, metallic sound. The surface is frequently washed or burnished purple-black or purple-brown colour. Turning marks can be seen on the interiors.

Shapes and forms : As mentioned above, the ware takes its name from the form of the vessel which is distinctive. The jars and crocks that belong to the class have an average thickness of 10 to 15mm. although in some sub-classes the thickness is far less. The shape most commonly found is the piriform jar with a flat base and sloping shoulders. The rim types in the class suggest high-necked jars (T.261, T.261.1) with internal projecting inverted rims, short straight-necked jars with externally projecting everted rim and an incised meander on the neck (T.429 D.199) and hole-mouth jars with rounded rim (T.427). A number of handles have been found in this class, suggesting that these were storage vessels meant for transportation. The bases are flat (Bs.43, Bs.43.1). The loop handles are small but sturdy. Some broken handles can be seen luted to the shoulders of some vessels (D.130, T.428, T.428.1). T.428 has a round bump on the curve and is an exact replica of the handle found in the excavations at Chaul (Gogte et al 2006: 62 – 80). Striation marks are frequently seen on the interior of the vessels.

Surface Treatment : The surface treatment is a thin wash or slip in most cases. Some vessels appear not to have any surface treatment at all. The slip or wash, when present, is a dark purple-black or purple-brown which gets a grey tinge due to weathering or deposition. The broad band-like ridges frequently have a thick cream slip applied to them, the pale colour standing out in contrast to the dark slip on the rest of the body. The interior surface is usually slipped but sometimes the slip is restricted to the exterior and the rim.

Decoration : The decoration on these jars is executed by incised meanders, horizontal lines, oblique lines and dots and rilling. The other common decoration is the presence of broad band-like flat ridges and the grooves between them. The ridges sometimes have a thick cream slip. The decoration is usually placed on the shoulder and body or on the neck or just below the rim.

Fabric : The fabric of this class is extremely compact, sturdy and hard-fired. It has large red platelet-like angular inclusions. Black and sometimes small white or buff inclusions are also noticeable. No mica is seen in this fabric when examined visually under lens. Four variations in the fabric and potting are noted, on the basis of which the following sub-classes have been made.

a) LISV-1: The fabric is MR(A). It is maroon-red to purple-red in colour and has large inclusions and what appears to be grog. This is the most common of the sub-classes. The surface of the vessels is slipped and often has decoration. These are the archetypical large storage jars reported from most sites. At Sanjan, they occur in large numbers in TT1 and TT2 but the context is doubtful. In TT4, they occur in the stratigraphic context, almost entirely in the lower part of layer 2 and upper part of layer 3.

b) LISV-2 : This sub-class has a fabric which is pale pink-red and has less black inclusions. This fabric, MR(B) has large red angular inclusions. The texture is hard and cement-like. They may or may not have a self-wash or a pale grey or black slip. Sometimes the surface is left untreated. In TT4, they occur in layers 2 and 3 with a few sherds in layer 4.

c) LISV-3 : The fabric of this sub-class is also MR(A). The vessels are smaller than the other sub-classes and thinly potted with an average thickness of 5 to 6mm. An incised design on the neck of a small high-necked jar (D.198) is the same as the one on the large storage jar in LISV-2 (D.19). The slip in these smaller vessels has a more pronounced brown tint. This sub-class also occurs mainly in layer 2 and is absent in layers 3 and 4 of TT4. It is again present in layers 5 and 6.

d) LISV-4 : The fabric of this sub-class is MR(C) which is much softer than the other fabrics in this class. It is orange-red in colour and has large white or buff inclusions and red platelets. It resembles RED(C) in its texture. The surface can have a greyish wash. The designs include incised meanders and horizontal lines placed on the neck under the rim (T.429 D.199) or broad band-like ridges (D.130). The occurrence of this sub-class is almost entirely restricted to layer 2 of TT4, with only one sherd occurring in layer 3.

e) LISV-5 : This is a very bright orange, hard-fired fabric with a stoneware-like compactness. It has fewer inclusions than the other fabrics in this class and can be either heavy-set or thinly potted. It usually has a self wash in the same bright orange colour. The only decoration seen in this sub-class are the broad ridges. Like LISV-4, this ware is also exclusive to layer 2 in TT4 with one sherd in layer 3.

Petrographic Analysis : Thin-sections of LISV-1, LISV-3 and LISV-5 were studied and they were found to belong to a distinct Petrofabric Group, I. No other samples from any other classes belong to this group. This distinct group has features unique to it, indicating a technological and perhaps a regional exclusivity. While it is possible that many centers probably existed for the manufacture of these unglazed utility wares, the similarity, if not uniformity, in the mineralogical profile indicates a common source at least for these three sub-classes.

The thin-sections falling under the category of fabric I had a medium to fine grained matrix. One of them was a micaceous matrix and showed slight pleochroism in PPL. The other samples are non-pleochroic in PPL and red to brown in XPL. The matrix looked ferruginous. The grains were angular to sub-angular in shape and in one case large number of flaky grains were found. The highest grain sizes in the samples were 50 μ, 60 μ, 80 μ, 100 μ, 120 μ and 400 μ. The grains were perfectly to well sorted and had a frequency of < 5 to 30%. The grain-size distribution pattern was unimodal. Grains did not show any orientation. Voids were very few to many and most of them were grain fallouts. The non-plastic inclusions included quartz, plagioclase feldspar and mica. There are a few hematite patches also. The geological terrain to which it belongs is probably II, with quartz and mica present in it.

Occurrence at Sanjan : This ware is very well represented in the Sanjan collection. It occurs in trench TT4 in the stratified context mostly in layers 2 and 3. The ware is poorly represented in layer 4. However it is interesting to note that LISV-3 occurs in layers 5 and 6 as well.

Discussion : This ware class is reported from most sites known to have contact with the Persian Gulf during the period under study. Kennet reports it from Ras al-Khaimah and suggests either Iran or a local manufacturing centre as the origin (Kennet 2004: 58). He dates the ware to the late eighth or early ninth century. Chittick calls these Massive Storage jars and compares them with those reported by Whitehouse from Siraf. He also mentions the occurrence of these at Hafun in Somalia in levels ascribed to the second to fourth century (Chittick 1984: 84, Plate 38, fig.40). At Unguja Ukuu, the ware is described by Juma as a type of unglazed pottery of the Near East. The illustrated sherd resembles LISV-1 at Sanjan (Juma 2004: 113, fig.6.17.11). Costa and Wilkinson report these large storage jars from 'Arja and suggest Siraf and Shirinu as the likely production centres. The date they assign to the ware is the 9th – 10th century (Costa and Wilkinson: 186, fig. 101).

LISV apparently has a long history which stretches from the pre-Islamic period to the Abbasid period. It becomes important to use this ware in the establishment of the chronology at Sanjan.

Mica Red Ware (Ware code:MRW)

Definition : This class of pottery at Sanjan is defined by the large amount of mica which coats the entire surface of the vessel and is present in the fabric as well, indicating a deliberate addition of mica to the red clay in order to achieve a metallic silver effect. The vessels are mainly thin walled, wheel-made pots, some of which may have been used for cooking as the darkening of the lip and rim indicate. The clay itself is red, fairly well levigated and sand tempered. The micaceous shine is the only decoration seen on these vessels, with ridges placed on the shoulder below the rim.

Shapes and forms : The vessels in this class are predominantly wide-mouthed cooking pots or handis as they are locally known. There are very limited variations in the rim forms. The basic shape of the vessel does not change. All the vessels have a constricted neck and a sharply defined mouth. The rims present in this class are externally projecting, extremely everted with square or rounded lips (T.77, T.78, T.78.1, T.79). The only variation noticed is in the placement or number of ridges. T.77 and T.78 have no ridge at all. T.78.1 is the most common shape with a ridge. T.79 has two blunt ridges on the shoulder and a carination on the body. T.80 is the only rim which is bilaterally projecting. The diameters of these pots ranges from 18 to 24cm. but most vessels have a diameter of about 22cm. The thickness of the vessels is between 3 and 5mm. Most rims are blackened or dark grey along the lip due to exposure to fire or heat.

Surface treatment : As mentioned earlier, the vessels have a heavy coating of mica which gives them a metallic silver shine. Mica is also present in the clay, though not in as much quantity. Two sub-groups can be identified on the basis of colour :

i) MRW-1 which is a deep red

ii) MRW-2 which is pale pink-red to pale orange-red.

The latter has a more pronounced silver tone due to the paleness of the body. Two small body sherds in this ware were unique in that the colour was violet purple. The mica covers the interior and the exterior surfaces of all vessels. The mica flecks are mostly fine but some slightly large particles are also seen on the surface.

Fabric: The fabric, RED(M) is well levigated with a heavy sand temper. Few inclusions or airholes are present. Mica is present in the fabric and was probably an addition by the potter. The colour of the fabric can range from red to pale pink and in some cases grey. This colour defines the colour of the vessel since no additional colour has been added as a slip or wash. No petrographic analysis has been done on these sherds but it is very likely that the ware is an indigenous, if not local, product.

Occurrence at Sanjan : This class was found in large numbers in TT1 and TT2. However its presence in TT4 is meagre. Most sherds belong to MRW-2 and come from the well deposit, from a depth of 4.45 to 5.15m only. Only one sherd occurs in TT4 NE in layer 3. It belongs to sub-class MRW-1. The only rim shape present in TT4 is T.78.1, whereas all the other shapes are present at TT1 and TT2.

Discussion : No known parallel wares are found to match these. Although micaceous red wares have been reported from various early historical sites in India, they are different from these in the shapes, fabric and treatment.

Slipped and unslipped Wares

Two sherds of Red Polished Ware were found in the excavations of season – 1. Their presence is of great significance even though the number is small and the context is suspect. They are dealt with in this section.

Indigenous wares which are functional and utilitarian are the most prolific at any site and Sanjan is no exception. These wares make up a bulk of the assemblage and have been sorted into two broad groups on the basis of whether they are slipped or unslipped. The distinct classes within these groups are discussed individually below. Since the preliminary sorting had taken fabric as a c onstant criterion, the colour of the fabric was taken as basis for categorisation. However this too was erroneously done in many cases since the colour identification was arbitrary. A large number of sherds had been identified as Slipped Red Ware in the first two seasons of excavations under the mistaken notion that the black surface was caused by firing or that the black fabric was a result of firing. This was found to be wrong when the corpus of material was re-examined under lens and with more care. The clays of the grey and red wares at Sanjan are no doubt similar, if not the same, and a large number of shapes are common to both. However the distinction between the deliberately created black vessels and red vessels needs to be addressed. It is also not specific enough to class all slipped wares with a grey fabric into a singular class since the slip colours could vary from black to red. The two are treated separately in this study in order to develop a more specific definition. Similarly, the slipped red fabric wares are classified as per the colour of the slip. The problem with using such wares for establishing a chronology for the site or for tracking the morphological changes over a period of time is compounded by the fact that these wares have a very long tradition in India and rarely change over time in any significantly noticeable way. In fact, these wares can begin in the Early historical period and continue well into the late medieval with almost no change. Most reports on pottery in India use the nomenclature Slipped Red Ware, Slipped Grey Ware, Coarse Red Ware, Coarse Grey Ware, Burnished Black Ware, Burnished Red Ware, etc regardless of whether the reference is to Chalcolithic, Early Historical or Medieval assemblages. This nomenclature has been deliberately avoided in this study so that some of the confusion arising out of the indiscriminate and generalized usage of terms may be overcome. The range of shapes in these classes is vast and frequently common. Hence the shapes are listed in detail in the section on illustrated and described sherds.

Red Polished Ware (Ware code: RPW)

Definition : Red Polished Ware is one of the most significant wares of the early Historical period. It is a matter of debate as to how long this pottery continued in use. But it is generally accepted that the ware belongs to the early centuries of the Christian era. It is also accepted that the origin for this ware is probably Gujarat. The vessels of this class have an extremely fine well levigated fabric which is pale red-pink in colour. The surface is has a deep red slip which is polished to a high gloss, which explains the nomenclature. Closed and open forms are present in this ware. However at Sanjan only two sherds were retrieved from the pit of a ring-well. One sherd is a sprinkler and the other is a spout.

Shapes and forms : As mentioned earlier, one sprinkler (T.263) and one spout (T.262) have been found at Sanjan. The spout has a ridged opening and six perforations on the base to act as a filter. The perforations have been pierced through the clay. The body of the vessel appears to be thin walled.

Surface treatment : The sherds have a dark red slip applied evenly on the external surface. This has been polished to achieve a shiny effect. No other attempt is seen at decoration or treatment.

Fabric : The fabric of this class is RED(N). Visual examination shows extremely well levigated clay in which small mica particles are visible. No inclusions or visible temper is seen. The fracture is smooth. The firing is even and the fabric has a pale pink-red colour. No petrographic analysis was done for these sherds.

Occurrence at Sanjan : Both the sherds of Red Polished ware have been retrieved from the ring-well pit and so nothing can be said about their stratigraphic context. It is surprising to find two sherds in such fine condition without any erosion or rolling when the rest of the collection shows a considerable amount of weathering. No other sherds of this ware have been found either in the excavations or explorations in the following seasons. It may be noted that the fill material for the pit could not have been transported from too large a distance. Logically speaking, the fill material for the pit probably belongs to an area in the vicinity of the site. It is possible that a site with RPW may be present in the vicinity of Sanjan Bandar and that it may even predate the occupation at Sanjan Bandar itself.

Discussion : RPW has been found from numerous sites both in India and outside. It has been found from the Early Historical context at most sites and has been seen by many scholars as being related to Roman trade. The ware first appears around the first century AD at sites such as Nevasa (Sankalia et al 1960: 69). However the termination date for this ware remains debatable. While a large number of scholars think that the date does not extend past the 5th century AD, there are others who think the ware continued for a longer period. This ware has also been reported from a number of sites in the Middle East and the Persian Gulf region as well as from sites on the East African coast. Kervran reports it from Suhar where it occurs from the earliest levels, i.e. from level 0 which starts in the middle or end of the second century to level IV which ends around 630 – 670 AD (Kervran 2004: 268, 271 – 86). Sedov dates it from the 2nd and the 4th centuries at Qana' (1996: 22, 26 – 27). Kennet reports it from Kush where it is most prominent in the 7th and 8th century context – a date which appears to be quite late (Kennet 2004 : 65 – 66). Kennet also mentions the occurrence of RPW at Sanjan as reported in the first preliminary report of the excavations and the possibility of an 8th century date for it. However the preliminary report did not mention the stratigraphic context of the sherds or that they were found in a pit.

Black Slipped Grey Ware (Ware code: BSGW)

Definition : This ware class forms the largest part of the Sanjan assemblage. The highest sherd count comes from this class which is defined by the deliberate application of a black slip to the surface of vessels with a grey semi-coarse to coarse fabric. The class includes closed as well as open forms. Most vessel forms suggest utility and storage. These sherds are present from layers 1 to 6, signifying a long tradition and a popularity which is not chronologically specific.

Shapes and forms : The range of shapes in this class is vast. Most shapes are common to both classes, BSGW as well as RSGW. The BSRW and RSRW classes also have many shapes in common with this class. High-necked pots, wide-mouthed pots or cooking pots, pots with constricted necks, carinated pots, water troughs, basins, dough plates, some bowls and dishes are represented in the collection. Pots number the highest in this class, the other shapes being less prolific. The rims of the wide-mouthed pots are mostly externally projecting and everted or bilaterally projecting, with a number of variations. They can have grooves on them (T.81, T.84, T.85, T.288) or ridges (T.87, D.191). A large number of the pots have constricted necks and some are carinated. The decoration is usually quite simple and is restricted to the shoulder of the vessel, the neck or the rim itself. The diameters of the rims can range from 32cm. to 12cm. but the average diameter found most commonly is 16 to 24cm. The thickness of most cooking pots does not exceed 5 to 6mm. While most shapes are common to all the slipped classes, spouts and handles are conspicuously missing from this class. The high-necked vessels have globular bodies suggested by the slope of the shoulders. The rims are externally projecting and everted with variations in the form. The bases of vessels are either contiguous flat or contiguous rounded with little variation. The basins or troughs have a thick rim, flat on the top, with a deeply impressed border (T.320 D.127) or are plain.

Surface treatment and Decoration : The external surfaces of these vessels have a black slip which often flakes and erodes. The closed form vessels have the slip on the exterior extending over the rim and stopping at the mouth of the vessel. Bowls, plates and other open forms have the slip on both surfaces. The decoration is restricted to the shoulder or neck or rim and does not display an imaginative range of designs. The most common forms of decoration are grooves and ridges (D.191), small puncture-mark triangles placed in horizontal rows (D.43, D.43.1, D.43.2, D.43.3 and D.43.4), small pendants formed with the same puncture-mark triangles (D.112, D.112.1) and a moulded or impressed border along the rim of a trough (D.127). The first two decorative features are also found on the other slipped wares. A few graffiti marks are present but these appear to be scratched on with no intention of decorating the vessel (D.193). The triangular puncture marks are either spaced or joined. In the well executed pieces, the joined triangles appear like the serrated edge of the upper part of the vessel which was luted on to the lower at the carination (D.46). D.42 has an appliqué band with oblique slashes cutting it.

Fabric : The fabric of this ware class is GREY(A). It is semi-coarse to coarse with a large number of inclusions and visible organic temper. It is sandy and has mica which can be seen clearly even without a lens. The texture is friable and the fabric is fired grey to black and, in some cases, fine red horizons can be seen. Many airholes are present. The compactness and the levigation in this fabric class vary. Not very frequently, the fabric can be GREY(B). This is a much more coarse and sandy fabric with a large number of inclusions, lime particles, and organic temper. The texture is harder and the firing is not even. There is less mica in this fabric.

Petrographic Analysis : The thin-section study done on samples of this class shows that it belongs to Petrofabric Group B. The matrix of the samples falling under fabric B is coarse grained, less compact and extremely ferruginous. It is non-pleochroic and has a brownish colour in PPL which turns to dark brown in XPL in XPL. The grain shape in this group varied from angular to sub-angular. The largest grains in this section were 200 μ, 300 μ, 360 μ, 400 μ, 600 μ and 1600 μ. The grains were well sorted. The frequency of non-plastic inclusions in various samples varied from 40 to 50%. The grain size distribution character in all samples was unimodal. The non-plastic inclusions in majority of the samples did not show any particular orientation except in two samples which showed more or less parallel orientation. The detritals in the thin-sections of the samples were constituted by quartz, plagioclase feldspar, microcline feldspar, augite, mica, bioclasts, cryptocrystalline silica, basalt fragments and argillaceous inclusions. It has a geological source designated I, i.e. a basaltic source with basalt fragments, basaltic components and argillaceous inclusions. The only wares that were found to belong to this Petrofabric group were the slipped red and grey ware classes and the unslipped red ware class. This confirms the visual observation made under lens that the sources for these fabrics are common and that it is the pottery traditions and cultural factors that cause variation.

Discussion : Black slipped or burnished wares have been reported from various medieval sites abroad and have been identified as Indian. Kennet reports them as SBBW or Black Burnished Ware at Ras al-Khaimah. His description of the ware mentions a lack of mica and vegetable temper. This is unusual since the Indian fabrics almost always contain mica in some proportion. Other than this, the description matches the common slipped or burnished black wares current in India during the early historic and medieval periods. He suggests that they started to circulate in the seventh or eighth century. The observation that most of these vessels represented in the collections outside India are cooking pots with everted rims and his speculation as to the reason for their trade over such long distances is significant in establishing the presence of Indian communities in foreign settlements with requirements that were probably defined by their ethnicity and cultural demands (Kennet 2004: 65, 66, fig 78).

Kervran reports Indian pottery, both RPW as well as the black or red slipped wares, from Suhar from pre-Islamic levels as well as Islamic levels. The pre-Islamic levels which she designates Level 0 to IV, dateable from the middle or end 2nd century to mid-7th century, contain notable quantities of Indian wares which continue in the later levels. The shapes described and illustrated by her as having a grey or black fabric and slip match the profiles of the cooking pots at Sanjan. The decoration on Indian pottery in Level II described by her is not seen at Sanjan. She notes the decrease of RPW in Level III and later, and the corresponding increase in other less fine Indian wares with a red or grey-black fabric and slip (Kervran 2004: 268, 271 - 275, fig.8, 27, fig.9, 9 - 10). Level IV continues to have the Indian grey slipped component (Kervran: 2004 285 – 86, fig.14, 11). In the Islamic levels, Level V and VI, the Indian ceramics increase tremendously, with much of the assemblage being black ware (Kervran 2004: 301, 316, fig.21, 21 - 23, fig.23, 25, fig.30, 10). She dates Level V from 670 to 890 AD and Level VI from 890 to 1150 AD.

Horton's description of Indian pottery from Shanga, Group 15, Grass-tempered greyware appears to bear resemblance to the Sanjan fabric. It occurs in both the ninth century context as well as the eleventh century. This is not surprising, given the long life of Indian pottery traditions (Horton 1996: 300, fig.224, a – g). The Manda assemblage also includes Indian wares, the most common of which, according to Chittick, are the grey fabric wares which have chaff temper and mica. His description closely matches that of the grey wares at Sanjan and the profiles of the cooking pots, which he calls 'chatties', are almost the same, except that a majority of the vessels appear to have a red slip instead of black. He dates them to Period I (mid-9th - early 11th century) and a few to Period II (mid-11th – late 12th century) (Chittick 1984: 101, fig.54, fig.55).

In India, most excavation reports mention Slipped Grey Ware or Burnished Black Ware but rarely are distinctions made between the colours of the slip or the chronological contexts. It is difficult to tell the Early Historical from the Medieval, more so because the medieval period has been treated with some disdain in archaeological studies. At Brahmapuri, Sankalia and Dikshi t report the wares belonging to the 'Post-Satavahana or Bahmani period' in some detail (Sankalia and Dikshit 1952: 57 – 83). But this period covers about a thousand years and no sub-division of the chronology has been attempted, which makes it difficult to trace any morphological or stylistic changes. At Maheshwar and Navdatoli, the periodisation is more detailed, with Burnished Black Ware and Coarse Black Ware suddenly showing a marked increase in Period VI (100 A.D to 500 A.D) and in Period VII (Muslim-Maratha) when they are associated with nineteen glazed ware and three celadon sherds. The pottery in Period VII is said to show a general decadence in fabric and the disappearance of shapes from the previous periods. Black Slipped Ware is reported to increase in this period. However, there are no assigned dates to Period VII, although it is mentioned that there is an absence of material from 500 AD to the Muslim-Maratha period which one presumes is datable to the 12th or 13th century at the earliest and possibly the 14th or 15th century, if the glazed wares are any indication. The black wares reported and illustrated closely match the grey wares at Sanjan, both slipped and unslipped, in profile and fabric (Sankalia et al 1958: 22 – 23, 150 – 170). The Shamalaji report describes and illustrates Burnished Black Ware and Plain Black Ware, along with their red counterparts. The descriptions do not include details about the fabric. According to Mehta, these wares occur at all levels from the beginning of the Christian era to 1800 AD (Mehta 1967: 17 – 23). The only data one can draw from this report is that some of the illustrated shapes match those from Sanjan. Similarly, the Nagara report gives the descriptions of Plain and Burnished Black Wares but does not mention the fabric or the exact chronology. Period III covers the beginning of the Christian era to the 8th – 9th century and Period IV covers the 14th – 18th century (Mehta and Shah 1968: 16 -19, 34 – 45). The gap of five hundred years in the chronology of the site makes it difficult to ascertain or assess the seriation of the pottery assemblage.

Red Slipped Grey Ware (Ware code: RSGW)

Definition : This ware class is different from the previous one only in the colour of the slip which is red instead of black. The colour can range from pale red to brown-red. The fabric remains the same, with the addition of a finer fabric in a few rare sherds. This ware occurs with far less frequency than BSGW.

Shapes and forms : The shapes in this class are similar to the previous class and to the RSRW class. Carinated and wide-mouthed cooking pots, jars, basins, dough plates, storage jars and bowls are present. However some shapes are present in this class which do not occur in the previous class. A lid finial (T.408) is found with a red slip but not with black. It is hand-made, conical and perforated on the tip. It occurs in layer 2, TT4 NE. A large basin (T.406) with straight sides and a square clubbed rim has been retrieved from layer 4 of the same quadrant. Spouts are present in this class. The dough plates have grey interior and red slipped exterior with flat bases which have an extremely rough gritty surface (T.252 and T.252.1). Conical spouts are present in this class, although the numbers are few. The cooking pots have constricted necks. Most rims are externally everted and may have grooves or an incised chain-like band on the top as decoration. Some bi-laterally projecting rims are also seen.

Surface treatment and decoration : The surface of these vessels is treated to a red slip which can vary from a dark brown-red tone to orange-red. The closed form vessels usually have the slip only on the exterior extending over the rim and stopping on the inside of the mouth. The open forms have the slip on both surfaces and, in the case of dough plates, the slip on the exterior may be different from that of the interior. Most vessels have a single coloured slip. The decoration of the vessels is not very innovative or imaginative. A single band of small puncture-mark triangles on the shoulders of some pots and an incised chain-like band along the top of the rim are the only attempts at decoration seen in this class.

Fabric: As mentioned earlier, the fabric of this class is in most cases the same as that of the previous class i.e. BSGW. GREY(A) appears to be the most common fabric in the grey ware classes, both slipped and unslipped. However, one other fabric has also been observed. GREY(B) is a fabric almost exclusive to the dough plate, basin and storage jar forms. The clay is coarse with a large number of large inclusions and vegetable temper. It is frequently ill-fired and has large grits which can roughen the surface. In the case of the dough plate, the entire flat base has been treated to a thick and even application of grit while the wall and interior of the vessel are both smoothened and slipped.

Discussion : None of the excavation reports accessed for this study mention a ware having a grey fabric and a red slip. Almost all reports describe red slipped pottery but the fabric has not been taken into account. It is therefore difficult to assess the occurrence of this ware at sites other than Sanjan. The similarity of shape or profile with red slipped wares reported from Nagara (Mehta, 1968), Shamalaji (Mehta and Patel 1967) and Nasik - Period IV (Sankalia and Deo 1955) may be noted. However, the fabric may not necessarily be the same. The same shapes occur in other classes as well and this makes it more difficult to find parallels. Kervran reports the Indian ceramics from Suhar. These include cooking vessels with a coarse grey-pink fabric with a polished orange or red slip on the exterior (Kervran 2004: 326 – 27, fig.35 no.18, 19, 20). They belong to Level VI, dated 880 – 1150 AD.

Grey Ware (Ware code: GW)

Definition : This pottery class consists of vessels having a grey to black fabric and no surface treatment. The vessel forms are utilitarian. Most forms are common to the above mentioned slipped wares. The fabric variations seen in the slipped wares are also found in this class, with the addition of an exceptionally fine fabric which has been seen in only a few samples. Three sub-classes can be demarcated in this category on the basis of fabric.

GW-1 is the more common sub-class with a majority of the vessels belonging to it. The fabric is GREY(A).

GW-2 is more rarely seen. The fabric is GREY(C), which is very fine textured, hard-fired and blue-grey in colour. Only a few shapes are seen in this sub-class.

GW-3 is represented by only one rim sherd, T.411 and is characterised by two clearly demarcated grey tones on the vessel. The interior of the vessel is a dark grey-black colour which extends up to the top of the rim. The exterior of the vessel and the lip of the rim are a pale shade of grey. The fabric is quite coarse in texture – GREY(B). This sample was found on the surface during exploration. Hence no stratigraphic context can be ascribed to it.

Shapes and forms : The shapes suggest utility and are mostly common to the slipped wares mentioned above as well as to the red wares. However some rim shapes appear to be specific to this class. Cooking pots with or without carination, jars, storage vessels, basins, bowls and dough plates are the common shapes. The cooking pots mostly have constricted necks, wide mouths and a variety of rim shapes, most of which are either externally projecting or bilaterally projecting (T.222, T.97, T.298, T.180, T.131, etc). The rims frequently have grooves on them, suggesting perhaps the use of lids. Pots with inverted rims are also present. These can have corrugation-like ridges on the shoulder (T.309). The jars have high necks and externally projecting rims. T.95 appears to be an extremely popular shape common to almost all the slipped and unslipped classes. A few lids and spouts are also represented in the class. The bases are contiguous flat, contiguous rounded and non-contiguous discoid (Bs.39, Bs.40, and Bs.80). All these vessels belong to GW-1. The range of shapes in the other sub-classes is not as varied. GW-2 consists of a small lamp or bowl with a diameter of only 6cm (T.409), a cooking pot which has a broad flanged rim having two deep grooves on top(T.410) and constricted necked pots (T.101, T.100.1, T.178). GW-3 consists of only one unique sample, T.411 which is a cooking pot having an externally projecting rounded rim. The vessel is two-toned, as described earlier. Most vessels are wheel made although a few handmade specimens are also seen.

Surface treatment and decoration : These vessels do not have any surface treatment. The decorations are also quite plain and consist of the ridges and grooves, bands of puncture triangle marks (D.43.3, D.74, D.76.1), bands of joined triangles to give an effect of luting (D.46) and bands of puncture dots (D.50). Only one small sherd was found with the traces of a painted design in black (D.54). The decoration is usually restricted to the shoulder of the vessel. No attempts at decoration have been noticed on sherds belonging to sub-classes GW-2 and GW-3.

Fabric : The lens observation of the body showed three fabric variations. GREY(A) was the most common fabric. It is semi-coarse clay with sand and organic temper, medium sized inclusions, visible mica particles and a number of airholes. It ranges from grey to black in firing. Impressions of the vegetable temper can sometimes be seen very clearly. GREY(B) is a more coarse clay with heavy sand temper and a large number of white inclusions that appear to be talc or lime. The firing is not always even and can sometimes show red clouds. There is less mica in this fabric. GREY(C) is extremely fine, well levigated clay which has been evenly fired to produce a bluish steel-grey colour. It is far more compact and has a very slight sand temper. There is very little mica in this clay.

No petrographic analysis was done on this ware class. However, GREY(A) fabric is common to BSGW and RSGW classes and has been analysed (see above).

Occurrence at Sanjan : Trench TT1 had a large number of these sherds distributed through the entire sequence. However, TT2 which was adjacent to TT1 had only two or three sherds and these were non-diagnostic. In TT4 the numbers were less than TT1 and were confined to layers 1, 2, 5 and 6. In contrast the Red Ware class was well represented in almost all trenches and at all levels.

Discussion : Most excavation reports refer to unslipped wares with a grey or black fabric as Plain Black Ware, Ordinary Black Ware, Coarse Black Ware, etc. Rarely are they given much prominence, especially if they belong to the Medieval period. In fact, the Nevasa report mentions the presence of Black Ware in Period VI and dismisses it in three sentences – "The black ware is relatively less than the red ware. The coarse variety is more abundant and the surfaces are drab black and core gritty and black. It gives commonplace shapes and deserves no further elucidation" (Sankalia et al 1960: 323). It is treated with more consideration in the excavation report of Maheshwar - Navdatoli. The excavators observe that the Coarse Black Ware increases in Periods VI and VII i.e. 100 AD – 500 AD and from c. 1400 onwards. The description of the fabric and shapes resemble those at Sanjan, in that the predominance of cooking pots and externally projecting rims are noted. However the prominence of lids in this ware class at Maheshwar is not seen at Sanjan. The gap in the chronology is also a deterrent in constructing a sequence (Sankalia et al 1958: 150 – 151, figs. 79 – 81, 162 – 163, fig.90). At Shamalaji (Mehta and Patel 1967) and Nagara (Mehta and Shah 1968) the Plain Black Ware vessels, especially cooking pots, closely resemble the Sanjan grey ware shapes. However the lack of fabric description and the non-chronological presentation makes it difficult to define which shapes belong to the Medieval period and which may be specific to the Early Historical. At Shanga, Horton reports Group 15, Grass-tempered Greyware in his description of Indian wares (Horton 1996: 300, fig.224 a – g). The fabric appears similar to GREY(A) and some of the vessels resemble the cooking pot shapes at Sanjan. However it is not certain that the two wares can be equated. He dates this ware to the 9th century.

Red Slipped Red Ware (Ware code: RSRW)

Definition : This class of pottery is the next most prolific at Sanjan after BSGW. The range of vessel forms is vast and mostly utilitarian in nature. The fabrics can vary, as can the colour of the slip. Two subclasses are noted in this category, RSRW-1 and RSRW-2, mostly on the basis of the slip colour but also fabric variation. Despite these variations, the class refers to vessels having a red body with a red slip as surface treatment. The similarity in fabric, shapes and decoration with the grey wares cannot be missed, although some marked differences are noted – the presence of small and medium sized bowls, spouted vessels, lids, painted decoration in the case of some vessels, etc. Pots, jars, bowls, dough-plates, basins and storage vessels are the common vessel forms.

RSRW-1:- This sub-class consists of red earthenware vessels having a dark red slip. The slip is usually quite thick and commonly shows signs of flaking or wear. This tone of colour is traditionally seen in vessels of the Early Historical period and in earlier samples from the Chalcolithic. Two fabric variations are noticed – RED(C) which is the most common red fabric at Sanjan and RED(D) which has a more coarse and sandy texture. Of the two subclasses, RSRW-1 is the more prolific.

RSRW-2:- This second group consists of similar vessels to the above mentioned subclass. The colour of the slip, in this case, is a pale orange-red. The surface is much smoother and more glossy, evidently due to more careful burnishing. Two fabric variations are noticed in this subclass as well – RED(C) and RED(E) which is more compact, better levigated and with visible quartz-like inclusions.

Shapes and forms : The most common shapes in this ware are carinated pots, rounded pots, pots with constricted necks and wide mouths, high-necked globular jars, bowls, narrow-mouthed jars with ledge rims, dough plates with or without lug handles, large storage jars, spouted vessels and basins. A bell shaped finial and a tea-cup handle also indicate shapes which are not found in other classes. A majority of the vessels are wheel-made, though some hand-made vessels are also found. The spout perforation is constantly maintained at a diameter of 8 – 10mm, regardless of whether the spout is bulbous, conical, grooved or plain. Most shapes in this ware class are common to the grey wares and appear to have a long tradition since most seem to continue into the Early Medieval period from the Early Historical, with a few variations. Other than the contiguous round and the contiguous flat bases found in the other red and grey classes, two unique base types are found in RSRW-1. Bs.81 is a high, complex pedestal base with a dark red slip and traces of white painted bands. This elaborate and ornate base is a unique piece and no sherds suggestive of the body have been retrieved. Bs.82 belongs to an extremely large thick storage jar and was retrieved from the Layer 3 in trench TT4. This vessel form is ovoid with a bulbous, nipple-base and a deep, sharply-defined well on the inside. The centre of this well contains a rounded bun-shaped feature. A similar shape is seen in BSRW as well but with some variation. These bases could not have been free-standing and must have been either sunk into the floor or had a support stand to hold them. Given the thickness and weight of the vessels, it is likely that they were sunk into the ground. T.163 is the fragment of a ring-stand or pot-rest with a dark red slip.

Surface treatment and decoration : The surface is treated to a red slip which can range from dark red to a pale orange-red, brown-red and even a yellowish red. The pots, jars and other closed forms have the slip applied to the exterior. It usually covers the rim and stops on the inside of the orifice or vessel mouth. The open form vessels such as bowls are slipped on both surfaces. The dark colour of RSRW-1 is seen in pottery of the Early Historical and Chalcolithic periods. The slip in the second sub-class, RSRW-2 is smoother and shows less signs of erosion or flaking. The decoration in this class is slightly more versatile than in the others. Other than the usual bands of triangular puncture mark, either single, double or multiple and bands of grooves and ridges, one also finds incised, painted and impressed decoration. The bands of puncture marks and the bands of grooves and ridges are almost always on the shoulder of the vessel. Sometimes the ridge or groove may be placed on the rim. The painted designs are restricted to RSRW-1 and consist of white dots, circles, loops or bands. In a few samples, traces of black bands are also seen. The painting is concentrated on the shoulder and the lower part of the body. In one case white painted loops are seen on the body around a spout. White bands are occasionally placed along the lip of the rim. Bowls never have any decoration. Bs.81 has white bands painted on it.

Fabric : The fabric of this ware class has several variations. However the most common fabric is RED(C). This is semi-coarse clay with sand and organic temper. It has a fair amount of mica and is fired red, sometimes with black horizons. The porosity of the fabric varies. Air-holes and inclusions are clearly visible. This fabric is present in both the sub-classes. RED(D) has a sand temper and is much more gritty in texture. Large white talc-like inclusions are seen in section. It lacks the vegetable temper of RED(C). The second sub-class, RSRW-2 can also have RED(E) as a fabric. This clay is better levigated and is finer than the others in this class. Mica and medium-sized quartz particles can be seen. This fabric is quite compact and does not have as many air-holes.

Petrographic Analysis : All samples of the red wares, slipped as well as unslipped, indicate that the clay belongs to Petrofabric group B. This is the same as the grey wares.

Occurrence at Sanjan : This ware is found in almost the

entire Sanjan sequence. However, it is noticeably absent in the uppermost levels of trench TT4, i.e. in layer 1. RSRW-1 begins to appear only at a depth of 0.49m onwards. RSRW-2, which is quantitatively far less, also appears only in layer 2 at a depth of about 0.53m onwards. In TT1 and TT2, this ware appears at all levels but the concentration in layer 1 is considerably less than the lower levels. Since trench TT4 and particularly quadrant NE is considered the index sequence, it is relevant that the ware has a high concentration in the earlier levels and is found in large quantities in layer 6 in the earliest occupation levels. The shapes are also indicative of change and evolution over time since the earliest shapes do not seem to appear in the later levels. However, it must be added that all shapes were not represented in the collection from TT4, NE and may therefore result in errors while reconstructing the sequence.

Discussion : Almost all excavation reports in India mention slipped red wares. The long tradition of this pottery continues from the Early Historical through the Early and Late Medieval period. However the pottery of the Early Medieval period rarely merits detailed documentation, making it difficult to identify typological markers. The excavation reports which document medieval pottery in Western India and Gujarat and which are relevant in the context of Sanjan are Nasik, Nevasa, Brahmapuri, Shamalaji, Nagara and Maheshwar, although the latter is in Central India. The problem is compounded by the fact that at a number of these sites the excavators report a break in occupation in the Early Medieval period, usually between the fifth and the tenth centuries or even later. Nevertheless, it is relevant to note that a number of shapes from the earlier periods at these sites and some from the supposedly later centuries of the medieval periods resemble those from Sanjan.

At Shamalaji, layers 4 to 7 represent Period II A and are dated 400 to 1000 AD. Layers 1 to 3 represent Period III, dated 1500 to 1800 AD. Plain red ware and Burnished red ware are mentioned as being very common and occurring in almost all layers (Mehta and Patel 1967: 8, 14, 23 – 33, fig.11 – 16). Unfortunately the lack of any morphological or contextual details makes it difficult to know which of the illustrated types belong to the different periods, or if these wares are indeed the same as the Sanjan class. At Nagara Periods I to III cover a long time span from the Iron Age to the early centuries of the Christian Era. However the habitation then falls into decay and "the town turned into the village, and did not grow for a few centuries. This was the time of the Chalukyas of Patan" (Mehta and Shah 1968: 9). Period IV extends from the 14th century to present day. The Plain and Burnished Red Ware occurs at all levels, as at Shamalaji. It is described as being "usually well fired pottery with gritty core and plain or burnished slip. The slip has different tones ranging from light red to deep red. This is also a utilitarian ware....Varieties in this ware are decorated by paintings or applied and incised design" (Mehta and Shah 1968: 45). Both sites, Shamalaji and Nagara mention the significant presence of bowls in this ware. While bowls are present in the Sanjan collection, the range and variety of bowl types from these sites is absent. While a large number of rim shapes, particularly those belonging to cooking pots, high-necked pots, jars and some bowls appear to be similar to the Sanjan samples, the problem of identifying them conclusively as belonging a particular period remains.

The chronology at Nevasa mentions an eight hundred year break – 200 AD to 1400 AD - in the occupation of the site. Period VI is called the Muslim-Maratha period and is said to extend from c.1400 – 1700 AD (Sankalia et al 1960: 64 – 70). It is relevant to note that despite the long gap between Period V and Period VI, a large number of rims, vessel forms and ware classes are said to continue into the latter period. In fact, a number of the shapes continue from Period III (c.1500 – 1000 BC) to Period VI, sometimes within the same ware class and sometimes in a different category. This continuation of forms over a span of two thousand years establishes the long pottery tradition in India and provides a convincing argument for the necessity of establishing stratigraphic context of excavated material. In Period VI, Coarse Red Ware, Red Slipped Ware and Ware with Red Wash are classes with presumably the same red fabric but with varying surface treatments. The Red Slipped Ware was less in quantity and is described as having a thin and unevenly applied slip on a core which contains degraissants such as sand and straw. Medium size storage pots are a peculiarity of this period (Sankalia et al 1960: 322 – 332). At Nasik a similar gap of nearly eight hundred years is seen between Period III (c. 50 – 200 AD) and Period IV or the Early Muslim-Mughal Maratha (c. 1400 – 1875 AD) (Sankalia et al 1955: 29). At this site too, the principal types survive from Period II, i.e. c. 200 BC – 200 AD well into Period IV. The Red Ware of Periods II and III is said to continue with some additional new shapes (Sankalia et al 1955: figs.32, 35, 36, 39). However no distinction has been made between the unslipped and the slipped wares. A large number of rim shapes closely resemble the Sanjan samples – ridged and ringed rims, carinated pots with everted rims, ridged-neck vessels, etc.

Brahmapuri has a continuous occupation with no break between the Early Historical and the Early Medieval and Medieval periods. The Late or "Post" (sic) Satavahana period is dated by the excavators from the 4th to the 9th centuries AD, followed by the Silahara and Bahmani period immediately after (Sankalia and Dikshit 1952: 57). While several pottery forms of the Satavahana levels are unique to that period, several others show continuity into the Late or Post Satavahana phase. Similarly, a number of forms from the Late or Post Satavahana period continue into the Bahmani period. The predominance of black slipped wares and a considerable quantity of red wares is mentioned as a feature of the Late Satavahana period while a chocolate slipped ware appears to be common in the Bahmani period. The red wares are more prominent during the Satavahana period. The Bahmani period has very little red slipped pottery (Sankalia and Dikshit 1952: 54 - 61). The carinated cooking pots and some of the jar shapes are similar to Sanjan. The "Eyes and Beads", "Maize" and "Bead and String" decorations which are so significant in their presence at Brahmapuri are totally absent at Sanjan.

The pottery of Period VI (c. 100 AD – 500 AD) at Maheshwar-Navdatoli has a Red-Slipped Ware which closely resembles the Sanjan shapes (Sankalia et al 1958: 150 – 156, figs. 81 and 82). However the pottery of Period VII, i.e. the Muslim-Maratha Period does not have any similarity to the Sanjan collection. There appears to be a gap yet again between Periods VI and VII. It is therefore difficult to tell when these earlier shapes and wares end in the site chronology.

Medieval Indian pottery found on Persian Gulf sites and in East Africa is frequently mentioned as being slipped red or black. At Ras al-Khaimah the Indian wares include at least three classes bearing a red slip – FIRE (Fine Indian Red), PAINT (Painted Indian Earthenware) and RSLIP (Coarse Red Slipped). Other than these, a general class called INDIA (Unclassified Indian Ware) mentions the presence of cooking pots but does not mention the slip or colour. IRAB (Indian Red and Black) on the other hand has red slip on the interior of the vessel and black or dark grey on the exterior. The predominance of cooking vessels in these classes is a relevant factor, indicating perhaps the settlement of an Indian community on these foreign shores (Kennet 2004: 65 – 67, 69 – 71, fig. 40). Kervran reports and catalogues a considerable collection of Indian pottery from Suhar. The presence of Indian wares from Level I (Mid to late 2nd century AD) to Level VI (1100 – 1150 AD) alongside the Pre-Islamic and Islamic wares gives a correlation to the shapes and fabrics. RPW predominates in the earliest levels, gradually decreasing while the less fine and more common red wares and grey wares increase. The rim shapes which appear in Level III, everted rims on constricted necked cooking pots are somewhat similar to Sanjan but the tenon around the shoulder of the vessel which Kervran mentions as a common feature is absent in the Sanjan collection. The slip of these vessels could be red or black, as could the fabric. But Kervran makes it a point to specifically mention the presence of mica in the Indian fabrics as an identification marker for wares which could have their origins in the region between the Narmada and the Indus. (Kervran 2004: 268, 278 – 315, figs.8, 10, 12, 14, 21, 23, 26).

At Shanga Horton reports Red-slipped orangeware which has an Indian origin and which he dates to from the 10th to the 12th centuries (Horton 1996: 302, fig. 224 k – n). The description of the ware and fabric could be similar to RSRW at Sanjan. It may be noted that the layers of white clay mentioned in the body of the vessels at Shanga are certainly not seen in the Sanjan wares. The Indian wares at Manda appear to resemble the Sanjan shapes of both the grey as well as the red wares. The fabric description also seems to indicate an Indian origin. The predominant shapes appear to be chatties or water-pots and some cooking vessels. Red slipped vessels number less than the black slipped ones. Chittick dates most of this pottery to Period I, i.e. mid 9th to early 11th century (Chittick 1974: 101, figs. 54, 55).

Almost all scholars note that the red slipped wares are usually found in large numbers in the Medieval period but are still less prominent than the grey and black slipped wares. However the similarity in fabric and shapes is also frequently noted, indicating a common tradition and possibly origin.

Black Slipped Red Ware (Ware code: BSRW)

Definition : This class of ware is very similar to BSGW, both in shapes, surface treatment and decoration. The fabric is RED(C) and the surface of the vessels has been treated to a black slip. Two sub-classes are demarcated on the basis of slip colour in this class:

BSRW-1 has a black slip which is identical to the slip on BSGW. The shapes and vessel forms are also very similar and, in many cases, the same.

BSRW-2 has a blackish slip which has either a yellow-red or brownish tone which borders on a chocolate colour. This sub-class numbers far less than the former.

Shapes and forms : Most shapes in this class are utilitarian. Some shapes are common in both sub-classes and can be differentiated only on the basis of slip colour. The most common shapes are the cooking pots with constricted necks and everted rims. These can be either carinated or rounded. The rims are usually externally projecting or bilaterally projecting with a number of variations. The presence of grooves and / or ridges on the rim is noticeable as in the case of the BSGW vessels. The luting of the external projection of the rim is sometimes obvious since the breakage has been along the joint section (T.136). This is also evident in the case of a deep wok-like bowl with an internally projecting rim (T.421). The carination of some vessels is extremely sharp and angled. The other shapes present in this class are high necked water jars, dough plates, woks, bowls with ledge rims, bowls with featureless square rims and vessels with spouts. Only one spout with a black slip has been found in the excavations and this sample belongs to BSRW-1 (T.255.2). It is a shape common to RW-1. It is conical with a rounded tip. No finials, lids or handles have been found. The bases appear to be contiguous round or contiguous flat. Bs.82.1 belongs to the sub-class BSRW-2 and is a variant of Bs.82 (described above in the section on RSRW). It is also a nipple base but more bulbous and the bun-like feature in the centre of the well is missing. A large number of body sherds of this extremely large storage vessel were retrieved along with the base at the bottom of the trench (TT4, NE. Depth 3.60 – 3.79m). Unfortunately no rim sherds of the vessel were found. In fact the excavators did mark a pit line into which this vessel was sunk. They noted that the pit extended from layer 5 into layer 6 (Gupta et al 2005: 59). However it must be noted that the pottery analysis shows a clear ceramic stratigraphy with almost no intrusive or mixed material from the earlier layers as would be expected in a pit. The wares and shapes form clear ceramic horizons with no evident disturbance.

Surface treatment and Decoration : The surface of BSRW-1 vessels is treated to a black slip very similar to BSGW. It is well smoothed and evenly applied in a thin coat. Horizontal burnishing streaks are sometimes obvious. The slip usually covers the exterior of the closed form vessels and stops just inside the mouth of the orifice. The slip does not preserve very well and is severely eroded in many cases. The surface of BSRW-2 is treated to a slip which is black to dark chocolate brown in colour, frequently with yellowish red patches. This sub-class shows even greater signs of erosion and wear. The decoration of these wares is similar to BSGW, RSGW and RSRW. Bands of ridges and grooves on the shoulder of the vessel or at the juncture of the neck and shoulder, ridge or groove on the rim, horizontal bands of triangular puncture marks either single or multiple along the shoulder or the carination and incised oblique lines are the common decorations found. A painted white band along the rim and on the neck were seen (T.439 D.202) in one case of BSRW-2.

Fabric : The fabric for both sub-classes is RED(C), the description of which has been given above in the section on RSRW.

Petrographic Analysis : The fabric of sub-class BSRW-1 shows that it belongs to Petrofabric Group B along with BSGW and RSRW. This once again supports the view that the origin of all these wares is common. The description of this fabric group has been given earlier and is therefore not repeated.

Occurrence at Sanjan : This ware has a long history and is present almost through the entire sequence in trench TT4, but with the highest concentration in layer 5. BSRW-1 occurs in layer 6 in small numbers but increases to more than double in layer 5. It continues in layers 4, 3 and 2 but with a decrease in the numbers. BSRW-2 is present in layer 6 but increases to more than three times in quantity in layer 5. Layer 4 has a small number of these sherds. They are present in layers 3 and 2 but in negligible quantities. Surprisingly, almost no sherds of this class appear in layer 1, with the exception of a few small non-diagnostic body sherds. In TT1 they appear in all layers and in TT2 they appear in layers 2 to 5. However these trenches were greatly disturbed and so do not provide the stratigraphic profile for the ceramics.

Discussion : The presence of black slipped utility wares has been noted at most sites. These have either a red fabric or grey but have not always been classified on this basis. At Brahmapuri, the excavators note the predominance of black slipped wares in the Bahamani and the Late or Post-Satavahana periods i.e. from around the 4th century AD to about the 16th century AD. They note that a large percentage of the pottery has a black slip which can range from jet black to grey and that the clay is usually porous with sand temper and "which burns red, pinkish and light red and often black to a light coloured grey"(sic) (Sankalia and Dikshit 1952: 55). While some shapes appear similar to the Sanjan pottery, a majority of the supposedly typical Bahamani and Post-Satavahana types such as the chocolate slipped wares, the slipped pottery with steatite decorative bands, the decorated Khujas, etc are absent at Sanjan. The Plain Black and Burnished Black wares at Shamalaji resemble the Sanjan shapes and decoration but the lack of fabric description makes it difficult to tell whether they can be considered as BSRW. There is also no chronological reference given to particular wares or shapes (Mehta and Patel 1967: 17 – 23, figs. 8 – 11). At Maheshwar the black slipped vessels in Period VI appear to have a grey or black fabric and are not present in Period VII (Muslim – Maratha). Moreover, there is a long gap of nearly a thousand years between these two periods of occupation (Sankalia et al 1958: 16, 151 – 152, 162).

White Slipped Red Ware (Ware code: WSRW)

This is a rare class at Sanjan, mostly seen in a few non-diagnostic body sherds. Only one rim sherd has been found (T.129) from TT1, SW in layer 2. The exterior of the vessel has a dark red wash with a white slip overlying it. The slip extends up to the lip of the rim. The interior of the vessel is a dark greyish-red. The rim has a wide ledge which is square in section. The lip is rounded and everted. The fabric of the vessel is RED(C), the description of which is given above. No excavation reports accessed for this study report any similar wares.

Red Ware (Ware code: RW)

Definition : This ware is one of the most prominent wares at Sanjan, both quantitatively as well as in the range of forms. The vessels in this class of utilitarian wares are semi-coarse to coarse in fabric and have no surface treatment. Both closed as well as open forms are seen. Some shapes are common to other slipped and unslipped classes. Both wheel-made as well as hand-made vessels are present. The firing is not always even. Three sub-classes are seen:

RW-1 is the more common sub-class. It is semi-coarse with RED(C) fabric. It is orange-red to deep red. Frequently there are black or grey clouds on the surface due to firing. Black horizons are also seen in section. There is a considerable amount of organic temper.

RW-2 is extremely coarse and often shows crude potting. Many vessels appear to be handmade in this sub-class. The fabric is RED(F). Large inclusions, airholes and heavy sand temper are seen. The firing is uneven.

RW-3 is dark brick-red colour in colour and has a softer and finer body than the earlier sub-classes. The fabric is RED(C) but it is better levigated and has less inclusions.

Shapes and forms : The vessel forms in this class include large storage jars, troughs, globular jars, carinated and rounded pots, high-necked pots, bowls with and without carination, small pots, dishes, dough plates, spouted vessels, etc. A number of spouts showing quite a lot of variation are found. Lug handles, finials and tea-cup handles are also present. The vessels are sometimes quite crudely made either by hand or on a slow wheel. The unique open-ended vessels (T.323, T.324) found exclusively at SJN-K in 2003 belong to this class. These unique vessels have no parallels in the excavation reports accessed for this study and their function remains unknown. They have an externally projecting, everted rim at the top end and a piecrust rim at the other. The jars in this ware class have externally projecting everted rims which are either rounded, square or tapering (T.167, T.168, T.169, T.170, T.322). The rims can also be inverted or bilaterally projecting (T.171, T.172, T.173). The troughs found in 2003 at SJN-K had bilaterally projecting rims (T. 320 D.127, T.321). The pots have rims very similar to the slipped classes and the Grey Ware class. The rims are mostly externally projecting and everted. They can be square (T.174, T.175, T.176, T.177, T.178, T.179, T.180, T.181, T.182, T.183, T.184, T.185, T.319), quadrangular (T.186, T.187, T.188, T.189, T.190, T.191, T.192, T.192.1), triangular (T.193, T.194, T.195, T.196, T.197, T.198, T.199, T.200, T.200.1), tapering (T.203, T.204, T.205, T.206, T.207, T.310, T.314), rounded (T.208, T.209, T.210, T.211, T.212, T.213, T.214, T.215, T.312, T.313, T.315 D.124, T.317, T.318 D.126) or bilaterally projecting (T.216, T.217, T.218, T.219, T.220, T.221, T.222, T.223, T.224, T.225, T.225.1, T.311). Most rims have grooves either on the lip, on the top or along the orifice. While many of these are shallow, some of the grooves may have served the purpose of holding a lid. A large number of the pots are carinated. Other than the externally projecting rims, inverted rims are also seen in the case of carinated vessels (T.226). Carinated bowls with inverted rims are also present, sometimes with a ledge or ridge on the exterior (T.234, T.235, T.236, T.236.1, T.237, T.240, T.241, T.242). Some straight-sided bowls are seen (T.243, T.243.1, T.243.2, T.244, T.304, T.305, T.306). A bowl with a small spout on the rim (T.307) is also seen in the collection. This is a single piece which comes from SJN-K. A few dough plates, basins and troughs are represented. The lug handles are either semi-circular (T.251) or angular (T.252, T.252.1, T.252.2, T.252.3). Finials are either flat with the appearance of discoid bases (T.253, T.253.1, T.253.2), solid conical (T.328), narrow-waisted with a knob on top (T.329), or bell-shaped with or without perforation (T.254, T.254.1). The spouts show variety. They are straight-sided (T.255, T.255.1, T.255.2, T.255.3, T.256, T.256.1, T.256.2), conical (T.257) or short (T.258, T.258.1, T.258.2, T.258.3, T.259, T.259.1, T.259.2). One short spout found in 2002 in TT1 was shaped like an elephant trunk with three horizontal grooves on the upper side and the lower lip extending downwards. The bases of the vessels are contiguous rounded (Bs.39, Bs.39.1, Bs.39.2), contiguous flat (Bs.40, Bs.40.1), discoid (Bs.41) and hour-glass shaped stand (Bs.42). Pedestal stands are also present (Bs.48, Bs.49).

Surface Treatment and Decoration : These vessels do not have any surface treatment. A few show signs of smoothening. The decoration is seen mainly on the pots, but also on some other vessels such as bowls and trough rims. The most common placement is along the shoulder of the vessel, neck or rim. The range of designs and techniques is not very vast. The execution is frequently crude. The decoration includes incised meanders and wavy lines (D.70, D.70.1, D.70.2), appliqué bands of oblique lines or chain-like thumb impressed design (D.71, D.71.1), appliqué bands with mesh-like impressed design (D.72), notched ledge under the rim (D.73), bands of triangles (D.74, D.74.1, D.74.2, D.74.3, D.75, D.75.1, D.75.2), rows of puncture dots (D.76, D.76.1), oblique lines (D.77, D.77.1) or oblique lines and inverted triangles (D.80). This is the only ware in which a similarity can be seen in decoration with TGW – the appliqué band with the chain-like thumb impressed design closely resembles D.28.

Fabric : Two fabric variations are seen in this class, as mentioned earlier. RED(C) is the more common fabric, also seen in the slipped red ware classes (see above for description). The second fabric is RED(F) which is a lot more coarse and has large inclusions which appear to be quartz particles. The firing is uneven and there is a heavy sand temper. The presence of large air-holes makes the vessels more porous. The surfaces are rough. Both fabric variations have mica in the clay.

Petrographic analysis : Two samples of this ware were subject to chemical analysis. Both belonged to RW-1 RED(C). They were found to belong to Petrofabric Group B along with RSRW-1, BSRW-1 and BSGW. The description of this ware group has been given in the sections above.

Occurrence at Sanjan : This ware occurs from the lowest levels to the uppermost in all trenches and localities at Sanjan. The concentration appears to be higher in layer 2 in TT4 as well as TT1 and TT2. But that may be in proportion to the other wares since the density of pottery in layer 2 is much greater than the other layers. RW-2 is better represented in layers 4 and 5 in TT4.

Discussion : Coarse Red Ware or Plain Red Ware has an extremely long tradition at most sites in India. These wares occur from the Chalcolithic and Early Historical periods to the Medieval, sometimes unchanged in fabric or form. As utilitarian wares which were probably mass-produced, affordable and easily available, these vessels appear to have formed a bulk of the pottery used by the population at Sanjan. The excavations at Shamalaji (Mehta and Patel 1967: 23 – 33, Figs.11 – 16) report Burnished Red and Plain Red Wares together, making it difficult to identify which shapes belong to the Burnished and which to the Plain categories. However the shapes are very closely reminiscent of those at Sanjan. At Nagara (Mehta and Shah 1968: 45 – 69, figs.25 – 38) again, a large number of shapes are very close to those at Sanjan but the lack of chronological details and the clubbing together of the burnished as well as unburnished types of red ware makes exact identification difficult. At Maheshwar, the Coarse Red flowers into a variety of shapes in Period VI (Sankalia et al 1958: 151, 156 – 58). However, Period VI ends around 500 AD. Only a few vessel shapes resemble the Sanjan collection. The Coarse Red in Period VII or the Muslim – Maratha period is said to occur in large quantity but no details are mentioned. Given the long gap between the two occupation periods, the coarse ware of the intervening period remains unknown. At Brahmapuri the Late or "Post" Satavahana period is dated to between the fourth and the ninth centuries, predating the Bahamani period (Sankalia and Dikshit 1952: 57 – 59). But the unslipped wares are not treated as a specific class making it difficult to differentiate between the slipped wares and the unslipped wares.

Miscellaneous Wares

A large number of sherds which do not seem to belong to any of the above mentioned categories have been retrieved in the three years of excavations. They vary in fabric, treatment, shape or decoration and are too few in number to be properly identified as distinct classes. A number of these are represented by non-diagnostic body sherds. No petrographic analysis has been done for these sherds. These wares are categorised as miscellaneous wares and have been recorded and described as under.

Miscellaneous-1 (Ware code: MISC-1)

These sherds have a dark grey core, brittle texture and sharp fracture. They are not weathered at all and look like modern sherds. They are wheel-turned and hard fired. The surfaces are smooth and polished with a red slip. The fabric is GREY(A). The only diagnostic sherd is a rim (T.430). The clay has a considerable amount of mica. Only two sherds of this type have been found, one from SJN-D and the other from TT1 NW.

Miscellaneous-2 (Ware code: MISC-2)

The sherds in this sub-class have a coarse texture. The fabric is GREY(B). it has a large number of inclusions and organic temper. The external surface has a red wash with some white inclusions showing through. The exterior is pitted. The internal surface has a black slip or wash. Both surfaces have particles of mica. Frequently the sherds appear purplish-maroon due to firing. One base sherd was retrieved (Bs.83) from layer 5 in TT4 NE at a depth of 2.16m.

Miscellaneous-3 (Ware code: MISC-3)

These sherds are unusually thin (2 – 2.5mm). Only one sherd had a thickness of 4 – 5mm. They are hard and brittle with a leathery texture. The firing is not even. A large number of quartz and calcritic inclusions can be seen in the section and sometimes also on the surface. The fabric is RED(F). The texture and finish of the pottery bears resemblance to Slipped Brittle Ware. The external surface is slipped and burnished. The colour of the slip can be red, black or chocolate brown. The identifiable shapes in this ware are T.264, T.431 and T.9. The only sherd showing any decoration is D.45. Most of these sherds were retrieved from TT1, with only one sherd found in TT4. They occur mostly in layer 1. The sherd found in TT4 was from the well deposit in the south-west quadrant. A large quantity of similar sherds has been found in the excavations at Chaul (Gogte et al 2006: 71), where it is has been classified as Red Slipped Pink Ware (in this study, Slipped Brittle Ware). Fine mica particles can be seen on the surfaces.

Miscellaneous-4 (Ware code: MISC-4)

This ware is of finer, better levigated clay. It is chalky and compact with a fine sand temper and has a laminated fracture. Mica is present in a considerable quantity. The fabric is RED(G). The sherds are fired from pale red-pink to grey and sometimes orange. A few sherds show traces of a pale self wash. Three identifiable shapes are seen – T.243, Bs.40 and T.432. A majority of the sherds come from TT1 and TT4. In the stratigraphic context of TT4 they occur in layers 2, 3, 4, 5 and one sherd in layer 6.

Miscellaneous-5 (Ware code: MISC-5)

The sherds in this class are different from the other classes both in fabric and in the decoration seen on them. The fabric is very fine and well levigated. It contains a large amount of mica but it is less than in MISC-4. The colour is pale red and salmon pink. The fabric is RED(G), same as MISC-4. The firing is even. A number of body sherds and one rim sherd which extends to the shoulder of the vessel show a moulded mesh design of little squares (T.433 D.200). These sherds come exclusively from the earliest levels of trench TT4 NE. They occur only in layers 5 and 6.

Miscellaneous-6 (Ware code: MISC-6)

The fabric of these sherds is RED(H). It is similar to RED(G) but it lacks mica and has white talc like inclusions. Red grog like particles and red platelets are also seen quite clearly even without a lens. T.434 and T.245 are the only diagnostic sherds found in this class. T.434 is an almost complete lid with a flat base, flaring rim and a deep well with a dome-shaped knob in the centre. As in the case of MISC-5, these sherds also occur in the earliest levels of TT4 NE. But they continue in the upper levels as well. The highest number occur in layers 6 and 4, with one sherd in layer 5 and one in layer 3. Layer 1 had two sherds. In TT1 they occur in very few numbers. The concentration is mainly in layer 2.

Miscellaneous-7 (Ware code: MISC-7)

This ware is slightly better represented than the other miscellaneous classes. The fabric is soft and slightly chalky. It is well levigated with some air-holes. Mica is present in the clay and the impressions of chaff, grass or straw and grains can be clearly seen in the surfaces. The colour is pale yellowish red to grey. The fabric has been designated RED(I). The potting is thick and appears to belong to large sturdy vessels. This ware occurs mainly in layers 2, 4 and 5 in TT4 NE, with one sherd in layer 6. In TT1 and TT2 it occurs only in layers 2 and 3. These sherds erode easily and are quite clumsily potted. The shapes seen in this class are varied. Four base types have been identified – Bs.39.1, Bs.39.2, Bs.51 and Bs.84. D.71 is the only sherd to show any design. This is an appliqué band with a chain-like design. The vessel shapes are T.168, T.169, T.170, T.171, T.172, T.173, T.251 and T.251.1. Of these, T.171 and T.172 may be fragments of the ringwells in TT1. The other forms are mostly large storage vessels. Some of these appear to be handmade. Some sherds appear to have either a wash or a slip which has eroded.

Miscellaneous-8 (Ware code: MISC-8)

This ware is pinkish red to pale purple in colour. The fabric is RED(J). It is somewhat similar to RED(H). One of the most noticeable characteristics of this fabric is the presence of white implosions which mark the surfaces. Fine red particles which may be grog are also seen. Some mica is present in the clay. The texture is not as fine as some of the other miscellaneous classes. Occasionally a red wash can be seen on some body sherds. Very few sherds of this ware have been found. They were found in layer 2.

Miscellaneous-9 (Ware code: MISC-9)

Only body sherds of this ware have been found. They are hard fired and brittle, ranging in colour from purplish pink to greyish red. The fabric is PINK(H). The fracture is laminated and particles of grog and white implosions are seen. Mica is present in the clay. The vessels appear to be both wheel turned as well as handmade. The thickness varies from 5 – 10mm. One sherd has a thickness of 14mm. most sherds have been retrieved from TT4. They are concentrated in layers 2, 3, 4 and one sherd in layer 5.

Miscellaneous-10 (Ware code: MISC-10)

These are very rare sherds found mostly in layer 1 in trench TT1. The fabric is RED(K). It is dark brick red in colour. A large number of black and white inclusions are seen. The fabric is coarse and has a large number of air-holes. It resembles MR and RED(B) in its hardness and texture. No diagnostic sherds have been found.

Miscellaneous-11 (Ware code: MISC-11)

Only two body sherds of this ware were found. They occur in layer 2 of trench TT4. The fabric is extremely coarse. It has been designated RED(L). It is grey-black on the interior and brown-red on the exterior with large black and white inclusions and a large amount of mica and quartz. The sherds have a sand-paper texture. No surface treatment is seen.

Chinese and Far Eastern Wares :

A considerable number of Chinese and Far Eastern Wares are represented in the Sanjan collection. Since the finer differentiation and identification of Chinese wares is a matter of specialization and is unavailable to the author, this study attempts at merely recording and describing the collection on the basis of observed features. In no way does the author claim this to be a definitive classification of Chinese ware types. Four broad categories have been considered based on the observations and general features of a glaze or fabric type and will definitely be subject to change in future when the data and expertise to study this body of work is available in India. Specific nomenclature has been used only in case of confirmed identification, for example Changsha Polychrome Underglaze Painted ware However, this is done with some hesitation. These wares are recorded but have not been used to suggest a chronology or for developing any kind of a sequence in the assemblage. The terms 'Chinese' and 'Far Eastern' are used since it is uncertain if all the wares are indeed of Chinese origin. The Chinese and Far Eastern wares at Sanjan have been broadly divided into four broad groups – Changsha Polychrome Underglaze Painted ware, Porcelain, Celadon and Stoneware for convenience.

The presence of these oriental wares at Sanjan is not surprising considering that the west coast of India was virtually a halfway-house for the east-west trade of the Indian Ocean in the Medieval period. Although quantitatively less prominent in the Sanjan collection, they are of great relevance in understanding the dynamics of the trans-oceanic trade between West Asia and the Far East. Trade between the two regions is known to predate Islam. However, it was in the 8th century and after that Chinese ceramics took precedence over silk, paper and other perishable goods which used to be the staple items of export. Large communities of Persian and Arab merchants are said to have settled in the ports of the South China Seas. Persian and Arab dhows made regular voyages to distant ports of China. Zhang Jun-Yan's study on the relations between China and the Arab lands from Han to Ming period is based on Chinese sources and traces the close links between the two (Zhang Jun-Yan 1983: 91 – 109). According to him, Chinese junks were supposed to have anchored in the Persian Gulf as early as the 5th century AD. Contact between the two regions appears to have intensified from the 7th to the 14th centuries, i.e. during the time of the T'ang (618 – 907 AD), Sung (960 – 1279 AD) and Yuan (1279 – 1386 AD) dynasties. Arabs were called Da-shi, derived from 'Tajik' and 'Tiao-zhi'. Jia Dan (730 – 805 AD), a famous geographer of the T'ang dynasty wrote 'The Route to the Foreign Countries across the Sea from Canton' which, it may be presumed, is the Chinese equivalent of the Periplus. This information is preserved in a work called Xin Tang Shu or 'The New History of the Tang Dynasty'. Canton, Quanzhou and Yangzhou were the main trading ports in South China. In 760 AD, the Da-shi are supposed to have sent thousands of troops to quell a rebellion of Arab and Persian merchants in Yangzhou. An account called Silsila al Tawarikh or 'The chains of History' by the Arab merchant Sulayman in 851 AD and complemented by Abu Zaid Hasan in 916 AD mentions the imposition of duties by the Chinese authorities on the foreign communities residing in Canton. The exact number of Muslims, Jews, Christians and Zoroastrians living there is placed at one hundred and twenty thousand. (Zhang Jun-Yan 1983: 91 – 109). This is a very large figure and if true, indicates an extremely high level of interaction and maritime trade. The presence of West Asian glazed earthenwares such as Turquoise Glazed Ware in South-east Asia and along the coastline of the South China Seas provides evidence for the commercial traffic between the two extremes of the Asian continent. The discovery of Chinese stonewares, porcelain and celadon at sites such as Samarra, Siraf, Suhar, etc further substantiate the demand for Far Eastern ceramics in Islamic lands.

Changsha Polychrome Underglaze Painted Stoneware (Ware code: CHANGSHA)

Definition : This ware is one of the earliest trade ceramic of the Early Medieval period. It takes its name from the Changsha region of Hunan Province where it was produced. It refers to a particular class of stoneware with a creamy buff to greenish grey fabric which is covered with a white slip, painted designs and a glaze. The potting is usually thick. The interiors of the bowls have paintings executed in green or green and brown on the cream or greenish cream background. Only one base sherd of this class which could be identified with certainty has been found at Sanjan.

Shapes and forms : Bs.54 D.132 is the base of a hemispherical bowl with a ring foot. The vessel is of medium thickness (5mm) although the thickness of the vessel floor increases to 10mm. The low foot ring has a diameter of 6cm. It is square in section with facets. A broad angular shallow groove within the ring suggests scraping or trimming. A deep narrow groove at the juncture of the vessel wall and the ring gives the foot sharper definition. The shape resembles the illustrated base no.1613 from Siraf (Tampoe 1989: fig.81). The ring and external surface of the lower part of the body are unglazed.

Surface treatment and Decoration : The surfaces have been treated to a white slip. The internal surface has the decoration painted on the floor of the vessel. The design consists of green and brown foliage painted in feather strokes and part of a curved design which may be floral. According to Kennet, the painting is done with cupric and ferric underglaze pigmentation (Kennet 2004: 46). The spear-like leaves present a simple fern-like motif, similar to that illustrated in the Siraf collection (Whitehouse 1968: Pl.VIb; also, Tampoe 1989: fig.81 no.1606). Some of the pigments appear depressed into the body. This has been noted by Tampoe as well. She explains this etched effect as being caused by "the use of both underglaze and overglaze techniques of painting in the ware, the overglaze pigments sinking through the glaze to give an e3tched appearance to the decoration. It is not always possible to tell the two apart since underglaze pigments sometimes move up towards the surface of the glaze. This technique is known as 'in glaze decoration'" (Tampoe 1989: 56). The translucent glaze which covers the vessel has a pale greenish cream colour. The internal surface is fully covered with glaze. The exterior has partial glaze which covers the rim and the upper part of the body, ending in a swag above the ring foot. The glaze is finely crazed.

Fabric : This fabric has been designated CHANG. It is a greyish buff stoneware with fine black and white particles which can be seen under lens. One small part in section appears to have turned pale pink in firing. The fabric is very similar to fabric P(B).

Occurrence at Sanjan : Only one sherd of this class has been found at Sanjan. It was found in Trench TT4, NW in layer 2 at a depth of 0.66m. This quadrant contained a large brick platform which abutted the well in the SW quadrant. The other wares associated with it are TGW-1 and 4, CPW, HsG(P), BW, BW(TJ), SBW-1 and LISV-3 and 5.

Discussion : Changsha wares have been reported from sites all along the ceramic trade routes going from West Asia to China. The numbers are always small but the presence of these wares is significant considering that they were some of the earliest Chinese wares to be traded along with Yeu and white wares of the kind found at Samarra. While a number of archaeological sites in West Asia and East Africa have recorded this ware, one of the largest and most significant finds has been the discovery of the ship-wreck off the island of Belitung in the waters of the western Java Sea. Guy's paper on this important find discusses the cargo and its relevance (Guy 2003: 13 – 27). Recovery of this vessel and its cargo was undertaken in 1998 and 1999. The first significant discovery was that it was an Arab dhow in Indonesian waters. The cargo was even more startling – 60,000 trade goods which included ceramics, gold and silver vessels, bronze mirrors and silver ingots, almost all of which were in a remarkably good condition. Of these, 56,500 vessels came from the Changsha kilns of Hunan. The sheer bulk of this consignment makes it obvious that the demand for these wares was high in Caliphal lands and that the Hunan kilns were meeting this demand. Large Dusun jars appear to have been containers for the Changsha bowls and other wares. Closed as well as open forms are mentioned in this cargo. "The predominant form among the Changsha wares is the shallow bowl, which appears with a rich variety of painted designs, including flowers, foliage, landscape, clouds, birds, fish, and fungus. All have a white slip ground over a typically grey-white body, with painted underglaze designs covered with a greenish-tinged clear glaze. The painting exhibits a rustic skill and a casual spontaneity, which results from the demands of factory-scale production" (Guy 2003: 23). On the basis of the calligraphic Changsha bowls, the cargo can be dated to the early ninth century, possibly 826 AD or soon after. Whitehouse discusses the Chinese stoneware from Siraf and dates them to the first quarter of the 9th century on the basis of their recovery from the Great Mosque. These include storage vessels of Dusun and the black glazed types, stoneware bowls with patches of glaze removed from the inside and Changsha wares which he calls 'painted stoneware' (Whitehouse 1973: 241 – 255). He mentions other find spots for the Changsha painted bowls – Susa, Sirjan, Minab, Dhahran and Brahminabad. To this list one may add Suhar (Rougeulle 1991: 21 – 25; also, Kervran 2004: 318 – 319, pl.35:7, pl.35:3 and pl.35:12), Ras al-Khaimah (Kennet 2004: 46), Shanga (Horton 1996: 303, fig.227 a – e), Aqaba (Whitcomb 1989: 182), Manda (Chittick 1984: 66, fig.32 c and pl.23 b) and Unguja Ukuu (Juma 2004: 107, pl.6 no.5). Mantai in Sri Lanka is another site to yield this ware (Carswell and Prickett 1984: 57, 64). At most sites the dating of this ware does not go beyond the middle of the 9th century.

Stoneware (Ware code:STONE)

Definition : The definition of stoneware by Medley and used by Kennet and Krahl is employed in this study as well – "Stonewares are clays that fire at temperatures between 1200°C and 1300°C. The clay and temper fuses completely to form an impermeable body. They are normally coated in a feldspathic or alkaline glaze (Medley 1976: 14). Stoneware first superseded earthenware during the Tang period (681 – 906 AD)" (Kennet 2004: 46).

This is a broad category which includes six sub-classes at Sanjan, the descriptions of which are mentioned below. The forms range from large storage jars to bowls and dishes.

STONE-1 : This sub-class has a glaze with two colours or shades. This is either seen in bands along the rim on both surfaces (T.340 D.137), or as a blurred line on the exterior just under the rim (T.340.1), or with fine incised lines on the inside of the rim (T.341 D.138). The glaze is usually creamy brown and has ochre or rust coloured band or line. The shapes are usually finely potted bowls with slightly everted delicate rims (T.4). Only one base sherd has been found (Bs.45) and this suggests a flat base of a jar. It has an olive green and brown glaze which has all but eroded. The glaze on the exterior stops short of the base. The thickness of most sherds is between 2.5mm – 4mm, except in the case of the jar, the thickness of which is 12mm. The fabric is S(A). It is not very fine, with some fine inclusions and a few airholes. The firing is grey to pale pink. Most sherds come from layer 2 in trench TT4 NE and NW, with two sherds occurring in layer 1.

STONE-2 : This sub-class consists of the type of stoneware possibly identifiable with 'Dusun'. These are large, sturdy storage jars which take their name from the Dusun tribe in Borneo who prize this ware and collect it (Whitehouse 1973: 246). They have a green glaze, thinly applied, which can sometimes appear mottled. The glaze on the exterior does not cover the base. The rims are everted or vertical and square or clubbed (T.342). More body sherds of this ware have been found than rims. No bases are seen in the collection. The fabric is extremely compact, well fired to steel-grey and fine, in most cases. It has been designated S(B). Occasionally black specks with white implosions can be seen on the external surface. The upper part of rim T.342 is dark as if unglazed or as if the glaze has worn out. Whitehouse notes this "shiny purple-brown finish" on the rim (Whitehouse 1973: 244). The average thickness of the sherds is about 10mm. Most sherds occur in trench TT4 in the lower part of layer 2 and one sherd in layer 3, i.e. from a depth of 0.75m to 1.09m. It is interesting to note that they do not occur in the upper layers or in the earlier levels even as intrusions. In trenches TT1 and TT2 where the context is disturbed, these sherds occur in layer 3 only.

STONE-3 : These sherds have an olive green to grey-green glaze. Stacking marks are sometimes seen. S(A) and S(B) fabrics both occur in this sub-class. S(C) is similar to S(A) but is much paler and has less inclusions. The shapes seen in this sub-class are T.7, T.8, T.21, T.343, and Bs.60. The base is a low foot-ring with the glaze covering all surfaces. Fine incised designs are seen on some of them (D.1, D.1.1 and D.2). These sherds occur mainly in layers 3 and 4, with one sherd in layer 5 of trench TT4, NE.

STONE-4 : Very few sherds of this sub-class have been found. None of them are diagnostic. They are unglazed and have a dark brownish grey to pinkish grey colour. Some of them are rough and appear crudely made. The fabric is S(A). These sherds have been retrieved only from trench TT4, mainly from layer 1.

STONE-5 : These sherds appear to belong to storage jars with a dark black or brown-black glaze on the exterior and a grey interior. The fabric S(D) is semi-coarse and grainy, fired a dark reddish-grey to grey. White inclusions are noticeable. Only three body sherds have been found in the excavations. All of them occur in trench TT4 in layer 1 within the first 30cm. Similar sherds occur in larger numbers at Chaul. Kennet's description of Martaban jars in the report on the pottery from Ras al-Khaimah appears very similar to this class (Kennet 2004: 50). However this needs confirmation.

STONE-6 : All the sherds in this sub-class are body sherds and no shapes can be inferred from them. They are distinct in that they are extremely coarse and rough. The fabric S(E) is full of large inclusions, air-holes and some shiny particles which could be quartz or glass. The fabric is fired to a dark pink. Some sherds appear to have a grey wash on the exterior. The average thickness of the sherds is about 7 – 8mm. No glaze is seen on these sherds. All the sherds come from TT4 NE, layer 1 and the upper part of layer 2.

Discussion : Stoneware sherds have been found at sites all along the coastline of the Indian Ocean, indicative of the great demand for these Far Eastern Wares. This is particularly true in the case of Dusun jars (STONE-2 in this study) which are commonly found throughout the maritime routes of the Indian Ocean. These vessels were not only items of trade but were also containers for other trade goods of a more perishable or fragile nature. This is borne out by the cargo of the Belitung ship-wreck. Large Dusun jars containing spices such as star-anis and smaller vessels such as Changsha bowls were recovered from the sunken dhow (Guy 2003: 18 - 19). Whitehouse reports Dusun jars as one of the earliest exported ceramic in the Persian Gulf region. Sherds of Dusun along with other stonewares such as Changsha painted bowls, black glazed jars and bowls with glaze removed from the interior have been founds in the platform fill of the Great Mosque at Siraf. On the basis of the stratigraphic evidence, Whitehouse suggests that Dusun jars were already in use and were being imported into the Persian Gulf by c.800 – 825 AD (Whitehouse 1973: 244 – 46). His description of Dusun closely matches the Sanjan samples. The other sites to report Dusun sherds are Manda (Chittick 1984: 66 pl.23a), Shanga (Horton 1996: 303, fig.227 f-h), Kilwa (Chittick 1974: 310), Unguja Ukuu (Juma 2004: 107, Pl.6 no.7), Mantai (Carswell and Prickett 1984: 57, 64), Ras al-Khaimah (Kennet 2004: 49 – 50) and Bhambore (anon. 1964: 51). It is difficult to identify the other sub-classes with certainty or to draw parallels for similar finds from other sites.

Celadon (Ware code:CEL)

Definition : The term 'Celadon' is said to have been taken from a character in a seventeenth century French play who always appeared on stage in green or grey-green attire. The term has come into popular use to refer to certain Chinese wares, presumably because of the colour of the glaze. Gompertz's description appears to be a good working definition – "Celadon may be held strictly to denote high-fired porcellanous wares with a feldspatic glaze of characteristic bluish or greyish green tone derived from iron" (Gompertz 1958: xv). Yeu wares or early celadon wares were imported into West Asia as early as the T'ang period. Celadon was prized and sought after for the mystical qualities attributed to it. It was believed that celadon vessels would reveal the presence of poison in the food. The green tone of the glaze and the hardness of the fabric aspired to a jade-like quality. While many scholars define the body of Celadon as stoneware, others consider it porcellanous. It could in actuality be described as a porcellanous-stoneware. A variety of forms both, closed as well as open, are seen. At Sanjan, most sherds belong to bowls and dishes. It is possible that some of the stoneware sub-classes with creamy brown or brownish-green glaze and finely incised designs may be pre-celadon wares. At the time of writing, the finer distinctions are unknown to the author. Hence only description of the ware classes has been attempted. Two sub-classes have been identified within the broader category of Celadon. CEL-1 is quantitatively much better represented than CEL-2 which is represented by only one sherd. The descriptions and defining traits of both sub-classes are given below.

CEL-1 : This sub-class consists of bowls and dishes having an olive green to grey-green glaze which covers the internal as well as the external surfaces. The glaze is evenly applied in a thin coating. It has fine crazing lines which can be seen under lens. The body is fine, pale grey and hard-fired – C(A). Occasionally a finer fabric is seen, which is white-grey in colour and almost appears to be porcelain – C(B). Most of the bowls and dishes have stacking marks or spur marks on the ring foot and on the internal surface of the vessel. The marks on the vessel interior are usually small, squarish or rounded and white against the green background. The rims are fine tipped and clearly defined. Foliate rims are also present (T.3). The ring bases have either a high graceful foot or a low ring. The rim forms are T.3, T.4, T.5, T.339, and T.438. T.3 also has a vertical 'scar' on the exterior, enhancing the lotus form. T.438 has a brown line along the lip which could be indicative of a metal rim which was fitted on. The base types are Bs.3, Bs.4, Bs.5, Bs.57, Bs.58, Bs.59 and Bs.60. Bs.57 has a shallow gadroon form on the exterior and a circular groove on the interior from which vertical grooves rise up the walls of the vessel. No stacking marks are seen. This sample is unique and has a paler green glaze than the others in this sub-class. This sub-class may be identified with Yeu ware (see discussion below). These sherds have been found mainly from layers 1 and 2 in trenches TT1 and TT2.

CEL-2 : Only one sherd of this subclass has been found at Sanjan. It is extremely fine in fabric as well as glaze. The fabric, C(B) is white-grey in colour and almost resembles porcelain. The glaze is a pale translucent apple green in colour. The application is slightly thicker than in the Yeu vessels. This sherd, Bs.2, could be the rim of a pedestal stand. Celadon of this refined and superior quality can be seen in the collections from Champaner and Chaul. At Sanjan, this single sherd occurs in trench TT2, SW in layer 2.

Discussion : CEL-1 can be tentatively identified as Yeu ware. The term applies to "most, if not all, the early celadon wares manufactured in Chekiang province, in the general area of Hangchou, as well as the later pi-se yao of Chinese literary fame, which became the special ware reserved for use by the Ruling House of the State of Wu-Yueh" (Gompertz 1958: 4). The ware was one of the earliest Chinese imports in West Asia and at other sites of the Indian Ocean trade network. It occurs at Samarra, Siraf, Suhar, Ras al-Khaimah, Fustat, Brahminabad and Mantai. Some of the Siraf sherds match the Sanjan shapes both in description and illustration. Tampoe's description of the Provincial Yeu rims and bases are especially relevant (Tampoe 1989: 51 – 54, figs. 76, 77, 78). The high foot of the foliate rim bowl (T.3) is one of the few such bases, a majority of them being low and broad. At Shanga they occur in a context which suggests a mid 9th – late 12th century date (Horton 1996: 307). At Suhar, they are identified as belonging to the Zhejiang province (Kerr 2002: 128; also, Kervran 2004: 319 – 323). The base sherd of a bowl from Mantai is almost an exact copy of T.5 from Sanjan (Carswell and Prickett 1984: pl.13B). This ware is reported from Ungujaa Ukuu as well (Juma 2004: 108, fig.6.16 no.2, pl.6 no.8). Excavation of the Intan shipwreck in the western Java Sea brought to light a varied cargo which indicated the complex nature of inter-regional trade along the maritime routes from West Asia to South-east Asia and China. Among the varied items which made up the Intan cargo were Chinese ceramics attributable to the 10th – 11th centuries. These include Yeu and Yeu-type green wares. The kilns of Guangdong which produced these wares during the Song period are thought to be the origin of these vessels (Guy 2003: 171 – 191). At Manda these sherds are found in the earliest deposits, i.e. Period I, mid 9th to early 11th century (Chittick 1984: 66). CEL-2 appears to be a more developed and finer type, though exact identification remains to be done.

Porcelain (Ware code:PORC)

Definition : The definition of Porcelain given by Medley and used by Kennet and Krahl is appropriate for this study. "Porcelain is made of a mixture of kaolin and white China stone, a refined non-feldspathic material derived from granite. When mixed and fired the materials form a vitrified body at temperatures between 1,280°C and 1,400°C. The colour of the body is pure white or very pale grey. They can be glazed with alkaline glazes or refired with lead glazes at a lower temperature (Medley 1976: 14). Porcelain was born out of a gradual improvement in the white earthenware and stoneware bodies in the late 8th or early 9th century (Medley 1976: 97)." (Kennet 2004: 46). To this definition may be added the characteristics of Porcelain as mentioned by Cox – "1. It is not porous even when unglazed. 2. The body cannot be scratched with a steel knife. 3. It is naturally of a whitish colour caused by the kaolin but the body may be stained many colours by mineral oxides. 4. It is generally more translucent than pottery or stoneware and less so than "soft paste porcelain", though translucency is not an absolute test for some of the denser porcelains are not visibly translucent even in fairly thin flakes. 5. It is fired in saggers at first for 24 hours at low temperature and then raised to about 1300°C or on some cases to about 1450°C for another 20 to 30 hours." (Cox 1945: xv).

A number of Porcelain sherds have been found in the excavations at Sanjan. They are varied in their quality, fabric and glaze. It is not possible at this time to identify the exact types. Hence they have been sorted and described as per their attributes to facilitate future identification. It is possible that different types of Porcelain may have been put into the same sub-class due to shared attributes.

PORC-1: This sub-class is quantitatively the best represented at Sanjan. It refers to a fine snowy white Porcelain body, P(A) with a white glaze overlying it. Sometimes the glaze can have a very pale bluish or greenish tinge though usually it is white. The shapes are almost without exception open forms – cups, bowls and dishes. They occur in trench TT4 in layers 1 and 2. In trenches TT1 and TT2 they are found in layers 1, 2 and 3, with higher concentration in layer 2. They have also been retrieved from the excavations at SJN-K and SJN-D. In the context of the Dokhma, one base sherd was found inside the central well in the deposit with the human bones. This base, Bs.55, has a square ring foot. The interior of the vessel has a broad, circular unglazed band with some impurities and grit adhering to it. Striation marks of a fast wheel can be clearly seen in this unglazed part of the vessel. The white glaze has a faint blue tinge. The shapes seen in this sub-class are T.1, T.2, T.6, T.3 D.97, T.265, T.333, T.333.1, T.333.2, T.334, T.335, T.336 and Bs.1, Bs.4, Bs.40.1, Bs.55, Bs.56 and Bs.56.1. Rolled or folded rims are common in this sub-class. T.3 D.97 is moulded on the exterior to form lotus petals. It is a small bowl with a white glaze having a very pale greenish tinge. T.336 is a fine small foliate rim of a small shallow bowl curving to a flat base. the glaze is white with a pale blue tinge. The vessel is finely potted and appears almost translucent. Most vessels have a thickness of 2 – 4.5mm, though they seem to be thicker at the base. The glaze rarely covers the base on the exterior. Bs.56 and Bs.56.1 are low foot ring bases with a shallow circular groove on the internal surface of the vessel. Some of these wares may belong to the T'ang period since they appear to have the characteristics of these wares – a porcelain body which is greyish in the earlier and white in the later samples, a white glaze which occasionally has a bluish tinge and which is thinly applied, unglazed foot and base, simple shapes with some bowls having foliate rims, and no attempt at decoration.

PORC-2: This subclass is represented by only four sherds but they are relevant for their morphological features. These sherds have a cream to ivory coloured body which is slightly granular compared to P(A). It has been designated P(B). This fabric is a little like the Changsha fabric, and strictly speaking, could well be described as porcellanous stoneware. The glaze overlying the body is also a creamy ivory colour. The bowls appear to be more thickly potted than PORC-1. The more relevant feature of this ware is the shape. The bowls have everted rounded rims with vertical ribs on the internal walls of the vessels, exactly like the SWGW sherds which are of West Asian origin. T.41 (SWGW) is exactly the same as T.331 (PORC-2), except that it is smaller. T.332 is a similar rim with a more rounded lip. These sherds occur in TT4, NE and NW in layers 1, 3 and 4. One sherd occurs in trench TT2, SE in layer 3 but this context is suspect. The sherd that occurs in layer 1 of TT4 NW is a small base fragment about which no diagnosis can be made. However it is evident that the ring foot was not glazed. Some glaze has dripped from the body on to the ring. Tampoe's description of Cream Stoneware in the Siraf collection fits the Sanjan samples perfectly (Tampoe 1989: 58, fig.84 no.400, fig.85 no.1702).

PORC-3 : Only one rim sherd makes up this sub-class. It must be noted that two other sherds of the same type were found in the excavation of the embankment tank (TT3) in 2003 but were discarded as modern material. The presence of this sherd in the undisturbed stratigraphic context of trench TT4, NE layer 3 at a depth of 1.43m makes it necessary to record it. This unique sherd has an extremely fine, milk-white body with no inclusions or air-holes. The glaze overlying this white body is a dark emerald green. The glaze is evenly applied and extremely glossy. No parallel sample has been found in any referred literature or accessed collection. The rim sherd T.337 belongs to a shallow saucer-like bowl. It has a diameter of 14cm and thickness of only 2mm.

PORC-4 : This sub-class consists of porcelain sherds with painted decoration. These include Blue and White Porcelain as well as sherds with decoration in dark green. They may belong to different traditions, regions or periods. The superior quality Ming Blue and White Porcelain found at Chaul, Khambat, Champaner or other sites is not seen at Sanjan. The few sherds with decoration come mainly from the excavations of 2003 at the Koli Khadi. Only four sherds occur in TT4 and these are mostly from the uppermost part of the well deposit. The designs are executed in dark blue, light blue and dark green against a white, grey, greyish-blue or greenish-white background. The designs include floral motifs, foliage, calligraphic or pseudo-calligraphic motif, etc. The painted designs are rarely sharp. They give a blurred impression. The fabric is P(A) but often shows inclusions and a slight coarseness which is lacking in the other porcelain classes. All the sherds in this sub-class are fragmentary and very small. D.87 which was found at SJN-K in 2003 is a small fragment with purple and dark green design on it. This sherd has a European provenance (Kennet, personal communication).

Discussion : Porcelain has been reported from almost all sites in the Indian Ocean littoral attesting to the great demand in the international market for this Chinese commodity. That a large part of the maritime trade was driven by the demand for speciality ceramics is clear from the far-flung regions which report them. From the East African coast to Japan, the demand for Chinese Porcelain appears to have been high during the T'ang period (618 – 960 AD) and reaching a peak during the Song and Yuan dynasties (960 – 1368 AD). Kerr mentions the maritime 'Porcelain Route' that transported the bulk of the ceramics westwards, by reason of their weight and fragility. This route stretched from the Southern Chinese coast through the Malacca straits or the Isthmus of Kra to Sri Lanka and the west coast of India to the Persian Gulf, the Red Sea and the coast of East Africa (Kerr 2002: 125-26). It is not surprising to find that Chinese Porcelain and stonewares have been reported from sites all along this route. Carswell explored the west coast of India, the Maldive islands and Sri Lanka and reported various sites with Chinese wares, particularly porcelain (Carswell 1977: 121 – 197; and Carswell 1979: 25 - 67). The large hoard of Chinese wares that were excavated on the Allaipiddy beach in Sri Lanka as a result of these explorations is an indication of the massive cargoes that must have regularly made their way to western markets (Carswell 1979: 25 – 67). At Siraf, a number of stoneware and porcelain types occur in the collection (Tampoe 1989: 47 – 62). The South Chinese White Stoneware of the Song period or GWW reported from Ras al-Khaimah by Kennet appears similar to some of the vessels classified as PORC-1 in this study. The rolled or folded rims are particularly noted. Kennet mentions that they first occur in the late T'ang period in Northern China and suggests a 10th / 11th century date (Kennet 2004: 47 - 48). Horton reports the presence of Qingbai, white porcelain, Blue-on-white porcelain, moulded whiteware or 'Marco Polo ware' and Ding ware at Shanga (Horton 1996: 309 – 10). Chittick suggests a date between 850 and 950 AD for the White porcelain at Manda (Chittick 1984: 65 – 66). At Kilwa only one sherd occurs in Period I, while most occur in the deposits of Period II. These are Ch'ing Pai or Qingbai type. In Period IIIa i.e. in the 14th century context, the presence of Blue-and –White ware and other porcelain types and stonewares is noticeable. This increases in the following periods (Chittick 1974: 310 – 312). Kervran attributes the White and Qingbai ware at Suhar to a period between the 9th and 12th centuries AD but suggests that some could have been traded as early as the end of the T'ang period (618 – 907 AD) (Kervran 2004: 323). Subbarayalu's report on Chinese ceramics in Tamil Nadu and Kerala mentions the site of Periyapattinam where Chinese wares belonging to the 14th century were found associated with a few West Asian glazed wares (Subbarayalu 1996: 109 – 114).

(Images at :_
http://www.wzcf.org/sanjan-excavations/pottery.html.)

* * *

Plate 1 - TGW Three-handled Jar

Plate 2 - TGW - 4 Barbotine Decoration

Plate 3 - TGW - Applique Decoration

Plate 4 - WGW Bowl

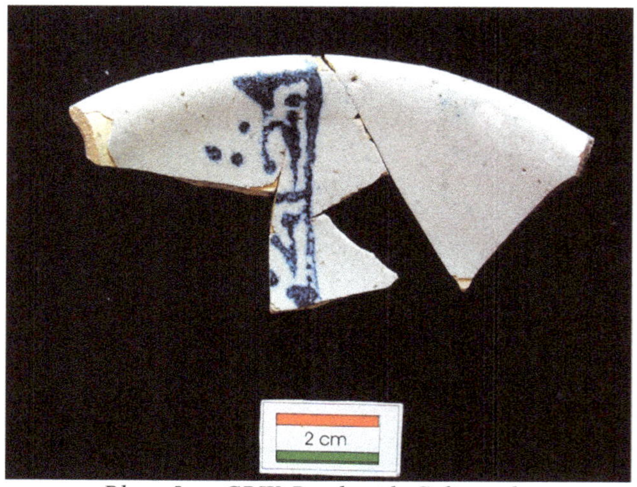

Plate 5 - CPW Bowl with Caligraphy

Plate 6 - CPW - Floral Painted Design

Plate 7 - SWGW - With Vertical Splashes & Indented base.

Plate 10 - HsGW - Bases of Bowls

Plate 8 - LPW-1 - 'Running Hare' Motif

Plate 11 - HsGW - Rims of Bowls

Plate 9 - LPW - Exteriors of Bowls

Plate 12 - CSGW - Conosegmental Bowl

Plate 13 - ESW(W) - Pseudocalligraphic Design

Plate 14 - MGPW - Bases of Dishes

Plate 15 - Porcelain Bowl Rims

Plate 16 - CHANGSHA - Base of Bowl

Plate 17 - STONE - Jars with Faux Handles & Bowls

Plate 18 - CEL 1 - Bowl Bases with Spur Marks

GLAZED EARTHENWARE

TURQUOISE GLAZED WARE	
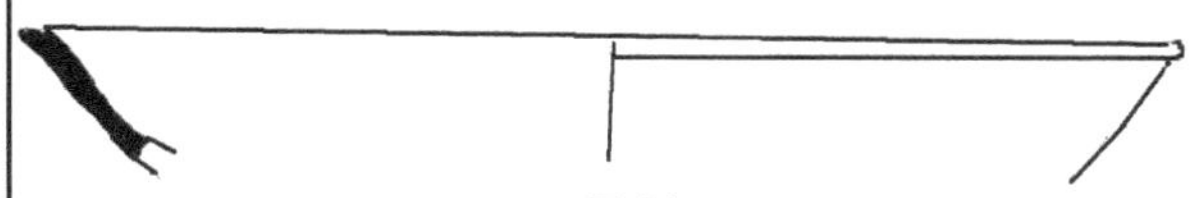 T.61	**T.61**:- Rim sherd of a small bowl with external projecting everted rounded rim. Creamy yellow fabric with blue-green glossy glaze. (I.D = 12cm). SJN-B, TT1, SE, 1, 0.3, 701, TGW-3, BUFF(A)
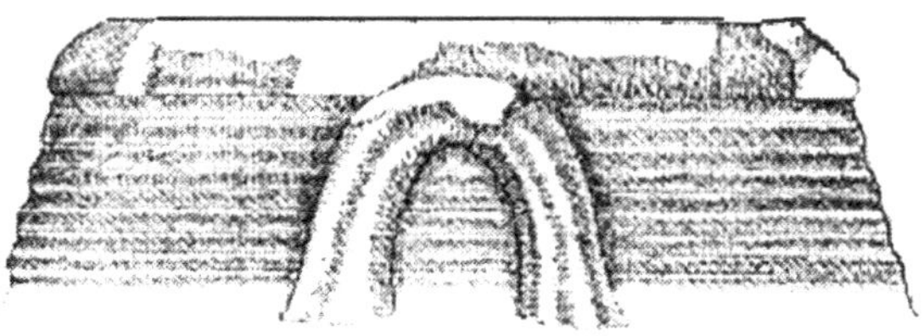 T.65.D.26	**T.65.D.26**:- Rim sherd of a storage jar with internal projecting beaked rim with a groove on the exterior. Corrugation marks are evident on the exterior surface of the neck. A large loop like a handle has been appliquéd to the neck and extends almost up to the rim. The fabric is yellow with a deep turquoise blue-green glossy glaze on both surfaces. (I.D = 14cm). SJN-B, TT1, NW, 2, 0.77, 308, TGW-4, BUFF(A)
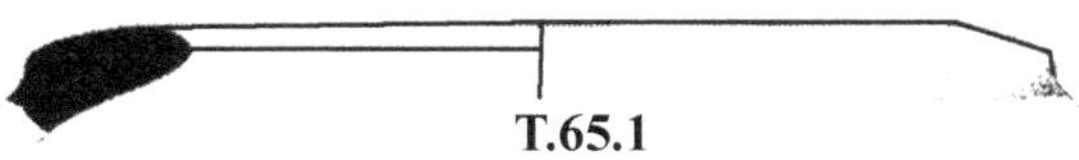 T.65.1	**T.65.1**:- Rim sherd of a jar having an inverted tapering rim and an external projection (I.D = 13cm)
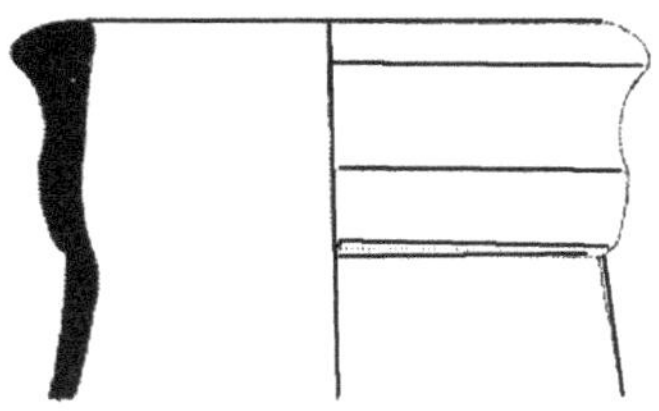 T.67	**T.67**:- Rim sherd of a high-necked jar with external projecting triangular rim. The neck is ridged. The jar appears to be having a narrow mouth. Dark turquoise green glaze on both surfaces. (I.D = 8cm)SJN-B, TT1, NW, 3, 1.07, 318, TGW-4, BUFF(A)
 T.76	**T.76**:- Fragmented Rim sherd and partial body of a high-necked jar with bilateral projecting rounded rim with external tilt. A large handle elliptical in section extends from just below the rim to the shoulder. This vessel probably had three such handles. The core is a pale yellow and the external surface has a deep blue-green glaze which extends to the inner side of the neck. The glaze has a translucent patina like gloss. A wavy horizontal line in a darker shade of the glaze is evident just under the rim. (I.D = 25 - 30cm)SJN-B, TT1, SW, 7, 3.95, 160, TGW-4, BUFF(A)
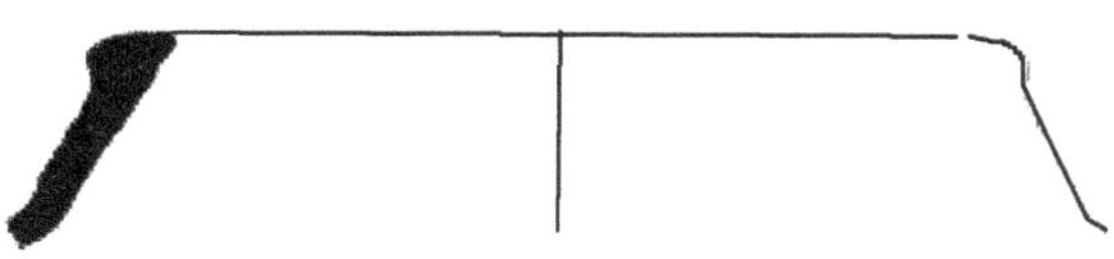 T.283	**T.283**:- Rim sherd of a high necked storage jar with corrugation on the neck. Turquoise glaze on internal and external surfaces. Corrugation on neck (I.D =14cm).SJN-B-TK, TGW-1, BUFF(A)

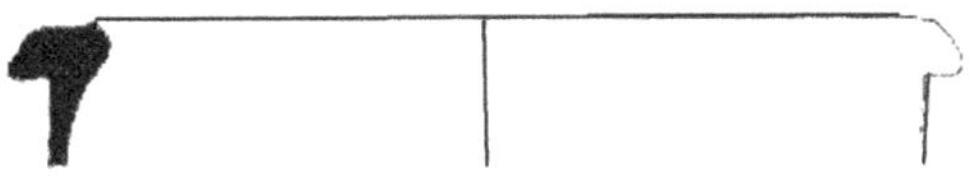 **T.284**	**T.284**:- Rim sherd of a vessel with internal projecting externally beaked rim. Glaze on both surfaces. (I.D =12cm).SJN-B-TK, TGW-1, BUFF(A)
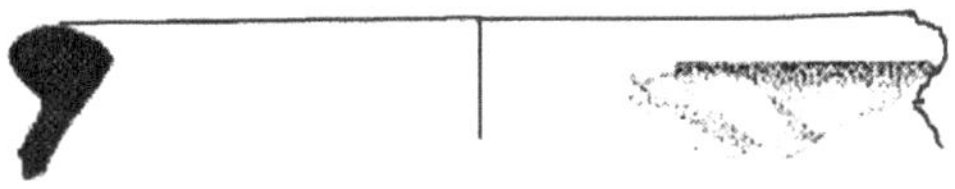 **T.373.1**	**T.373.1**:- Rim sherd of a high-necked jar, with external projecting short-elliptical rim. No grooves or bands are seen. (I.D = 16cm)SJN-B, TT4, SE, 2, 1.16, 51723, TGW-1, BUFF(A),
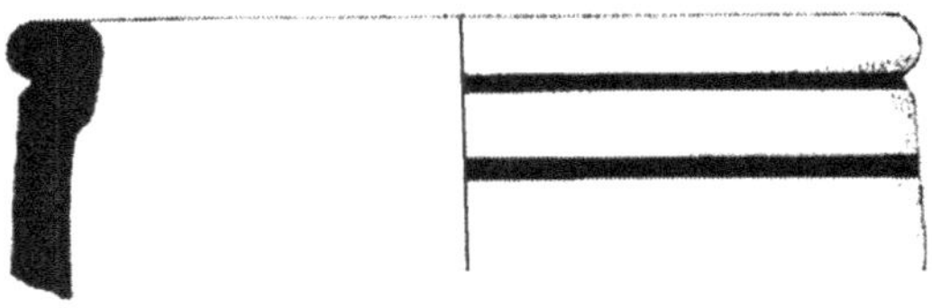 **T.373 D.168**	**T.373 D.168**:- Rim sherd of a high-necked jar, with external projecting short-rounded rim, with a deep groove just under the rim and a similar groove lower down on the neck. The glaze has pooled in the grooves and gives the appearance of two dark-green bands. (I.D = 10cm) SJN-B, SURFACE, TGW-1, BUFF(A)
 **T.374**	**T.374**:- Rim sherd of a jar, having internal projecting, externally overturned rim. A broken spur mark is seen on the flat upper surface. (I.D = 9cm)SJN-B, TT4, NE, 2, 1.03, 51528, TGW-1, BUFF(A)
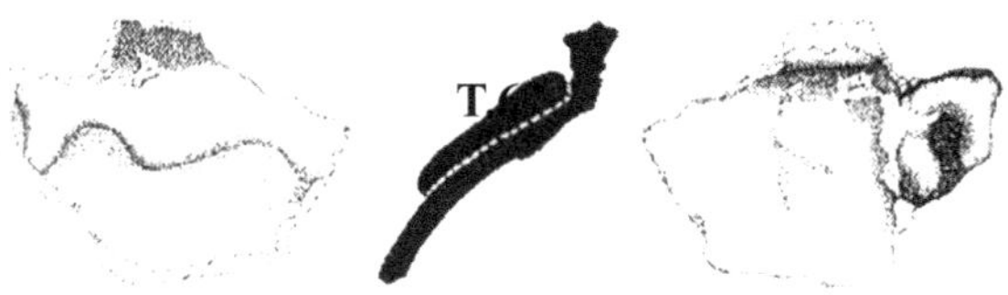 **T.375 D.169**	**T.375 D.169**:- Rim to shoulder of a globular jar. The rim is broken but appears to be externally or bilaterally projecting. There is a raised bump on top. A deep groove is present just under the lip. The glaze has pooled and appears like a dark band. There is evidence of a moulded or impressed design just under on the neck. The shoulder has a rounded curve on which a small faux handle has been appliquéd. The glaze is dark green. SJN-B, TT4, NE, 2, 0.88, 51523, TGW-1, BUFF(A)
 **T.376**	**T.376**:- Rim to base of a straight-sided bowl having slightly in-turned square rim and a non-contiguous discoid base with a convex molding. A very slight carination, just above the base, gives the vessel a slight curvature before going straight up to the rim. Both surfaces are glazed. (I.D = 18cm)SJN-B, TT4, NW, 2, 0.78, 51321, TGW-2, BUFF(B)
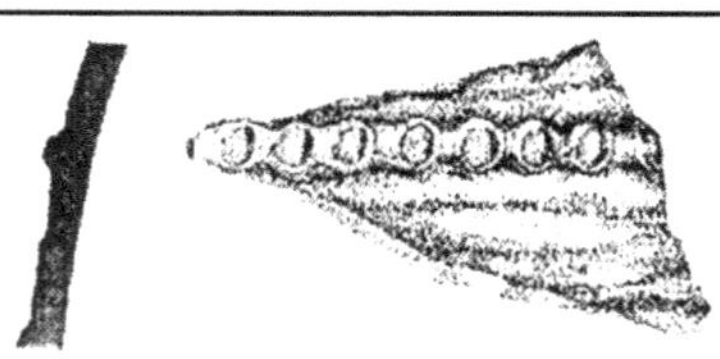 **D.28**	**D.28**:- Horizontal bands with rounded chain-like design appliquéd on shoulder of large storage jar. (Th = 15mm) SJN-B, TT1, SW, 1, 0.57, 103, TGW-1, BUFF (A)

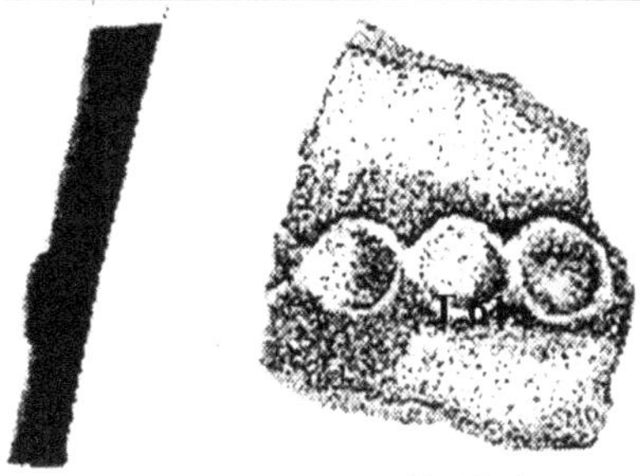 **D.28.1**	**D.28.1**:- Horizontal bands with square chain-like design appliquéd on shoulder of vessel. (Th = 12-15mm)SJN-B, TT1, NW, 1, 0.74, 307, TGW-1, BUFF (A)
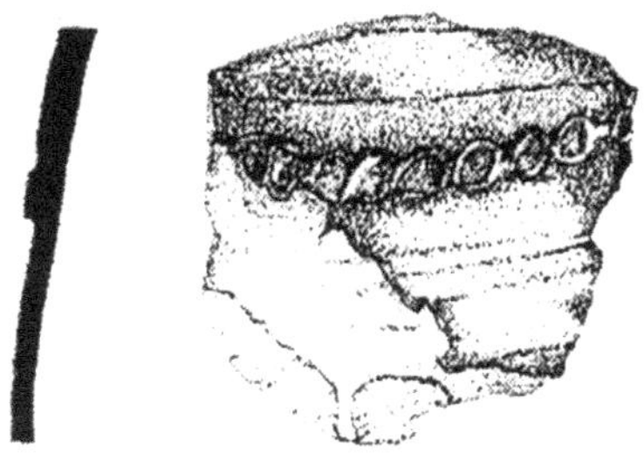 **D.28.2**	**D.28.2**:- Horizontal bands with square chain-like design appliquéd on shoulder of vessel. th = (Th = 13mm) SJN-B, TT2, SW, 2, 1.05, 910, TGW-1, BUFF
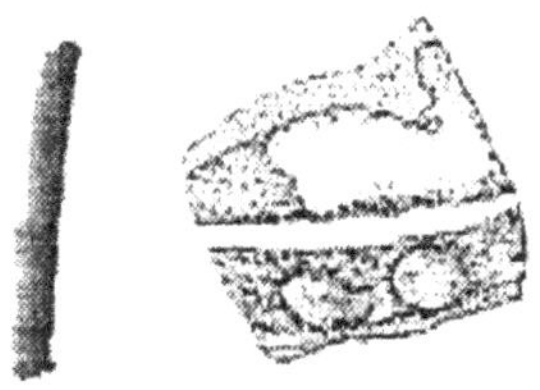 **D. 29**	**D.29**:- Vertical bands with dots. th = 7mmSJN-B, TT2, SE, 2, 1.05, 1514, TGW-1, BUFF(A)
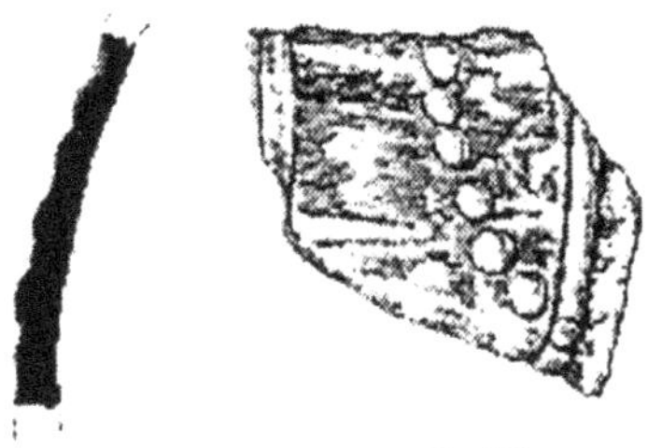 **D. 30**	**D.30**:- Barbotine sherd, Curved bands and rows of dots. (Th = 10mm)SJN-B, SURFACE, TGW-4 BUFF(A)
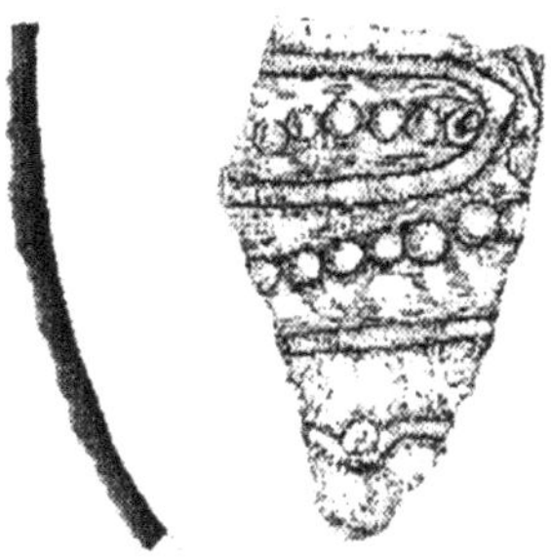 **D.30.1**	**D.30.1**:- Curved bands and rows of dots. (Th = 10mm) SJN-B, TT1, NE, 2, 0.71, 504, TGW-3, BUFF(A)

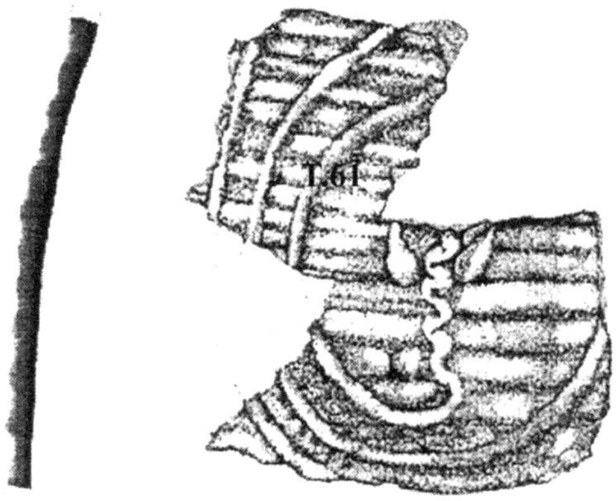 **D.31**	**D.31**:- Coiling bands continuing into a long vertical spiral with leaf- like design on either side. (Th = 8 – 10mm) SJN-B, SURFACE, TGW-4, BUFF (A)
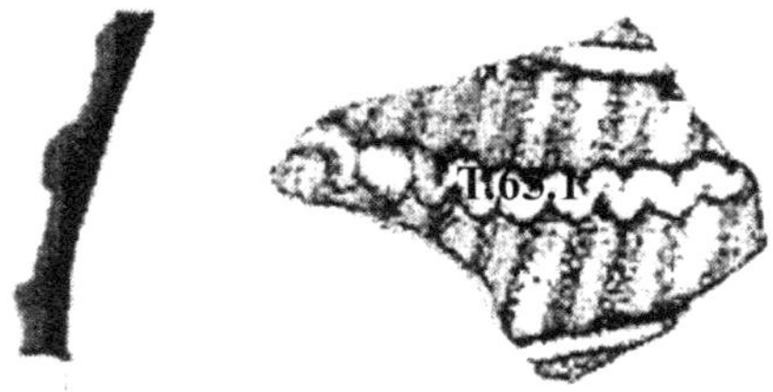 **D.31.1**	**D.31.1**:- Coiling bands with spiral and dots. SJN-B, SURFACE, TGW-1, BUFF(A)
 **D.31.2**	D.31.2:- spiral band between two straight-bands. SJN-B, TT1, NE, 3, 0.86, 508, TGW-1, BUFF(A)
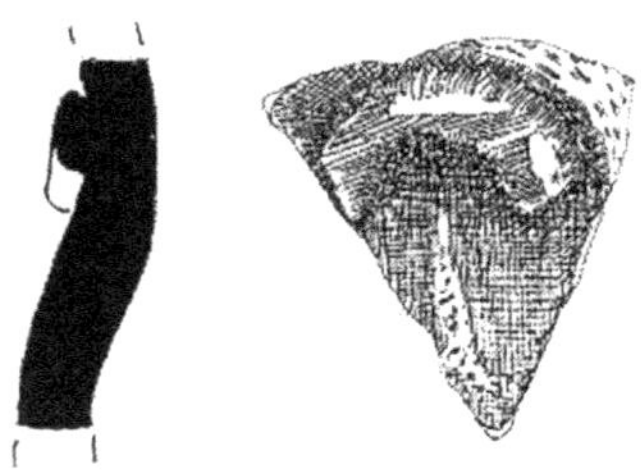 **D.32**	**D.32**:- Faux handle. Loop-like.SJN-B, TT1, SW, 1, 0.35, 101, TGW-1, BUFF (A)
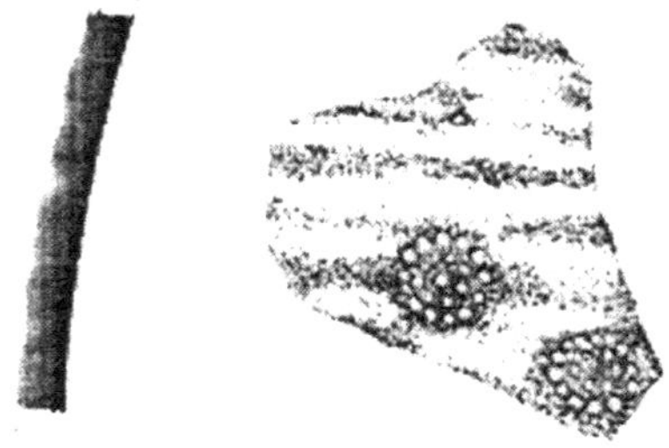 **D.33**	**D.33**:- Stamped – dots arranged to form a floral design. Th = 10mm)SJN-B, TT1, SW, 2, 0.83, 110, TGW-4, BUFF (A)

 **D.34**	**D.34**:- Combination of stamped and appliqué design. (Th = 7mm)SJN-B, TT1, NW, 3, 1.38, 324, TGW-1, BUFF (A)
 **D.35**	**D.35**:- meander incised just below rim. (I.D = 4cm) SJN-B, TT1, NW, 2, 0.82, 310, TGW-1, BUFF(A)
 **D.35.2**	**D.35.2**:- meander incised just below rim.
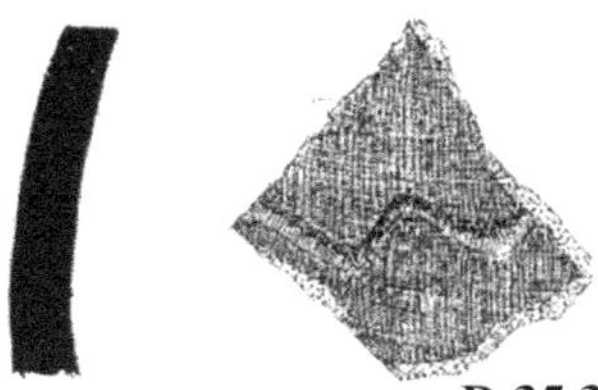 **D.35.3**	**D.35.3**:- meander incised on the shoulder of the vessel. (Th = 15mm)SJN-B, SURFACE,TGW-1, BUFF(A)
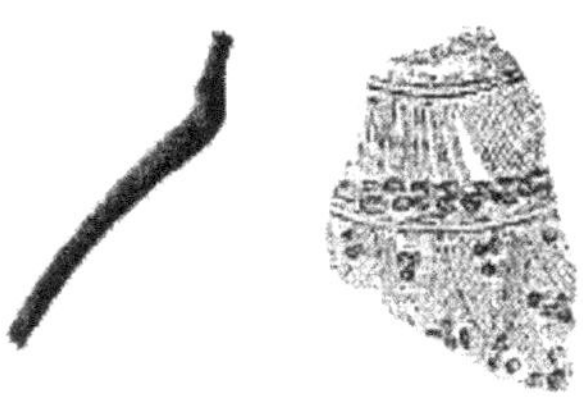 **D.36**	**D.36**:- Diamond design formed by bands of fine lines with dots in the centreSJN-B, TT2, SE, 4, 1.67, 1525, TGW-4, BUFF(A) (also in BW)
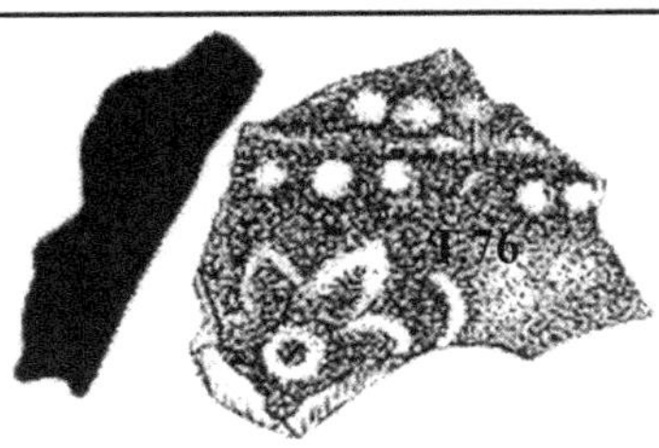 **D.110**	**D.110**:- Body sherd with stamped flower having three extant pointed petals and a circle in the centre. Two horizontal rows of dots can be seen above the flower.SJN-B-TK, TGW-1, BUFF(A)

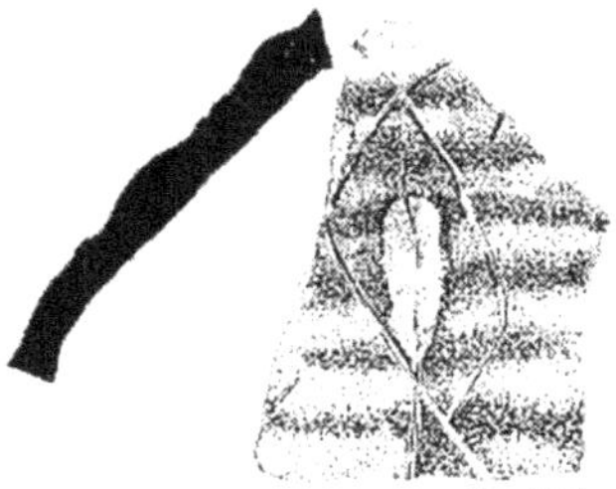 D.111	**D.111**:- Body sherd having vertical intersecting incised lines forming a chain of loops. In the centre of the loop can be seen an appliqué vertical line. An incised line dissects the appliqué and part of the loop.SJN-B-TK, TGW-1, BUFF(A)
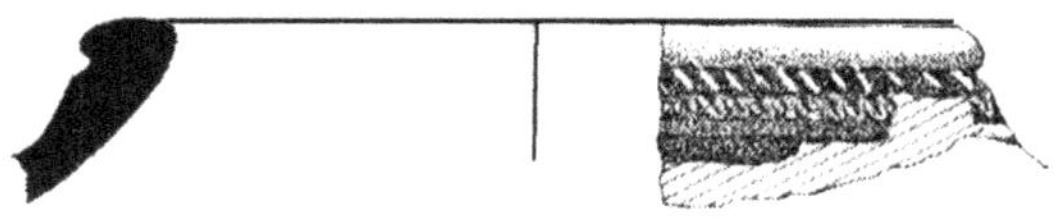 D.167	**D.167**:- Rim sherd of a large storage jar, with inverted external projecting rim. On the exterior, just under the rim is a design consisting of two bands of oblique lines. The lower band has S-shaped oblique incised lines.SJN-B, TT4, NE, 2, 1.19, 51534, TGW-1, BUFF(A)
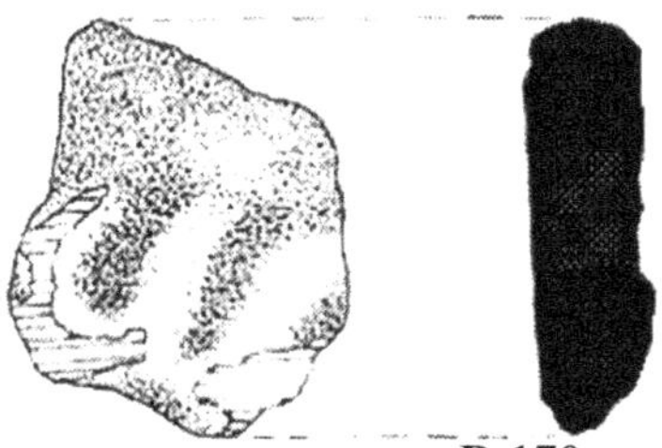 D.170	**D.170**:- Body sherd of a thick storage jar, with an incised design, which appears to be calligraphic. SJN-B, SURFACE, TGW-1, BUFF(A)
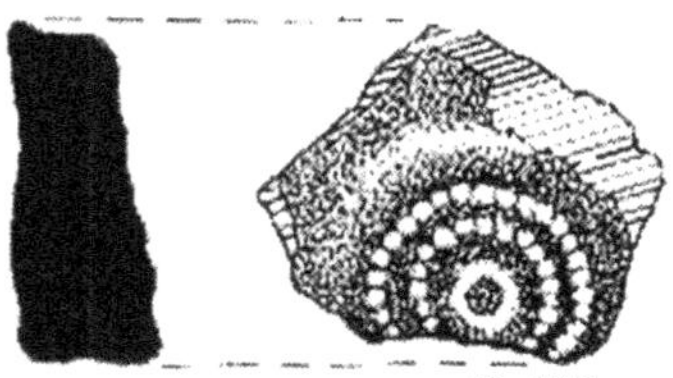 D.171	**D.171**:- Body sherd of a thickly potted jar having a boss consisting of concentric circles of dots; it has a diameter of 8mm. The glaze on the exterior is turquoise blue-green and on the interior it is grey and turquoise. SJN-B, TT4, NE, 2, 0.39, 51508, TGW-4, BUFF(A)
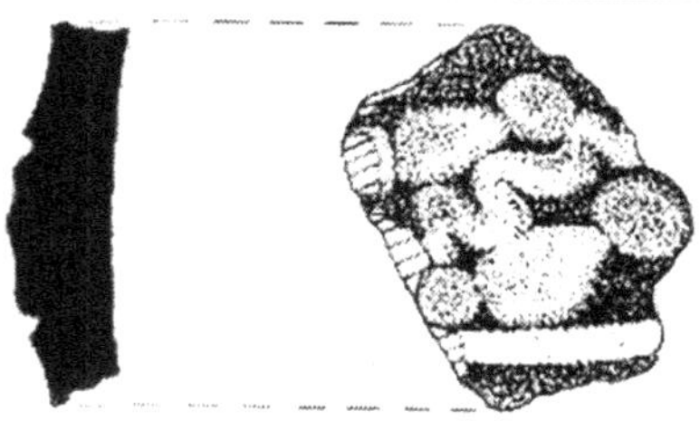 D.172	**D.172**:- Body sherd of medium thick vessel having a complex decoration of dots, bands and wavy lines; the appliqué design appears to be elaborate. The glaze on the exterior is very dark green and on the interior it is speckled dark green and black. The interior glaze is almost 1mm thick. (Th = 7-8mm)SJN-B, TT4, NW, 2, 0.51, 51311, TGW-1, BUFF(A)
 D.173	**D.173**:- Part of a moulded appliqué band or possibly part of a vessel. The band consists of a sharp edge or rim upon which two horizontal bands of raised dots can be seen. In between these bands is a ridge. SJN-B, TT4, NW, 3, 1.04, 51331, TGW-1, BUFF(A)

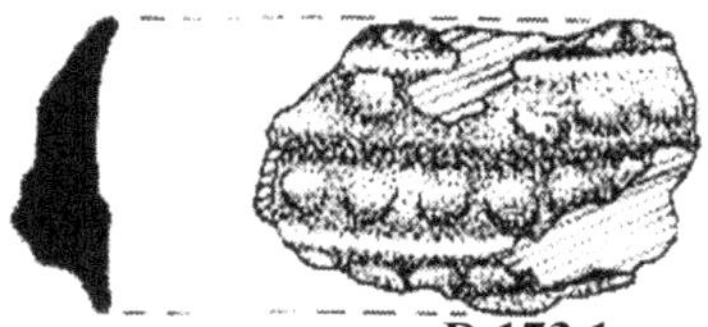 D.173.1	**D.173.1**:- Very similar to D.173. At least two of these bedded 4 ridge borders are adjoined and curved implying that they may have circled some other central motif. These are definitely appliquéd sections that have come off the body. SJN-B, TT4, NE, 5, 1.86, 51547, TGW-1, BUFF(A)
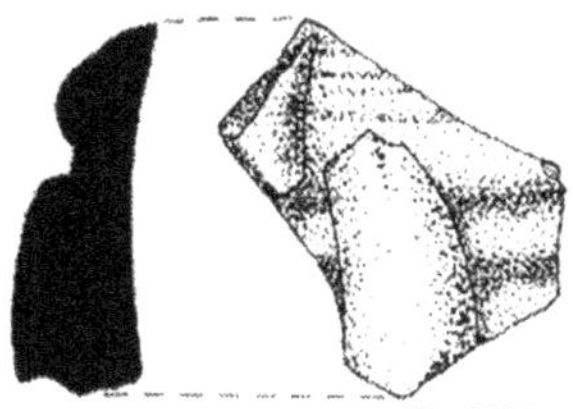 D.174	**D.174**:-Body sherd of a small or medium size vessel and appliquéd faux handle part can be seen tapering at one end. SJN-B, TT4, NE, 4, 1.73, 51544, TGW-2, BUFF(B)
 T.72	**T.72**:- Handle. Oval / elliptical in section.
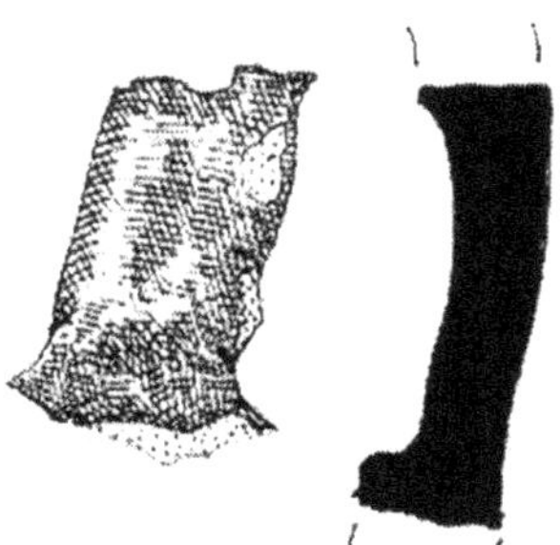 T.72.1	**T.72.1**:- Handle. Oval / elliptical in section. Dark green glaze. (L = 4cm)SJN-B, TT2, SE, 4, 2.01, 1532, TGW-1, BUFF (A)
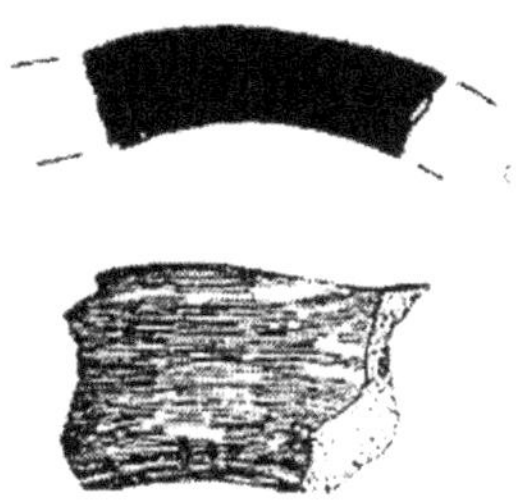 T.72.2	**T.72.2**:- Handle. Oval / elliptical in section. Dark green glaze. (L = 3cm)SJN-B, TT1, NW, 1, 0.52, 301, TGW-1, BUFF(A)
 T.282	**T.282**:- Handle, oval in section. Broad and flattened at one end and where it was probably luted to the vessel. Green glaze.SJN-B-TK, TGW-1, BUFF(A)

 **T.73.2**	**T.73.2**:- Handle. Round in section. Bright green glaze. (L = 2.5cm)SJN-B, TT1, NW, 3, 1.07, 318, TGW-3, BUFF(A)
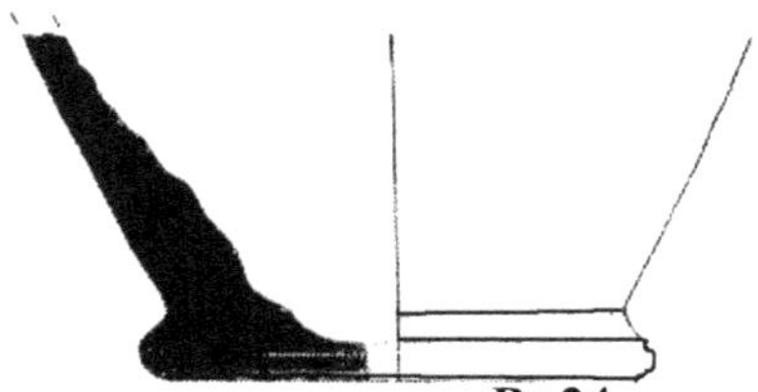 **Bs.24**	**Bs.24**:- Discoid base. (D = 9cm)SJN-B, TT1, NW, 3,1.43, 325, TGW-1, BUFF (A)
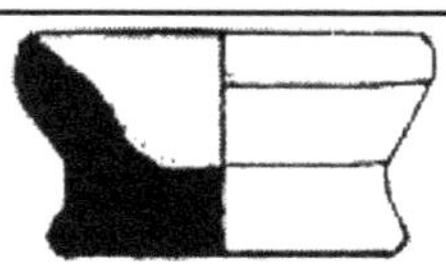 **Bs.27**	**Bs.27**:—small base. discoid. . (D = 3cm)TT1, NE, 2,0.83,507, TGW-1, BUFF (A)
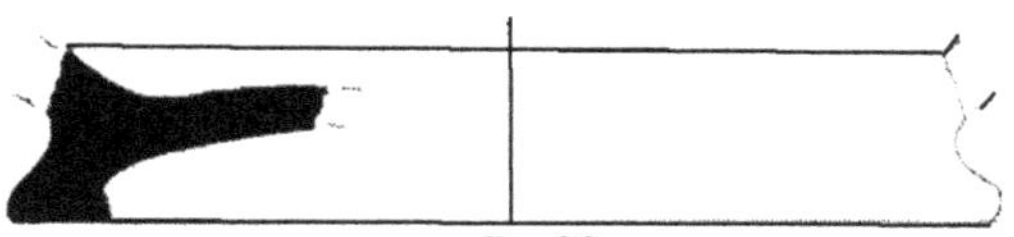 **Bs.32**	**Bs.32**:- Ring base. (D = 20cm) SJN-B, TT2, SW, 2, 0.84, 907, TGW-1, BUFF(A)
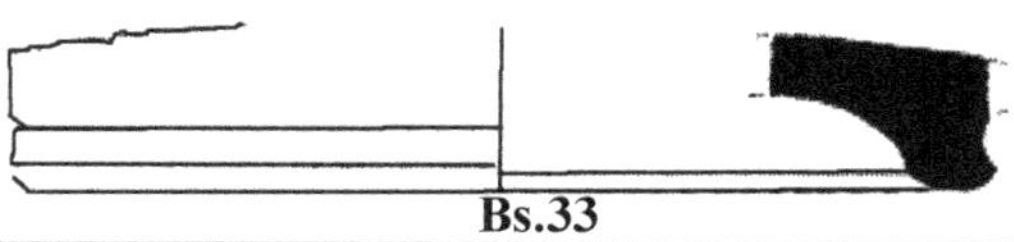 **Bs.33**	**Bs 33**:- Very thick, medium, coarse ring base with dark green blue glaze. Thick uneven glaze (up to 7mm) in places. (D = 30cm) SJN-B, TT1, NE, 3, 1.23, 516, TGW-4, BUFF(A)
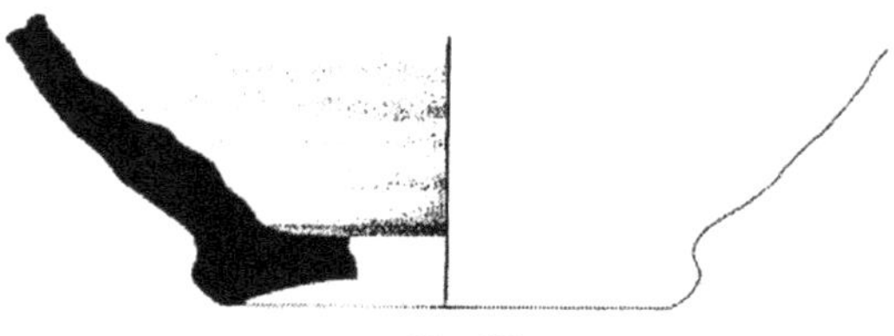 **Bs.67**	**Bs.67**:- Non contiguous ring base slightly concave in the centre of thickly potted and crudely made bowl or pot. Deep ridges and grooves on the inside of the vessel give it a corrugated effect. The body has uneven thickness and the base is very crude. There is a lot of extra clay and glaze adhering to the base.SJN-B, TT4, NE, 6, 3.46, 51580, TGW-1, BUFF(A)
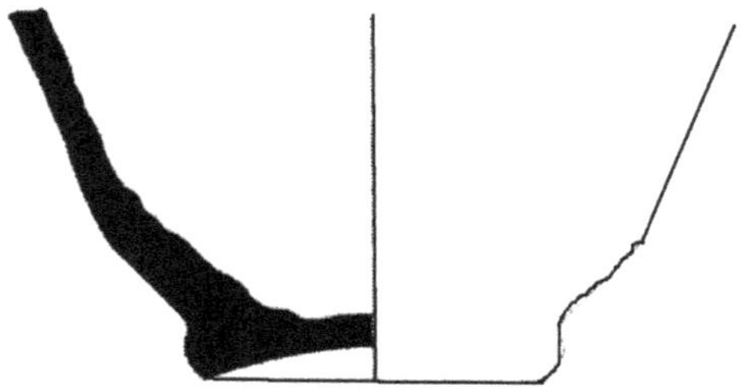 **Bs.68**	**Bs.68**:- Noncontiguous discoid base with convex moulded and a concave centre. This vessel is also crudely potted with a bright green crackled glaze. A broken spur mark can be seen on the underside of the base. The inner glaze is black and grey and very thick in the centre of the vessel. (D = 9cm) SJN-B, TT4, NE, 6, 3.46, 51581, TGW-5, BUFF(A)

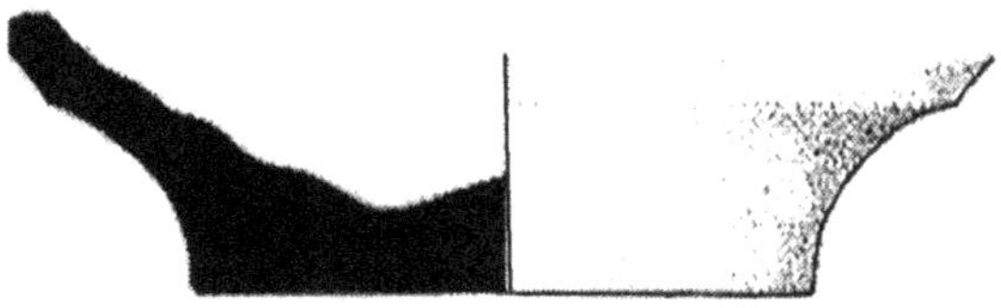 **Bs.70**	**Bs.70**:- Contiguous shallow concave base of a convex sided bowl having three spur marks on the under side and a slight carination like ridge on the external surface of the wall. Three eroded marks on the interior correspond with a spur mark on the exterior of the base. (D = 6cm)SJN-B, TT4, NE, 2, 0.34, 51507, TGW-1, BUFF(A)
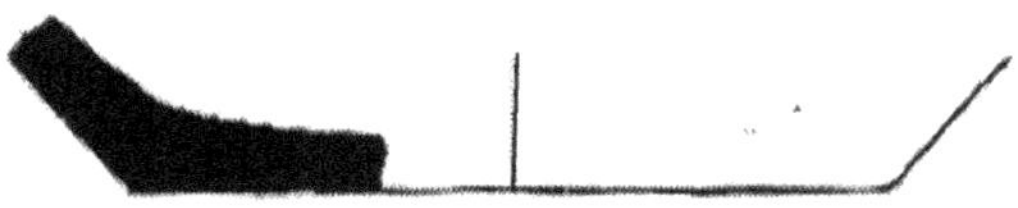 **Bs.69**	**Bs.69**:- Contiguous flat base of a convex sided/straight sided small bowl. (D = 4cm)SJN-B, TT4, NW, 2, 0.63m, 51315, TGW-1, BUFF(A)
WHITE GLAZED WARE	
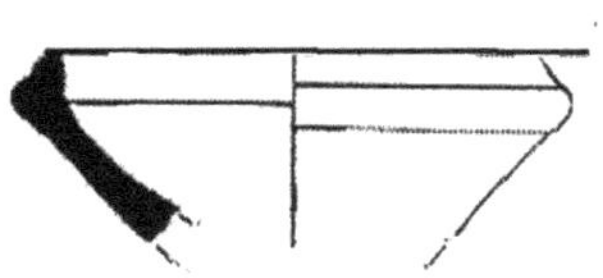 **T.45**	**T.45**:- Rim sherd of a bowl with inverted tapering rim with sharp carination at the base of the rim. Uneven surface. Creamy white fabric with glaze that is the same colour as the fabric (I.D = 10cm).SJN-B, TT2, SE, 3, 1.36, 1520, WGW, BUFF(A)
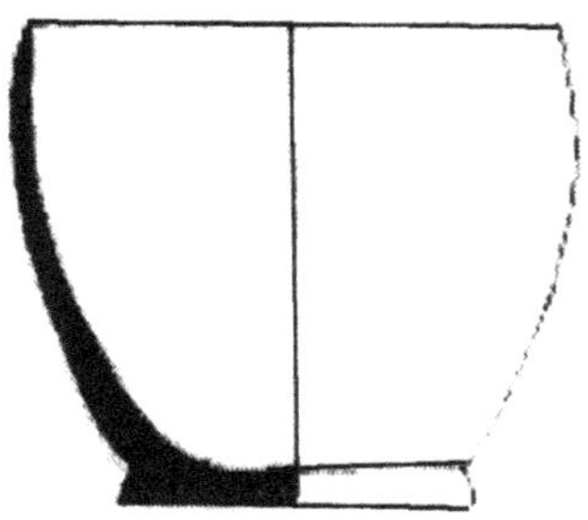 **T.46**	**T.46**:- Almost complete convex-sided bowl with slightly inverted tapering rim. The bowl has a completely eroded surface which bears evidence of glaze. It has a contiguous ring base. (I.D = 10cm)SJN-B, TT1, SW, 2, 0.83, 110, WGW, BUFF(A)
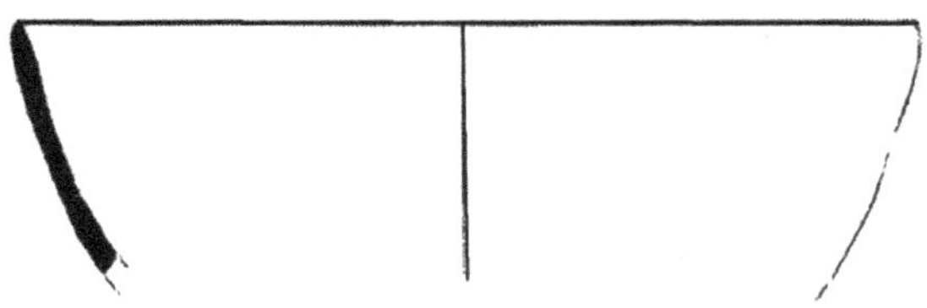 **T.47**	**T.47**:- Rim sherd of a bowl with rounded rim. Creamish yellow fabric with white glaze which has degraded to brown. The glaze appears to be unevenly applied. (I.D = 22cm). SJN-B, TT2, SW, 2, 1.05, 910, WGW, BUFF(A)
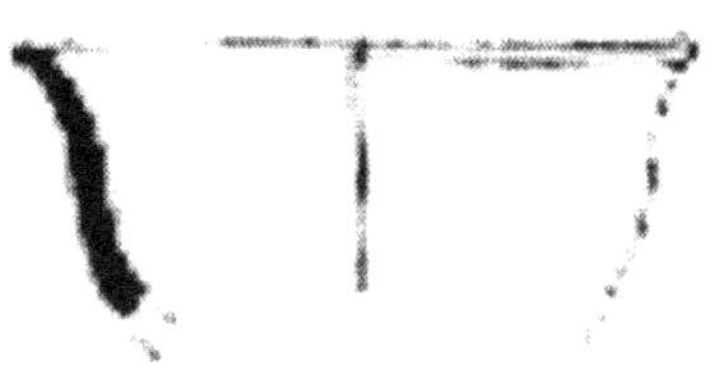 **T.56**	**T.56**:- Rim sherd of a small cup with external projecting tapering rim. Concave body and a carination at the base. White glaze.SJN-B, TT2, NW, 1, 0.56, 302, WGW, BUFF(A)

 T. 63	**T.63**:- Rim to base of a bowl with external projecting everted triangular rim. There is a blunt carination first above the base. The base appears to be a contiguous ring base. There is a white glaze on both surfaces, which has been highly eroded. (I.D = 20cm).SJN-B, TT1, SE, 3, 1.03, 714, WGW, BUFF (A)
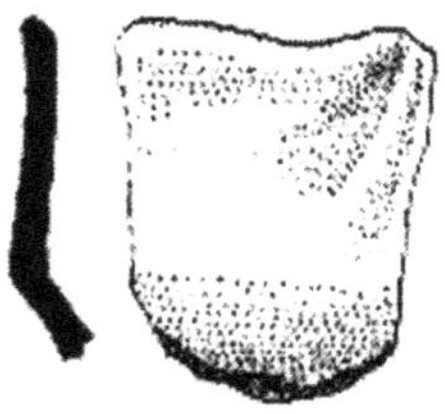 T. 281	**T.281**:- Rim sherd of a small carinated cup with everted slightly curved tapering rim. White glaze all over. The rim is foliateSJN-B-TK, WGW, BUFF(A)
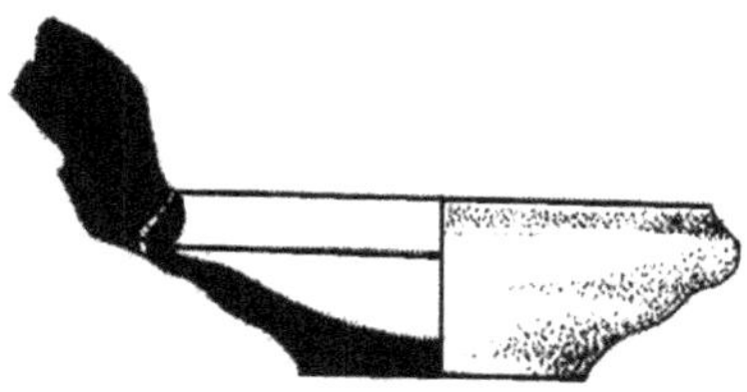 T. 381	**T.381**:- Rim to base of a small saucer lamp having an inverted tapering rim with a sharp carination at the base of the rim. The vessel slopes down to a contiguous flat base. A small hand made crudely shaped and pinched handle is luted to one side of the lamp. The channel is broken. The handle is flat on the top. The glaze is white. (I.D = 7.3cm)SJN-B, TT4, SW, 2, 5.51, 51148, WGW, BUFF(A)
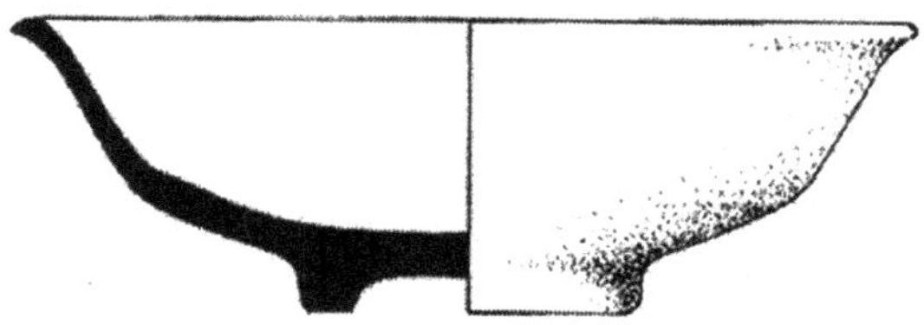 T. 382	**T.382**:- Rim to base of a conosegmental bowl having an everted tapering rim and a non-contiguous ring base. The glaze has not degraded and is a milky white. (I.D =12cm) SJN-B, TT4, SW, 2, 3.94, 51142, WGW, BUFF(A)
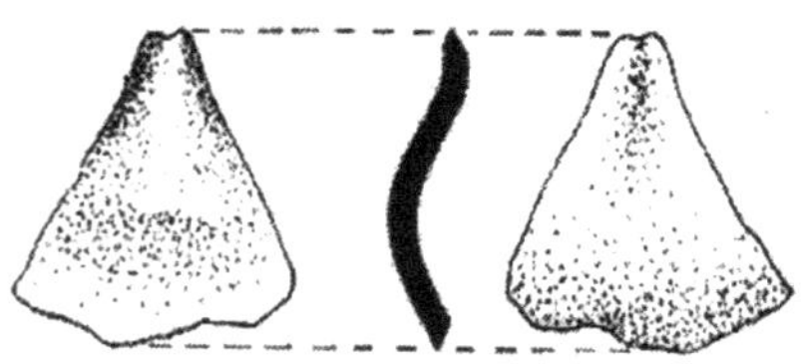 T. 383	**T.383**:- Rim sherd of a small cup or bowl with a pinched/ foliated rim and a blunt carination. White glaze on both surfaces. The rim is too small to measure.SJN-B, TT4, SW, 2, 3.42, 51139, WGW, BUFF(A)
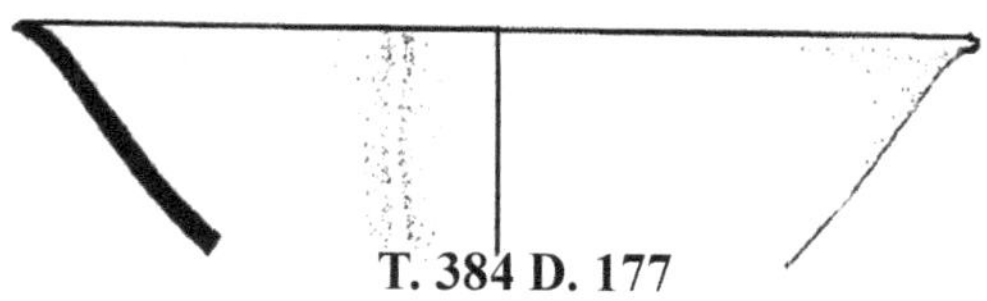 T. 384 D. 177	**T.384 D.177**:- Everted tapering rim having a white glaze. A vertical ridge or rib extends from the base to the rim on the interior of the vessel. The vessel probably had three to five such ribs. (I.D – 26cm)SJN-B, TT4, SW, 2, 5.51, 51148, WGW, BUFF(A
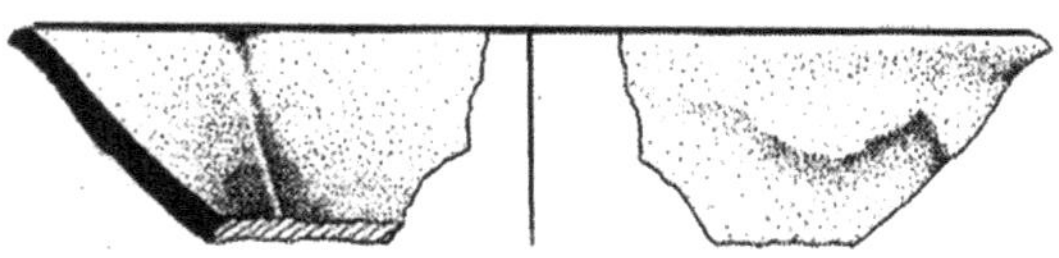 T. 384 D. 177.2	**T.384.1 D.177.2:-** This fragment of a rim is very similar to T.384 and has the same characteristics as Bs 72.1. It may belong to the same vessel.SJN-B, TT4, SW, 2, 5.51, 51148, WGW, BUFF(A)

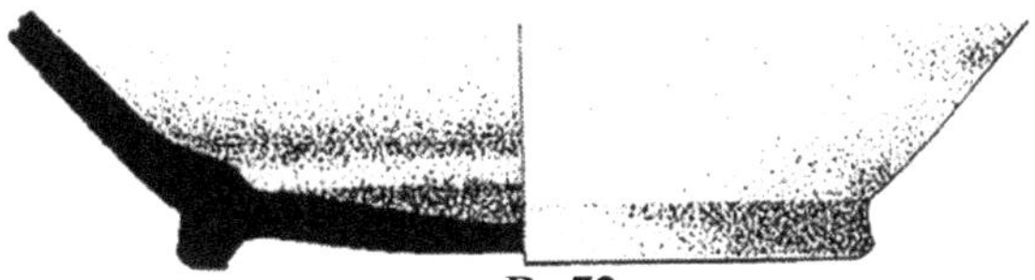 **Bs.72**	**Bs.72:-** Noncontiguous ring base very similar to Bs.72.1. The glaze is vitrified and grey. But it may originally be white. The inside of the vessel has a sharp circular ridge and a stepped floor from which five vertical ribs rise towards the rim.(D = 10cm)SJN-B, TT4, SW, 2, 3.94, 51142, WGW, BUFF(A)
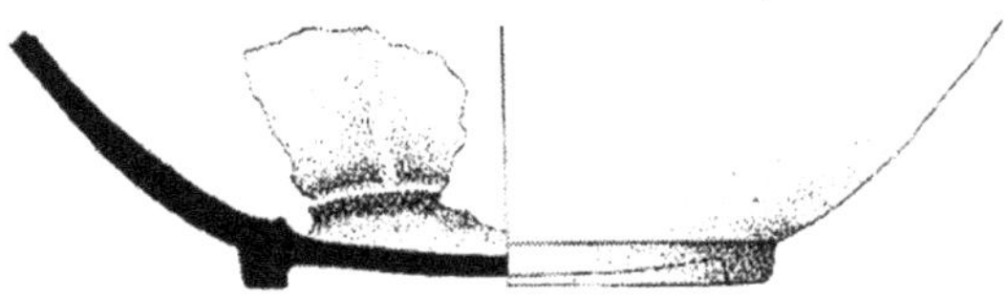 **Bs.72.1 D.177.1**	**Bs.72.1 D.177.**1:- Noncontiguous ring base of a large dish or bowl having a stepped base on the interior. From this step two vertical ridges can be seen rising towards the rim. The glaze has completely crumbled. (D = 12cm)SJN-B, TT4, SW, 2, 3.94 - 4.45, 51142/44, WGW, BUFF(A)
 **Bs.22**	**Bs.22**:- Contiguous flat. (D = 14cm)SJN-B, TT1, SW, 2, 0.86,111, WGW, BUFF (A)
SPLASHED WHITE GLAZED WARE	
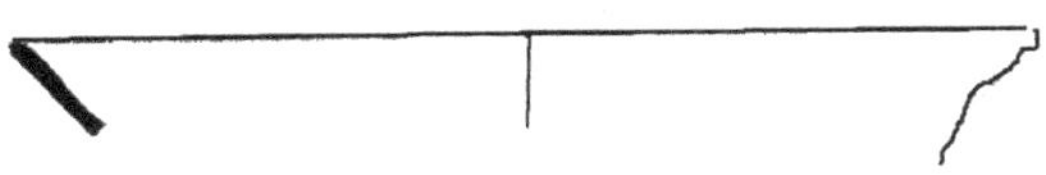 **T.42**	**T.42**:- Rim sherd of a dish with everted rounded rim. Light yellow fabric. Manganese black and grey streaky glaze. (I.D =30cm)SJN-B, TT1, SW, 1, 0.57, 103, SWGW, BUFF(A)
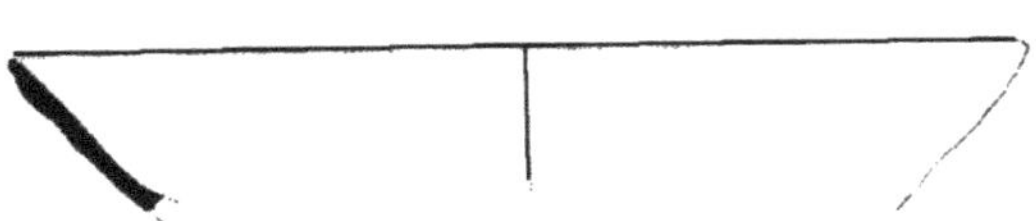 **T.42.1**	**T.42.1**:- Rim sherd of a dish with everted rounded rim similar to T.42. Sky blue glaze eroded. Creamy white fabric. (I.D = 34cm).SJN-B, TT1, SW, 3, 1.27, 915, SWGW, BUFF(A)
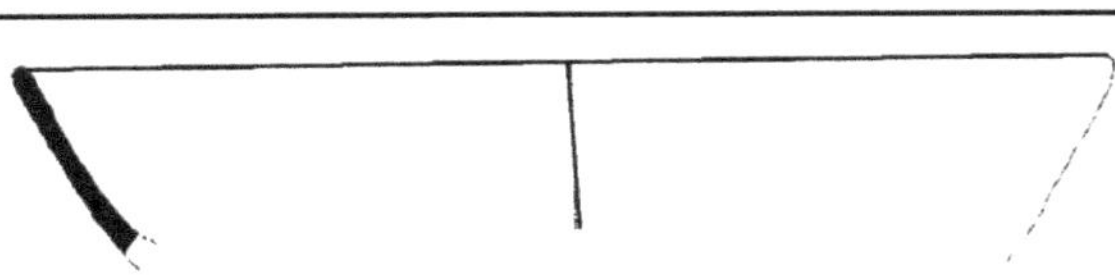 **T.42.2**	**T.42.2**:- Rim sherd of a dish with everted rounded rim similar to T.42. This dish has a blunt carination as it sloped down to the centre. Yellowish fabric. Over burnt glaze. The inner surface has vertical bands of grey and black. (I.D = 34cm) SJN-B, TT1, SW, 3, 0.4, 102, SWGW, BUFF(A)
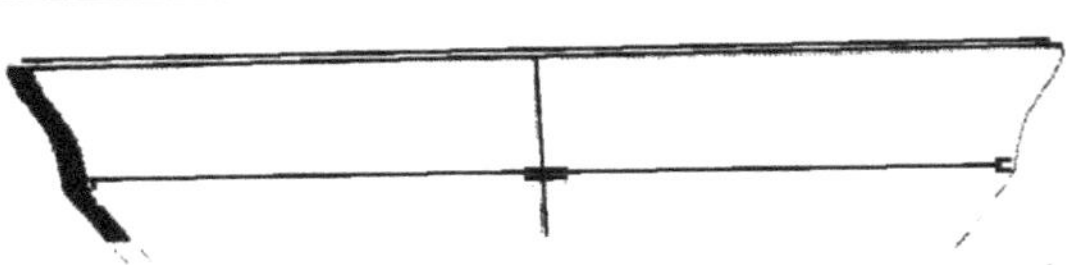 **T.59**	**T.59**:- Rim sherd of a blunt carinated bowl with external projecting everted tapering rim. Yellowish fabric. The greyish blue glaze has vertical green-blue streak of colour. The external surface glaze has been eroded. (I.D = 20cm). SJN-B, TT2, SW , 3, 1.27, 915, SWGW, BUFF(A)

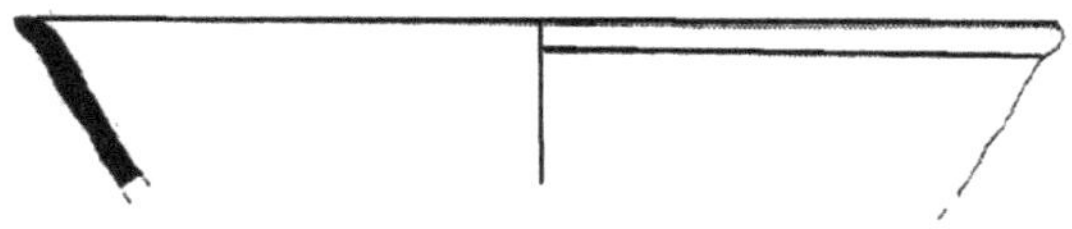 T.62	**T.62**:- Rim sherd of a bowl with external projecting everted rim. Creamy yellow fabric with light blue glaze and vertical bands of the same blue in a darker shade (I.D = 26cm). SJN-B, TT1, NW, 2, 0.96, 315, SWGW, BUFF(A)
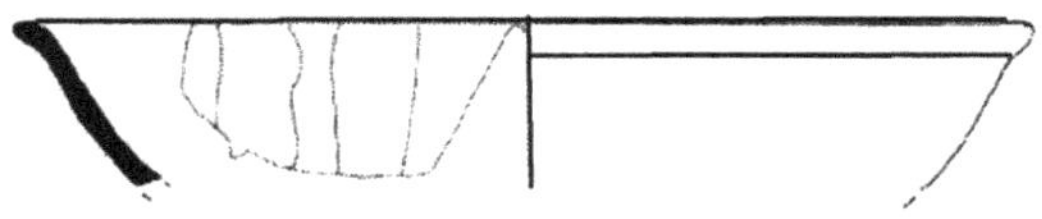 T.62.1 D.25	**T.62.1.D.25:-** Rim sherd of a bowl with external projecting everted rounded rim. This bowl is slightly convex sided. The external surface as a greyish coloured glaze. The inner surface has the same background glaze with vertical bold bands of a dark turquoise colour. (I.D = 24cm). SJN-B, TT1, SE, 2, 0.51, 705, SWGW, BUFF(A)
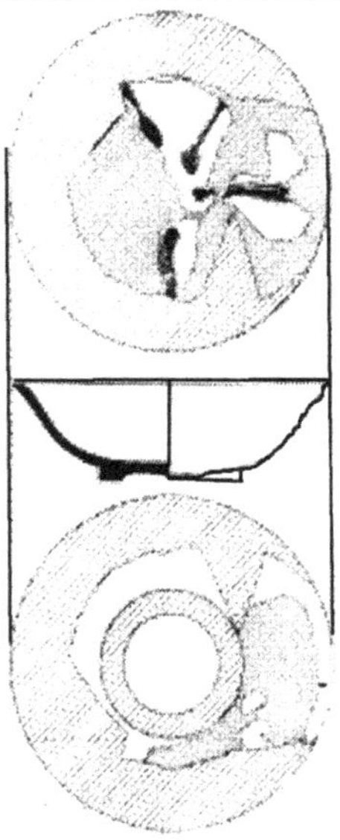 T.67	**T.386 D.180**:- Rim to base of a large bowl or dish having an everted tapering rim. The external surface has a white glaze with some blue traces. The base is non-contiguous ring. The interior of the vessel has a white glaze with sky-blue and aqua vertical streaks radiating from the centre to the rim. The colours appear to be splashed and the colours are eroded. (I.D = 27cm) SJN-B, TT4, SW, 2, 3.94m, 51142, SWGW, BUFF(A)
COBALT PAINTED WARE	
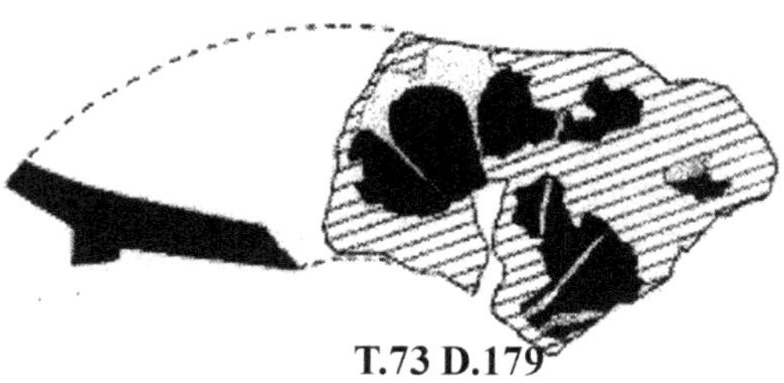 T.73 D.179	**Bs.73 D.179:-** Contiguous low ring base of a dish having a brownish coloured eroded glaze. On the interior surface a dark blue painted design can be seen. The design may have been a floral motif. SJN-B, TT4, NE, 2, 0.80m, 51521, CPW, BUFF(A)
 D.179.1	**D.179.1**:- Base of a dish very similar to Bs 73. The originally white glaze has degraded to brown. A dark blue painting which may be part of a larger motif can be seen. The outline of the painting is somewhat blurred. SJN-B, TT4, NE, 3, 1.24, 51535, CPW, BUFF(A)

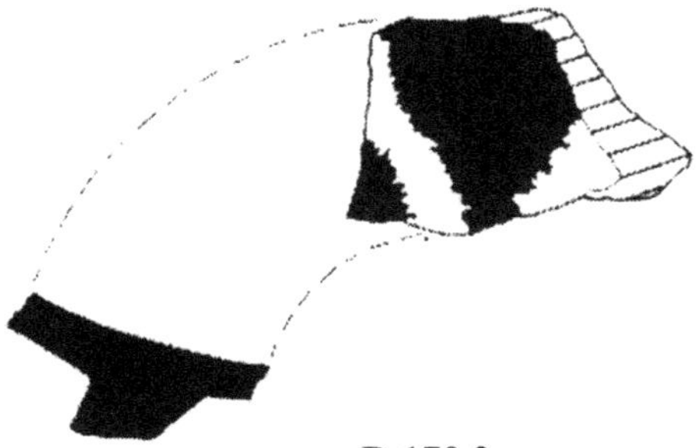 **D.179.2**	**D.179.2**:- Base of a dish similar to Bs.73. The background glaze is now degraded to brown. On the interior of the vessel a blackish blue leaf like design is painted. SJN-B, TT4, SE, 2, 1.16, 51723, CPW, BUFF(A)
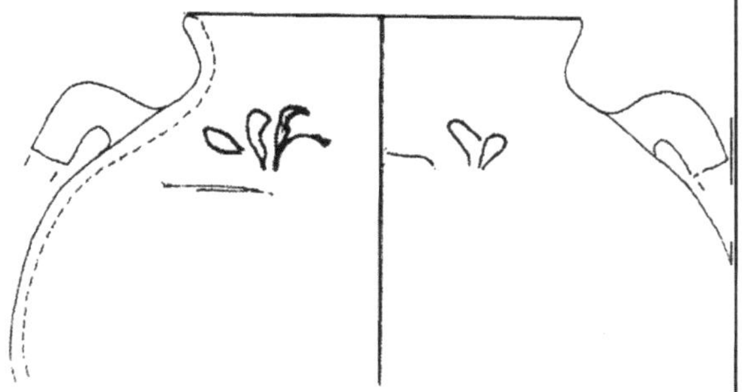 **T.71 D.27**	**T.71 D.27**:-.Rim sherd of a globular pot with external projecting everted rounded rim. White glaze with blue painted design. Floral motifs. SJN-B, TT1, SW, CPW BUFF(A)
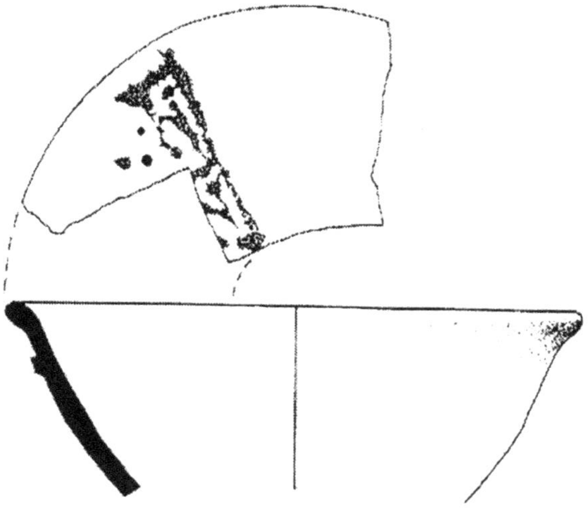 **T.385.D.178**	**T.385.D.178**:- Everted tapering rim of a convex sided bowl having white glaze on both surfaces. The interior of the vessel has the remnants of a dark blue painted inscription extending from rim to rim across the vessel. Only part of the inscriptions can be seen. (I.D = 10cm) SJN-B, TT4, SW, 2, 5.42, 51149, CPW, BUFF(A)
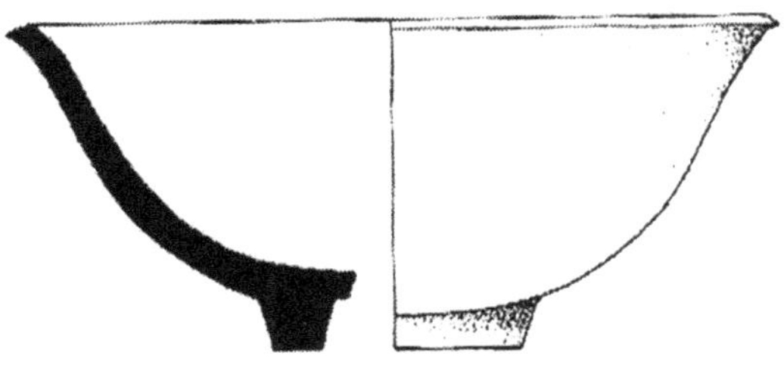 **T.385.1**	**T.385.1**:- Rim to base of a small convex sided bowl having an everted tapering rim and a non-contiguous ring base. White glaze covers the exterior and interior.Blue tinge. But there no attempt at decoration is seen on the sherd (I.D = 10cm) SJN-B, TT4, SW, 2, 5.42, 51149, CPW, BUFF(A)

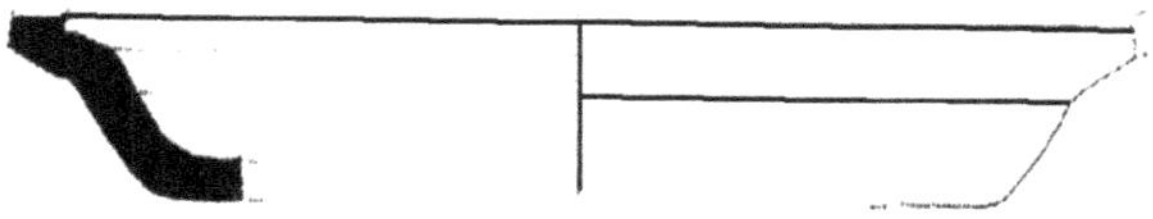 **D.40**	**D.40**:- Ivory white with yellowish green horizontal almond-shaped dots. (Th = 7 – 10mm)SJN-B, SURFACE , LPW-1, BUFF(A)
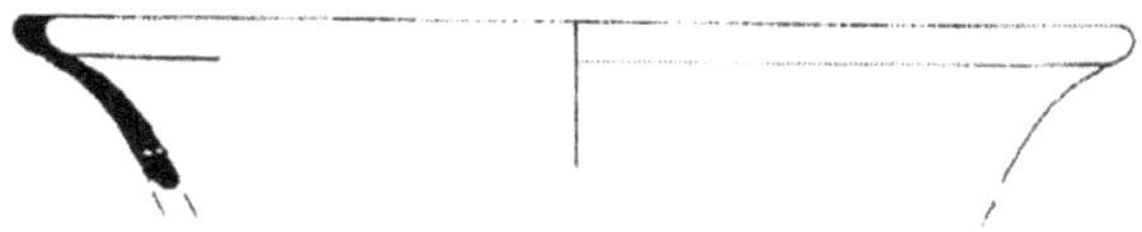 **T.60**	**T.60**:- Rim sherd of a bowl with external projecting extremely everted (almost horizontal) rounded rim. Yellow fabric with a brownish glaze and faint traces of yellow design on the lip of the rim. (I.D = 18cm). SJN-B, TT2, SE, 3, 1.24, 1517, LPW-1, BUFF(A)
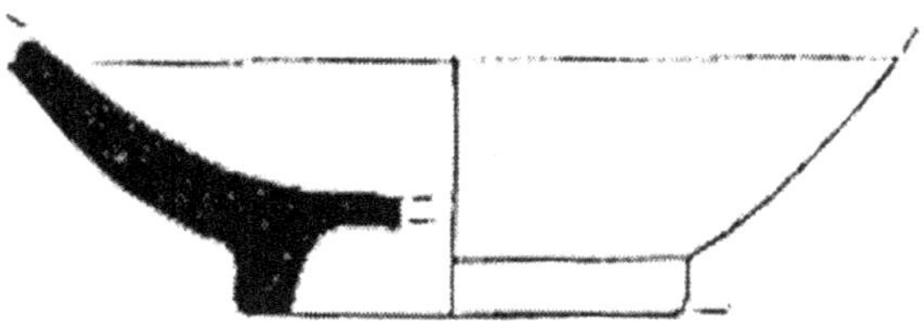 **Bs.30**	**Bs.30**:- Ring base. (D = 4cm)SJN-B, TT1, NW, 1, 0.68, 305, LPW-1, BUFF(A)
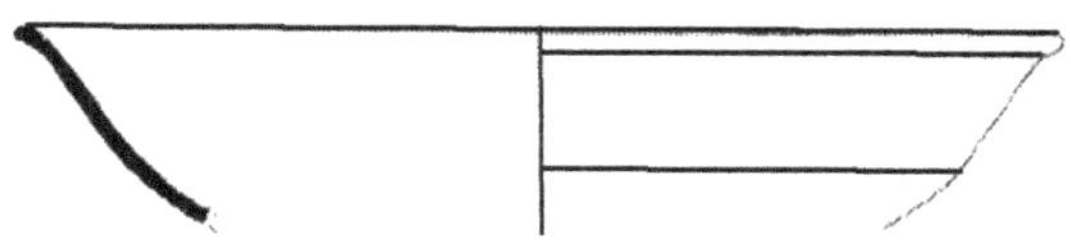 **T.43**	**T.43**:-Rim sherd of a dish with external projecting extremely everted tapering rim. Creamish white fabric. Glossy glaze with painted designs on exterior surface – brown background with red painting. Glaze is eroded (I.D = 32cm).SJN-B, TT1, SW, 3, 1, 114, LPW-3, BUFF (A)
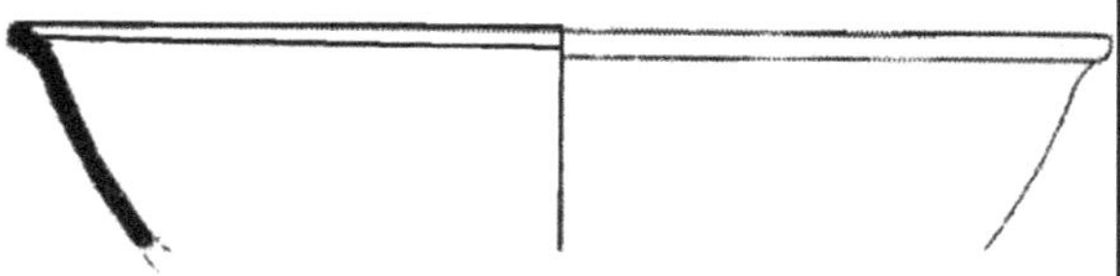 **T.43.1**	**T.43.1**:- Rim sherd of a dish with external projecting extremely everted rounded rim. Creamish white fabric. Surface has yellowish brown glaze with red painting in evidence on the exterior surface, and mustard yellow painting on the white background of the interior. (I.D = 34cm). SJN-B, TT1, NW, 3, 1.07, 318, LPW-3, BUFF (A)
 **T.50 D.22**	**T.50 D.22**:- Rim sherd of a bowl with external projecting everted rounded rim with a blunt ridge on the external side. Creamish yellow fabric with creamish brown glossy glaze with olive green painted design on the exterior surface. The design consists of a broad green circle within which is a smaller circle filled in with wavy lines and inverted 'C's There is also a border-like design along the rim which has oblique wavy lines ending in dots. The inner surface has pale yellow dots all over. (I.D = 28cm). SJN-B, TT1, SW, 2, 0.83, 110, LPW-1, BUFF(A)

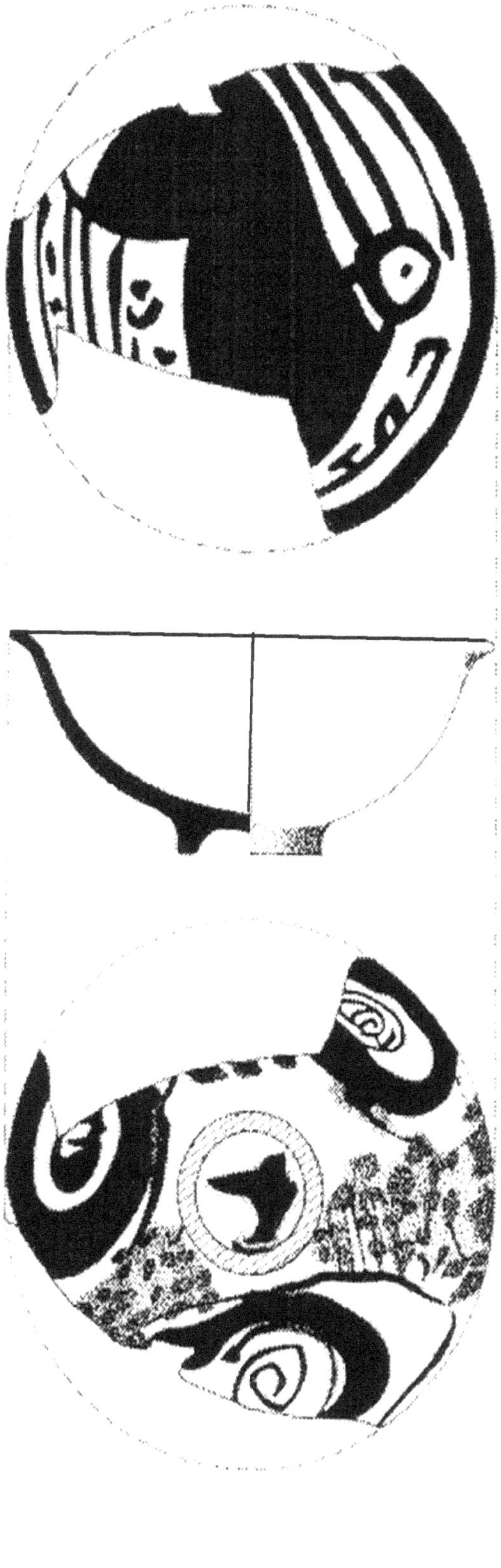

T.283

T.379 D.175.1:- More than 3/4th of a convex sided bowl having a non-contiguous ring base which is high and vertical-sided. The rim is external projecting, everted and tapering. The glaze of the vessel is white. The lustre painting on the interior is similar to D.175, with the same central motif. The pseudo calligraphic symbols are different. The painting is mustard gold with a greenish-tinge and the exterior three broad pinkish-red circles are seen with a coiling swirl within each. The swirls are mustard in color. Between these circles the space is filled with a smudged design of oblique lines and dots. (I.D = 12cm)
SJN-B, TT4, SW, 2, 2.91, 51137, LPW-1, BUFF(A)

 T.61 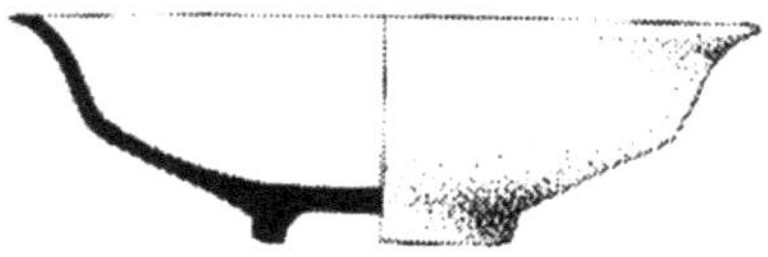 T.67  T.378 D.175	**T.378 D.175**:-Almost complete conosegmental bowl having an external projecting everted tapering rim and a non contiguous vertical sided ring base. The bowl has a white glaze covering the exterior and interior. The lip and upper part of the rim has a mustard yellow band painted in lustre. A similar band runs along the interior of the vessel. Within this circle is a stylized motif of a running hare. Some pseudo calligraphic symbols are also seen. The centre of the bowl has a small circular groove. On the exterior three large swirls can be seen. These are reddish- green in color. This lustre painted bowl is quite well preserved. (I.D = 12cm)SJN-B, TT4, SW, 2, 2.91-3.11, 51137/38, LPW-1, BUFF(A)
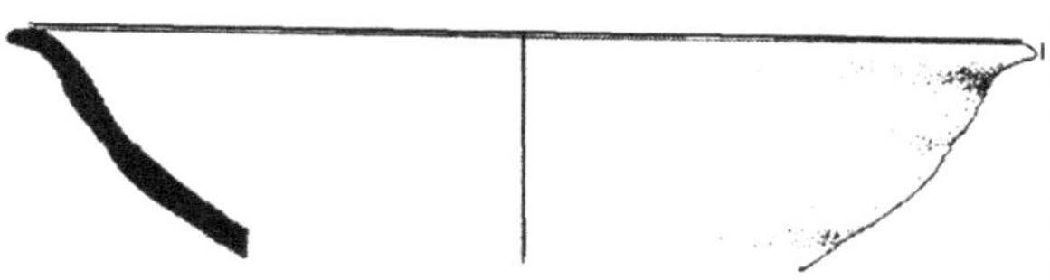 T.379.1	**T.379.1**:-Everted tapering rim, very similar to T.379. The eversion of the rim forms shallow groove on the exterior. The lustre painting has completely faded. On the interior it faintly resembles the motif of D.175 and has a yellowish-green color. On the exterior a reddish band is faintly visible just under the lip in on part. (I.D = 12cm)SJN-B, TT4, SW, 2, 2.91, 51137, LPW-1, BUFF(A)
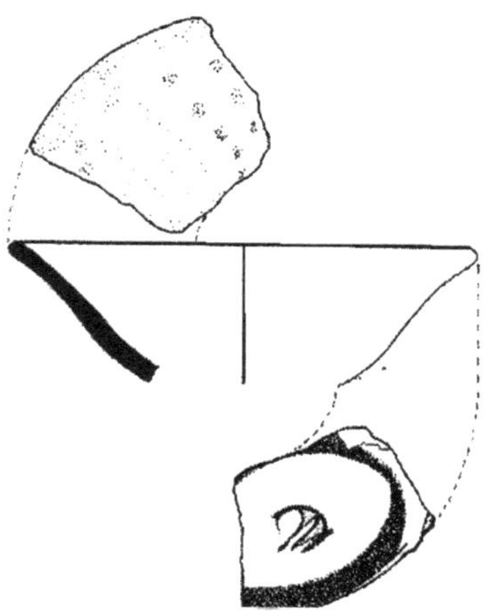 T.380 D.176	**T.380 D.176**:-Slightly everted tapering rim of a bowl with a blunt carination. The interior of the vessel has a completely faded design painted in yellow. The motif appears to be leaf like with small dots around it. A yellow band may run along the lip. On the exterior is a large red circle containing a small ochre-yellow squiggle. (I.D = 15cm) SJN-B, TT4, SW, 2, 5.42, 51149, LPW-1, BUFF(A)

HATCHED SGRAFFIATO WARE	
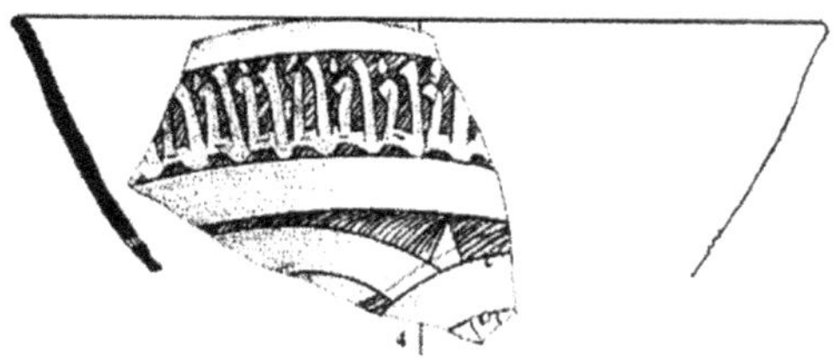 **D.99**	**D.99**:- Rim sherd of a convex sided bowl The glaze is in shades of light creamish green and dark green on the internal and external surface. A dark green band runs along the rim. The inner surface has incised design which consists of a border contained within two horizontal lines. The design is a repetition of a single motif which gives a calligraphic effect. Hatched fine oblique lines in the background bring the main motif into relief. Traces of some other design on the body of the vessel can be seen under this border. This consists of plain broad horizontal bands and hatched lines. One perforation can be very clearly seen on the body of the bowl. (I.D = 16cm). SJN-B-TK, HsG(P), PINK(B)
 **D.100**	**D.100**:- Rim sherd of a convex sided bowl (three pieces refit) with light creamish green and green glaze. There is a dark green band running along the rim. There is no evidence of glaze on the external surface. Two horizontal parallel lines run along the inside of the rim. An abstract motif which may be part of a larger design is evident Two perforations, one on the rim and one on the body of the bowl (vertically aligned) are extant. (I.D = 22cm).SJN-B-TK, HsG(P), PINK(B)
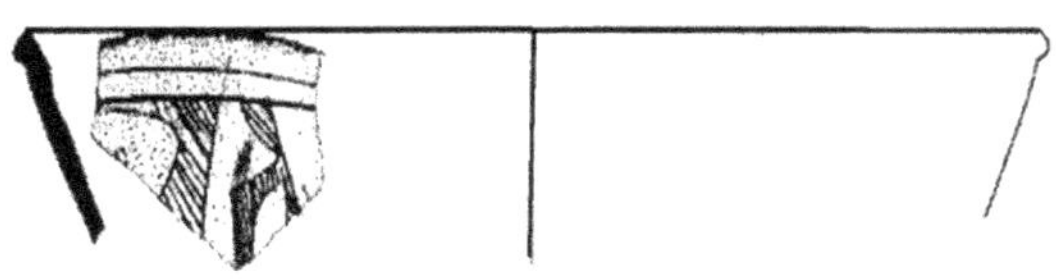 **D.101**	**D.101**:- Rim sherd of a convex sided bowl. Dark green glaze on internal surface. The glaze appears to be stop just below the rim on the external side. One incised horizontal line and one horizontal band runs parallel to the rim. The design is part of some motif and is brought into relief by hatched oblique lines (I.D = 21cm).SJN-B-TK, HsG(M), PINK(B)
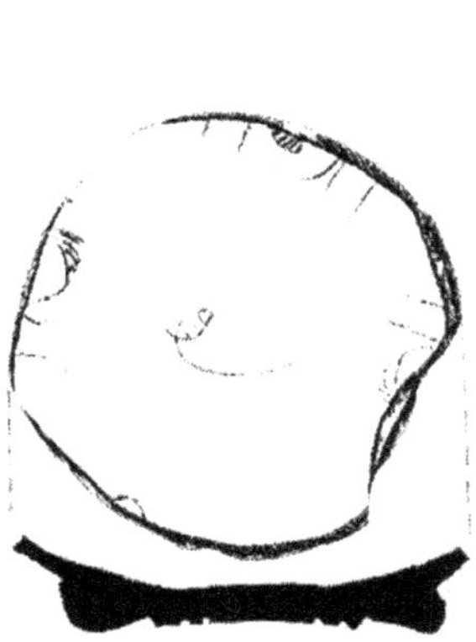 **D.103**	**D.103**:-Base of a bowl/dish. Light creamish green glaze with ochre yellow splash in the centre and four dark green splashes opposite each other. The ochre yellow splash has a curved inscribed line with a closed loop at one end. The inside of the loop is hatched. The green splashes each have incised extant semi-circles and are flanked by two parallel incised vertical lines. Evidence of glaze on external surface. Ring base with discoid centre. Contiguous flat base with design on inner surface. Yellow and green glaze. A floral/abstract design is very sharply incised on the base and the body of the vessel. Some curved lines can be seen. At the junction of the body and the base two deep

	grooves and a sharp ridge inbetween can be seen. There are further emphasized by horizontal incised lines that run along them. SJN-B-TK, HsG(P), PINK(B)
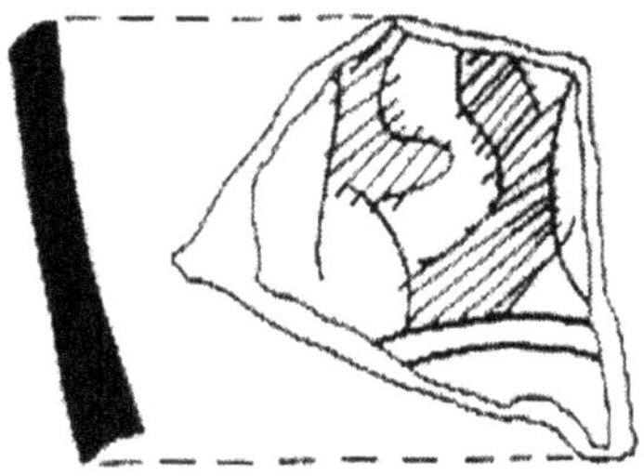 **D.149**	**D.149**:-Rim sherd of a convex sided bowl of a simple rounded rim. The interior has whitened-green glaze. (I.D = 14cm) SJN-BTT4, NE, 2, 0.34, 51507 HsG(B), PINK(B)
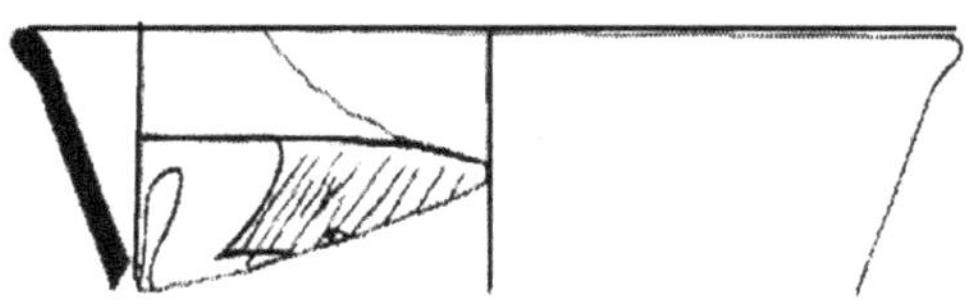 **D.150**	**D.150**: Rim sherd of a convex-sided bowl or a dish having external projecting everted rounded rim. A green band runs along the inside of the rim, forming a border. An incised line separates the border from the white body of the vessel which has the decoration. SJN-B, TT4, SW, 2, 0.56, 51113, HsG(B), PINK(B)
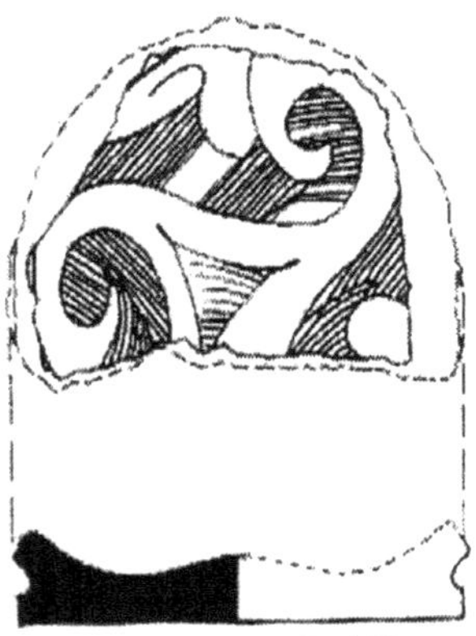 **D.152**	**D.152**:- Discoid base. Paisley decoration incised within a circle on interior. Small hatched triangle in the center. Eroded glaze. SJN-B, NE, 2, 0.31, 51506, HsG(B), PINK(B)
 **D.153**	**D.153**:- Body sherd of a bowl with wide spaced hatched design. Green and white glaze. SJN-B, SW, 2, 0.75, 51118, HsG(B), PINK(B)

 **D.154**	**D.154**:- Small body sherd with glaze on both surfaces. Two concentric circlular motifs with hatching in-between them. SJN-B, SW, 2, 0.58, 51114, HsG(B), PINK(B)
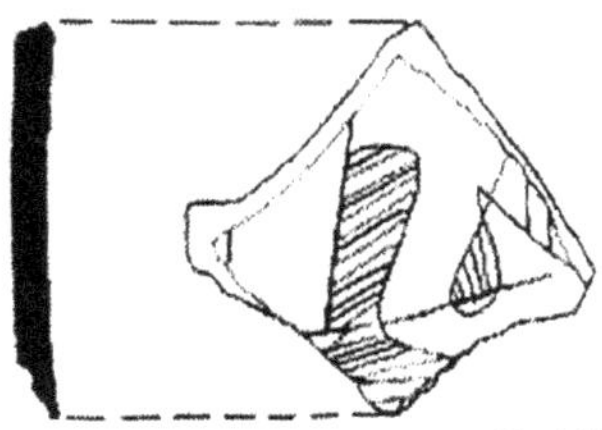 **D.155**	**D.155**:- Small body sherd with hatched design. white and green glaze.SJN-B, TT4, SE, 2, 1.04, 51702, HsG(B), PINK(B)
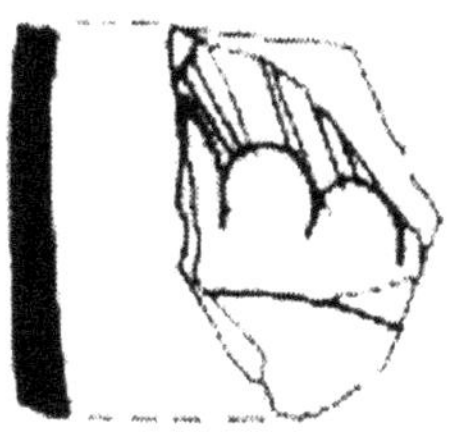 **D.156**	**D.156**:- Body sherd with rust-brown glaze on the interior and greenish-gold on the exterior. Hatched design of loops.SJN-B, TT4, NW, 2, 0.44, 51309, HsG(B), PINK(B)
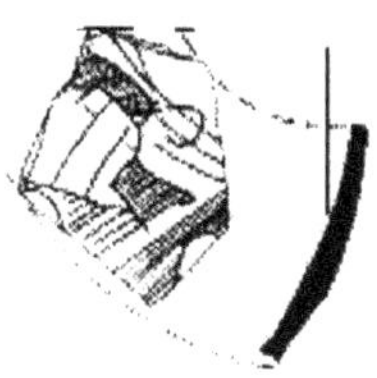 **D.157**	**D.157**:- Complex design. Pseudo-calligraphic. Three black manganese dots on a double-line border which is under the motif. SJN-B, TT4, NW, 2, 0.44, 51309, HsG(B), PINK(B)
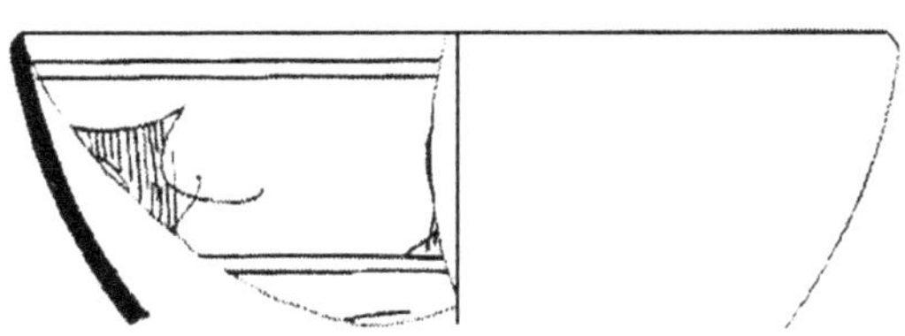 **T.353 D.147**	**T.353 D.147**:- Rim sherd of a convex sided bowl having a simple rounded rim white and green glaze on both surfaces. The interior surface has a hatched widely spaced design contained within a broad border of double lines (I.D = 17cm).SJN-B, TT4, SW, 2, 0.93, (WELL), 51123, HsG(B), PINK(B)
 **T.354 D.148**	**T.354 D.148**:-Rim sherd of a dish having simple tapering rim. A slight undulation of surface is suggestive of a carination or a curve. The inside of a lip has a border of double lines and finely hatched design is seen under it. I.D = 26 cm) A perforation like hole is seen on the interior. (I.D = 26cm) SJN-B, NW, 2, 0.51, 51311, HsG(B), PINK(B)

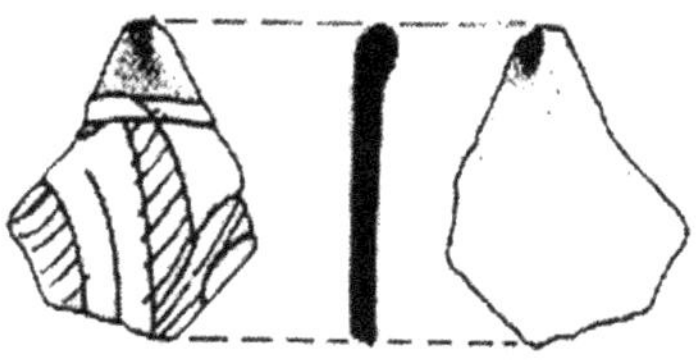 **T.355 D.151**	**T.355 D.151**:- Very small fragment of a bowl having a fluted rim. White and green glaze on both surfaces. Double row of incised lines form a border on the inside of the rim. Hatched decoration on the interior wall under this border. SJN-B, TT4, NW, 2, 0.82, 51324, HsG(B), PINK(B)
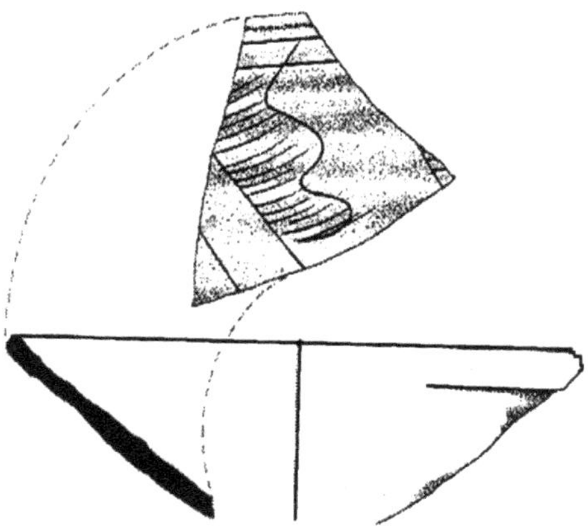 **T.356 D.158**	**T.356 D.158**:- Rim having glaze on both surfaces. White and green glaze. Double-line border along rim interior and part of hatched design underneath. SJN-B, NE, 2, 0.39,51508, HsG(B), PINK(B) Bs.65 D.144:-This is a complex base having a non contiguous rounded ring. The inner part of the base is concave but a circular disk can e seen in the centre. The inside of the vessel has a brown-green glaze and an incised design with hatching. It is a convex sided bowl. The centre of the bowl has a raised point. A triangle (hatched) with a circle insde it is the central motif. SJN-B, TT4, NW , 2, 0.51, 51311, HsG(M), PINK(B) (SEE BELOW)
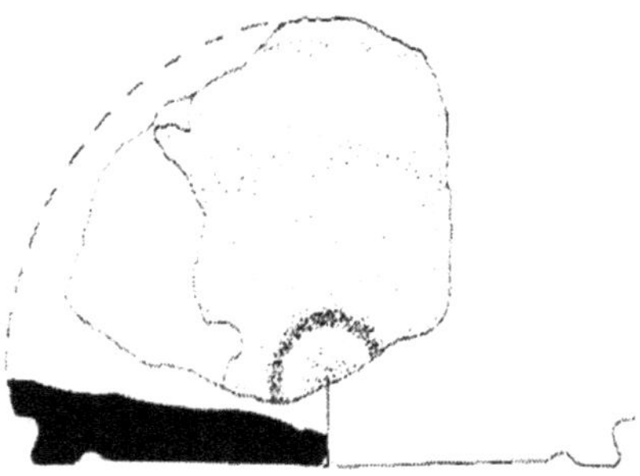 **Bs.65**	**Bs.65 D.144**:-This is a complex base having a non contiguous rounded ring. The inner part of the base is concave but a circular disk can e seen in the centre. The inside of the vessel has a brown-green glaze and an incised design with hatching. It is a convex sided bowl. The centre of the bowl has a raised point. A triangle (hatched) with a circle insde it is the central motif. SJN-B, TT4, NW , 2, 0.51, 51311, HsG(M), PINK(B) (SEE BELOW)
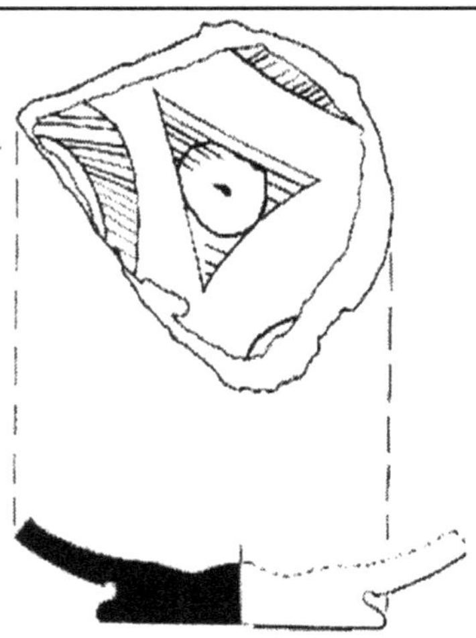 **Bs.65 D.144**	**Bs.65 D.144**:-This is a complex base having a non contiguous rounded ring. The inner part of the base is concave but a circular disk can e seen in the centre. The inside of the vessel has a brown-green glaze and an incised design with hatching. It is a convex sided bowl. The centre of the bowl has a raised point. A triangle (hatched) with a circle insde it is the central motif. SJN-B, TT4, NW, 2, 0.51, 51311, HsG(M), PINK (B)

 T.272	**T.272**:- Rim sherd of a dish with everted straight rounded rim with a sharp ridge on the inner side. Green glaze on internal surface (I.D = 31cm). SJN-K, A5, SW, 1, 0.33, 4903, MGB, BUFF(A)
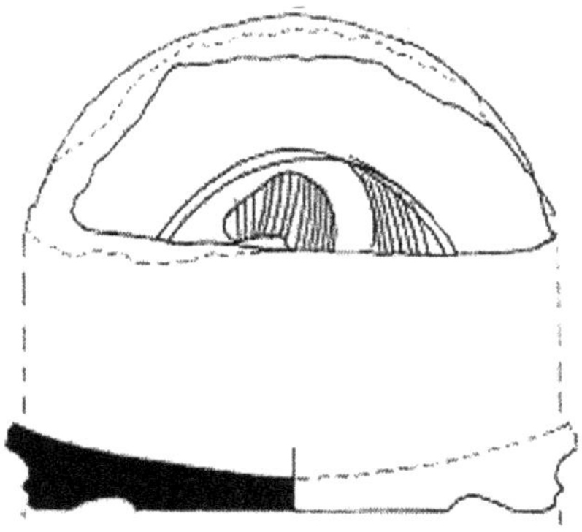 D.143	**D.143**:- Contiguous discoid base. Green glaze overlying a white slip. The colour has filled the incisions of the design. The design is contained within a circle. The hatched fillers can be seen. (D = 9cm) SJN-B, TT4, NW, 2, 0.51, 51311, HsG(M), PINK(B)
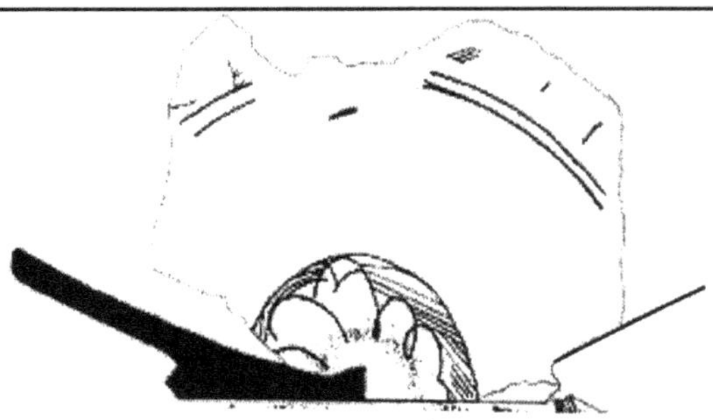 Bs.66 D.145	**Bs.66. D.145**:- Non contiguous discoid base with carinated disk having a ring like appearance. The external wall of the vessel has transparent glaze and white slip unevenly applied. The exterior has obliquely placed shallow vertical grooves. The internal surface has a hatched design contained within a circle and a border. (D = 11cm) SJN-B, TT4, SW, 2, 0.42, 51109, HsG(P), PINK(B)
 D.102	**D.102**:- Body sherd Probably part of a floral motif can be seen incised on a cream and green glaze. SJN-B-TK, well / tank, HsG(B), PINK(B)
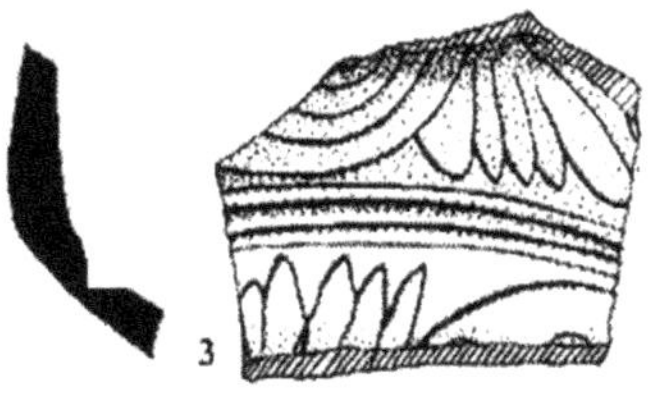 D.103.1	**D.103:**-Fragment of a sherd, with very deeply incised design. No evidence of hatching. Yellow, green ochre glaze SJN-B-TK, HsG(P), PINK(B)
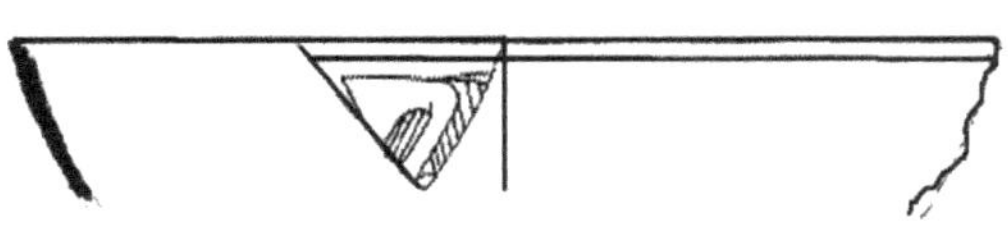 T.13	**T.13**:- Rim sherd of a bowl with simple external projecting everted tapering rim. Pink fabric with incised design on inner surface green and yellow glaze (I.D = 16cm). SJN-B, TT1, SW, 3, 1.16, 137, HsG (P) PINK(B)

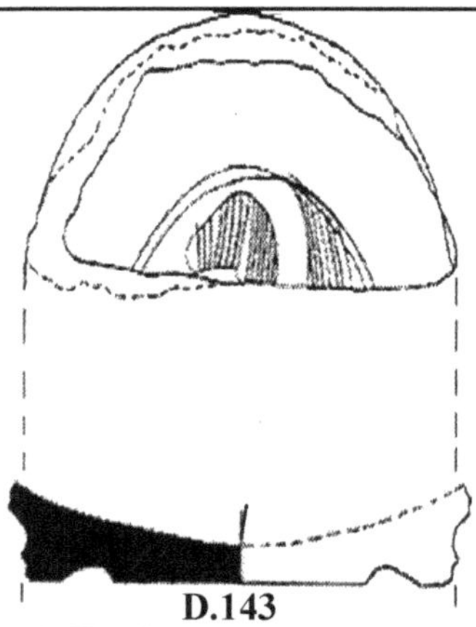 **D.143**	**D.143**:- Contiguous discoid base. Green glaze overlying a white slip. The colour has filled the incisions of the design. The design is contained within a circle. The hatched fillers can be seen. (D = 9cm)SJN-B, TT4, NW, 2, 0.51, 51311, HsG(M), PINK(B)
MONOCHROME GLAZED WARES	
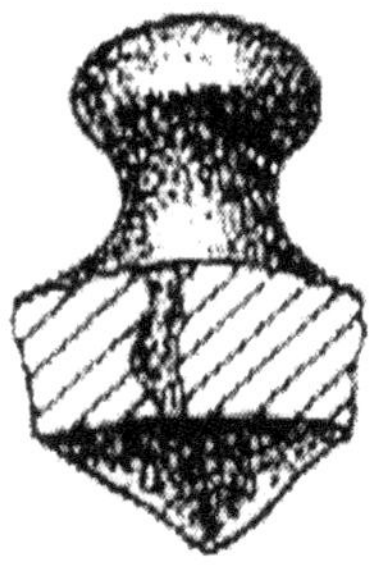 **T.344**	**T.344**:- Finial of a lid having a norrow-waisted stud handle. Dark brown glaze. The underside of the finial tapers to a point. The stud handle itself has a flat top. A broken perforation, diagonally placed beside the stud handle can be seen, giving this finial a resemblance to modern tea pot lids. The fabric is a reddish-pink. The glaze is not eroded. SJN-D, LX1, NE, 1, 1.27, 21901, MGP, PINK(A)
 **T.9**	**T.9**:- Rim of a convex sided bowl with simple everted rounded rim. Leaf green glossy glaze on interior surface and on rim. The glaze comes down to about two cm on the exterior surface. The rest of the exterior surface is left untreated (I.D = 20cm). Pale creamish pink fabric. SJN-B, TT2, SW, 2 & 3, 1.05 - 1.27, 915 & 910, MGB, BUFF(A)
MONOCHROME GLAZED WARES	
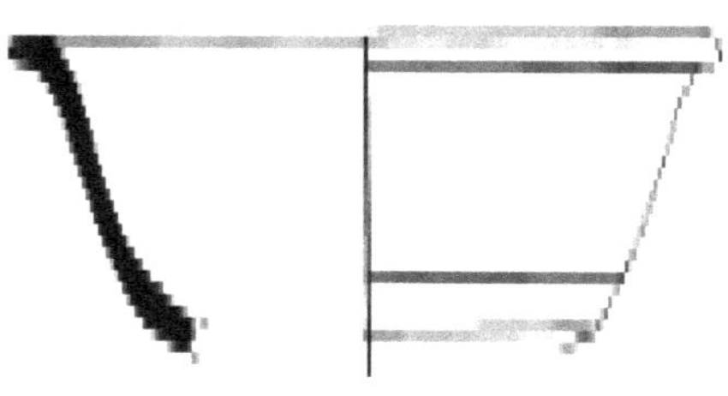 **T.14.1**	**T.14.1**:- Rim sherd of a bowl with simple external projecting extremely everted rounded rim. Pale pink fabric and white glaze streaky colours (I.D = 18cm). SJN-B, TT1, NW, 2, 0.96, 315, CSGW, PINK(F)

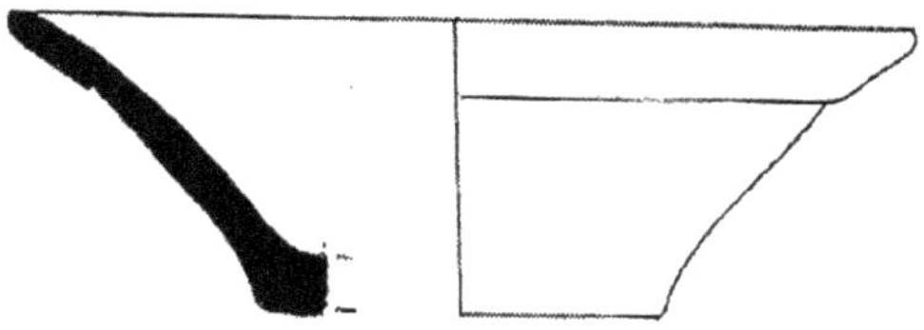 **T.14**	**T.14**:-Rim sherd of a bowl with simple external projecting extremely everted tapering rim with a deep groove on the external side. Light pink fabric with a white glaze (I.D = 14cm).SJN-B, TT1, SW, 1, 0.35, 101, CSGW, PINK(F)
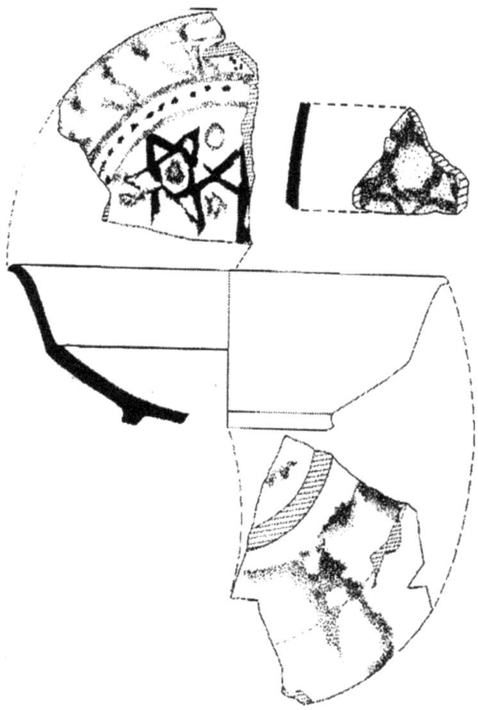 **T.351 D.141**	**T.351 D.141**:- Rim to base of a cono-segmental bowl having external projecting, extremely everted tapering rim. The base is a non-contiguous ring. The exterior of the vessel has glazed splashes of colour-ochre, white, blue and black. It appears to be a smudged panel design. It stops short of the foot. The interior of the vessel shows a combination of decoration techniques. The glaze on the walls of the bowl and the rim are smudged and appear to have had a design. White, ochre, black and blue splashes can be seen and the border which is eroded along the lip. On the carination can be seen a border of black and blue bands within which are placed black dots. The floor of the bowl is dramatically covered in a geometric design of inter connected six pointed stars. The centers of the stars contain dots that are black and blue. The stars are executed in black. The entire effect is that of Islamic tile work. This is a singular sample. No other similar complex motif has been encountered. (I.D = 21cm) SJN-B, TT4, SW, 2, 3.11, 51138, CSGW, PINK(B)
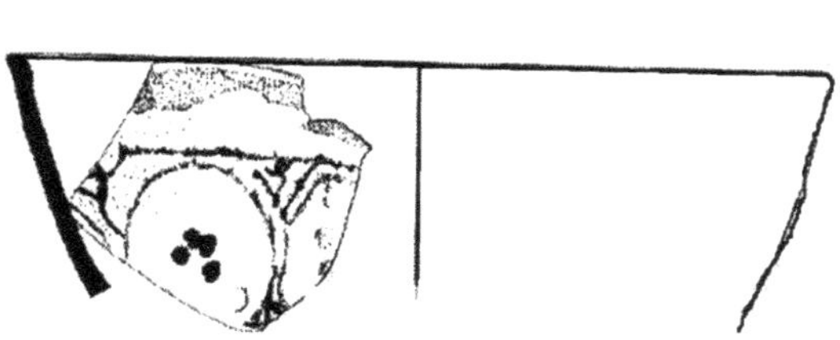 **D.104**	**D.104**:-Rim sherd of a convex sided bowl or dish having white glaze on internal surface and evidence of same glaze on the external surface. The design is painted in a dark purple-black on the internal surface. The design consists of a horizontal line under which can be seen a row of circles each containing four dots. Between the circles and the horizontal line is filled with an inverted triangle (I.D = 17cm) .SJN-B-TK, WGPW, PINK(B)
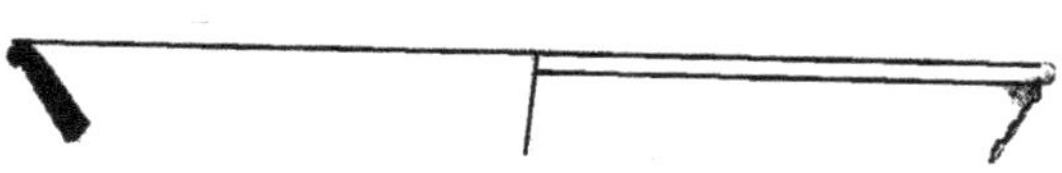 **D.139.T.345**	**D.139.T.345**:- Externally projecting everted rounded rim of a convex sided bowl having a white glaze on both surfaces and a black band placed on the interior just below the lip. (I.D = 23cm) SJN-B, TT4, NW , 2, 0.51, 51311, WGP, PINK(B)

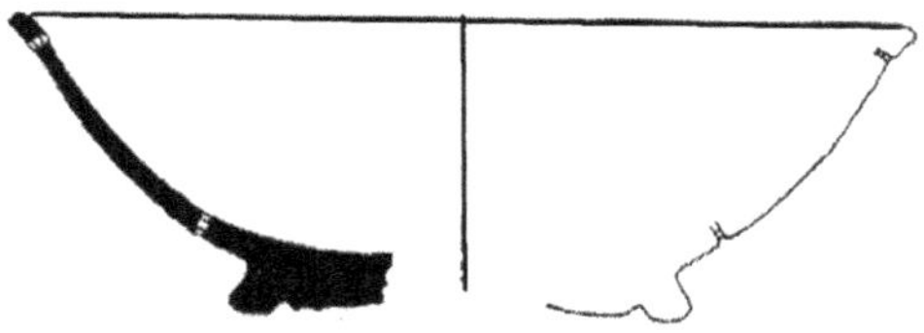 **T.15**	**T.15**:- Rim sherd of a dish with simple external projecting rounded rim. Fabric is pink. Traces of a whitish glaze can be seen .Mending marks in pairs along olf break lines. (I.D = 24cm). SJN-B, TT2, SE, 2, 0.81, 1509, WGP, PINK(B)
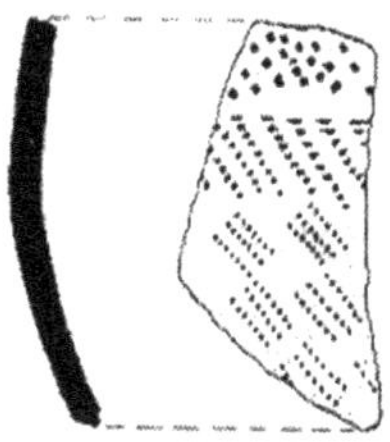 **D.142**	**D.142**:- Body sherd having a white glaze on the exterior but no glaze. This may be due to erosion. Deep square punctation marks form a band. After a small space another band of obliquely placed lines of dots are similarly punctuated. The lines of dots have been probably made with instrument. Under these lines can be seen a chess board pattern of small squares. The squares are found of three oblique dotted lines each. The fabric is reddish pink. SJN-D, LX1, NE, 1, 1.27, 21901, IGP, PINK(D)
 **D.105**	**D.105**:- Body sherd of a vessel having glaze on the external surface with painted hatched diamonds. Highly eroded. SJN-B-TK, UGPW, PINK(C)
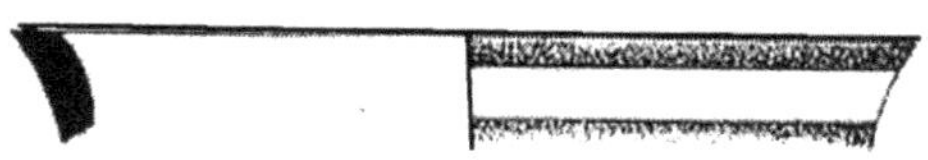 **D.140**	**D.140**:- Very small fragment of a rim having no glaze on the inner surface. The exterior has a white glaze with two ochre-brown horizontal bands. One band is just below the lip and the second is about 5 mm under it. SJN-B, TT4, SW, 2, 0.46, 51110, WGPW, PINK(B)
GLAZED RED WARE	
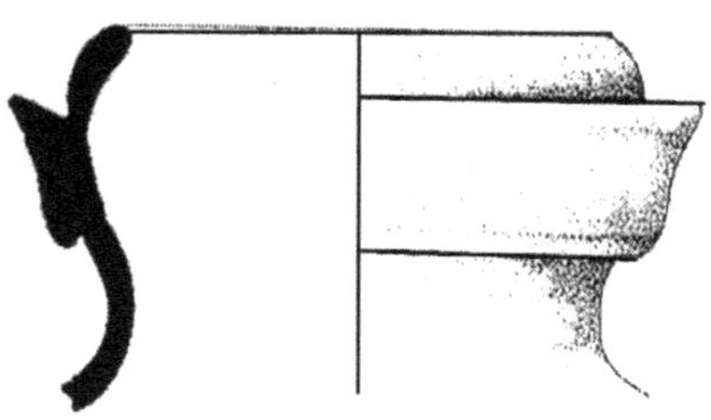 **T.401**	**T.401**:- Complex rim of a large jar having a double rim. The vessel form resembles modern glaze jars or "burnis". The vessel has a short neck which rises to form an inverted rounded rim. The neck has a convex profile (s-shaped). The glaze stops inside the mouth of the vessel though the slip continues on the interior. The second rim is a faux rim which is like a broad thick ledge applied on the exterior mid-way between the shoulder and the mouth of the vessel. The ledge widens as it goes up. It has two shallow grooves running along the exterior at ends. This external rim stops short of the mouth and a deep groove is found between the ledge and the internal rim. I.D of the inner rim is 7.5 cm and diameter of the outer rim is 11.5 cm) SJN-B, TT4, SW, 2, 4.60, 51145, GRW, RED(B)

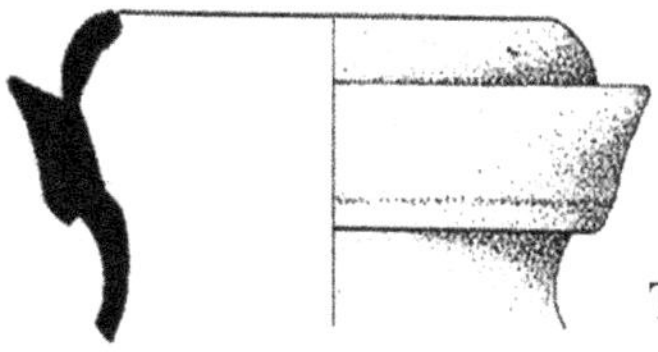 **T.401.1**	**T.401.1**:- A complex rim exactly the same as T.401. This one is broken in half. (I.D = 7.5cm)SJN-B, TT4, SW, 2, 4.80, 51146, GRW, RED(B)
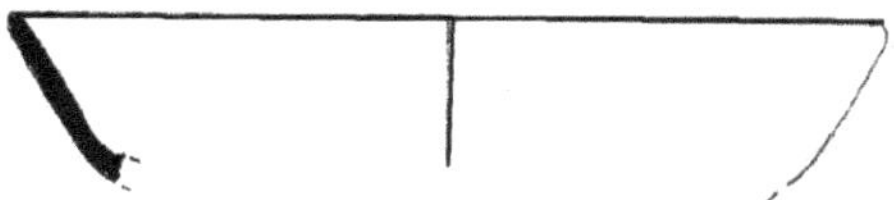 **T.9.1**	**T.9.1**:- Rim sherd of a bowl with simple everted rounded rim very similar to T.9. The only difference in shape is the sloping sides and blunt carination. The fabric is deep reddish-pink with an eroded greenish glaze (I.D = 20cm) SJN-B, TT1, NW, 2, 0.81, 309, GRW, RED(B)
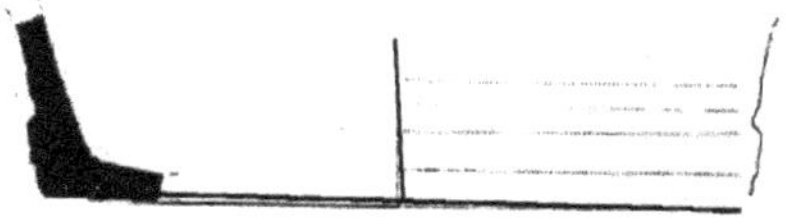 **Bs.6**	**Bs.6**:- Contiguous flat base of a straight-sided vessel. Olive green glaze on both surfaces. (D = 16) SJN-B, TT1, NE, 3, 1.63 - 2.23, 522, GRW, RED(B) (Also seen in MGP but with a smaller diameter of 9 – 10cm)
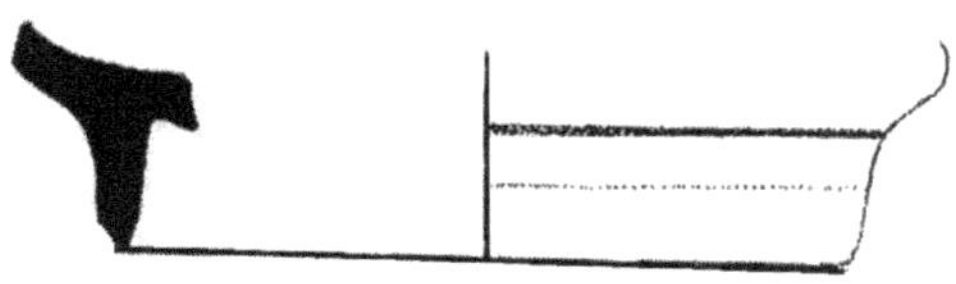 **Bs.44**	**Bs.44**:- High ring base with greyish white background and two pale blue horizontal bands on the outer surface of the ring. Crackled effect.SJN-K, B4, NE, 1, 0.08, 8501, FRIT, FRIT
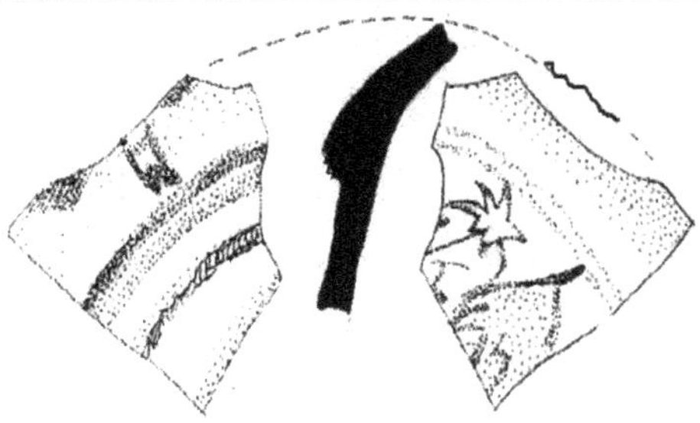 **D.96**	**D.96**:-Base fragment of a bowl. Pale bluish white background. Design of inner and outer surface. One other surface two horizontal blue bands are seen just above the ring base and some vertical design/lines above the band. One inner surface a blue palm tree like motif is painted next to a dune/spire like depiction (could be part of a landscape/scenery). SJN-K, A4, SE, 1, 0.27, 4701, FRIT, FRIT

UNGLAZED EARTHENWARE

EGGSHELL WARE	
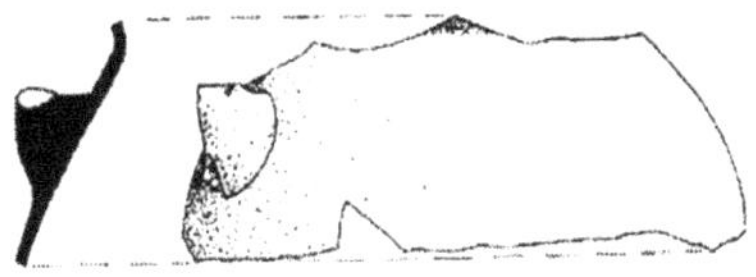 **T.396**	**T.396**:- Shoulder of a globular pot / pitcher having a luted handle. The exterior is not smooth and the handle appears to have been oval or circular in section. There have been a creamy white wash on the vessel but it has been eroded. The fabric is pale pink and not very fine.SJN-B, TT4, SW, 2, 3.94, 51142, ESW(P), PINK(G)
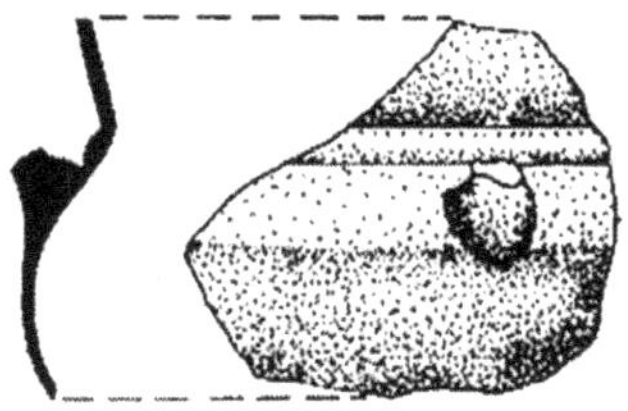 **T.398**	**T.398**:- Shoulder and neck of a straight sided globular pitcher or flask having a blunt carination and part of a luted handle. A shallow ridge can be seen just above the luted handle on the shoulder.SJN-B, TT4, SW, 2, 3.74, 51141, ESW(R), RED(A)
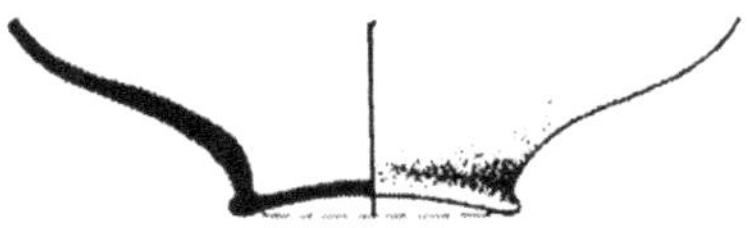 **Bs.76**	**Bs.76**:- Non-contiguous discoid base with a sharp rim and slightly concave in the centre. The body is globular and flares outwards. The centre of the vessel is slightly deep. SJN-B, TT4, SW, 2, 4.45, 51144, ESW(W), BUFF(G)
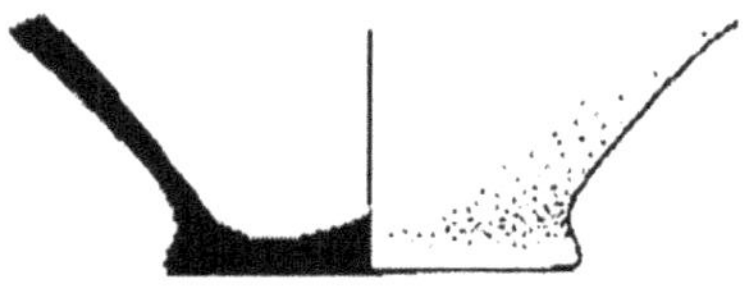 **Bs.76.1**	**Bs.76.1**:- Very similar to Bs.76 but this base is flat on the underside. The centre of the interior is raised like a cake. The body of a vessel curves immediately above the base SJN-B, TT4, SW, 2, 4.45, 51144, ESW(W), BUFF(G)
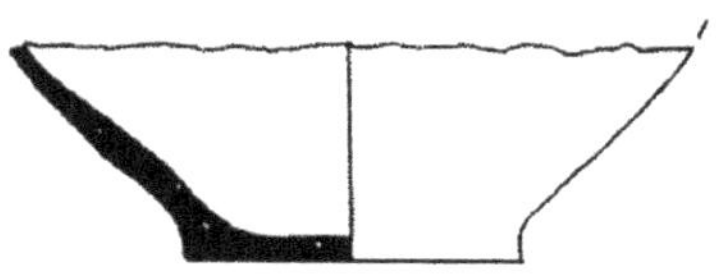 **Bs.23**	**Bs.23**:- Discoid base. (Th = 2mm) SJN-B, TT1, NE, 2,0.81,506, ESW (W), BUFF (G)
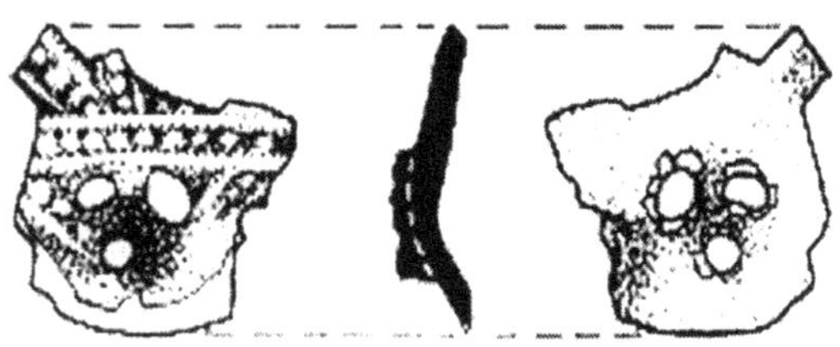 **D.187**	**D.187**:- Fragment of a filter having six extant perforation and a moulded complex design consisting of a beads/dots and lines. Raised clay can be seen on the rivers due to the perforations. The thin body itself appears to be slightly curved or depressed. The breakage pattern of this sherd is angular. The design appears moulded. SJN-B, TT4, SW, 2, 3.74, 51141, ESW(W), BUFF(G)

 **D.187**	**D.187**:- Probably part of the same filter as above.
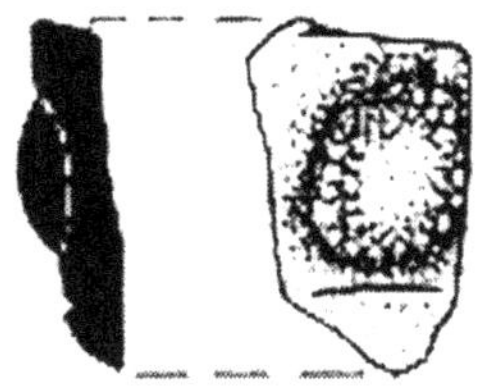 **D.188**	**D.188**:- Small body fragment having one incised line and a composite circular dot which is appliquéd. It appears to have smaller dots and some eroded design which may have been moulded. SJN-B, TT4, NE, 3, 1.34, 51537, ESW(W), BUFF(G)
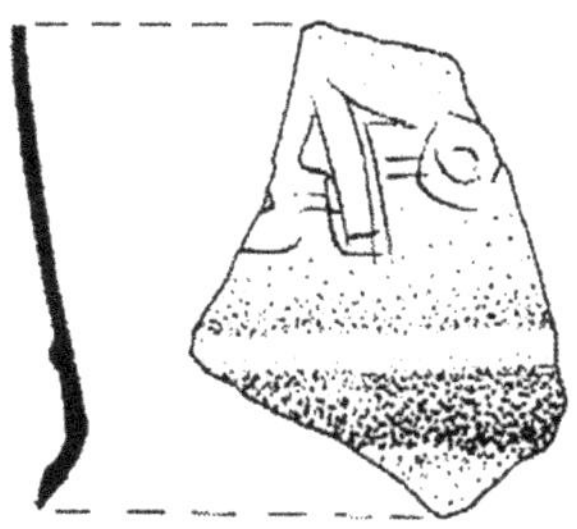 **D.190.2**	**D.190.2**:- Body sherd having part of an incised Kufic inscription similar to the previous sample. SJN-B, TT4, SW, 2, 4.45, 51144, ESW(W), BUFF(G)
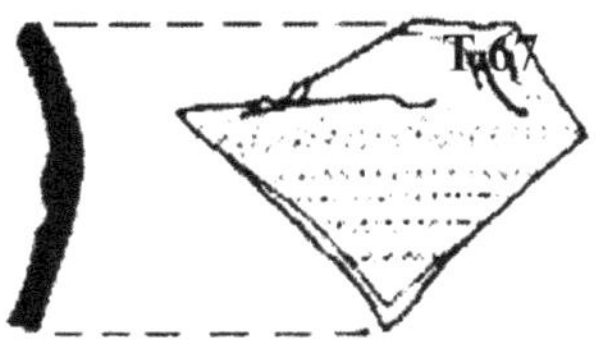 **D.190.3**	**D.190.3**:- Fragment of neck and shoulder having traces of same inscription but fine ridges and grooves are present below it. SJN-B, TT4, SW, 2, 4.60, 51145, ESW(W), BUFF(G)
 **T.399**	**T.399**:- Tea cup handle having a long slim stem, circular in section. The handle was luted on the body and extending from the rim to the shoulder. A small thumb rest has been attached to the top of the curve started back a little. SJN-B, TT4, SW, 2, 3.94, 51142, ESW(R), RED(A)
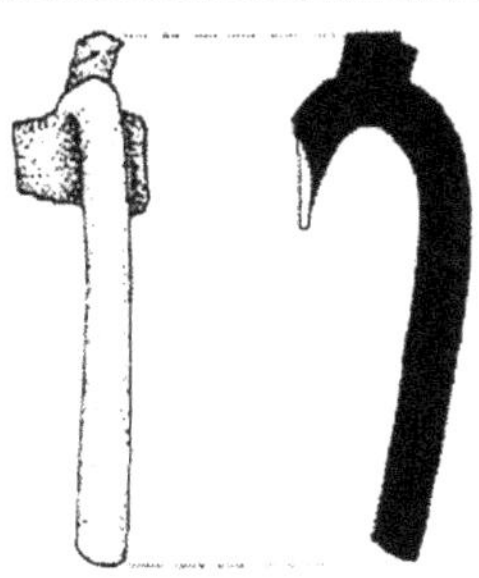 **T.399.1**	**T.399.1**:- Tea-cup handle almost exactly like T.399, difference being that the thumb rest being pressed on to the top slightly somewhat crookedly. SJN-B, TT4, SW, 2, 3.94, 51142, ESW(R), RED(A)

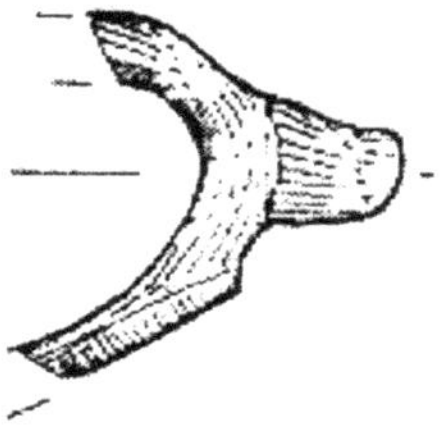 **T.74**	**T.74**:- Tea-cup handle SJN-B, TT2, SW, 3, 1.22, 914, ESW(W), BUFF(G)
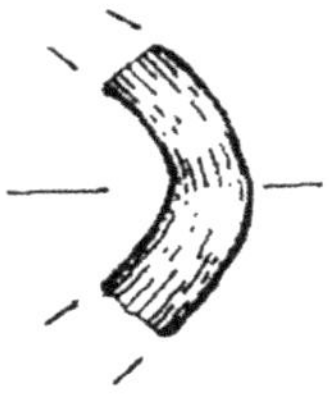 **T.373.1**	**T.74.1**:- Tea-cup handle SJN-B, TT1, NW, 3, 1.07, 318, ESW(W), BUFF(G)
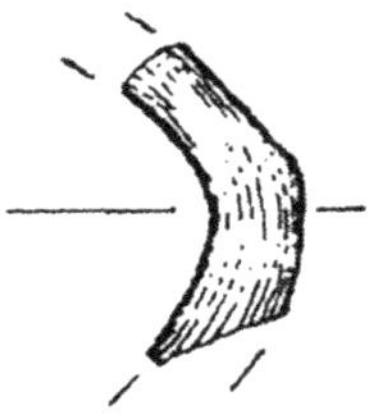 **T.74.2**	**T.74.2**:- Tea-cup handle SJN-B TT1, NW, 2, 0.87, 312, ESW(W), BUFF(G)
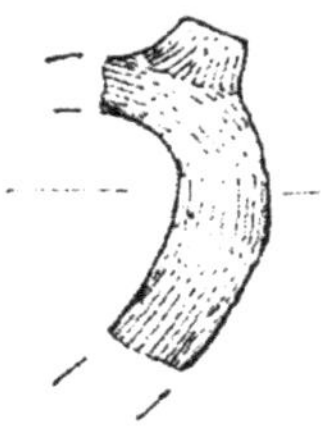 **T.74.3**	**T.74.3**:- Tea-cup handle SJN-B, TT1, NW, 3, 2.1, 335, BW, BUFF(F)
 **T.395**	**T.395**:- Small curved fragment of a tea-cup handle with a thumb rest. The handle is round in section and the thumb rest has narrow triangular points towards the vessel. It has a cut on the top giving it a appearance of animal head with horns and ears. It appears to have been pinched into this shape. SJN-B, TT4, SW, 2, 3.94, 51142, ESW(W), BUFF(G)
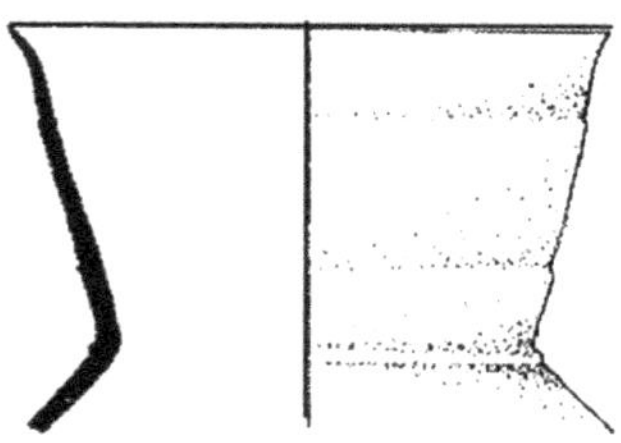 **T.393.2**	**T.393.2**:- Rim and neck of a straight sided pitcher with a globular body. A low ridge is seen at the junction of neck and shoulder. The exterior of the neck has two offsets. In between these a small fragmentary incisions can be seen which may have been of an inscription or design. The shoulder is slightly thicker than the neck. (I.D = 8cm) SJN-B, TT4, SW, 2, 4.45, 51144, ESW(W), BUFF(G)

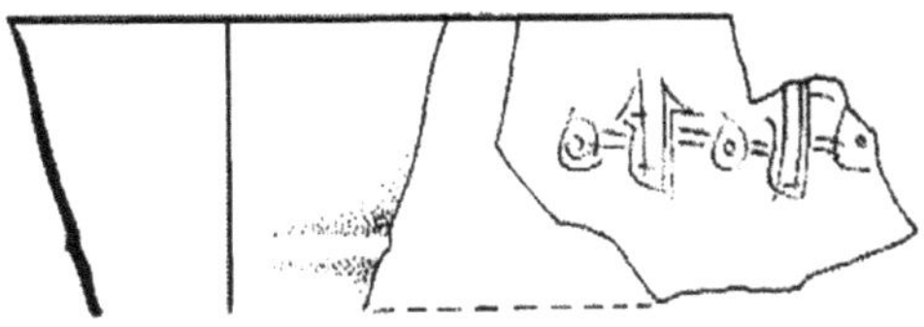 **T.393.5 D.190.1**	**T.393.5 D.190.1**:-Rim sherds of a straight sided globular bodied pitcher having tapering sharp rim. The sides of the vessel have an effect on the exterior and the broad angular ridge just above the shoulder. An incised inscription very similar to T.393.4 and D.190 can be seen between the ridge and the offset. The upper part of the inscription touches the offset. (I.D = 8cm) SJN-B, TT4, SW, 2, 4.88, 51147, ESW(W), BUFF(G)
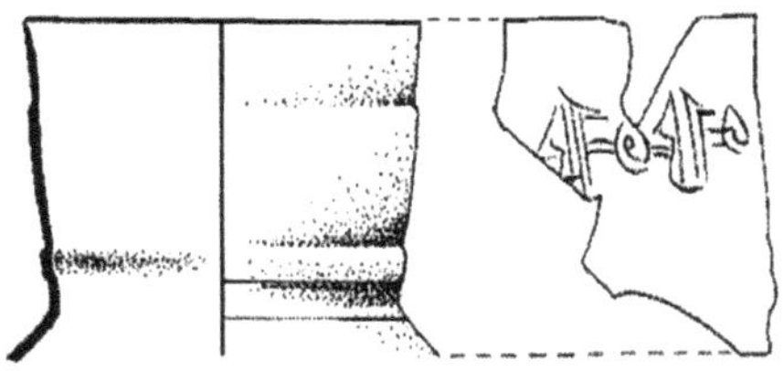 **T.393.5 D.190.1**	**T.393.5.D.190.1**:-Rim sherds of a straight sided globular bodied pitcher having tapering sharp rim. The sides of the vessel have an effect on the exterior and the broad angular ridge just above the shoulder. An incised inscription very similar to T.393.4 and D.190 can be seen between the ridge and the offset. The upper part of the inscription touches the offset. (I.D = 8cm) SJN-B, TT4, SW, 2, 4.88, 51147, ESW(W), BUFF(G)
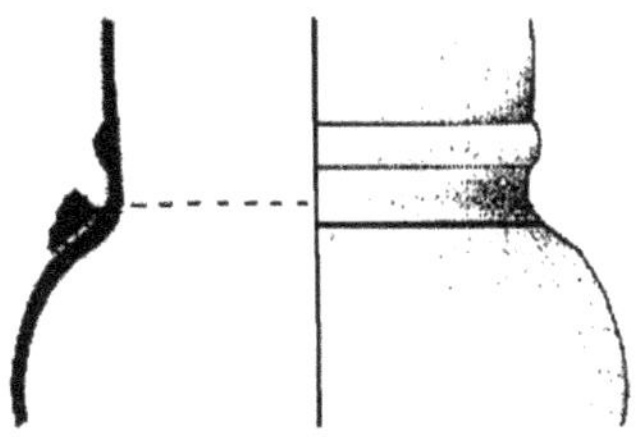 **T.394**	**T.394**:- Neck and shoulder of a pitcher or jar similar to the previous samples. This body sherd has two shallow grooves and a ridge at the juncture with the neck. The shoulder of the body has a shallow groove near the neck and a blunt carination. On this blunt carination can be seen part of the luted handle extending upwards. SJN-B, TT4, SW, 2, 4.88, 51147, ESW(W), BUFF(G)
 **T.67.1**	**T.67.1**:- Rim sherd of a small vessel with an external projecting rounded rim similar to T.67. No surface treatment is evident. (I.D = 5cm) .SJN-B, TT1, SW, 4, 2.25, 143, BW, BUFF(F)
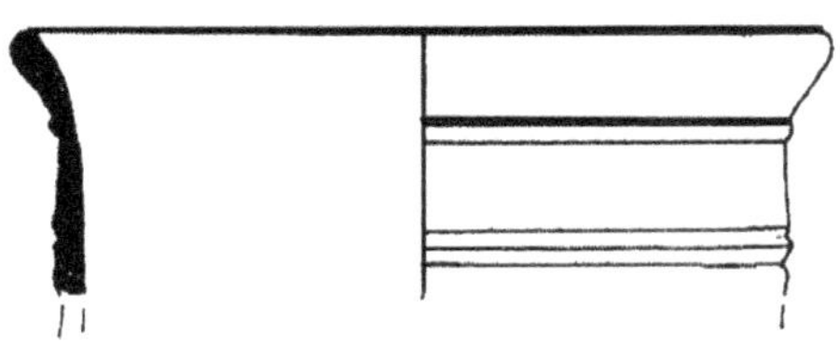 **T.68**	**T.68**:- Rim sherd of a pot with external projecting everted rounded rim which broadens at the tip. It has an uneven ridge on the neck. Untreated surface. The shoulder has some deep grooves (I.D = 12cm). SJN-B, TT1, NW, 3, 1.07, 318, BW, BUFF(F)

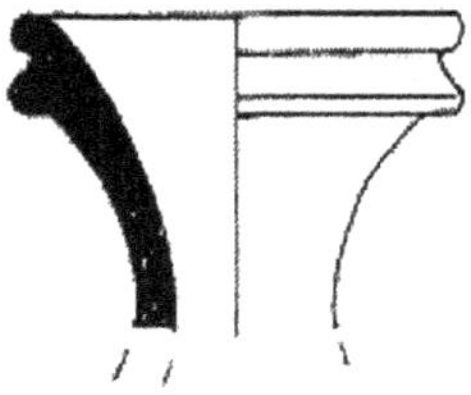 **T.70**	**T.70:-** Rim sherd of a narrow mouthed pot with external projecting extremely everted rounded rim with a bevelled ledge just below. Untreated surface. White fabric (I.D = 6cm). SJN-B, TT1, SE, 2, 0.72, 708, BW, BUFF(F)
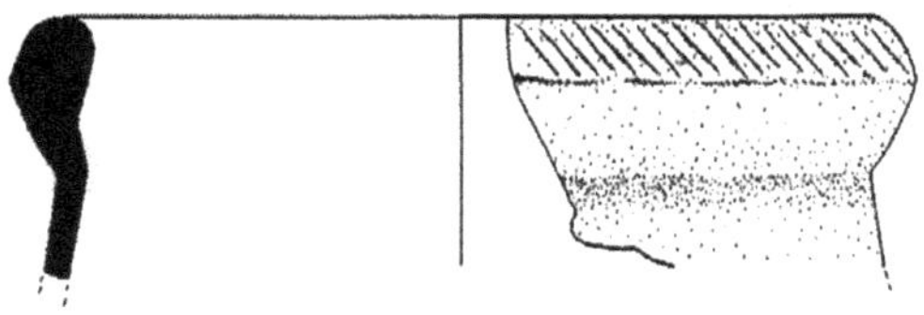 **T.387 D.181**	**T.387 D.181**:- Rim sherd of a high necked vase or storage jar having a slightly inverted profile with a rounded clubbed rim which projects externally. A band of oblique lines has been incised on the exterior of the rim on the upper part of the curve. (I.D = 20cm) SJN-B, TT4, NE, 5, 2.02, 51550, BW, BUFF(E)
 **T.387.1**	**T.387.1:-** It is a smaller rim similar to T.387. SJN-B, TT4, NE, 5, 1.86, 51547, BW, BUFF(E)
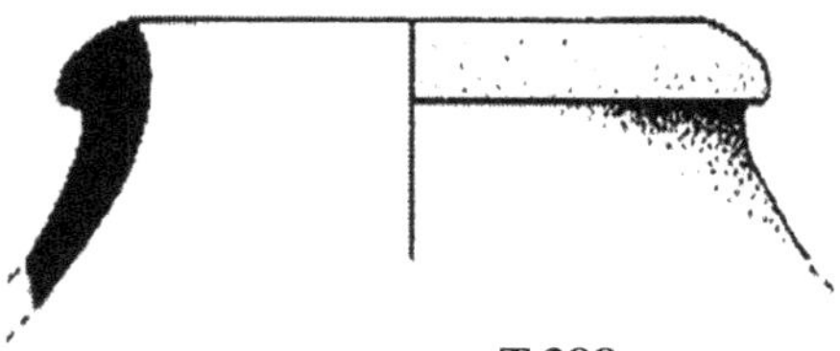 **T.388**	**T.388**:- Rim sherd of a small pot having an external projecting triangular beaked rim which seems to be overturned. (I.D = 6cm) SJN-B, TT4, NE, 3, 1.50, 51539, BW, BUFF(F)
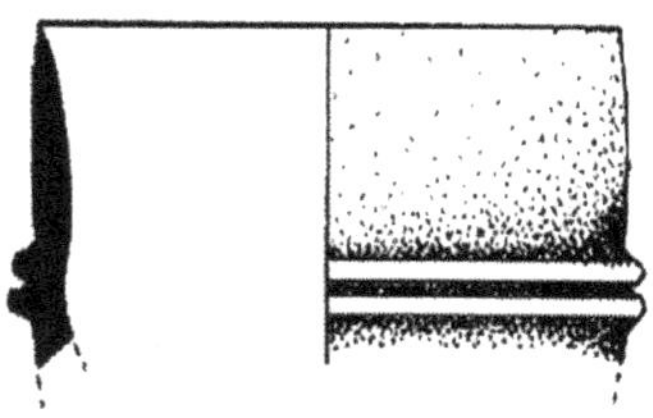 **T.389**	**T.389**:- Rim sherd of a beaker having a vertical tapering rim and two sharp ridges and placed on the exterior lower down the body. (I.D = 6cm) SJN-B, TT4, NE, 2, 1.03, 51528, BW, BUFF(F)
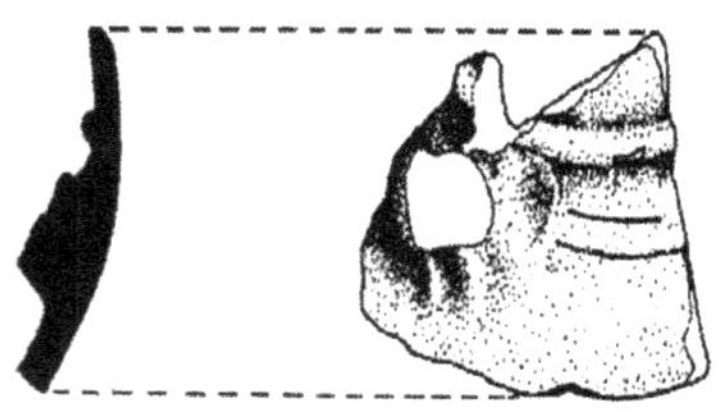 **T.390**	**T.390**:- Neck and shoulder of a globular pot, having a high ridge at the juncture of neck and shoulder. A broken luted handle can be seen just below the ridge. The luting has been done quite crudely. SJN-B, TT4, NE, 5, 2.16, 51553, BW, BUFF(F)

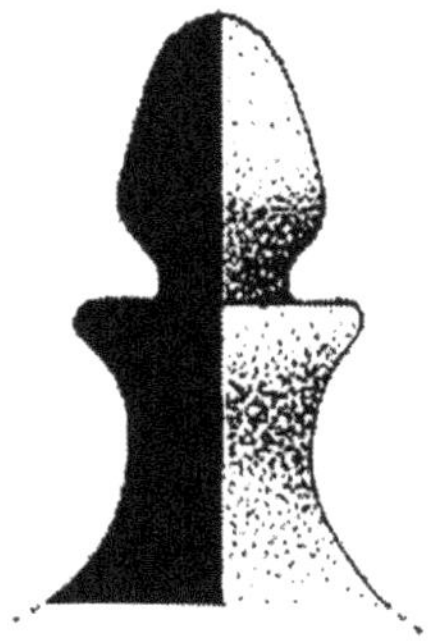 T.391	**T.391**:- This is a turban handle with a tall dome like finial having a narrow waist. It is fixed to a flat circular platform which is also waisted. It broadens downwards to the point where it was attached to the actual stem of the handle. SJN-B, TT4, SW, 2, 4.45, 51144, BW, BUFF(A)
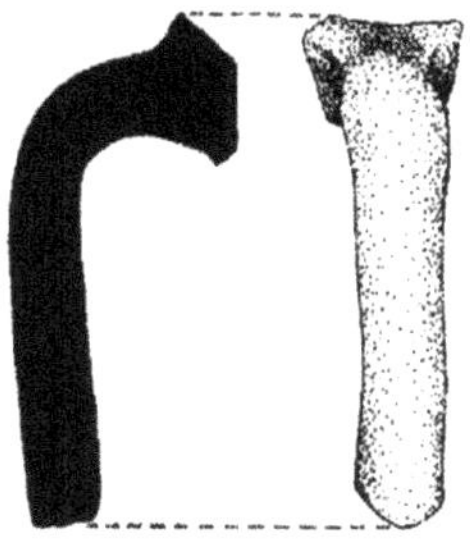 T.392	**T.392**:- Very small rim sherd having an everted tapering rim. A large curved handle is luted to the rim. The vessel was probably a high vertical necked jar or vase having a handle extending from the rim to the shoulder. The handle is rectangular in section with rounded edges. The vessel may have been quite large. The rim is too small to measure. SJN-B, TT4, NE, 6, 3.32, 51577, BW, BUFF(A)
 T.73	**T.73**:- Handle. Round in section. (L = 5.5cm) SJN-B, TT2, SE, 3, 1.31, 1519, BW, BUFF(A)
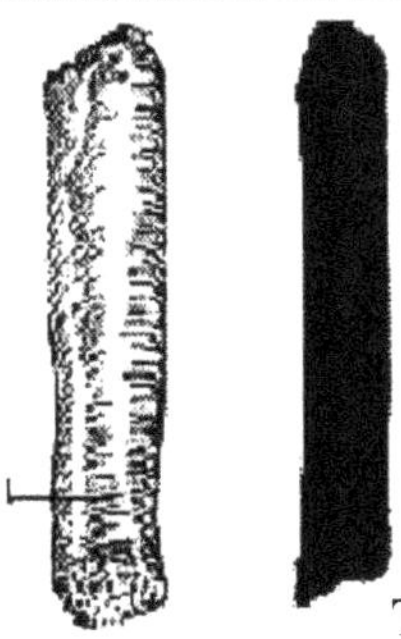 T.73.1	**T.73.1**:- Handle. Round in section. (L = 4cm) SJN-B, TT2, SE, 4, 1.81, 1528, BW, BUFF(F)
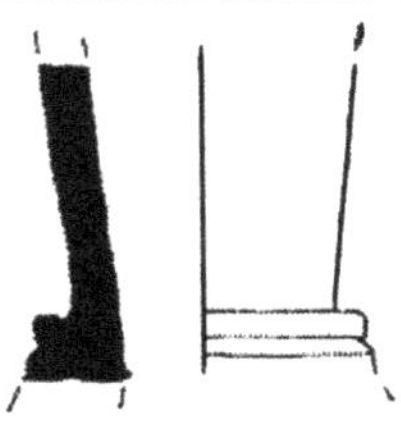 Bs.36	**Bs.36**:- Neck of a flask or bottle, mistakenly noted as stem of a tubular stand. (I.D = 2.5cm) SJN-B, TT2, SE, 3, 1.52, 1522, BW, BUFF (A)

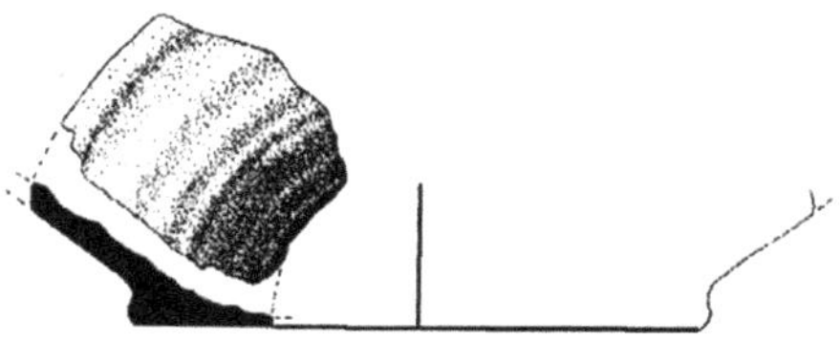 **Bs.74**	**Bs.74**:- Non-contiguous discoid base having broad turning ridges on the interior. SJN-B, TT4, NE, 2, 0.95, 51526, BW, BUFF(E)
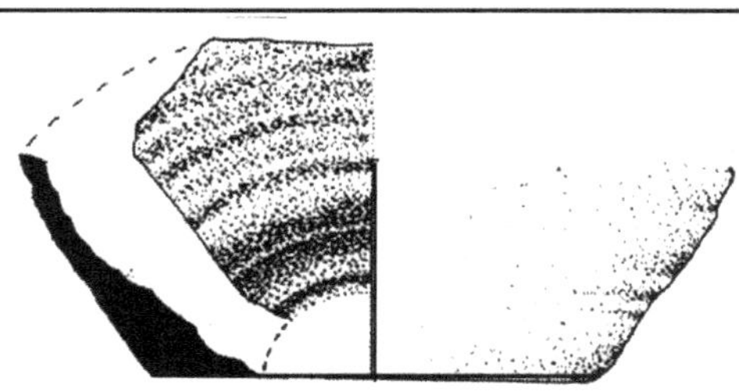 **Bs.75**	**Bs.75**:- Contiguous flat base, slightly convex sided with turning ridges on the interior and a very thin base floor. SJN-B, TT4, NE, 3, 1.43, 51538, BW, BUFF(F)
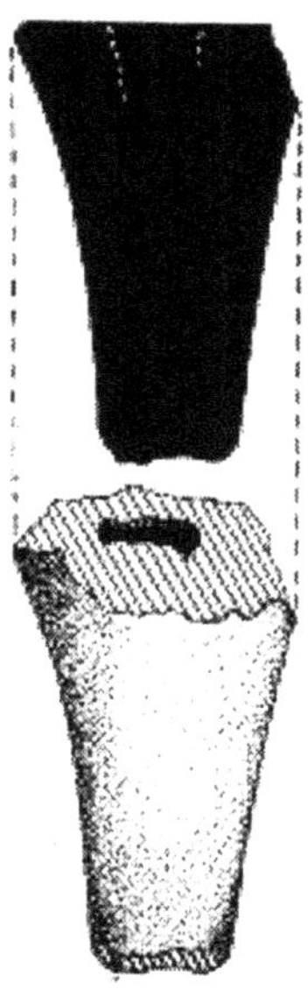 **Bs.77**	**Bs.77:-** Elongated toe of an amphora or Torpedo jar. The fabric is yellowish-white. The surface is uneven and the end is broken or damaged but it is round in section. It is probably hand made and is hollow on inside. The interior was lines with bitumen. Residue can be seen coating the inside and on the section of the vessel after it broke. SJN-B, SURFACE, BW(TJ), BUFF(A)
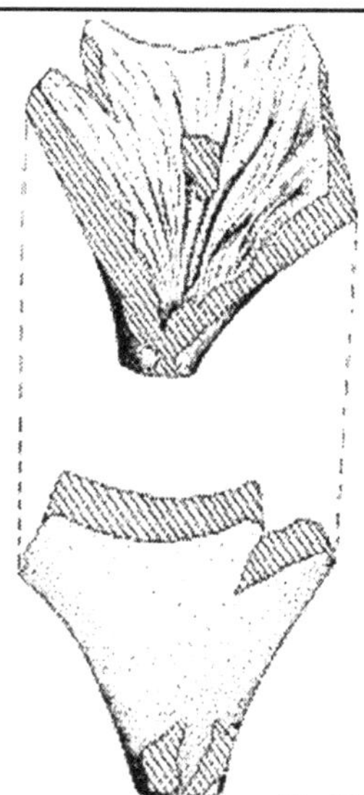 **Bs.77.1**	**Bs.77.1**:- Toe of an amphora or Torpedo jar. It is not elongated like Bs.77 but it also has a round toe. The body flares upwards immediately from it. Residue or bitumen can be seen on the interior vessel wall. SJN-B, TT4, NE, 5, 2.43, 51559, BW(TJ), BUFF(A)

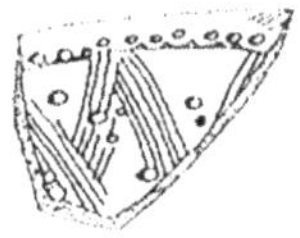 **D.36.2**	**D.36.2**:- Diamond design formed by bands of fine lines with dots in the centre. (Th = 5m) SJN-B, TT1, NW, 3, 2.23, 337, BW, BUFF(A) (ALSO SEEN IN TGW)
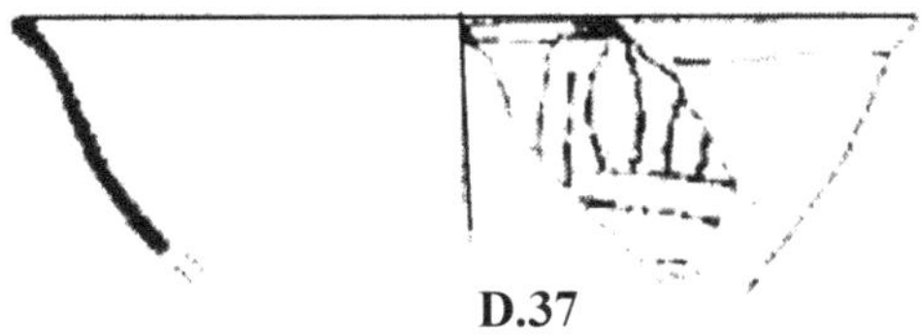 **D.37**	**D.37**:- Oval incised within two parallel horizontal and vertical lines. (I.D = 8cm) SJN-B, TT2, SE, 4, 1.57, 1523, BW, BUFF(F)
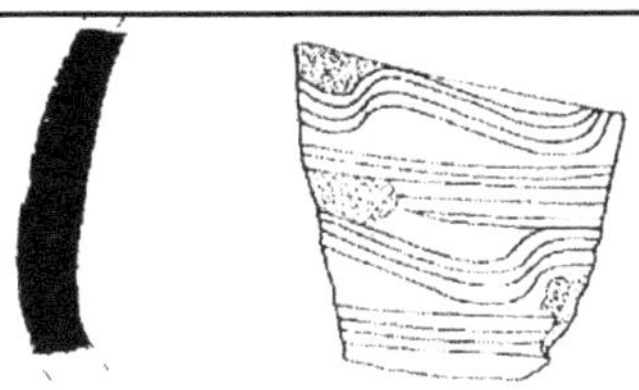 **D.38**	**D.38**:- Horizontal, alternate bands of horizontal and wavy lines
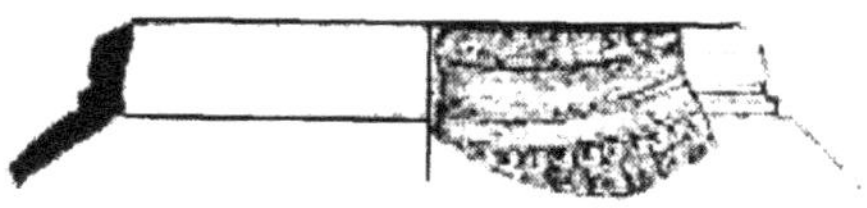 **D.39.1**	**D.39.1**:- stamped design of square and circle. (Th = 4mm) SJN-B, SURFACE, BW, BUFF(F)
 **D.108**	**D.108**:- Neck and shoulder portion of a small pot There is an appliqué horizontal band with pinched design on the shoulder. The neck and body is corrugated. SJN-B-TK, BW, BUFF(A)
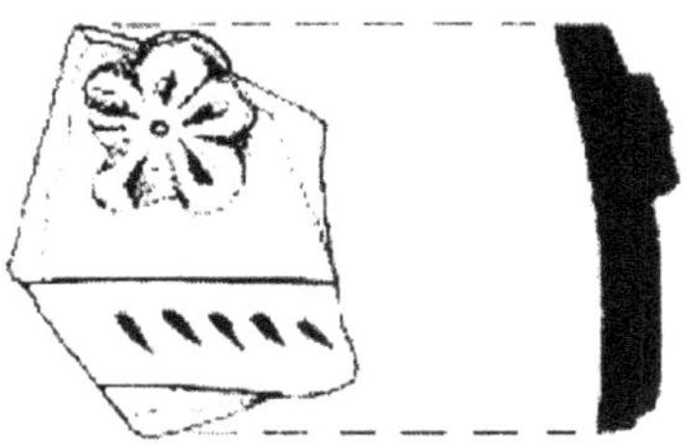 **D.182**	**D.182**:- Body sherd of a jar or a vase having a floral stamped motif appliquéd on to it. The flower has six petals with scooped centres and a hollow dot in the middle. Under this appliqué motif is a band of two incised lines containing five oblique incised lines. SJN-B, TT4, NE, 5, 1.80, 51546, BW, BUFF(F)
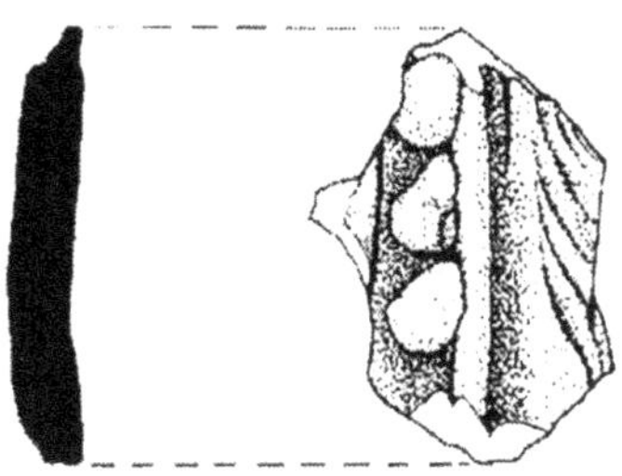 **D.183**	**D.183**:- Body sherd with appliqué lines and dots having single incised line on one side of the dots and a band of incised oblique lines beside the appliquéd lines. SJN-B, TT4, NE, 5, 1.96, 51549, BW, BUFF(E)

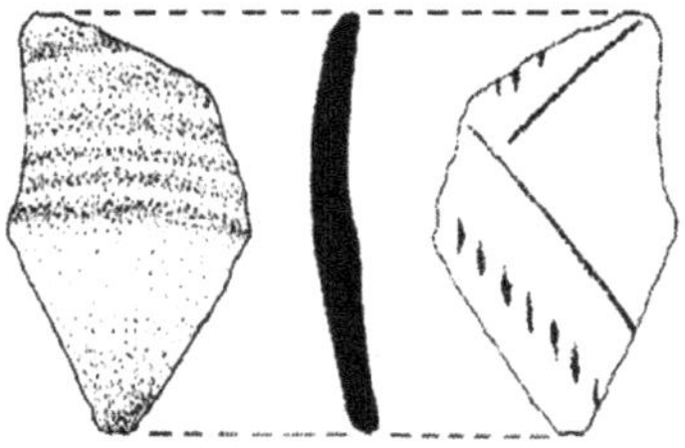 **D.184**	**D.184**:- Body sherd having incised designs of two perpendicular lines and oblique small incisions (shaped like rice grains). SJN-B, TT4, NE, 5, 1.96, 51549, BW, BUFF(E)
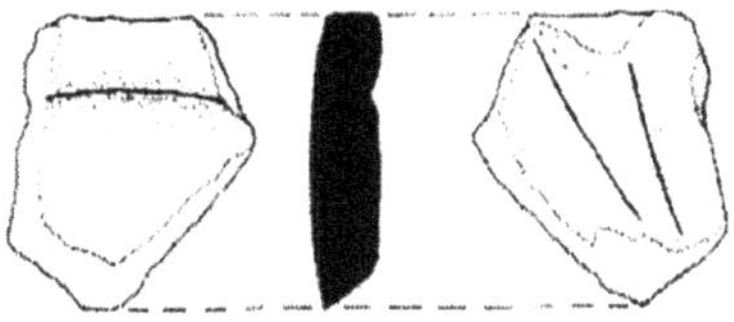 **D.184.1**	**D.184.1**:- Body sherd having a shallow incised design similar to D.184. SJN-D, LX1,NW, 1, 1.35, 21701, BW, BUFF(E)
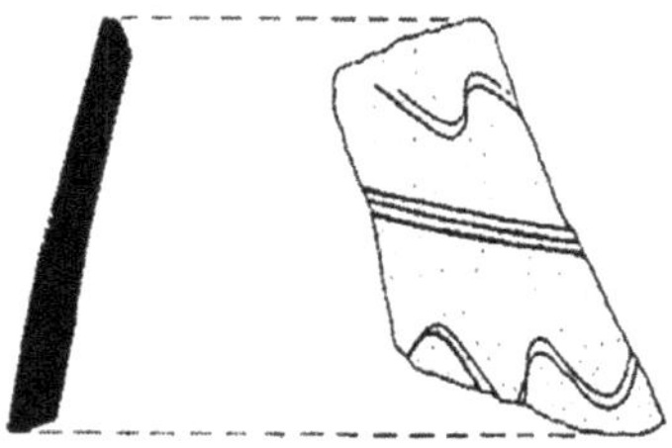 **D.185**	**D.185**:- Neck fragment of a jar or vase having a band of three grooves and a double lined wavy band under it. The incisions may have been done with an instrument. SJN-B, TT4, NE, 5, 2.11, 51552, BW, BUFF(E)
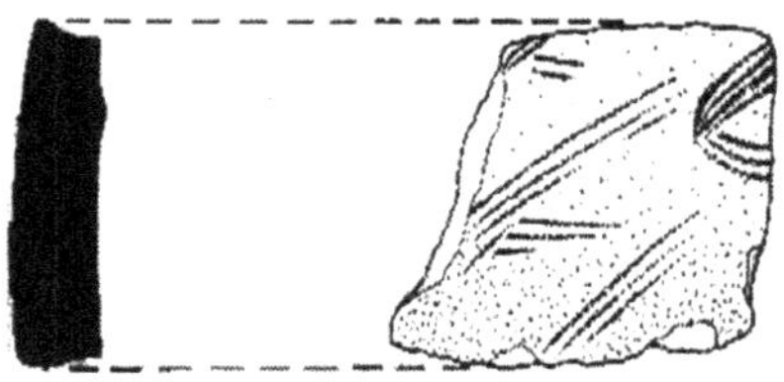 **D.186**	**D.186**:- Body sherd of a jar having a design incised on the exterior using a comb like instrument or tool. SJN-B, TT4, SW, 2, 2.91, 51137, BW, BUFF(E)
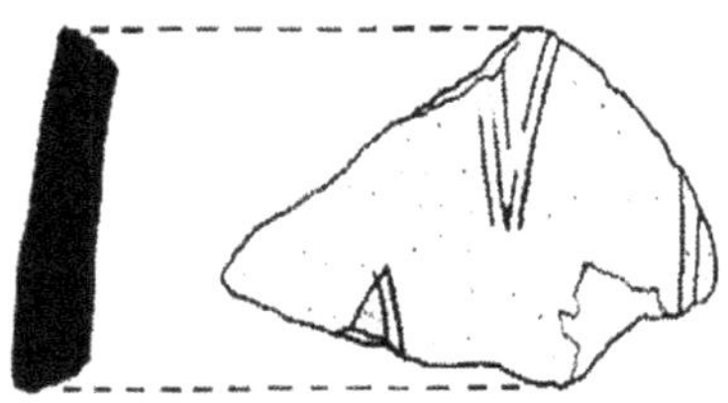 **T.186.1**	**D.186.1**:- Design similar to D.186 but shallow incisions. SJN-B, TT4, SW, 2, 2.91, 51137, BW, BUFF(E)
UNGLAZED PINK WARE	
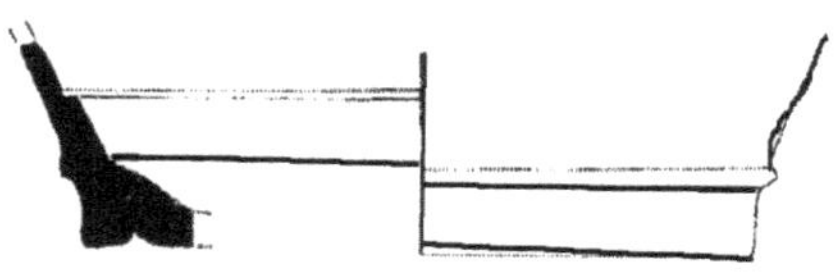 **Bs.18**	**Bs.18:-** Flat base with a sharp groove scraped underneath to create a ring. Sharp ridge on the exterior. SJN-B, TT1, NW, 2, UGP PINK(B)

WHITE SLIPPED PINK WARE	
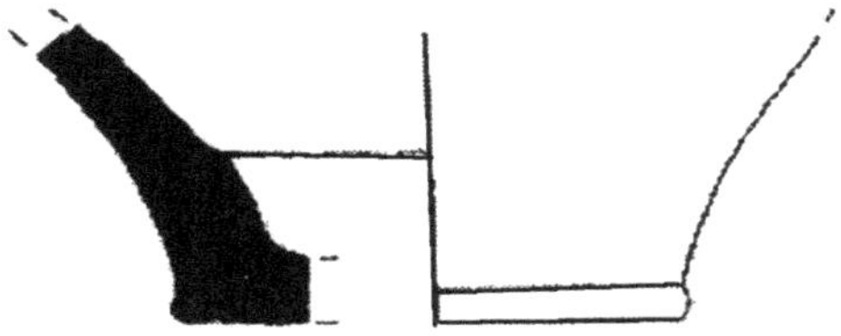 **Bs.20**	**Bs.20**:- Contiguous flat base. (D = 10cm) SJN-B, TT1, NE, 2, 0.71, 504, WSPW-1, PINK (F)
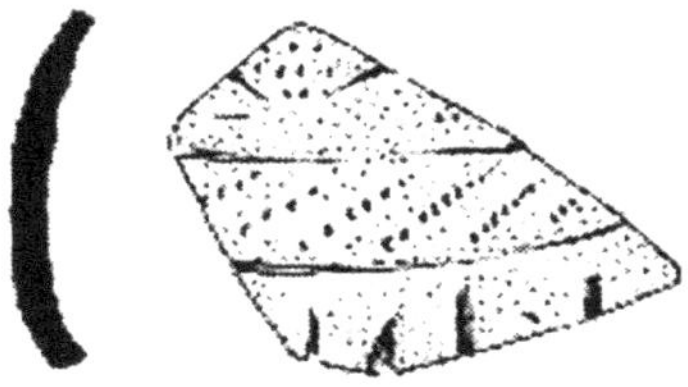 **D.106**	**D.106**:- Unglazed body sherd. Design consists of two horizontal incised lines containing oblique lines of punctate dots. Some punctate dots are also seen on the rest of the body. SJN-K, B5, SW, 1, 0.26, 8903, WSPW-3, PINK(F)
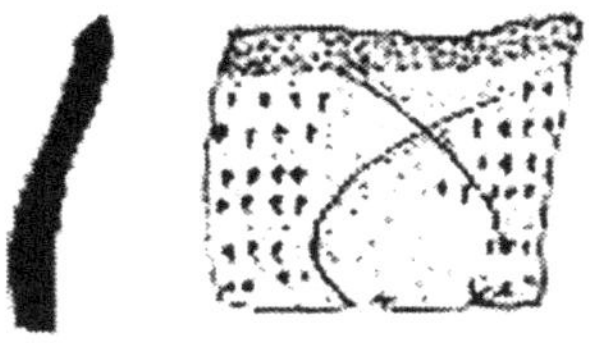 **D.106.1**	**D.106.1**:- Similar to above sample (probably same vessel). An incised curving line forms a loop on either side of which can be seen horizontal rows of punctate dots. SJN-K, B5, SW, 1, 0.26, 8903, WSPW-3, PINK(F)
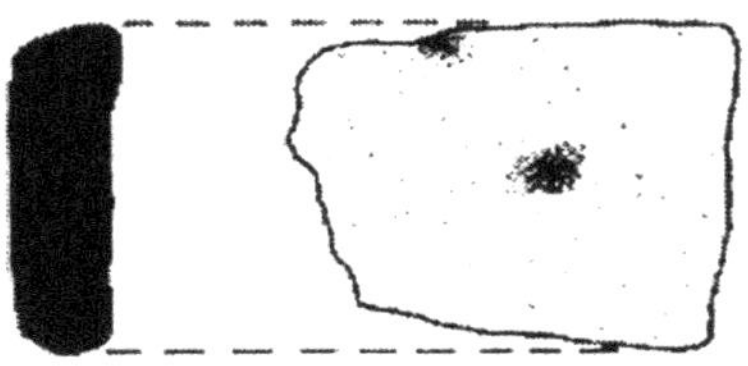 **D.159**	**D.159**:- Body sherd with partial-perforation. Black coating on the interior. SJN-B, TT4, NE, 4, 1.73, 51544, WSPW-2, PINK(E)
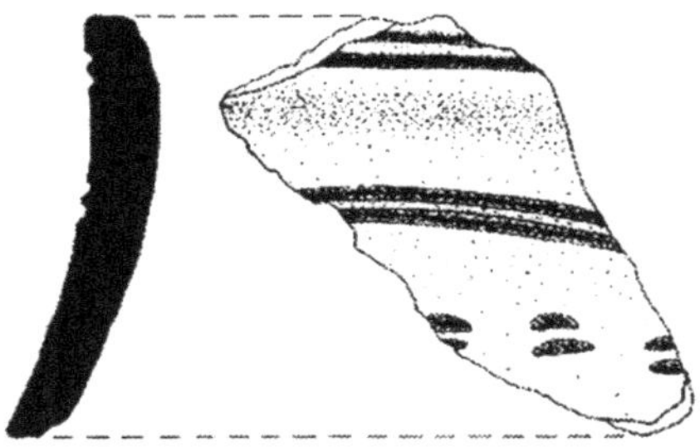 **D.160**	**D.160**:- Shoulder of a thick storage jar with two horizontal bands widely spaced, each consisting of two grooves and one sharp ridge. The lower part of the neck has a band of thumbnail impressions in sets of two. SJN-B, SURFACE, WSPW-1, PINK(E)
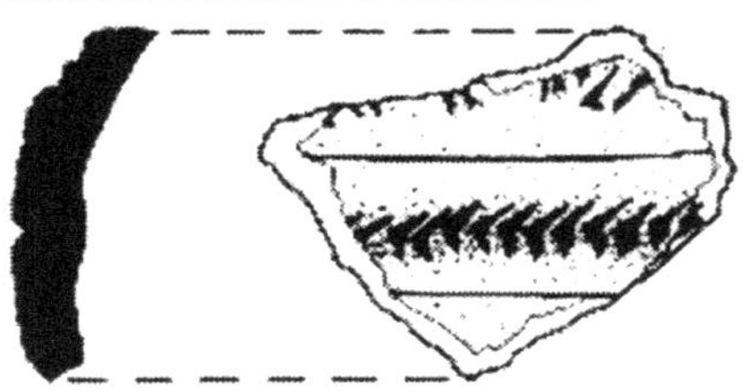 **D.161** **T. 384 D. 177.2**	**D.161**:- Body sherd of a small jar. The exterior has a thick white slip. A sharp and deep rouletted design is seen. It looks like an angular chain or band. SJN-B, TT4, NE, 5, 2.16, 51553, WSPW-3, PINK(B)

 **T.27**	**T.27**:-Turban Handle SJN-B, TT1, NE, 2, 0.76, 505, WSPW, PINK(E)
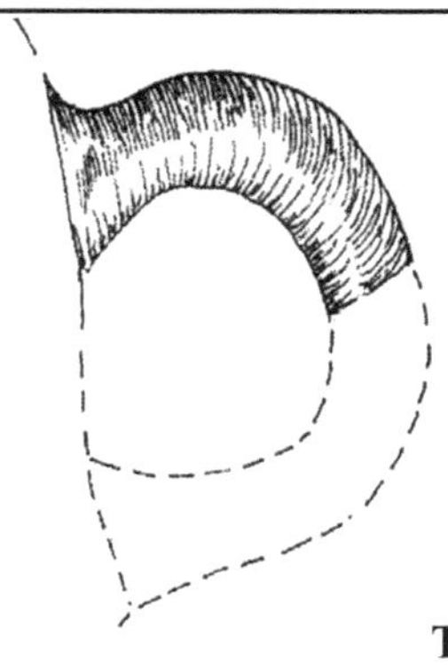 **T.29**	**T.29**:- Knuckle-shaped, semi-circular handles with two high, sharp ridges and a deep groove in-between. (L = 7cm) SJN-B, TT1, NW, 3, 1.15, 320, WSPW-1, PINK(E)
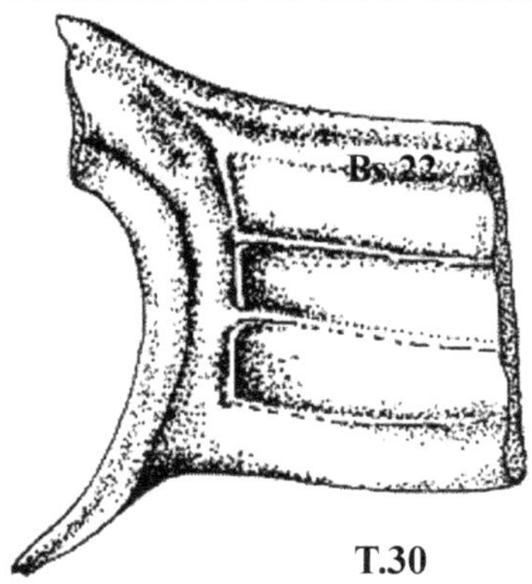 **T.30**	**T.30**:- Flat, very wide handle with three grooves within two high ridges and divided by two shallow ridges. This specimen has a rim under which the handle is luted. The rim is of the T.25 type with an I.D of 9cm) SJN-B, SURFACE, WSPW-1, PINK(F)
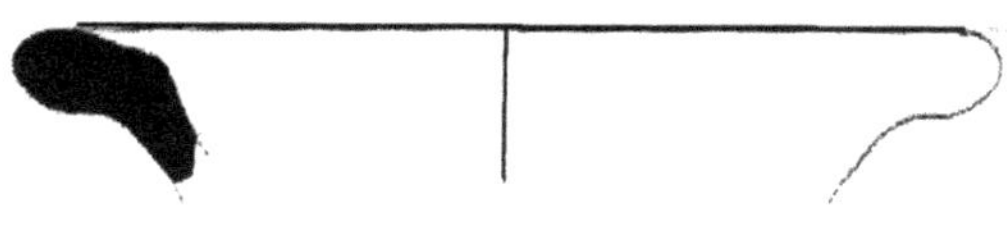 **T.357**	**T.357**:- Rim sherd of a thick storage jar having external projecting extremely everted rounded rim. Thin white coating like a wash on the exterior and upper part of the rim. SJN-B, TT4, SW, 0.42, 51109, WSPW-1, PINK(D)
 **T.358**	**T.358**:- featureless rim of a storage vessel or basin. Internal surface has a coating of a black substance which could be bitumen / asphalt and some residue. Creamy white exterior. Compact fabric. (I.D = 43cm) SJN-B, TT4, NE, 4, 1.73, 51544, WSPW-2, PINK(E)

SLIPPED BRITTLE WARE	
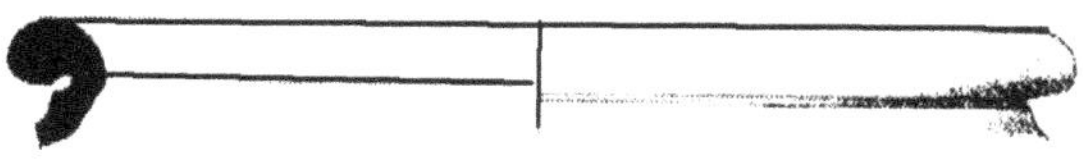 **T.362**	**T.362**:- External projecting, extremely everted, elliptical rim of a pot having a constricted neck. A deep and sharp groove at the juncture of the neck gives the rim a beaked appearance. (I.D = 25cm) SJN-B, TT4, NW, 2, 0.66, 51317, SBW-1, PG(A)
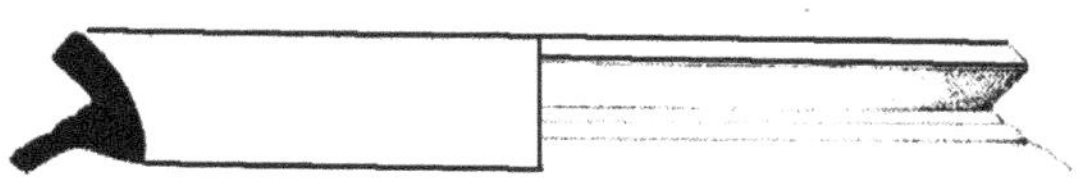 **T.363**	**T.363**:-Bilaterally projecting, everted, wide quadrangular rim of a pot with a constricted neck. A low ridge runs along the exterior of the body just below the rim. (I.D = 18cm) SJN-B, TT4, NW, 3, 0.92, 51327, SBW-2, PG(A)
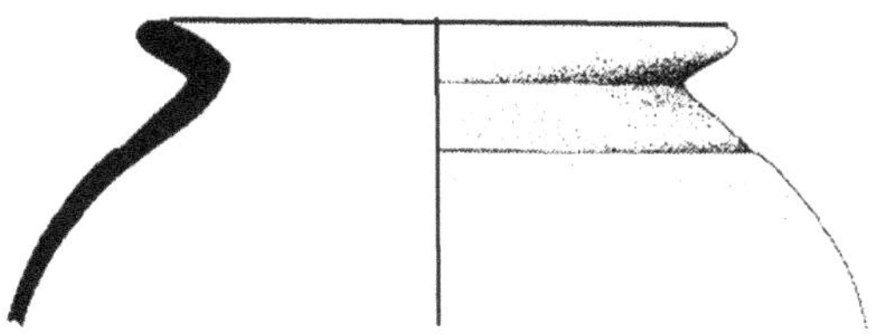 **T.364**	**T.364**:- Rim sherd of a pot having an external projecting tapering rim with a wide flange. A low ridge runs along the shoulder. (I.D = 20cm) SJN-B, TT4, NE, 5, 2.05, 51551, SBW-2, PG(A)
 **T.364.1**	**T.364.1**:- External projecting, everted rim, similar to T.364. (I.D = 16cm) SJN-B, TT4, SW, 2, 3.11, WELL, 51138, SBW-3, PG(A)
 **T.365**	**T.365**:- External projecting everted rim of a pot. The rim is perpendicular to the neck and has a deep groove. The lip turns up from this groove and is rounded. A low ridge runs along the neck just below the rim. (I.D = 15cm) SJN-B, TT4, SE, 2, 0.73, 51711, SBW-2, PG(A)
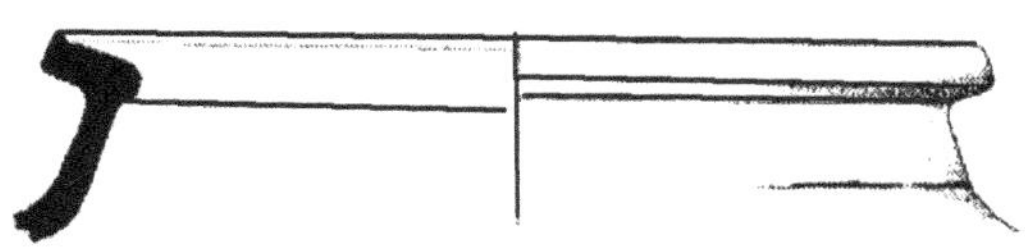 **T.366**	**T.366**:- Rim sherd of a high-necked pot having external projecting quadrangular rim with a shallow groove along the top of the lip. The rim is almost perpendicular to the vessel. It has a rounded edge at the mouth, like a ridge. (I.D = 16cm) SJN-B, TT4, SW, 2, 3.67, 51140, SBW-2, PG(A)
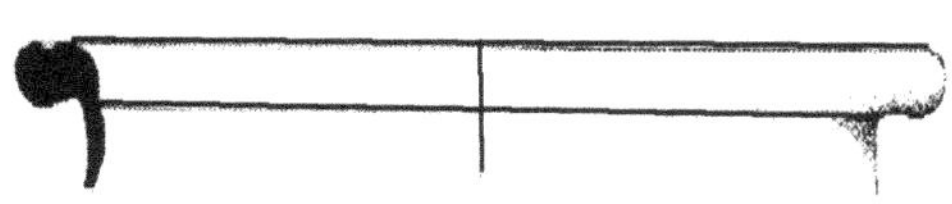 **T.367**	**T.367**:- Rim sherd of a pot having averted external projecting wide rim, flat on the top and bulging slightly on the underside,. The rim has a dark band, along the lip and under it, perhaps due to firing. A shallow groove runs along the top of the rim and another wider groove runs along the inside of the neck, just under the mouth. (I.D = 20cm) SJN-B, TT4, SW, 2, 4.04, 51143, SBW-1, PG(A)

 **T.367.1**	**T.367.1:-** Very similar to T.367, but this rim does not have the groove on the top. (I.D = 25cm) SJN-B, TT4, SW, 2, 3.94 - 4.45, 51142 / 44, SBW-1, PG(A)
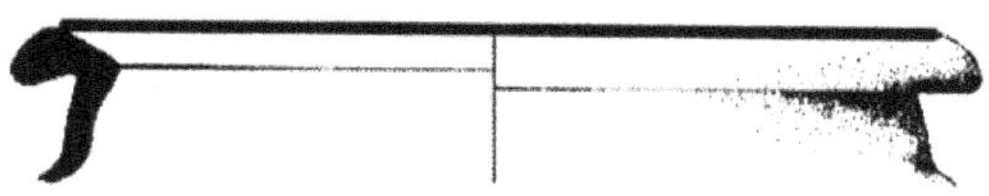 **T.368**	**T.368**:- Rim sherd of a globular pot, having external projecting beaked rim, with a pointed lip. The internal and external junctures are angular. (I.D = 16cm) SJN-B, TT4, NW, 2, 0.23, 51303, SBW-2,PG(A)
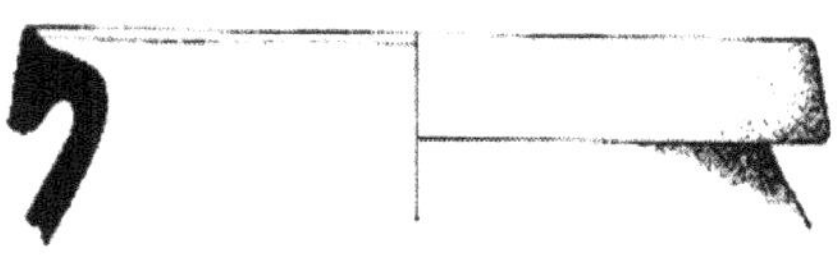 **T.369**	**T.369**:- Rim sherd of a small globular pot, having an extremely averted external projecting rim, which is so sharply beaked, that it appears overturned. The rim is triangular and the upper lip of the rim has been pinched into a sharp ridge. (I.D = 8cm) SJN-B, TT4, NE, 5, 2.47, 51560, SBW-2, PG(A)
 **T.370**	**T.370**: Rim sherd of a large globular pot, having external projecting triangular rim, which appears to broaden above the lip to form a flat ring. The rim is wide and flaring. (I.D = 24cm) SJN-B, TT4, SW, 2, 3.67, 51140, SBW-2, PG(A)
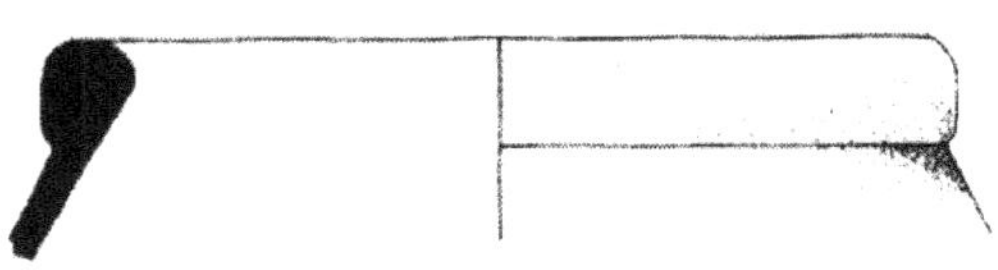 **T.371**	**T.371**:- External projecting inverted rim, which is triangular. A shallow groove on the exterior gives it the impression of a folded rim. (I.D = 16cm) SJN-B, TT4, NE, 5, 2.63, 51563, SBW-2, PG(A)
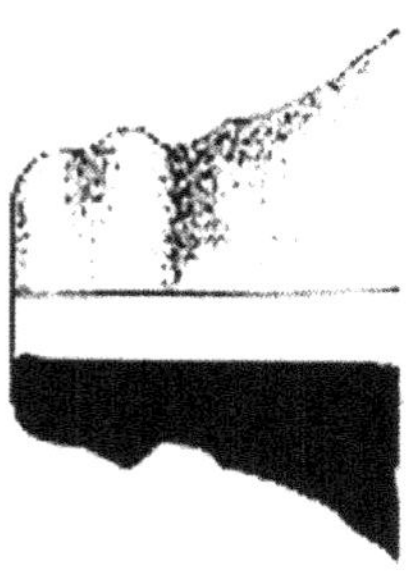 **T.372**	**T.372**:- Conical spout with a broad angular bulge at the tip. The perforation is even (10mm). A small ridge can be seen at the base, which has been caused by pinching. SJN-B, TT4, SW, 2, 3.74, 51141, SBW-1, PG(A)

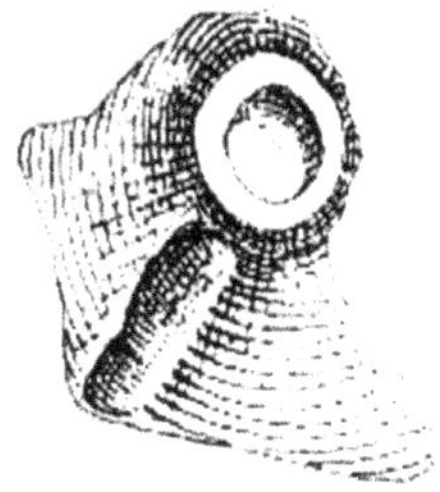 **T.40**	**T.40**:- Broken spout showing evidence of a serrated design where it is luted to the body. Red slip.(L = 2.5cm) SJN-B, SURFACE, SBW-2, PG(A)
 **T.39**	**T.39**:- Handle. Quadrangular in section. Flat at one end, probably where it was luted to the vessel. (L = 6cm) SJN-B, TT1, SW, 2, 0.86, 111, SBW-2, PG(A)
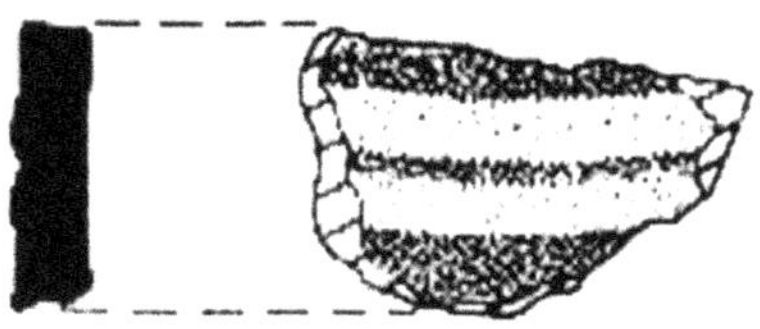 **D.164**	**D.164**:- Body sherd, part of a shoulder, having two low ridges and two shallow grooves which appear to circle the vessel. The ridges are light orange red and the grooves are dark red. SJN-B, TT4, NE, 6, 3.06, 51572, SBW-2, PG(A)
 **D.165**	**D.165**:- Body sherd of a vessel having a sharp carination. The shoulder and body appear luted to form a sharp ridge which has been nicked at intervals along the edge. SJN-B, TT4, SW, 2, 3.74, 51141, SBW-2, PG(A)
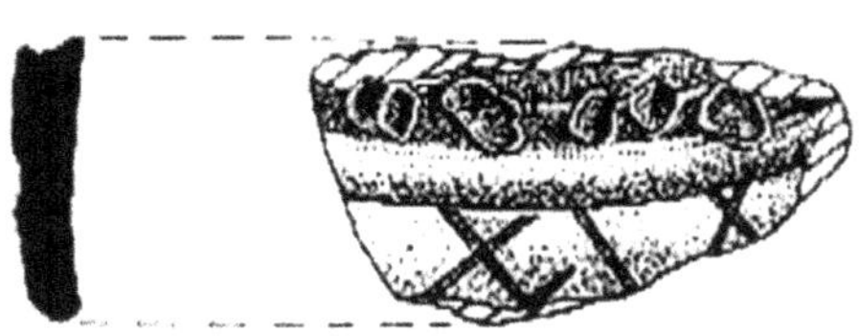 **D.166**	**D.166**:- Body sherd decorated with impressed, irregular-sized dots, which are connected by a narrow line, forming a sort of chain. These are placed over a broad groove, under which crisscross lines are incised. SJN-B, TT4, SW, 2, 3.11, 51138, SBW-1, PG(A)
LARGE INCISED STORAGE VESSELS	
 **T.50 D.22**	**T.23**:- Rim sherd of a high-necked storage jar with inverted rounded rim with flat tip. It has two deep parallel horizontal grooves on the exterior. The fabric is reddish pink, medium coarse and has a white slip on the exterior. (I.D = 14cm) SJN-B, TT2, SE, 3, 1.17, 1516, LISV-4, MR(C)

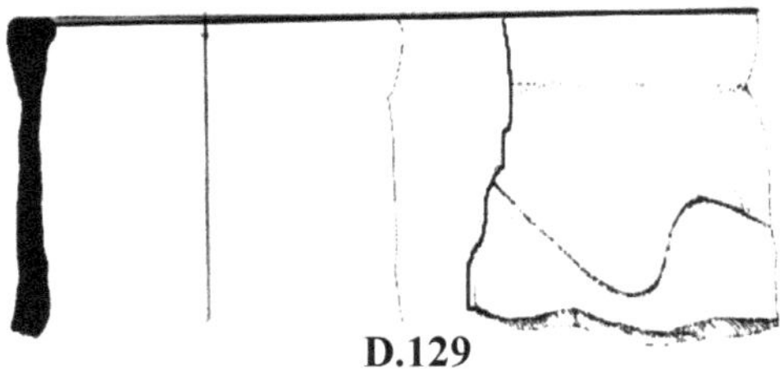 D.129	**T.330 D.129:-** Rim sherd of a high necked jar with external projecting triangular rim. With a groove under the rim on the external side. On the body is seen an incised wavy line (I.D = 13cm). SJN-B-TK, LISV-1, MR(A)
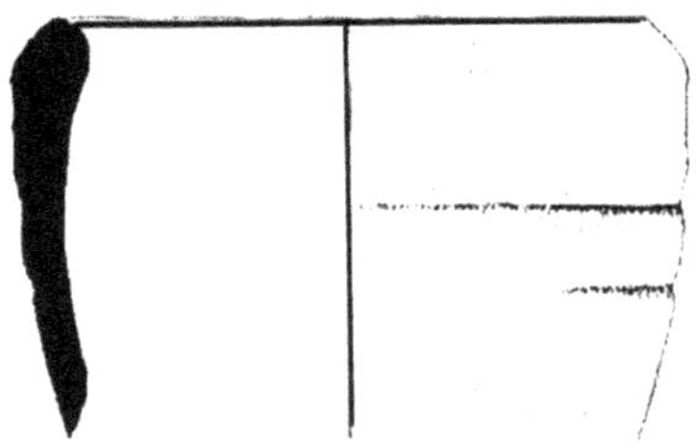 T.427	**T.427**:- Rim sherd of a large storage vessel . On the exterior, just below the rim can be seen a broad groove (I.D = 14cm) SJN-B, TT4, SW, 2, 0.78, 51139, LISV-1, MR(A)
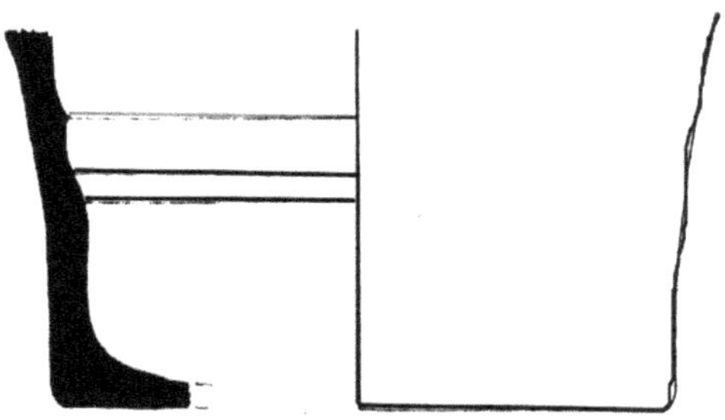 Bs.43	**Bs.43**:- contiguous flat base SJN-B, TT1, NE, 2, 0.81, 506, LISV-1, MR(A)
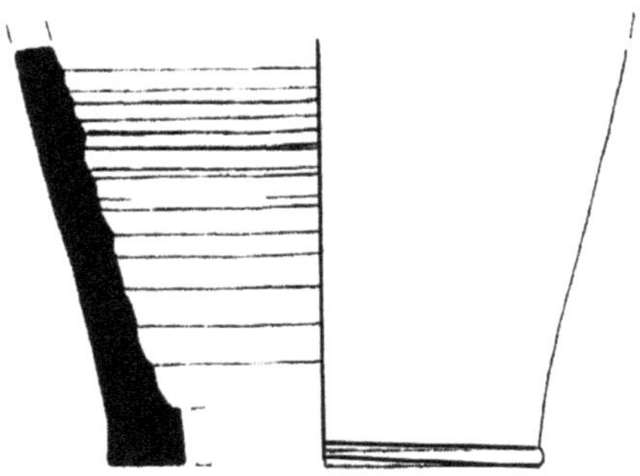 Bs.43.1	**Bs.43.1**:- contiguous flat base SJN-B, TT2, SE, 3, 1.17, 1516, LISV-1, MR(A)
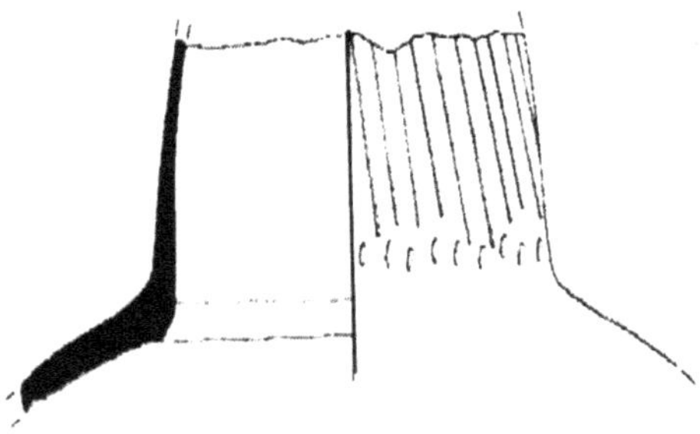 D.19	**D.19**:- Neck portion of a storage jar with long oblique parallel lines incised with deep dots below each of them. The fabric is medium coarse and reddish pink. The design is restricted to the neck portion and not on the shoulder. (Th = 8-11mm) SJN-B, TT1, NW, 2, 0.96, 315, LISV-2, MR(B)

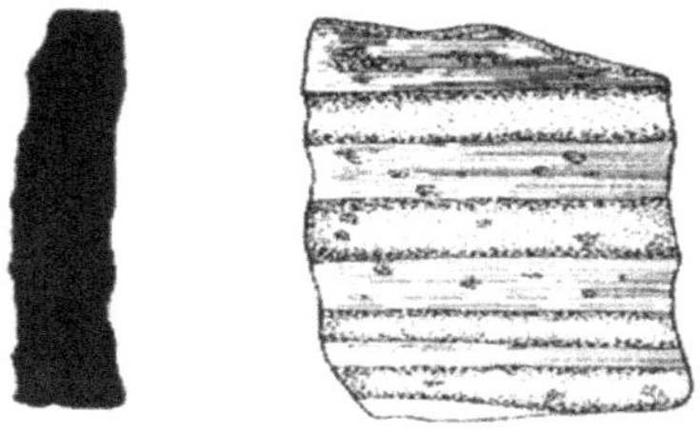 **D.82.1**	**D.82.1**:- Broad ridges on the body, painted white, SJN-B, LISV-1, MR(A)
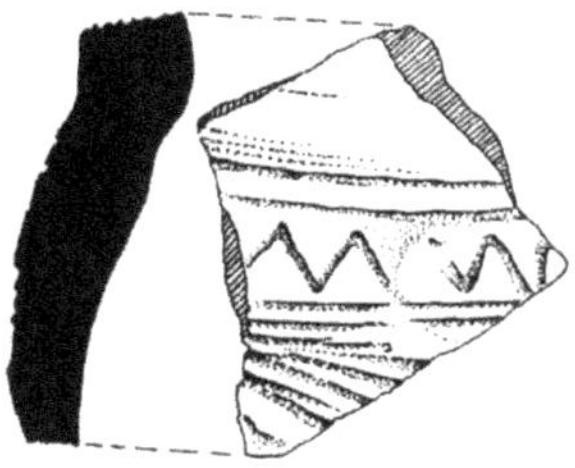 **D.131**	**D.131**:- Body sherd with incised horizontal bands under which is a zig-zag line. Under this line can be seen some incised lines fanning out. Evidence of a handle that may have been luted after the incised design was made. SJN-B-TK, LISV-1, MR(A)
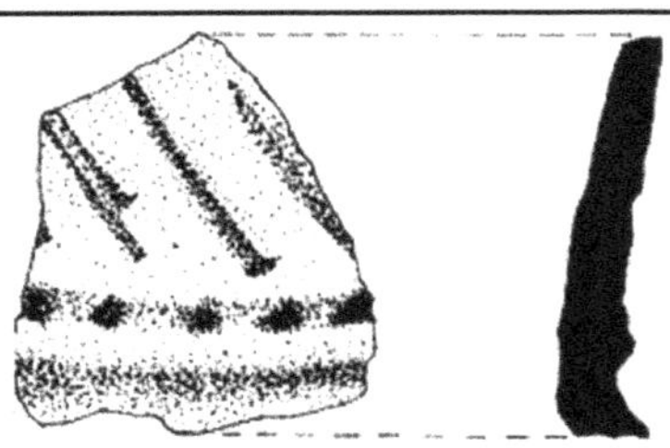 **D.198**	**D.198**:- Body sherd of a small jar or vase (shoulder and neck) having a broad rounded ridge running along the juncture. On this ridge five deeply impressed dots form a border. On the high vertical neck three oblique shallow and broad grooves extend upto but stop just short of ridge and dots border. SJN-B, TT4, NW , 2, 0.78, 51321, LISV-3, MR(A)
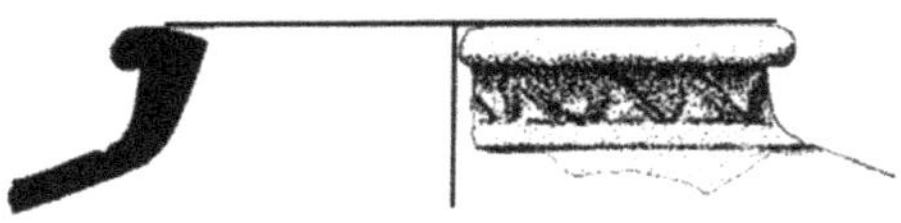 **T.429 D.199**	**D.199 T.429**:- Rim sherd of a pot with a short vertical neck and external projecting short rounded rim. The exterior of the neck below the rim has crudely incised broad wavy lines and short groove.(I.D = 10cm) SJN-B, TT4, NW, 2, 0.51, 51311, LISV-4, MR(C)
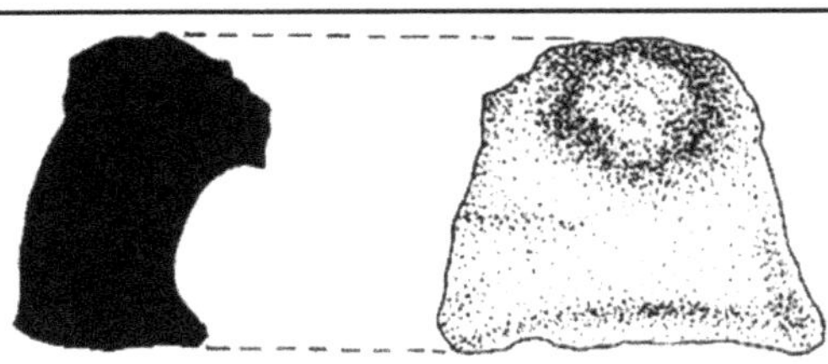 **T.428**	**T.428**:- Upper part of a thick large strap-handle oval in section. On the curve of which an appliquéd or moulded bump can be seen. SJN-B, TT4, SE, 2, 1.04, 51720, LISV-1, MR(A)
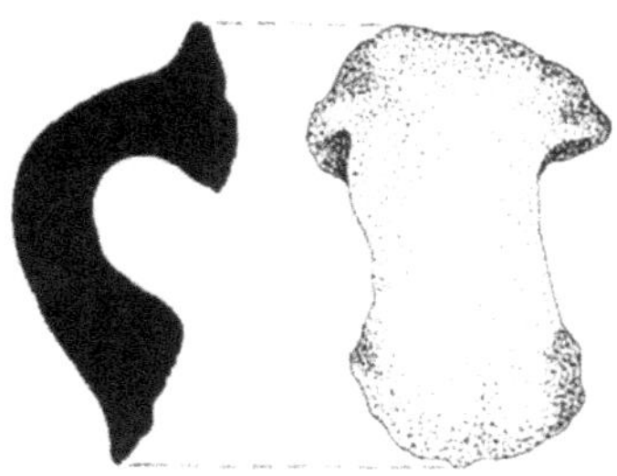 **T.428.1**	**T.428.1**:- Strap handle very similar to T.428, but the handle has no bump on it and the curvature is small. The handle is complete. The vessel may have had multiple such handle .SJN-B, TT4, SW, 2, 3.11, 51138, LISV-2, MR(B)

MICA RED WARE	
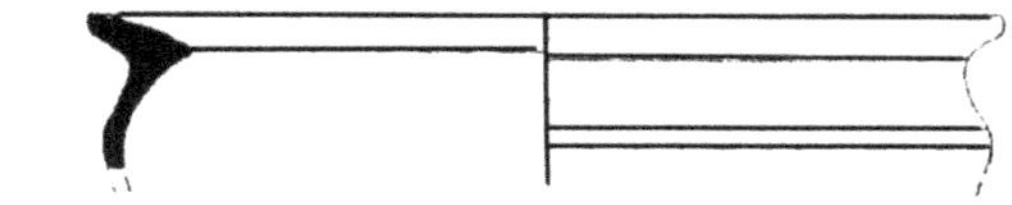 **T.77**	**T.77**:- Rim sherd of a cooking vessel with external projecting everted tapering rim. Constricted neck. (I.D = 22cm) SJN-B, TT1, SW, 1, 0.57, 103, MRW-1, RED(M)
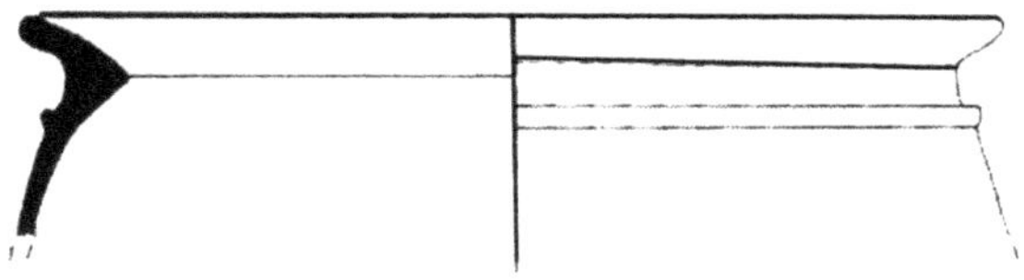 **T.78**	**T.78**:- Rim sherd of a handi with external projecting extremely everted rounded rim (I.D = 22cm). SJN-B, TT1, NW, 2, 0.82, 310, MRW-1, RED(M)
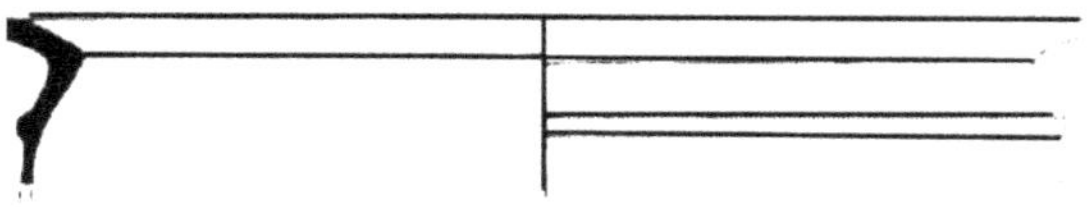 **T.78.1**	**T.78.1**:-Similar to T.78, with the difference in location of the ridge (I.D = 24cm). Constricted neck. SJN-B, TT1, SE, 3, 0.77, 709, MRW-1, RED(M)
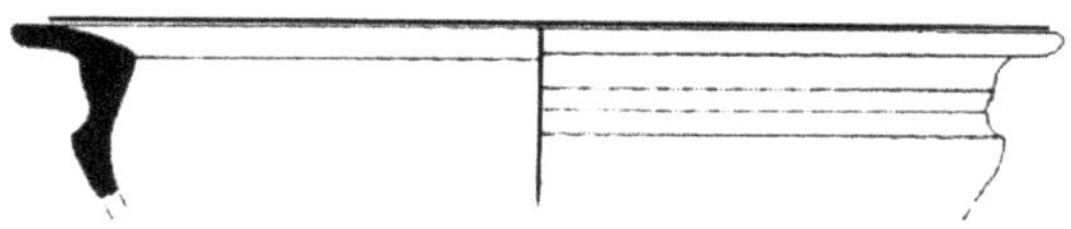 **T.79**	**T.79**:- Rim sherd of a handi with external projecting extremely everted rounded rim with two blunt ridges under the rim and a carination on the body. Constricted neck. (I.D = 24cm). SJN-B TT1, NE, 3, 0.96, 509, MRW-1, RED(M)
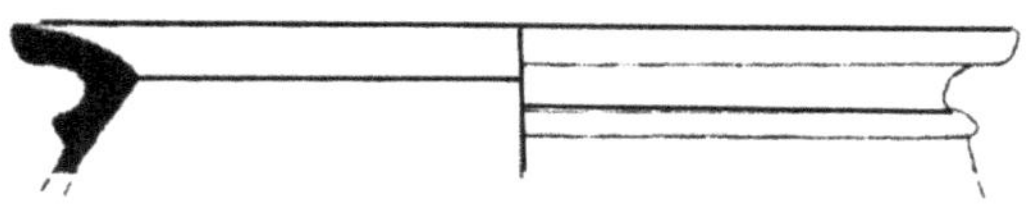 **T.80**	**T.80**:-Rim sherd of a handi with constricted neck and bilaterally projecting rim with more external projection (I.D = 19cm). SJN-B, TT1, NW, 1, 0.63, 304, MRW-1, RED(M)
RED POLISHED WARE	
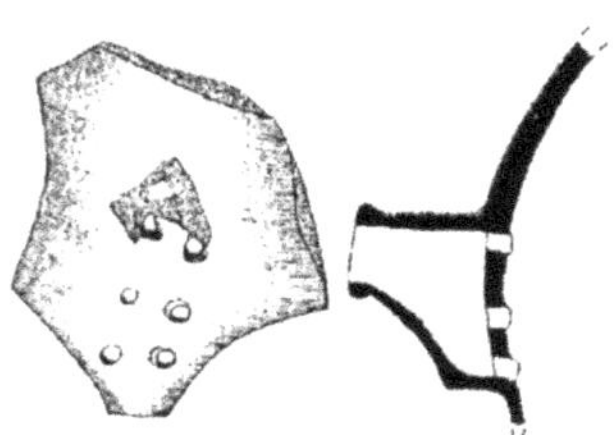 **T.262**	**T.262:-** Spout with a ridged opening and six perforations at its base. The spout was luted on to the body of the vessel and a part of the body can be seen. SJN-B, TT1, SW, 3, 2.51, 147, RPW, RED(N)
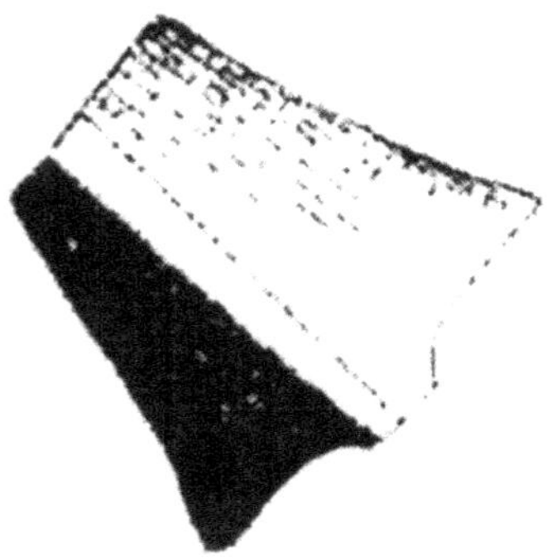 **T.263**	**T.263:-** Sprinkler. Chipped on the edges. SJN-B, TT1, NW, 3, 2.32, 339, RPW, RED(N)

BLACK SLIPPED GREY WARE	
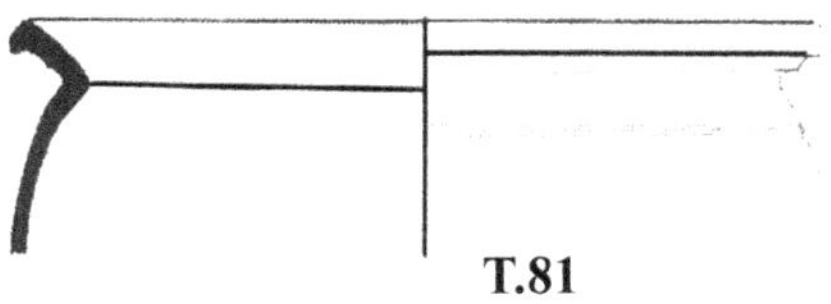 **T.81**	**T.81**:- Rim sherd of a wide-mouthed vessel with external projecting everted rim with an overturned tip. It has two shallow grooves on the upper side of the lip and on the body (I.D =24cm). SJN-B, TT1, SW, 3, 1.06, 115, BSGW, GREY(A)
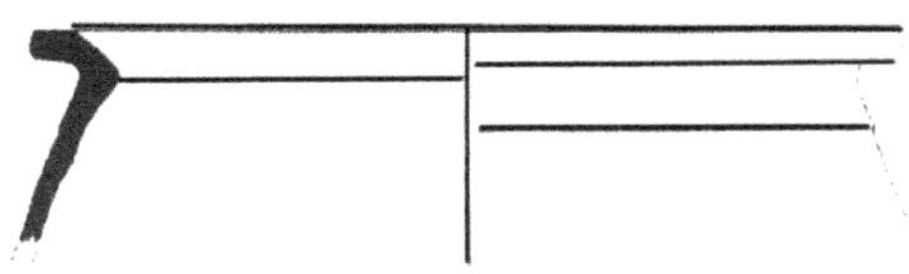 **T.82**	**T.82**:-Rim sherd of a wide-mouthed vessel with an external projecting everted rounded rim (I.D = 21cm). SJN-B, BSGW, GREY(C)
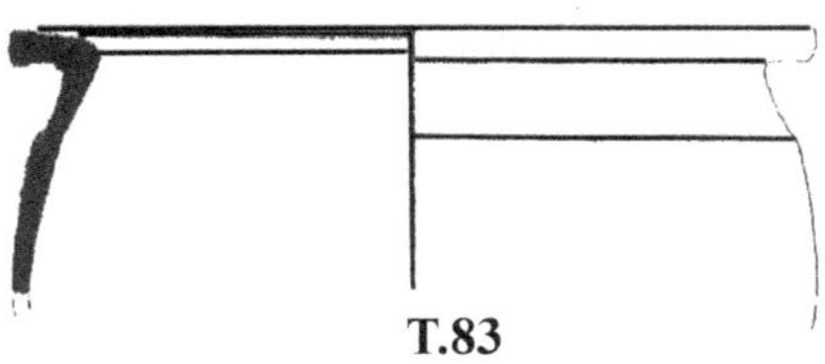 **T.83**	**T.83**:-Rim sherd of a wide mouthed pot with external projecting everted squared rim with constricted neck (I.D = 15cm). SJN-B, TT2, SE, 3, 1.12, 1515, BSGW, GREY (A)
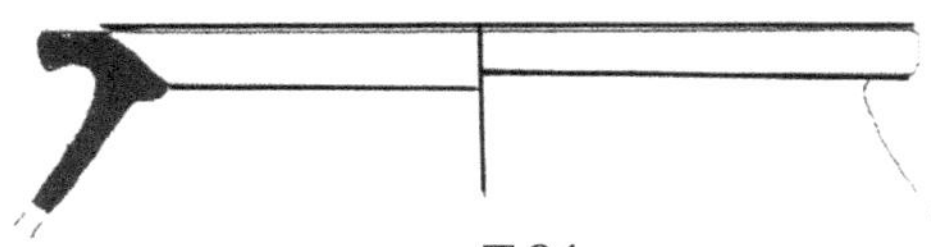 **T.84**	**T.84**:- Rim sherd of a pot with bilateral projecting rim with squared external projecting and tapering internal projection. The rim has a groove on the upper side of the lip. Constricted neck (I.D = 16cm). SJN-B, TT2, SW, 2, 0.74, 905, BSGW, GREY(A)
 **T.85**	**T.85**:- Rim sherd of a pot with an external projecting everted squared rim with a deep groove running along the tip. (I.D = 22cm). SJN-B, TT1, SW, 3, 1.83, 129, BSGW, GREY (A)
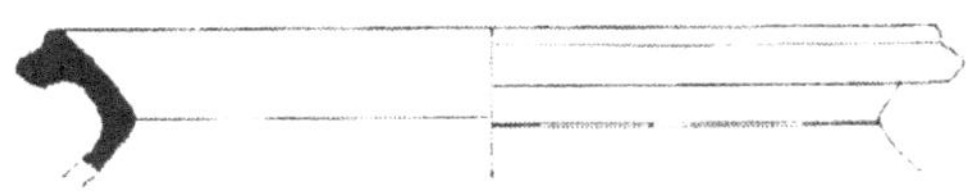 **T.87**	**T.87** - Rim sherd of a pot with external projecting extremely everted rounded rim with a high ridge on the lip (I.D = 22cm). SJN-B, BSGW GREY(A)
 **T.88**	**T.88**:-Rim sherd of a pot with external projecting everted tapering rim. The neck is not constricted (I.D = 14cm). SJN-B, TT1, SW, 3, 2.34, 144, BSGW, GREY (A)
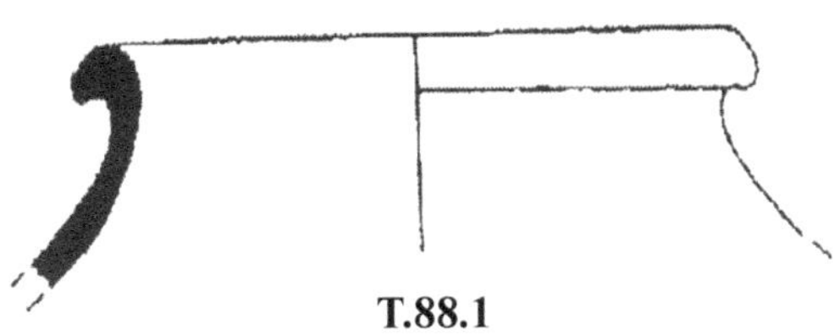 **T.88.1**	**T.88.1**:-Rim sherd of a pot with external projecting everted rounded rim.. (I.D = 16cm). SJN-B, TT2, SE, 4, 1.98, 1531, BSGW, GREY(A)

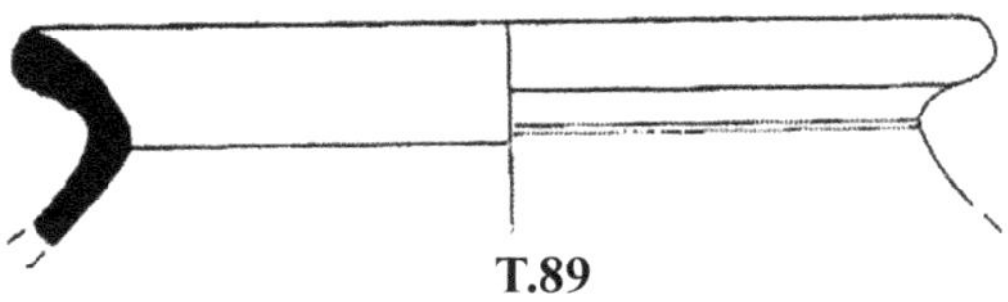 T.89	**T.89**:-Rim sherd of a pot with external projecting everted squared rim with rounded edges. (I.D = 18cm). SJN-B, TT2, SW, 3, 1.42, 917, BSGW, GREY (A)
 T.90	**T.90**:-Rim sherd of a pot with external projecting squared rim so everted that its lower lip has fused with the body. (I.D = 22cm). SJN-B , TT1, SE, 2, 1521, BSGW, GREY(A)
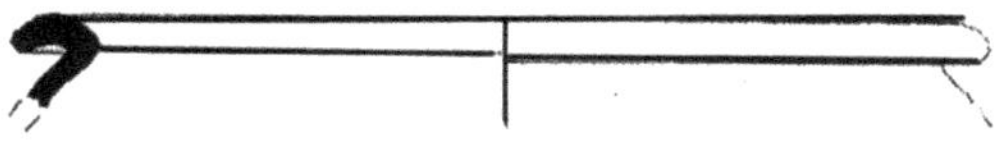 T.91	**T.91**:-Rim sherd of a pot with external projecting extremely everted rim with an extremely constricted neck (I.D = 18cm). SJN-B, BSGW, GREY(A
 T.92	**T.92**:-Rim sherd of a pot with external projecting everted rounded rim (I.D = 14cm). SJN-B, TT1, SW, 3, 1, 114, BSGW, GREY(A)
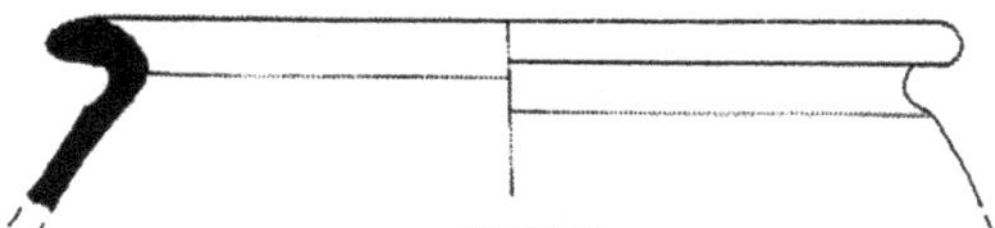 T.92.1	**T.92.1**:-Rim sherd of a pot with external projecting everted rounded rim, similar to T.92 (I.D = 14cm) SJN-B, TT2, SE, 3, 1.42, 1521, BSGW, GREY(A)
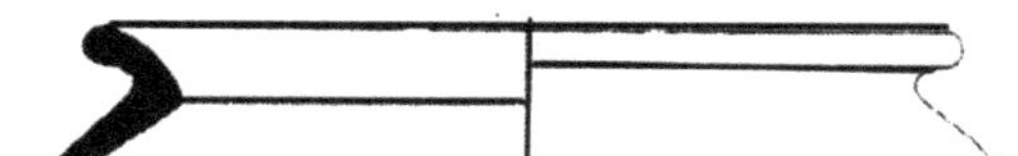 T.92.2	**T.92.2**:-Rim sherd of a pot with external projecting everted rounded rim. Thick in section (I.D = 20cm). SJN-B, TT2, SE, 3, 1.42, 1521, BSGW
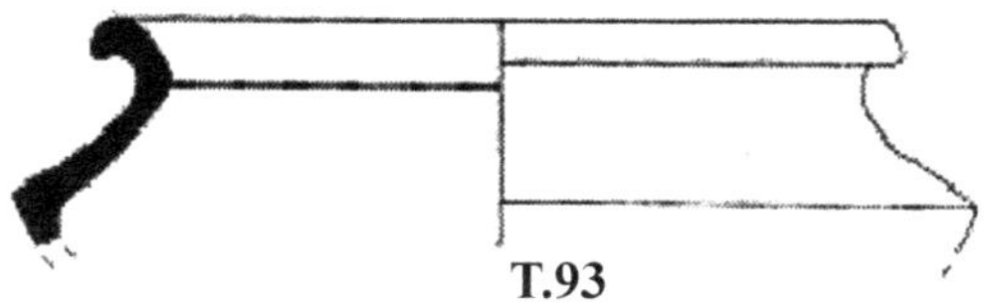 T.93	**T.93**:-Rimsherd of a carinated pot with external projecting extremely everted rounded rim (I.D = 17cm). SJN-B, TT2, SE, 4, 1.6, 1524, BSGW, GREY(A)
 T.95	**T.95**:-Rim sherd of a high necked pot with external projecting, triangular rim (I.D = 16cm). SJN-B, TT1, NE, 3, 1.41, 520, BSGW, GREY(A)
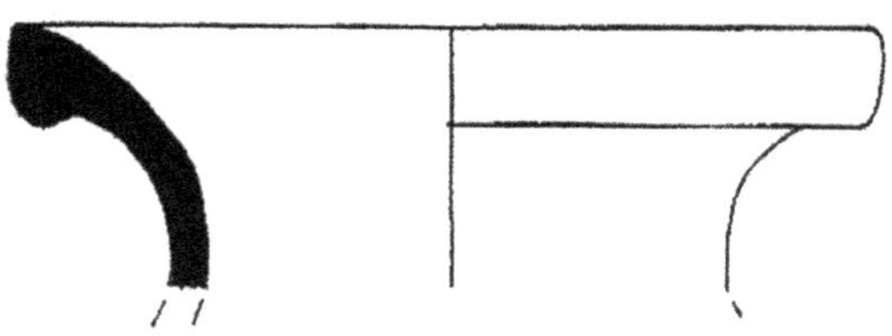 T.95.1	**T.95.1**:-Rim sherd of a high necked pot with external projecting triangular rim similar to T.95 (I.D = 16cm) SJN-B, TT1, NW, 3, 1.5, 326, BSGW, GREY(A)

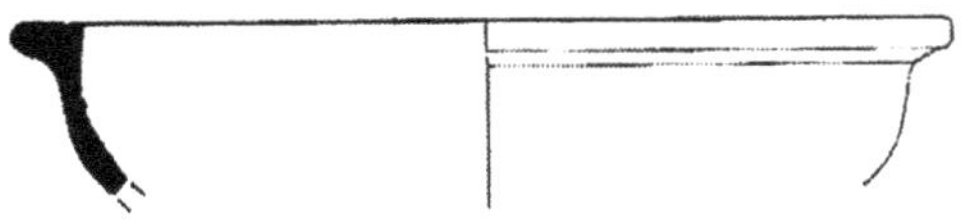 T.96.1	**T.96.1**:- Rim sherd of a convex-sided bowl with external projecting triangular rim. (I.D = 26cm). SJN-B, TT2, SW, 3, 1.1, 911, BSGW, GREY(A)
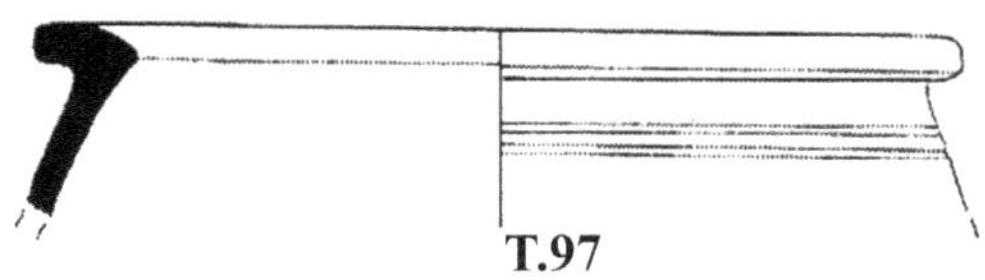 T.97	**T.97**:- Rim fragment of a wide-mouthed pot with constricted neck. The rim is external projecting, extremely everted and square (I.D = 19cm). SJN-B, TT2, SE, 2, 0.98, 1512, BSGW, GREY (A)
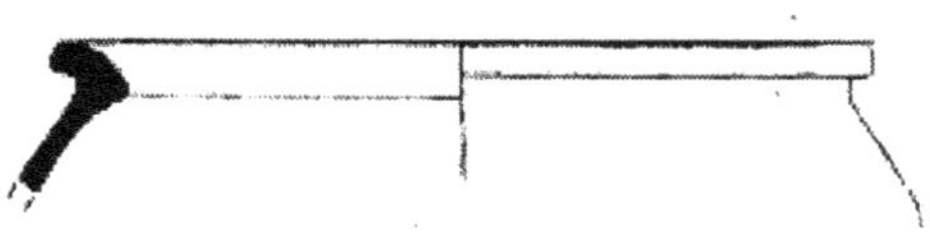 T.98	**T.98**:-Rim sherd of a pot with constricted neck having an external projecting extremely everted squared rim with a shallow groove on the tip and a deeper groove on the orifice. (I.D = 17cm) SJN-B, TT1, NE, 1, 0.57, 503, BSGW, GREY(C)
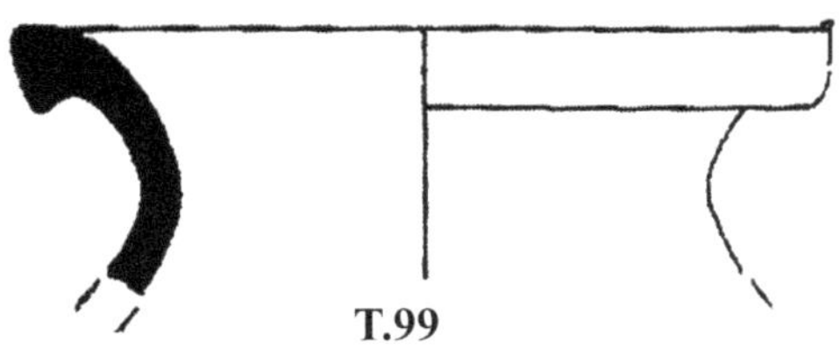 T.99	**T.99**:-Rim sherd of a pot with external projecting extremely everted squared rim with an undercut. (I.D = 9cm). SJN-B, TT2, SE, 3, 1.12, 1515, BSGW, GREY (A)
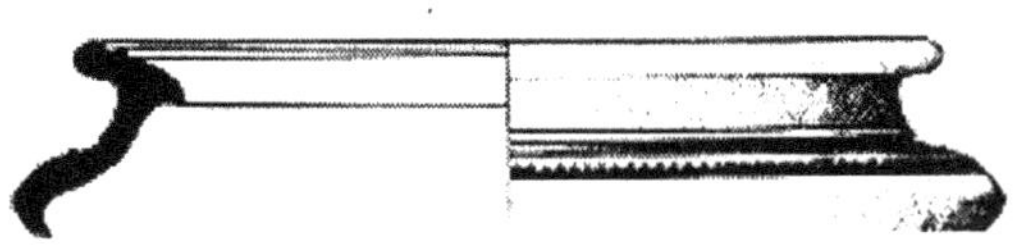 D.191	**D.191**:-Rim sherd of a pot with a bilaterally projecting rim almost perpendicular to the neck. The external projecting of the rim is wide and tapering having two grooves and a ridge on top near the lip. The neck and upper part of the shoulder up to the carination has a series of shallow ridges and grooves. The horizontal band of punctated triangles runs under the last groove. (I.D = 21cm) SJN-B, TT4, SW, 2, 1.03, 51125, BSGW, GREY(A)
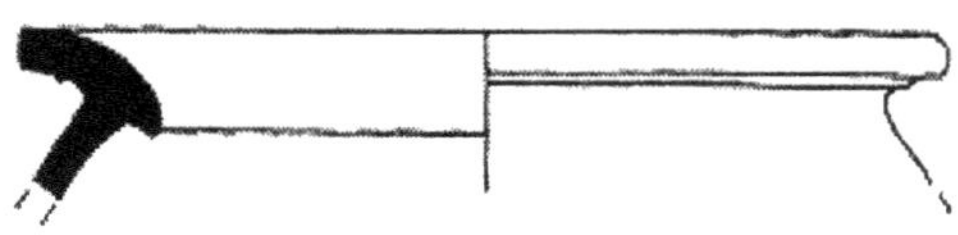 Bs.65	**T.100**:-Rim sherd of a wide-mouthed pot with constricted neck having a bilateral projecting rim with a square external projection and a beaked internal projection There is a groove on the external projection on the lip. (I.D = 17cm). SJN-B, TT1, SE, 2, 0.58, 706, BSGW, GREY (A)
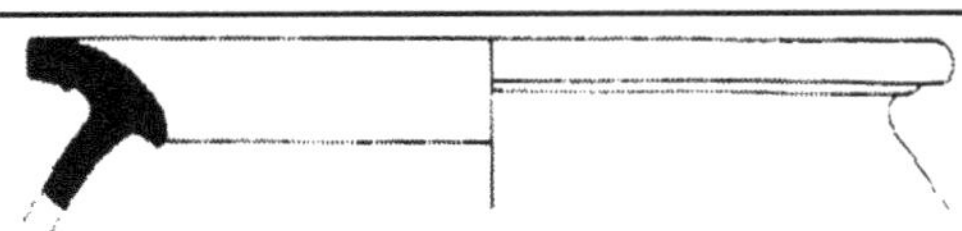 T.100.1	**T.100.1**:- Rim sherd of a wide-mouthed pot with bilateral projecting rim with a square external projection and a beaked internal projection, similar to T.100. This rim does not have a groove on it (I.D = 17cm). SJN-B, BSGW, GREY(A)
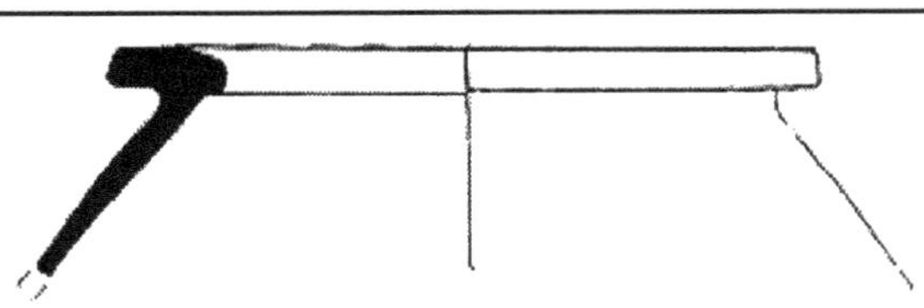 T.101	**T.101**:- Rim sherd of a wide mouthed pot with constricted neck having an external projecting everted squared rim (I.D = 18cm). SJN-B, TT1, NW, 1, 0.68, 305, GW-2, GREY(C)

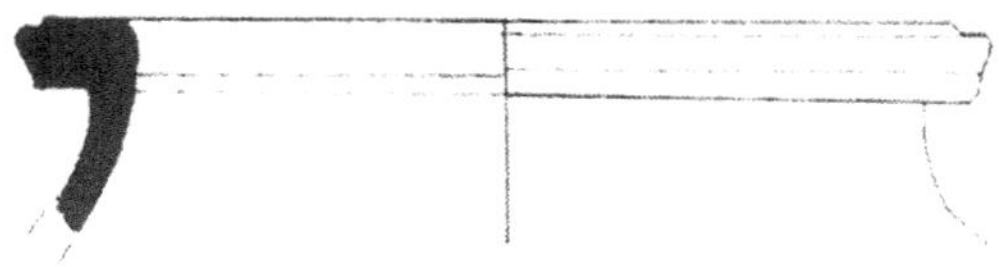 **T.102**	**T.102**:- Rim sherd of a wide-mouthed pot with constricted neck having an external projecting extremely everted quadrangular rim with a deep groove on the tip as well as on the lip (I.D = 13cm). SJN-B, TT1, NE, 3, 1.24, 517, BSGW, GREY (A)
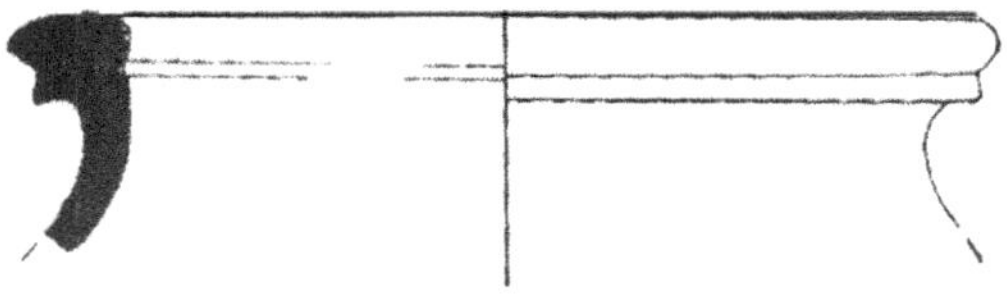 **T.103**	**T.103**:- Rim sherd of a pot with external projecting quadrangular rim with a deep groove on the tip exterior and another on the mouth. (I.D =14cm). SJN-B, TT1, NW, 1, 0.59, 303, BSGW, GREY(C)
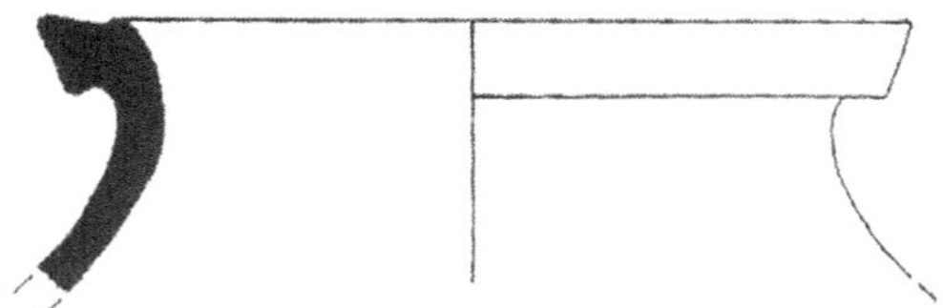 **T.104**	**T.104**:-Rim sherd of a pot with external projecting quadrangular rim with a groove at the mouth. (I.D = 15cm). SJN-B, TT1, SE, 2, 0.63, 707, BSGW, GREY (A)
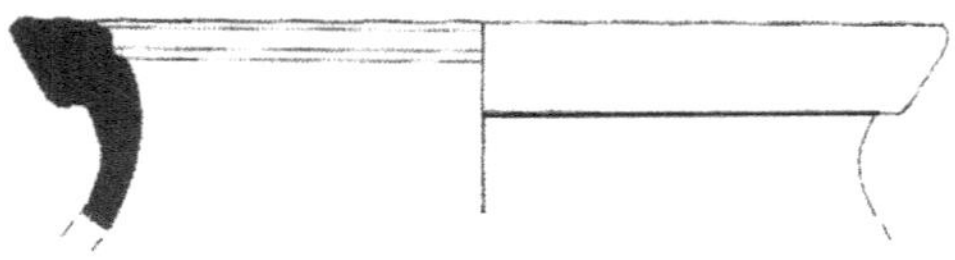 **T.105**	**T.105**:-Rim sherd of a pot with external projecting quadrangular rim with a groove on the mouth (I.D = 20cm). SJN-B, TT1, NE, 4, 1.63, 521, BSGW, GREY(A)
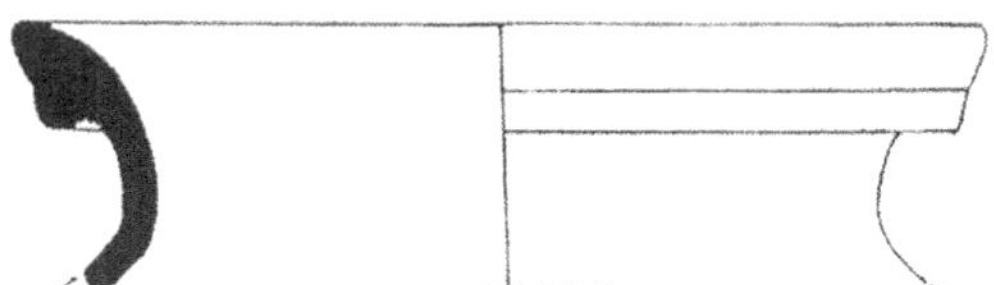 **T.106**	**T.106**:- Rim sherd of a pot with external projecting triangular rim with a groove on the external side (I.D = 20cm). SJN-B TT1, SW,2, 0.86,111, BSGW, GREY(A)
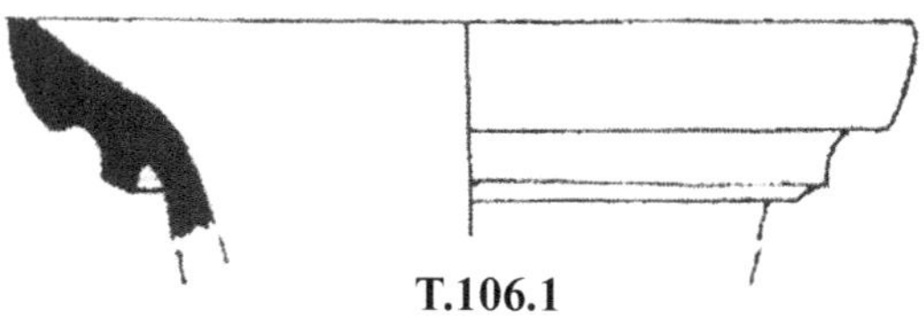 **T.106.1**	**T.106.1**:- Rim sherd of a pot with external projecting triangular rim with a deep groove just below the external side, which appears to form a ledge (I.D = 16cm). SJN-B, TT1, NW, 1, 0.68, 305, BSGW, GREY(A)
 **D.103.1**	**T.107**:- Rim sherd of a pot with constricted neck and external projecting extremely everted tapering rim (I.D = 20cm). SJN-B, TT1, SW, 1, 0.4, 102, BSGW, GREY(A)
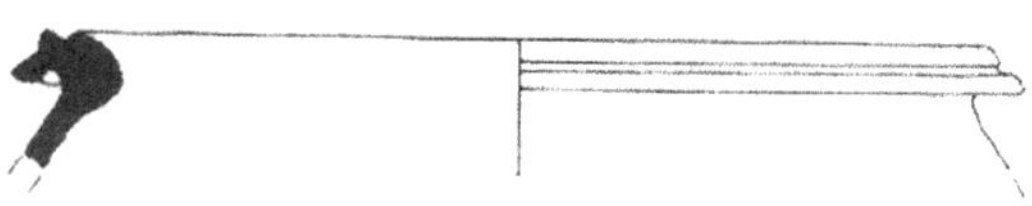 **T.109**	**T.109**:- Rim sherd of a pot with extremely projecting extremely everted tapering rim with a very high and prominent ridge on the upper part of the lip. (I.D =20cm). SJN-B TT1, NW, 1, 0.63, 304, GW-1, GREY(A)

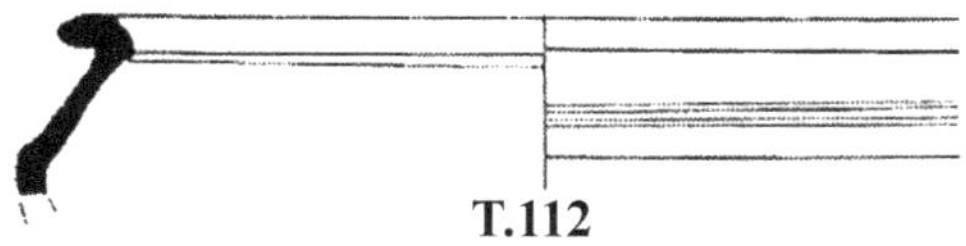 T.112	**T.112**:- Rim sherd of a carinated pot with external projecting extremely everted tapering rim. (I.D = 26cm). SJN-B, TT2, SE, 4, 2.05, 1533, BSGW, GREY (A)
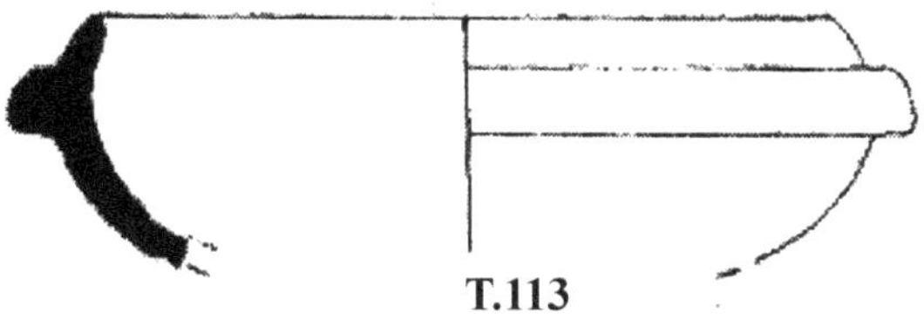 T.113	**T.113**:- Rim sherd of a bowl with reverted rounded rim with a ledge just below on the exterior (I.D = 9cm). SJN-B, TT1, SW, 1, 0.35, 101, BSGW, GREY(A)
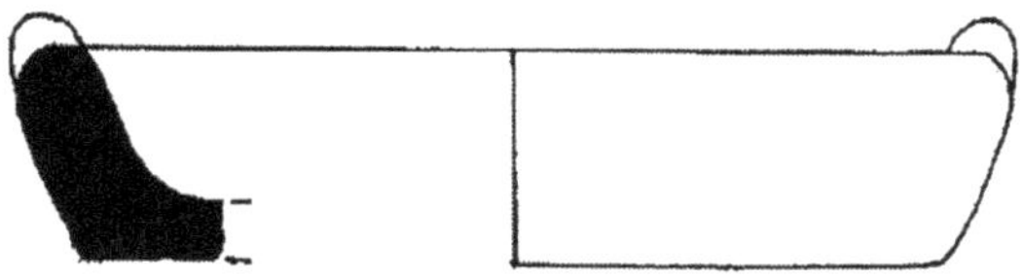 T.115.1	**T.115.1**:- Rim sherd of a plate; rounded, probably a lug. It is not possible to measure the diameter. SJN-B, TT1, SW, 3, 1.02, 138, BSGW, GREY(A)
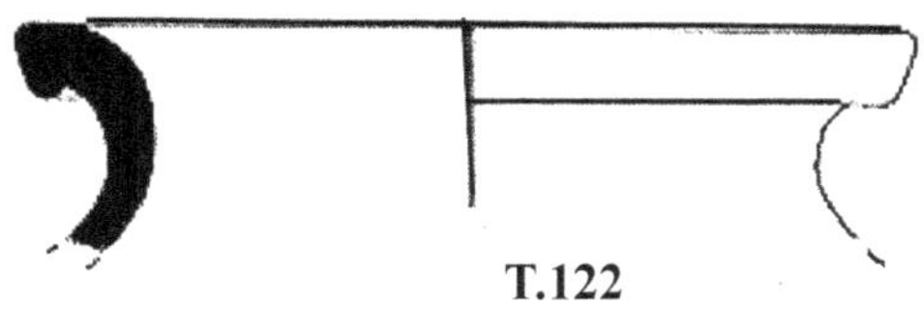 T.122	**T.122:-** Rim sherd of a high necked pot with external projecting quadrangular rim with an under cut. Slip is mainly on the exterior surface and upto the orifice on the inside. (I.D = 10cm). SJN-B, TT1, SW, 2, 0.92, 112, BSGW, GREY(B)
 T.143	**T.143**:- Rim sherd of a pot with external projecting triangular rim. Grey slip (I.D = 12cm). SJN-B, TT2, SE, 4, 2.31, 1538, BSGW, GREY(A)
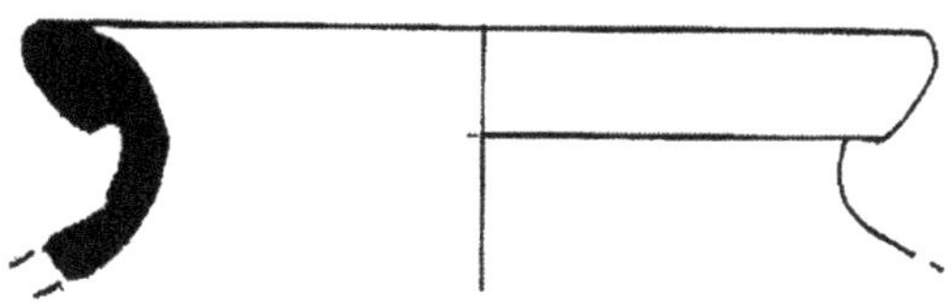 T.144	**T.144**:- Rim sherd of a pot with external projecting triangular rim. Grey slip (I.D = 12cm). SJN-B, BSGW, GREY(A)
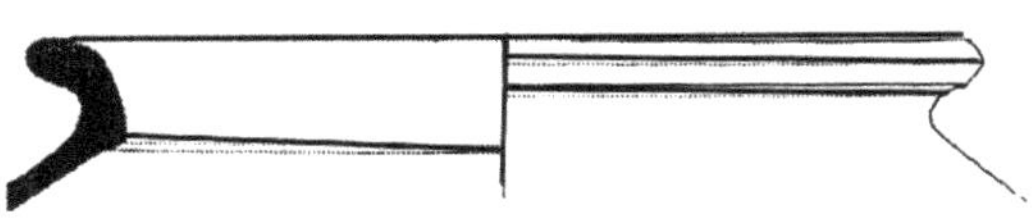 T.147	**T.147**:- Rim sherd of a pot with external projecting everted rounded rim. Grey slip (I.D = 22cm). SJN-B, TT1, SW, 3, 1.08, 116, BSGW, GREY(A)
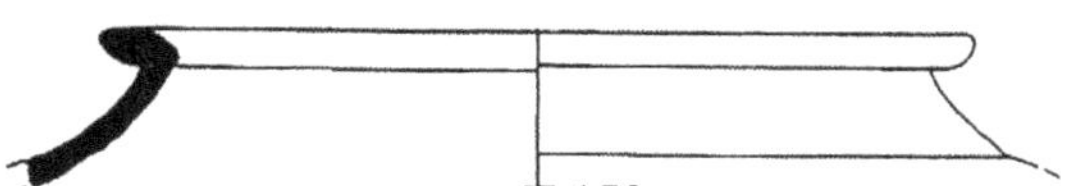 T.150	**T.150**:-Rim sherd of a pot with external projecting extremely everted tapering rim (I.D = 18cm). SJN-B, TT1, SW, 1, 0.57, 103, BSGW, GREY(A)
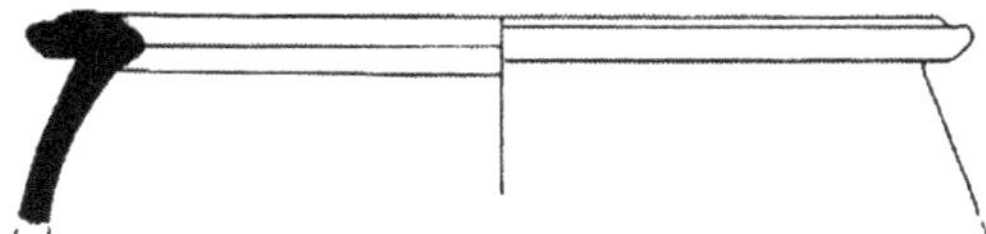 T.151	**T.151**:- Rim sherd of a pot with external projecting everted rounded rim with a broad lip. Grey slip. (I.D = 13cm). SJN-B, TT2, SE, 4, 2.26, 1537, BSGW, GREY(A)

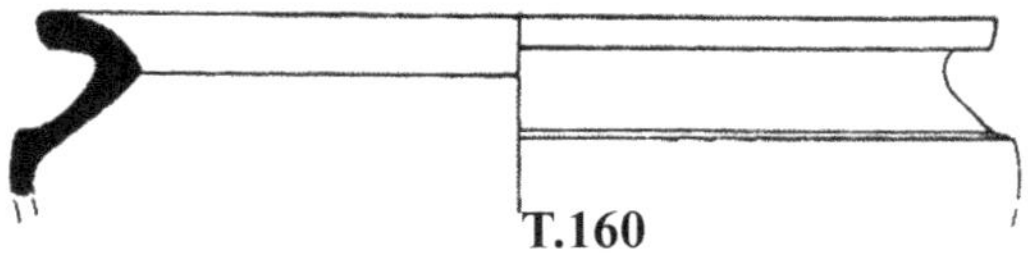 T.160	**T.160**:- Rim sherd of a carinated pot with external projecting everted squared rim. Grey slip (I.D = 18cm). SJN-B, TT2, SE, 3, 1.36, 1520, BSGW, GREY(A)
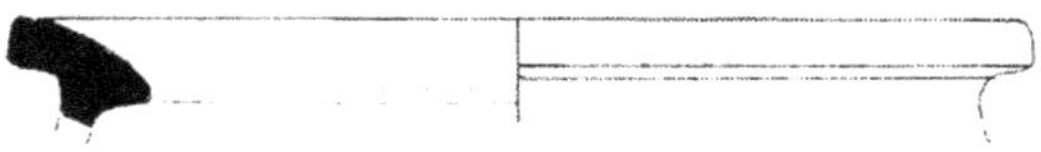 T.178	**T.178**:- Rim sherd of a pot with bilateral projecting everted square rim with a groove on the upper side of the lip (I.D = 20cm). SJN-B, TT2, SW , 3, 1.22, 914, BSGW, GREY(A)
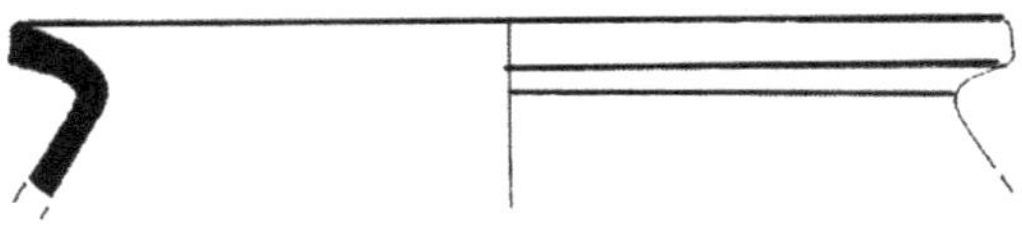 T.179	**T.179**:- Rim sherd of a pot with external projecting everted square rim with a groove on the tip (I.D = 18cm). SJN-B, TT1, SW, 1, 0.57, 103, BSRW-1, RED(C)
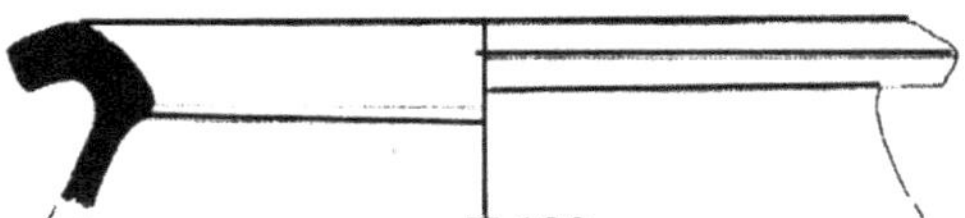 T.180	**T.180**:-Rim sherd of a pot with external projecting everted square rim (I.D = 18cm). SJN-B, TT1, NW, 1, 0.52, 301, BSGW, GREY(A)
 T.186	**T.186**:- Rim sherd of a pot with external projecting quadrangular rim (I.D = 11cm). SJN-B, TT1, NW, 2, 0.96, 315, BSGW, GREY(A)
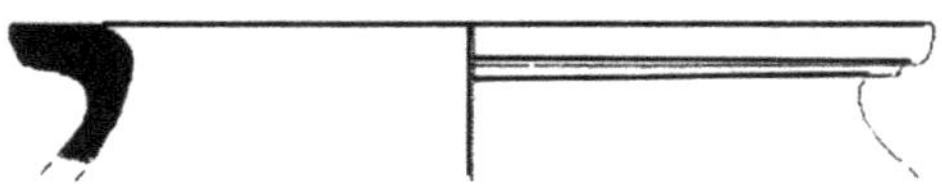 T.190	**T.190**:- Rim sherd of a pot with external projecting quadrangular rim (I.D = 12cm). SJN-B, TT1, SW, 1, 0.57, 103, BSGW, GREY(A)
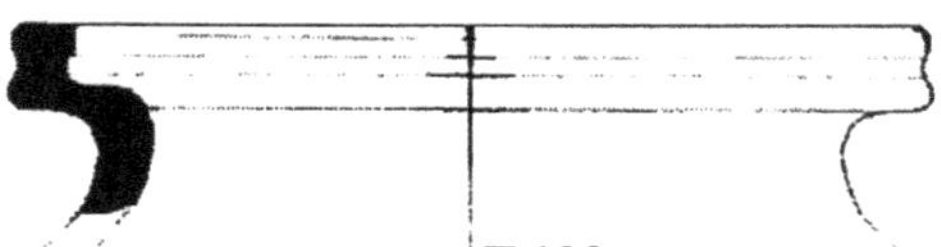 T.192	**T.192**:- Rim sherd of a pot with external projecting quadrangular rim with a deep groove on the tip (I.D = 15cm). SJN-B, TT1, NW, 1, 0.68, 305, BSGW, GREY(A)
 T.197	**T.197**:- Rim sherd of a pot with internal projecting triangular rim with a groove on the external side (I.D =11cm). SJN-B, TT2, SE, 3, 1.12, 1515, BSGW, GREY(A)
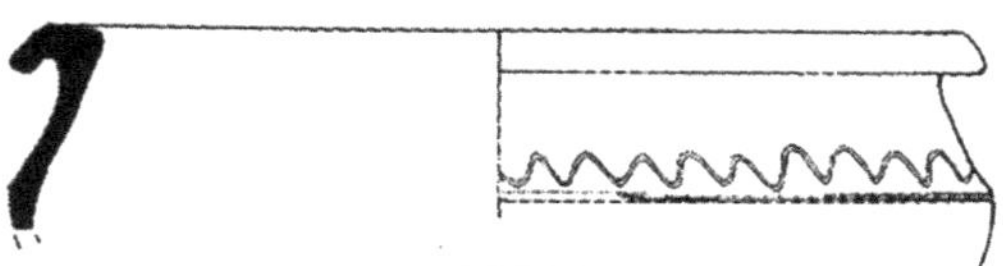 T.204	**T.204 D.66**:- Rim sherd of a pot with external projecting extremely everted tapering rim with an incised design on the shoulder. The rim has a groove on the upper side. The design has a horizontal wavy line with a straight horizontal line incised below it. (I.D = 20cm). SJN-B, TT1, NE, 2, 0.81, 506, BSGW, GREY(A)
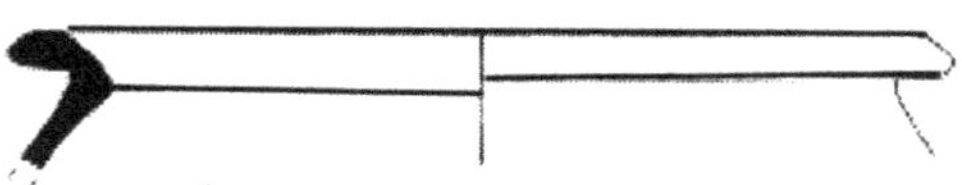 T.206	**T.206**:- Rim sherd of a pot with external projecting extremely everted tapering rim with a groove on the upper side (I.D = 15cm). SJN-B, TT1, SW, 2, 0.8, 108, BSGW, GREY(A)

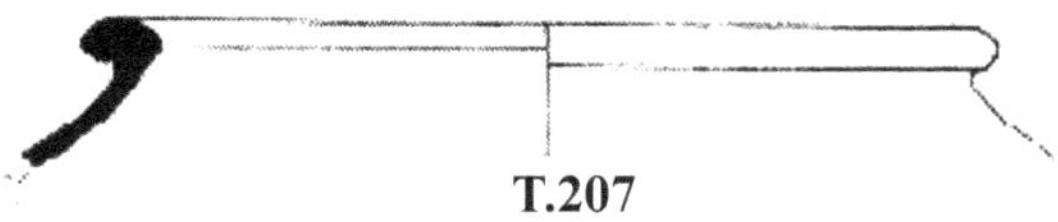 T.207	**T.207**:- Rim sherd of a pot with external projecting extremely everted tapering rim with thickening in the middle of the lip (I.D = 24cm). SJN-B, TT2, SE, 2, 0.85, 1510, BSGW, GREY(A)
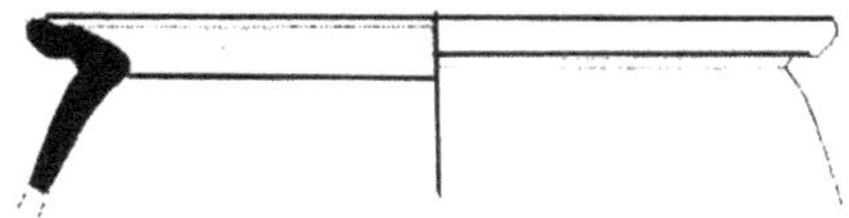 T.211	**T.211**:- Rim sherd of a pot with external projecting extremely everted rounded rim. Shallow groove on the upper side of the lip. (I.D = 16cm) SJN-B, TT1, SW, 1, 0.57, 103, BSGW, GREY(A)
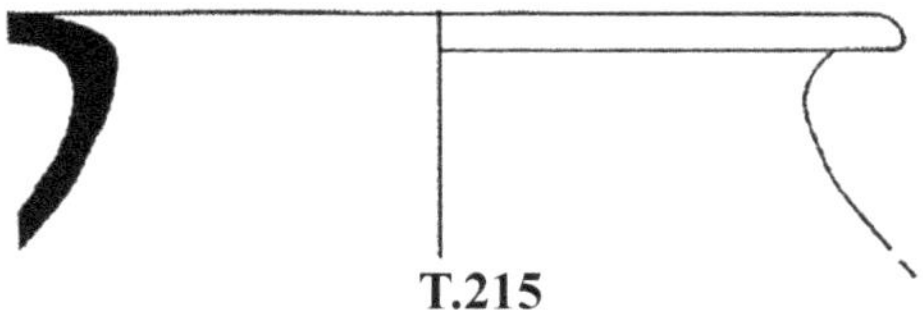 T.215	**T.215:-** Rim sherd of a pot with external projecting everted rounded rim (I.D = 13cm). SJN-B, TT1, SE, 2, 0.44, 704, BSGW, GREY(A)
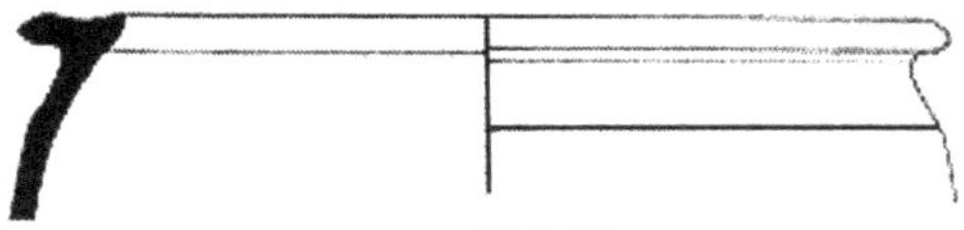 D.142	**T.216**:- Rim sherd of a pot with bilateral projecting rim with both edges tapering and slightly raised internal projection (I.D =18cm). SJN-B, TT1, SW, 1, 0.4, 102, BSGW, GREY(A)
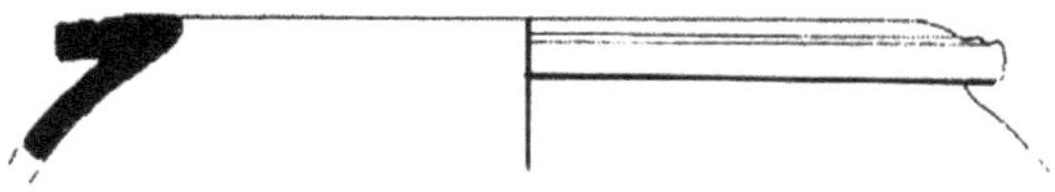 T.220	**T.220**:- Rim sherd of a pot with bilateral projecting rim with a squared external projection and two grooves on the upper side (I.D =18cm). SJN-B, TT2, SW, 2, 0.94, 908, BSGW, GREY(A)
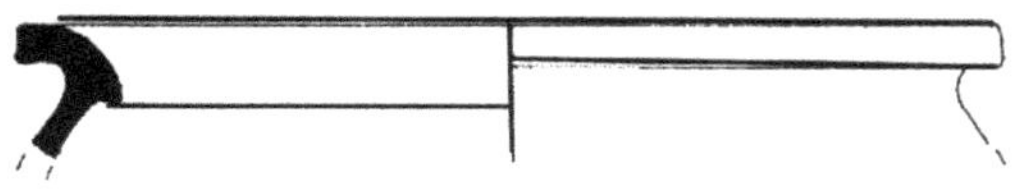 D.140	**T.224**:- Rim sherd of a pot with bilateral projecting rim with square external projection and a groove along the tip and on the upper side (I.D = 24cm). SJN-B, TT1, NW, 1, 0.71, 306, BSGW, GREY(A)
 D.140	**T.229 D.67**:- Rim sherd of a carinated pot with external projecting everted squared rim with a groove on the inner side and a horizontal incised line under which is a row of punctate marks. The design is on the shoulder. This vessel is squat with blunt carination (I.D = 12cm). SJN-B, TT1, NE, 3, 1.23, 516, BSGW, GREY(A)
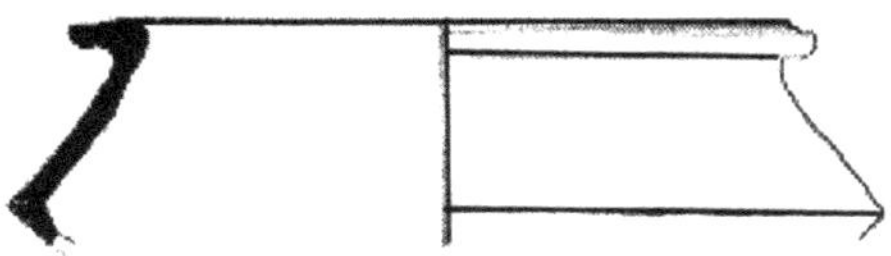 T.232	**T.232:-** Rim sherd of a small carinated pot with external projecting extremely everted tapering rim (I.D =11cm). SJN-B, BSRW-1, RED(C) slip,
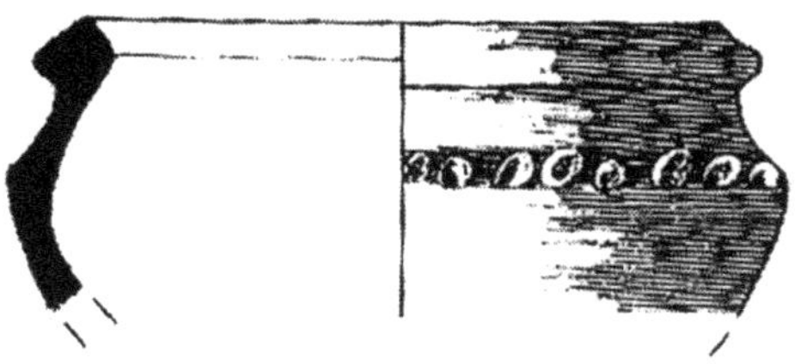 T.233	**T.233 D.68**:- Rim sherd of a small carinated pot with external projecting everted square rim with a row of punctate dots on the carination (I.D = 7cm). SJN-B, TT2, SE, 3, 1.27, 1518, BSGW, GREY(A)

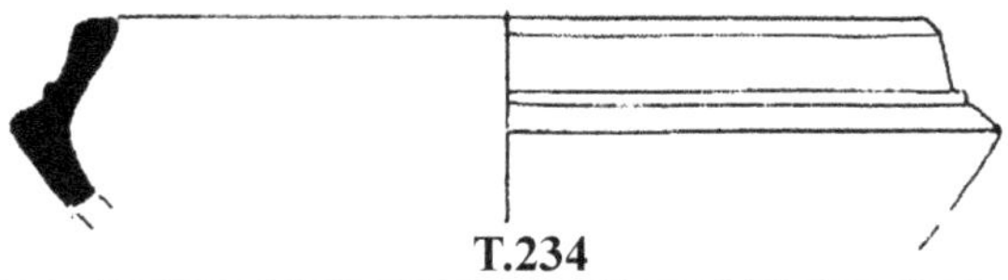 **T.234**	**T.234**:- Rim sherd of a carinated bowl with inverted rounded rim and a ridge on the carination. (I.D =16cm). SJN-B, TT1, SW, 1, 0.57, 103, BSGW, GREY(A)
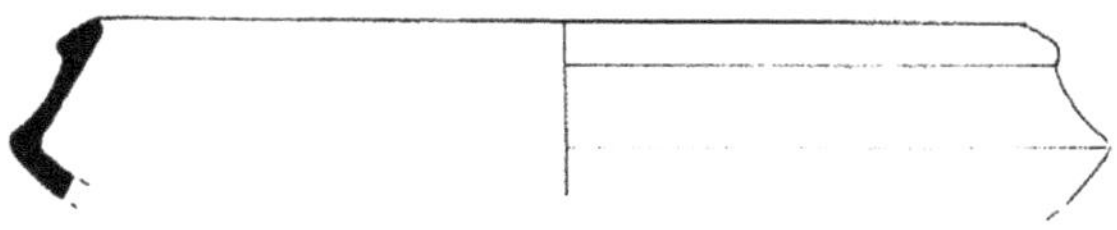 **T.236**	**T.236**:- Rim sherd of a carinated bowl with an inverted tapering rim externally and with a sharp lower lip edge. (I.D = 24cm). SJN-B, TT1, SW, 2, 0.92, 112, BSGW, GREY(A)
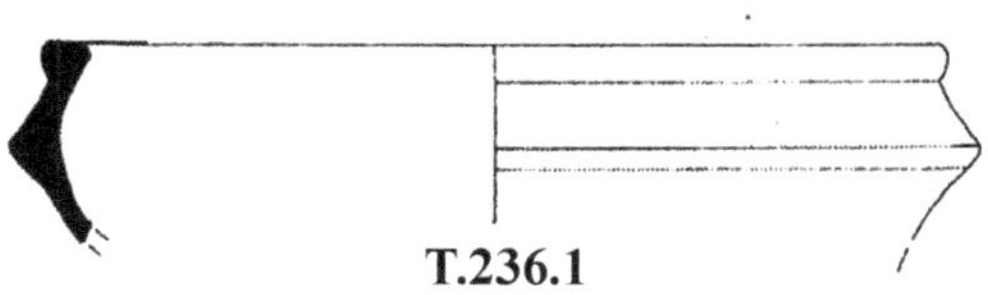 **T.236.1**	**T.236.1**:- Rim sherd of a carinated bowl, similar to T.236 (I.D =18cm). SJN-B, TT1, SE, 2, 0.58, 706, BSGW, GREY(A)
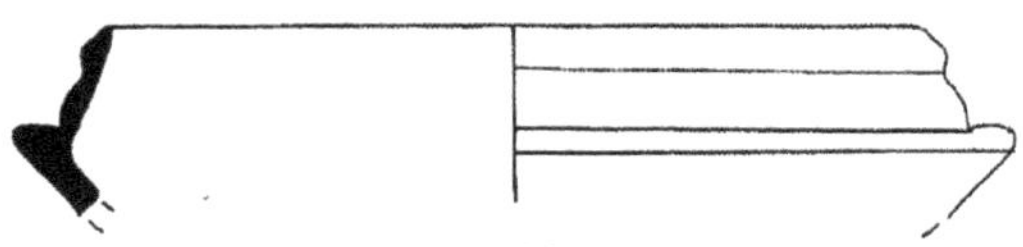 **T.237**	**T.237**:- Rim sherd of a carinated bowl with an inverted rounded rim and with a ridge above the pronounced ledge on the carination. There is a groove just above the ledge. (I.D = 18cm). SJN-B, TT1, SE, 3, 0.81, 710, BSGW, GREY(A)
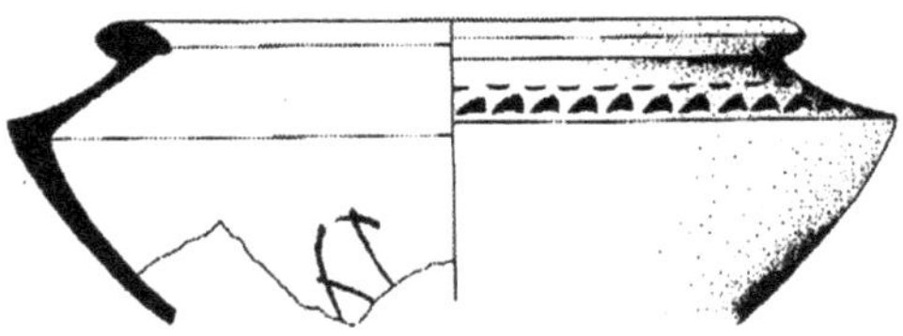 **T.402**	**T.402**:- Rim sherd of a cooking pot or handi having a sharp carination. The rim is external projecting extremely everted and tapering. On the carination a band of triangles can be seen cut into a clay to give an impression of luting (I.D = 20cm) SJN-B, TT4, SW, 2, 5.42, 51149, BSGW, GREY(A)
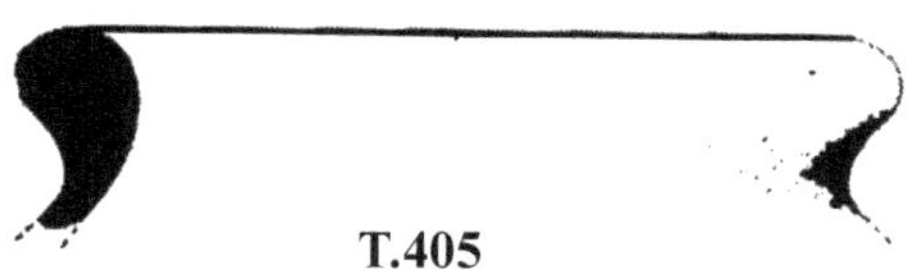 **T.405**	**T.405**:- Rim sherd of a jar with external projecting everted rounded rim heavily potted. The rim is somewhat clubbed and heavy (I.D = 32cm) SJN-B, TT4, SW, 3, 3.74, 51141, BSGW, GREY(A)
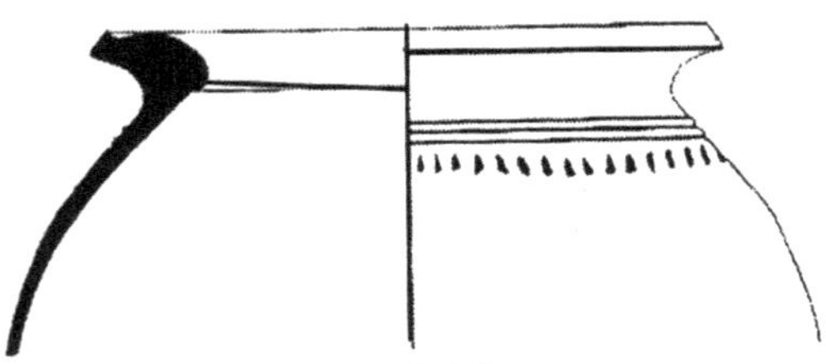 **D.96**	**T.440 D.71.1**:- Rim sherd of a pot having external projecting extremely everted triangular rimwith a deep groove like off-set on the top near the lip (perhaps to hold a lid) There are two grooves on the shoulder with a band of short vertical (deeply incised) lines underneath. (I.D = 18cm).
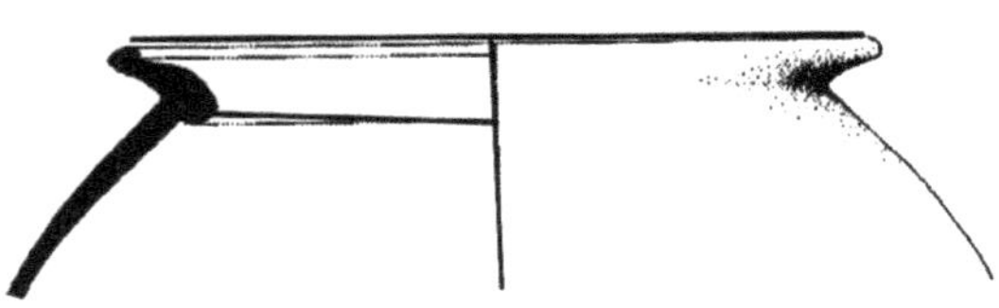 **T.440.1**	**T.440.1**:- Rim sherd very similar to the above except that this form has a bilateral projection, the internal end of which is rounded. (I.D = 18cm)

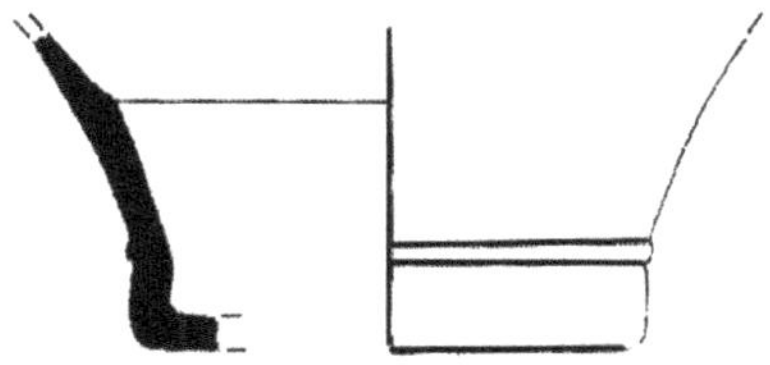 **Bs.37**	**Bs.37**:- This base is contiguous flat and has a ridge on the outer surface just above it. The edge of the base is rounded. SJN-B, TT1, NW, 3, 2.23, 337, BSGW, GREY(A)
 **D.42**	**D.42**:- horizontal appliqué band with oblique lines cut into it. SJN-B, TT2, SW, 3, 1.27, 915, BSGW, GREY(A)
 **T.373 D.168**	**D.43**:- Horizontal band of punctate triangles SJN-B, TT2, SE, 3, 1.42 , 1520, BSGW, GREY(A)
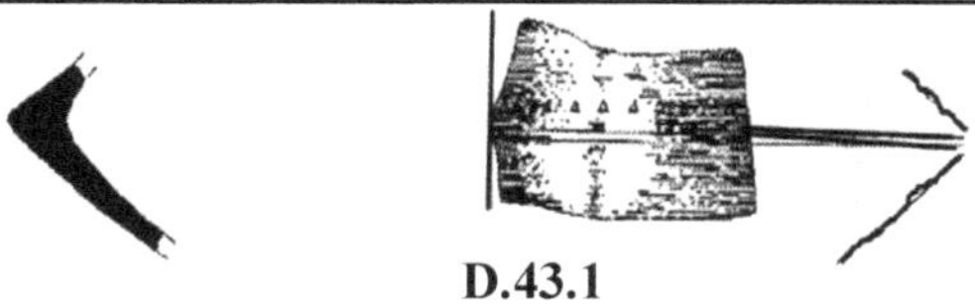 **D.43.1**	**D.43.1**:- Horizontal band of punctate triangles SJN-B, TT2, SE, 3, 1.52, 1522, BSGW, GREY(A)
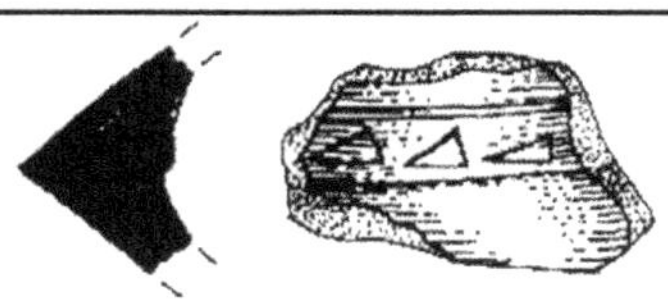 **T.43.2**	**D.43.2**:- Horizontal band of punctate triangles SJN-B, TT2, SE, 4, 1.84, 1529, BSGW, GREY(A)
 **D.43.3**	**D.43.3**:- Horizontal band of punctate triangles SJN-B, TT2, SE, 3, 1.17, 1516, BSGW, GREY(A)
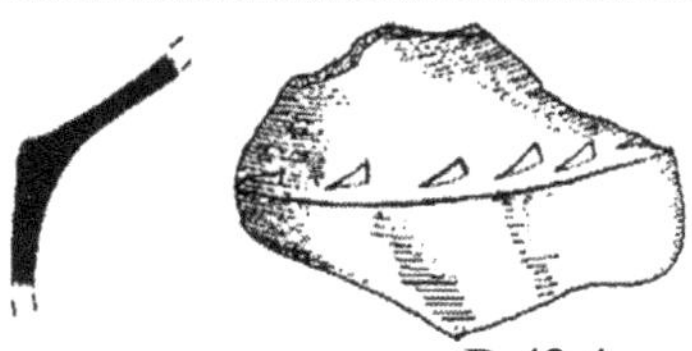 **D.43.4**	**D.43.4**:- Horizontal band of punctate triangles SJN-B, TT2, SE, 3, 1.27, 1518, BSGW, GREY(A)
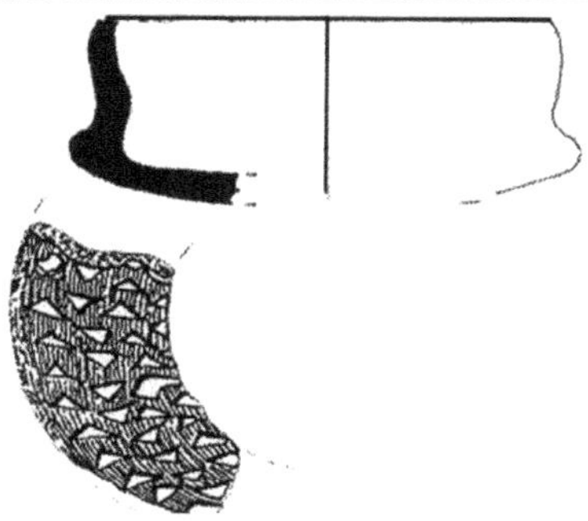 **D.44**	**D.44**:- Semi circular punctate marks in concentric rows; alternate 'c' and inverted 'c' rows SJN-B, TT1, SW, 4, 2.25, 143, BSGW, GREY (A)

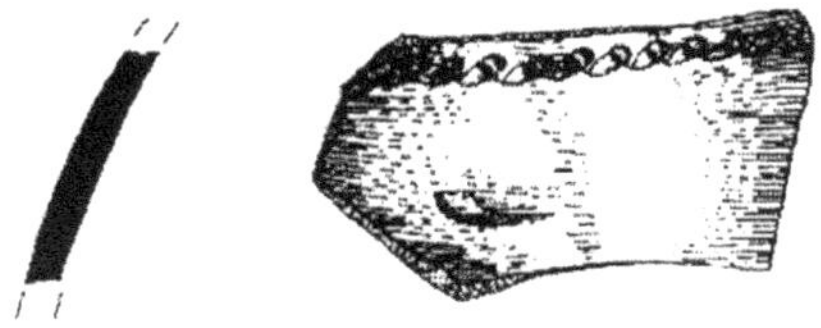 D.45	**D.45**:- Punctate marks with raised clay on one side SJN-B, TT2, SE, 4, 2.31, 1538, MISC-3, RED (F)
 D.46	**D.46**:- Shoulder fragment with a triangular design cut onto the shoulder so as to give an effect of luting. SJN-B, TT1, SW, 3, 1.9, 131, BSGW, GREY (A)
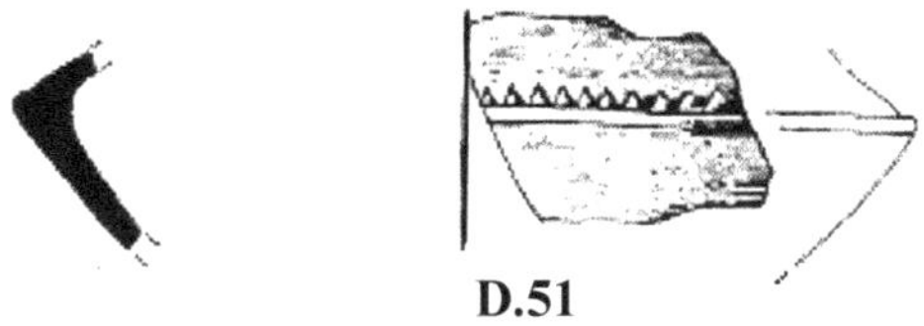 D.51	**D.51**:- Triangular design cut into the shoulder so as to give the impression of luting SJN-B, TT2, SE, 3, 1.12, 1515, BSGW, GREY(A)
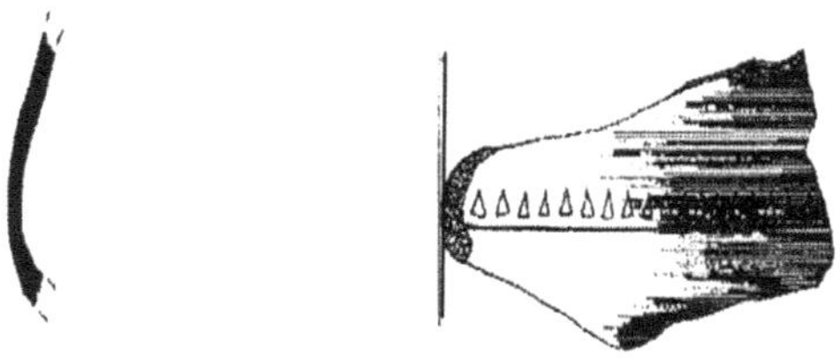 D.55	**D.55**:- Band of triangular punctate marks SJN-B, TT1, NW, 3, 121, 322, BSGW, GREY(A)
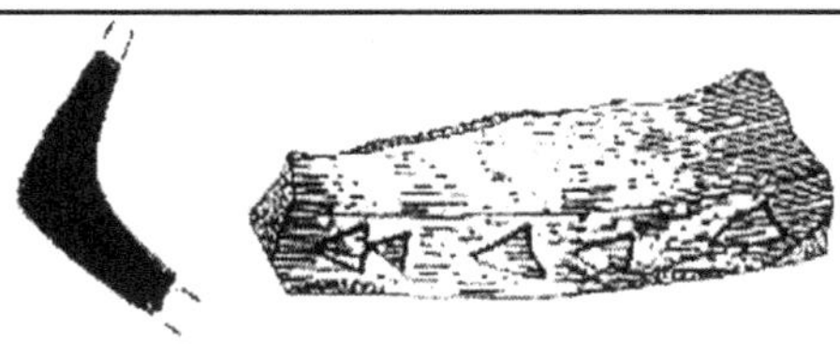 D.55.2	**D.55.2**:- Bands of triangular punctate marks SJN-B, TT2, SE, 3, 1.31, 1519, BSGW, GREY(A)
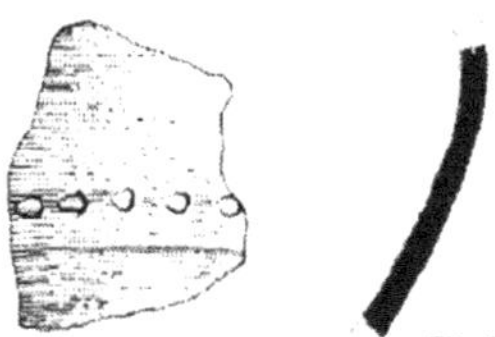 D.55.3	**D.55.3:**- Bands of triangular punctate marks SJN-B, TT2, SE, 3, 1.31, 1519, BSGW, GREY(A)
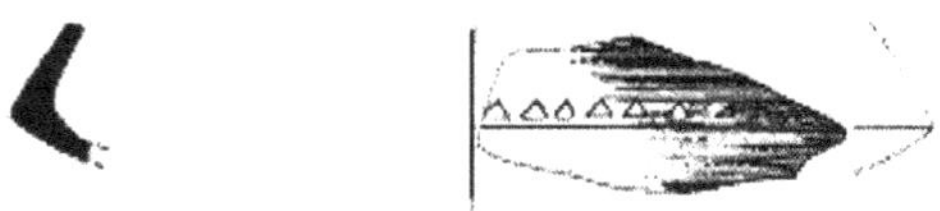 D.55.4	**D.55.4:**- Bands of triangular punctate marks SJN-B, TT1, SW, 3, 1.24, 139, BSGW, GREY(A)
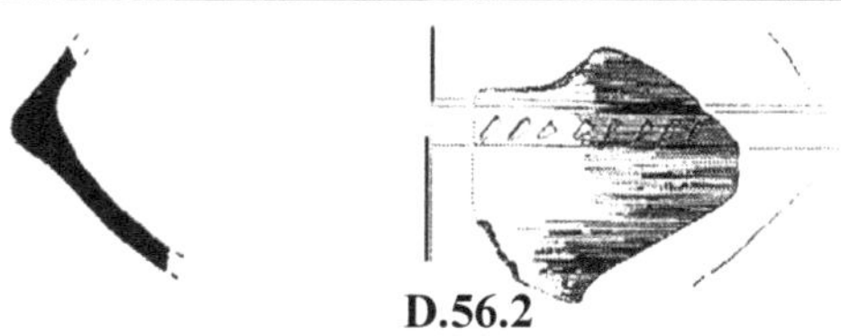 D.56.2	**D.56.2**:- Bands of triangular punctate marks with grooves SJN-B, TT1,NW, 3, 1.02, 316, BSGW, GREY(A)

 **D.63**	**D.63**:- Incised horizontal lines and lines of punctate triangles SJN-B, TT2, SE, 3, 1.31, 1519, BSGW, GREY(A)
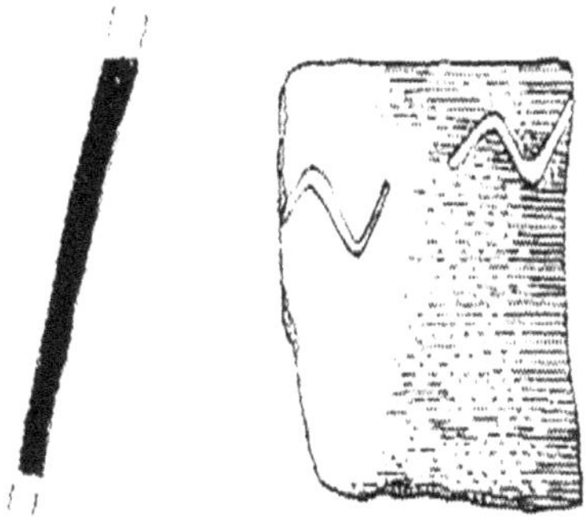 **D.62**	**D.62**:- Incised wavy lines SJN-B, TT1, NW, 3, 1.79, 330, BSGW, GREY(A)
 **D.75.2**	**D.75.2**:- Double rows of triangles
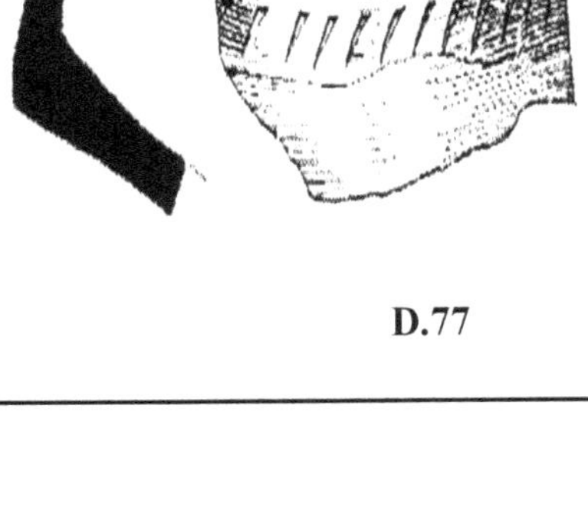 **D.77**	**D.77:-** Rows of oblique lines

RED SLIPPED GREY WARE	
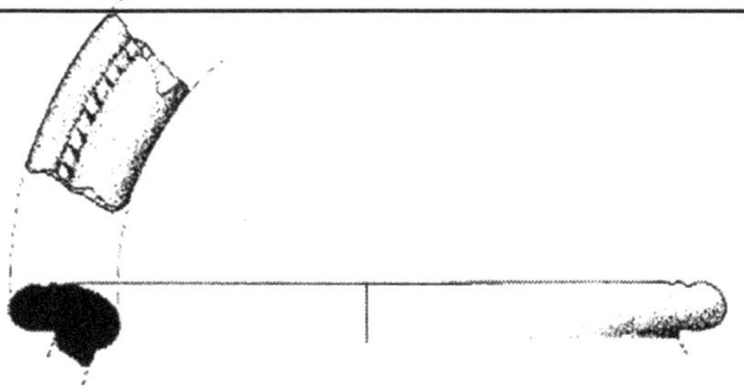 **D.192**	**D.192**:- Small rim sherd of a pot having external projecting rounded rim. On the rim along the lip can be seen two grooves within which a series of oblique lines has been incised. The vessel has a red slip. SJN-B, TT4, SE, 2, 0.73, 51711, RSGW, GREY(A)
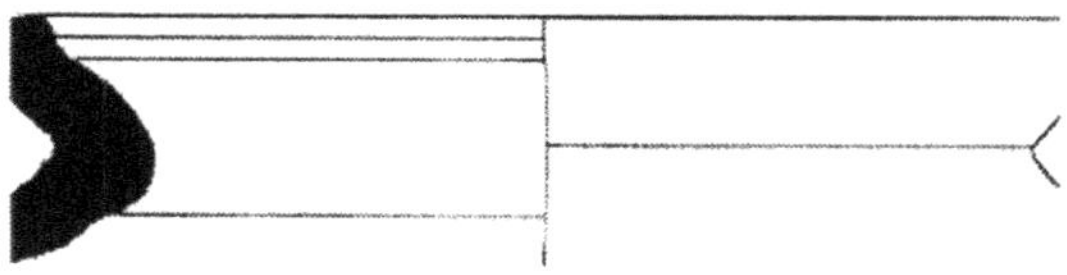 **T.373 D.168**	**T.121**:- Rim sherd of a large storage jar with everted squared rim with a groove on the inner side of the lip. (I.D = 24cm). SJN-B, TT1, SW, 2, 0.86, 111, RSGW, GREY(B)
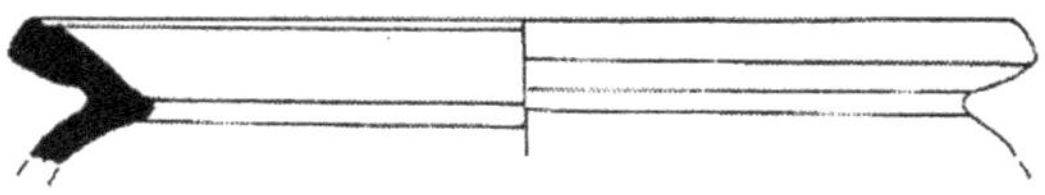 **T.135**	**T.135**:- Rim sherd of a wide-mouthed pot with external projecting extremely everted square rim similar to T.134 with a groove on the upper side of the lip. Red slip (I.D =18cm). SJN-B, TT1, NE, 3, 1.29, 518, RSGW, GREY(A)
 **T.157**	**T.157**:- Rim sherd of a pot with bilateral projecting rim with external projection being square. It has a groove on the upper side of the lip (I.D = 20cm). SJN-B, TT2, SE, 4, 2.31, 1538, RSGW, GREY(A)
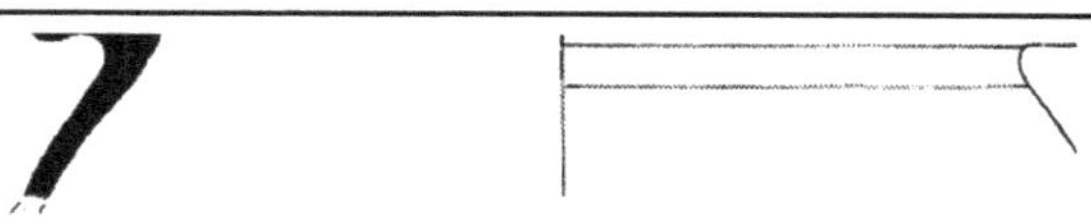 **T.185**	**T.185**:- Rim sherd of a pot with external projecting extremely everted rim with a groove on the inner side (I.D = 20cm) SJN-B, TT1, NE, 2, 0.76, 505, RSGW, GREY(A)
 **T.406**	**T.406**:- Rim sherd of a straight sided basin having a square rim which bulges slightly on the exterior. This vessel is illfired and coarse. The exterior has a pale red wash or slip. (I.D = 42cm) SJN-B, TT4, NE, 4, 1.78, 51545, RSGW, GREY(B)
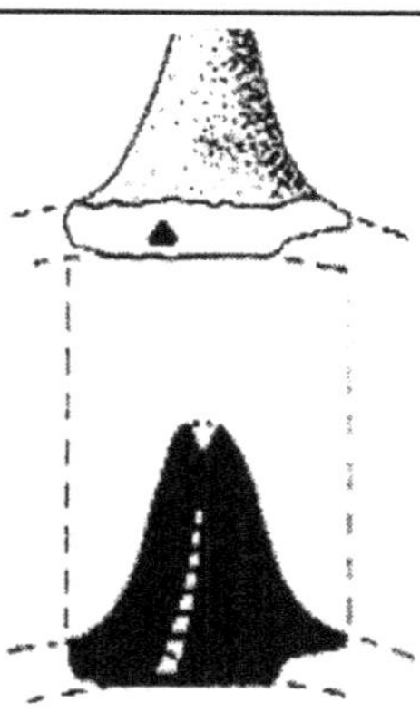 **T.408**	**T.408**:- Finial of a lid crudely made having a conical shape. It appears to have been pinched upwards. There is a perforation at the top but it is unclear if it is complete. SJN-B, TT4, NE, 2, 0.39, 51508, RSGW, GREY(A)

 T.227	**T.227**:- Rim sherd of a pot with everted beaded rim (I.D = 24cm). SJN-B, TT2, SE, 2, 0.98, 1512, RSGW, GREY(A)
 T.228	**T.228**:- Rim sherd of a pot with an external projecting triangular rim (I.D = 20cm) SJN-B, TT2, SE, 3, 1.12, 1515, RSGW, GREY(A)
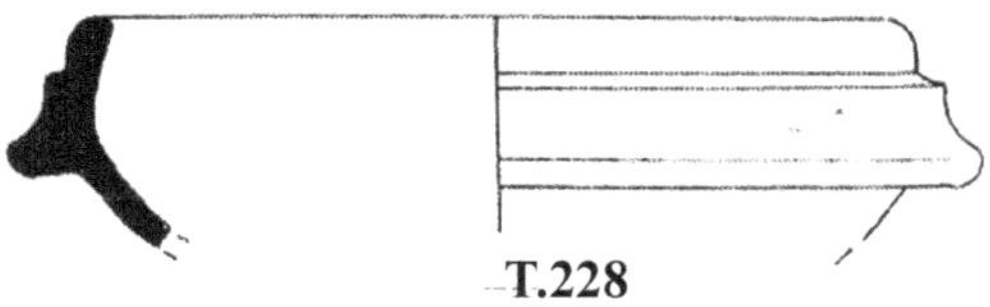 T.228	**T.235**:- Rim sherd of a carinated bowl with inverted squared rim with a corrugation above the ledge at the carination (I.D =16cm). SJN-B, TT1, NE, 3, 1.31, 519, RSGW, GREY(A)
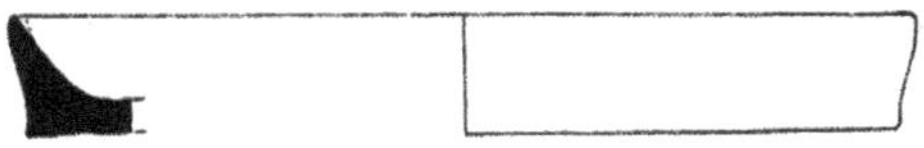 T.252	**T.252**:- Angular lug handle SJN-B, TT1, NE, 2, 0.76, 505, RSGW, GREY(B)
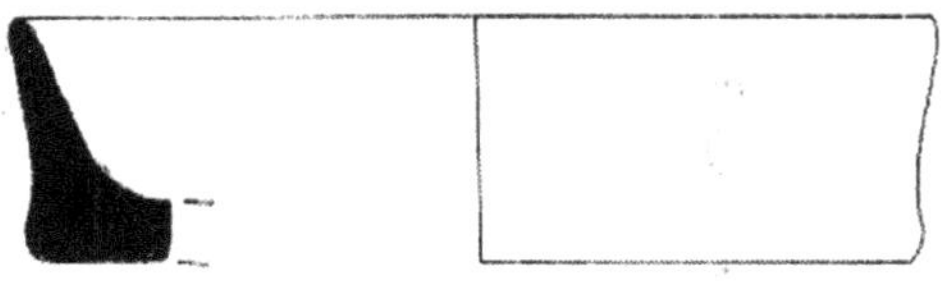 T.252.1	**T.252.1**:- Angular lug handleSJN-B, TT1, SE, 2, 0.58, 706, RSGW, GREY(B)
GREY WARE	
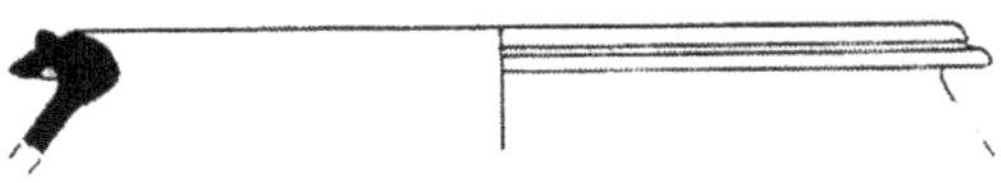 T.109	**T.109**:- Rim sherd of a pot with extremely projecting extremely everted tapering rim with a very high and prominent ridge on the upper part of the lip. (I.D =20cm). SJN-B, TT1, NW, 1, 0.63, 304, GW-1, GREY(A)
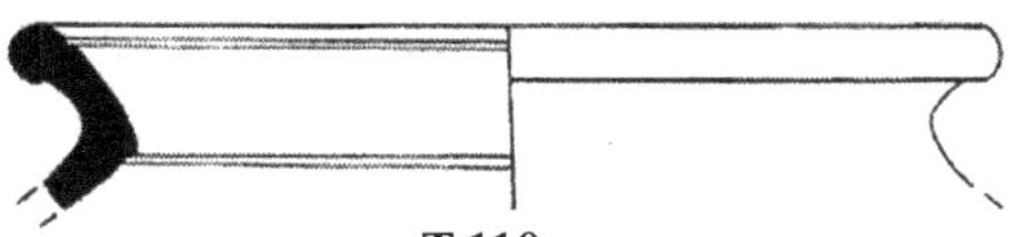 T.110	**T.110**:- Rim sherd of a pot with external projecting extremely everted rounded rim with two grooves on the upper side of the lip and a groove at the mouth. (I.D = 16cm). SJN-B, TT1, SE, 2, 0.44, 704, GW-2, GREY(C)
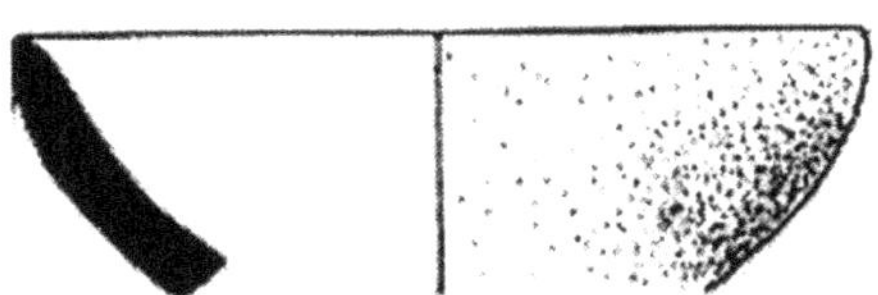 T.409	**T.409**:- Rim sherd of a very small lamp or shallow bowl having a simple tapering rim and convex sides. I.D – 6 cm) The fabric is quite fine and well fired. SJN-B, TT4, NW, 2, 0.41, 51308, GW-2, GREY(C)
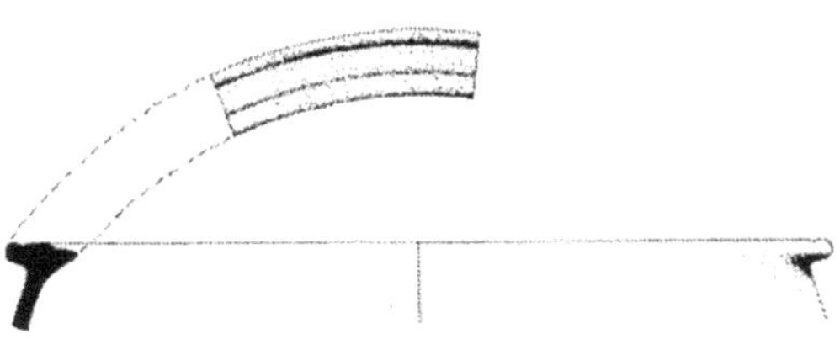 T.410	**T.410**:- Rim sherd of a pot or handi having external projecting wide rounded rim with two broad grooves running along the top, one near the mouth and other near the lip. The fabric is fine well levigated and evenly fired. (I.D = 38cm) SJN-B SURFACE, GW-2, GREY(C)

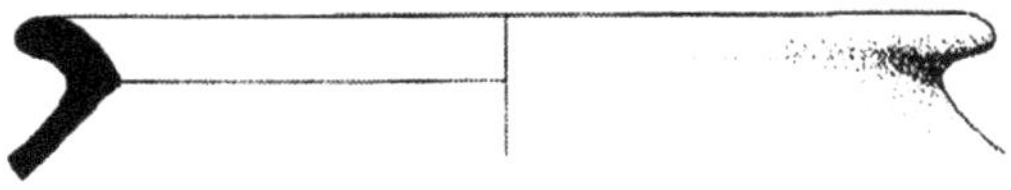 T.411	**T.411**:- Rim sherd of a pot having external projecting rounded rim. The inner surface and the mouth of the vessel (upto half of the upper part of the rim) is black and the half part of the rim and the exterior is light grey. This is the only two-toned sample to be found (I.D = 19cm)SJN-B SURFACE, GW-3, GREY(B)
RED SLIPPED RED WARE	
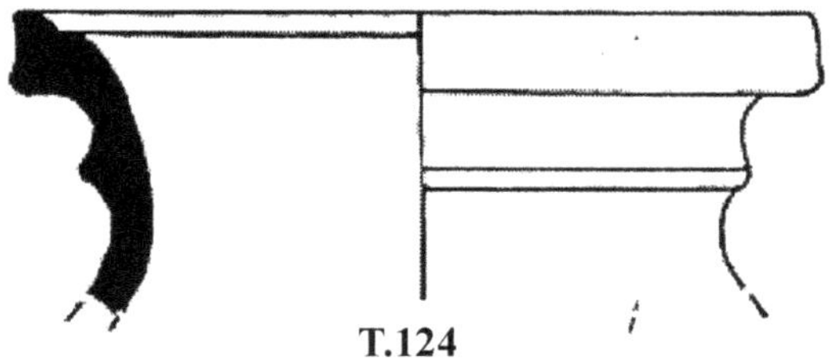 T.124	**T.124**:- Rim sherd of a high necked pot with external projecting triangular rim with a groove on the inner side and a bevelled ridge on the neck. The slip is shiny red. (I.D = 12cm). SJN-B, TT2, SE, 3, 1.42, 1521, RSRW-1, RED(C)
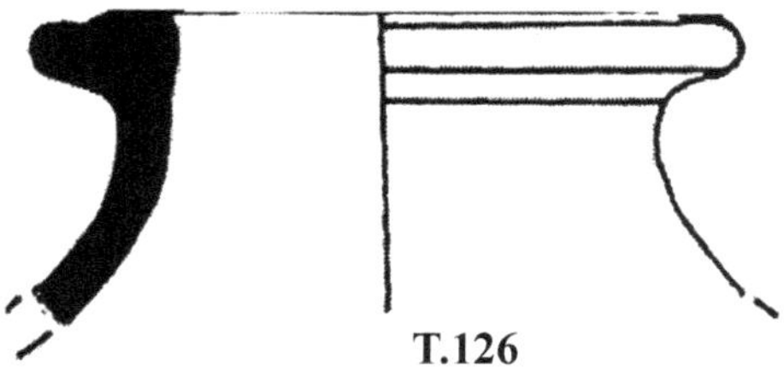 T.126	**T.126**:- Rim sherd of a high-necked pot with external projecting beaked rim with a groove on the upper side of the lip. The lip tapers sharply to the tip. Red slip. (diameter or orifice = 8cm). SJN-B, TT2, SE, 2, 1.05, 1514, RSRW-1, RED(C)
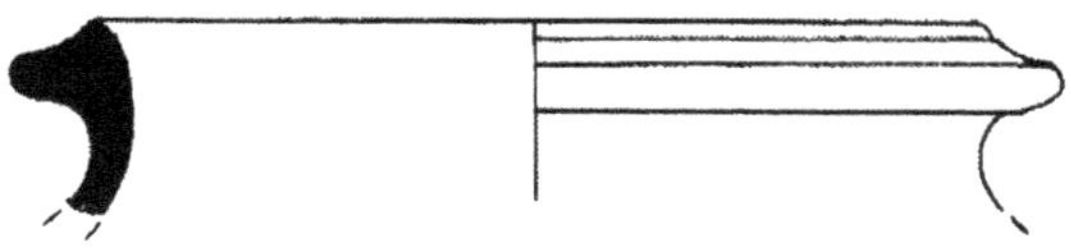 T.127	**T.127**:- Rim sherd of a pt with extremely wide ledge with triangular section and small rounded upper lip. Red slip applied externally and upto the orifice inside the rim (I.D =16cm) SJN-B, TT1, SW, 3, 1.64, 126, RSRW-1, RED(C)
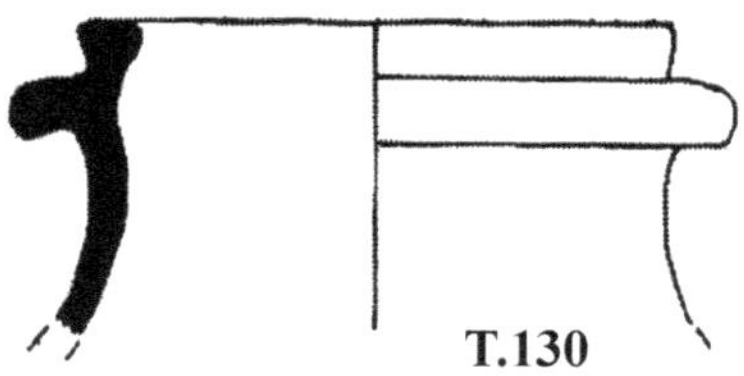 T.130	**T.128**:- Rim sherd of a pot with wide ledge with square section and small square upper lip with a small external projection and an inner groove. (I.D = 9cm). SJN-B, TT1, SW, 3, 1.06, 115, RSRW-1, RED(C)
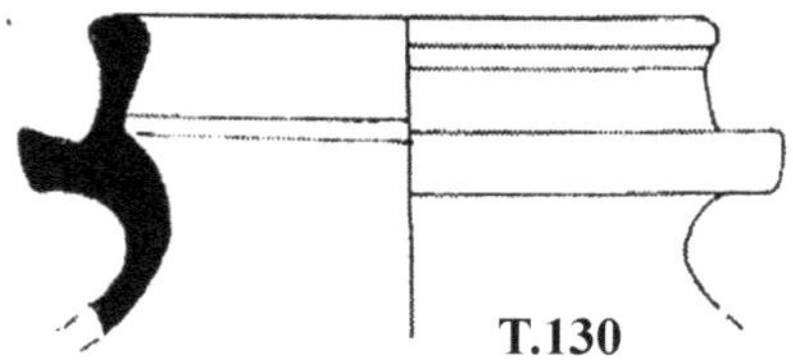 T.130	**T.130**:- Rim sherd of a pot with inverted squared rim with a small internal projection and a wide ledge, square in section just below and a subsequent deep groove on the inner side. Red slip (I.D = 9cm). SJN-B, TT1, NW, 3, 1.43, 325, RSRW-1, RED(C)
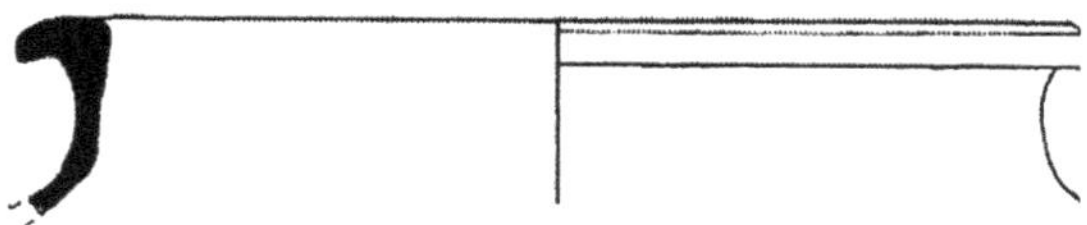 T.131	**T.131**:- Rim sherd of a wide mouth pot with external projecting extremely everted square rim with a groove on the upper side of the lip (I.D = 20cm). SJN-B, TT2, SW, 4, 1.58, 921, RSRW-1, RED(D)

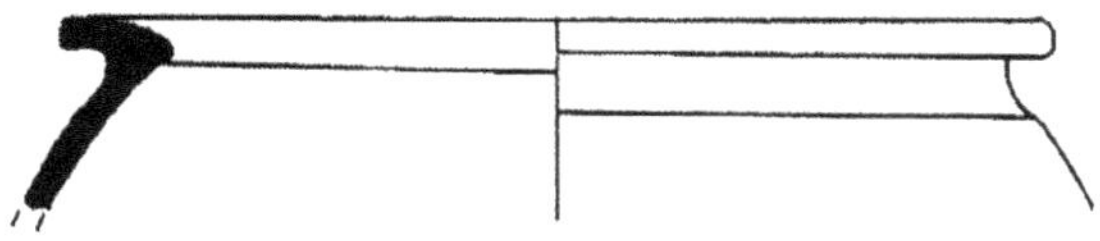 T.132	**T.132**:- Rim sherd of a wide-mouthed pot externally projecting extremely everted square rim with a groove on the upper side of the lip and an extremely constricted neck (I.D = 16cm). SJN-B, TT1, SW, 3, 1.16, 137, RSRW-2, RED(C)
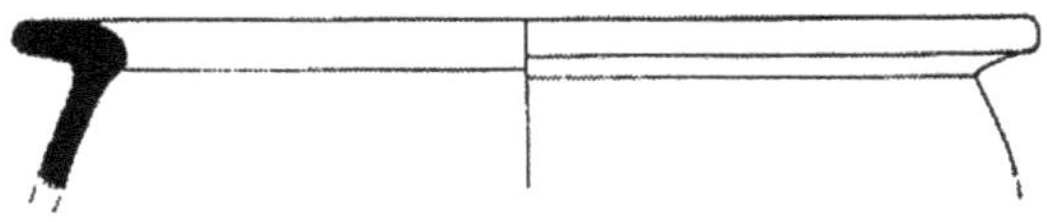 T.133	**T.133**:- Rim sherd of a wide-mouthed constricted necked pot with an external projecting extremely everted square rim (I.D = 18cm). SJN-B, TT2, SE, 3, 1.52, 1522, RSRW-1, RED(C)
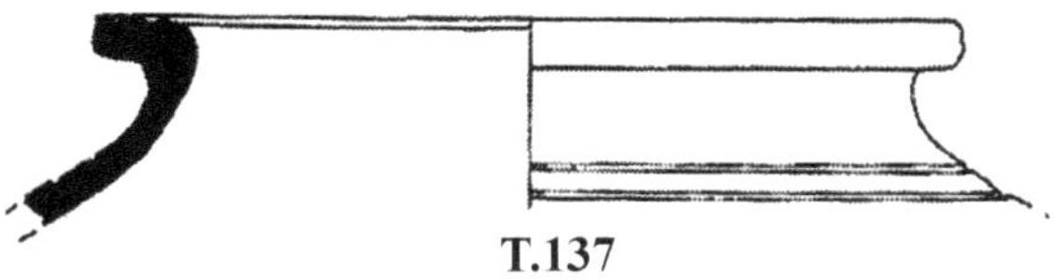 T.137	**T.137**:- Rim sherd of a wide mouthed pot with external projecting extremely everted squared rim with two grooves on the upper side of the lip. Red slip (I.D = 13cm). SJN-B, TT2, SE, 2, 1.05, 1514, RSRW-2, RED(C)
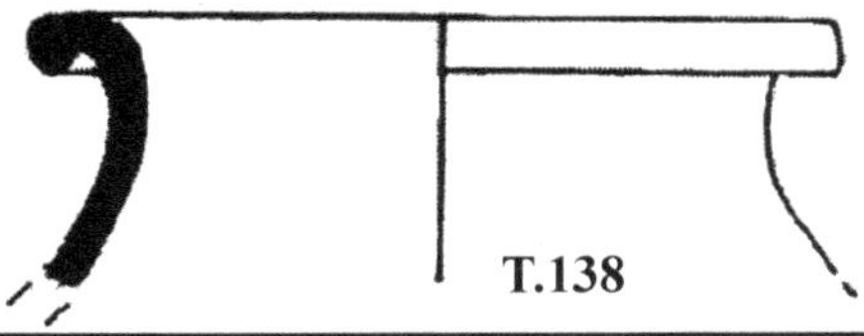 T.138	**T.138**:- Rim sherd of a small pot with external projecting extremely everted square rim with undercut. Grey slip (I.D = 8cm). SJN-B, TT1, SW, 2, 0.86, 111, BSGW, GREY(A)
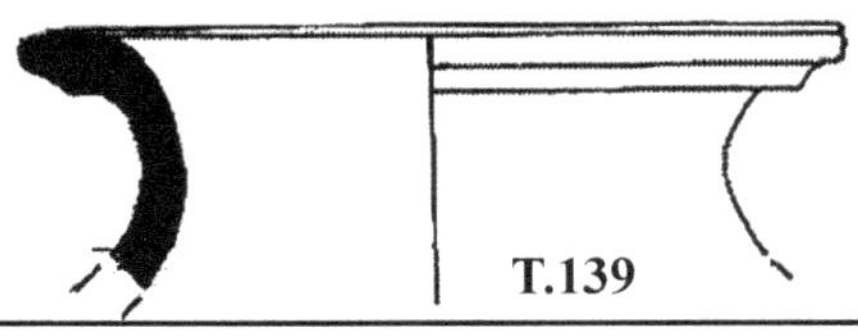 T.139	**T.139**:- Rim sherd of a small pot with external projection extremely everted triangular rim (I.D = 7cm). SJN-B, TT1, NE, 3, 1.18, 515, RSRW-1, RED(C)
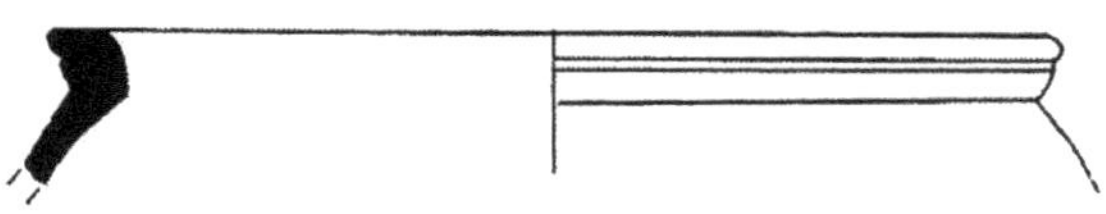 T.140	**T.140**:- Rim sherd of a wide-mouthed pot with constricted neck and an external projecting everted square rim and grooved on the underside of the lip. Grey slip. (I.D = 20cm). SJN-B, TT2, SE, 3, 1.31, 1519, BSRW-1, RED(C)
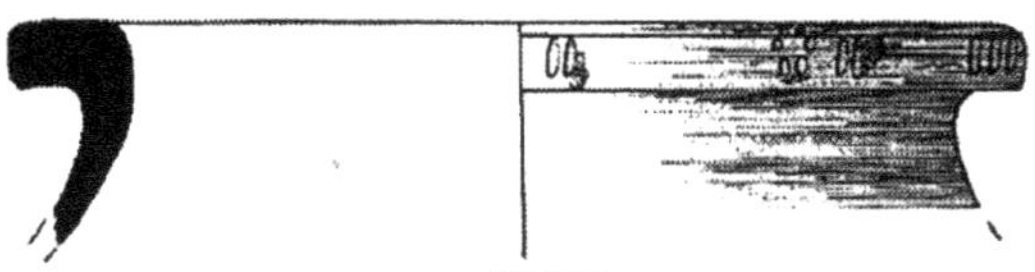 T.376	**T.142.1**:- Rim sherd of a pot with external projecting everted quadrangular rim. Grey colour. Three notches of wear marks on the tip, evenly spaced and an even horizontal band of slip erosion on the mouth (I.D = 18cm). SJN-B, TT1, SW, 2, 0.77, 107, BSGW, GREY(A)
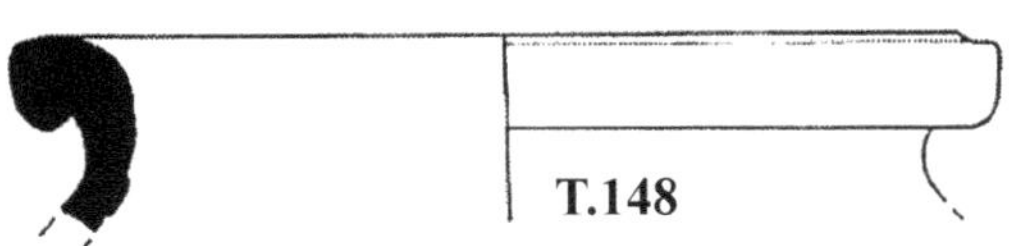 T.148	**T.148**:- Rim sherd of a pot with external projecting everted rounded rim. Grey slip (I.D = 18cm). SJN-B, TT2, SW, 4, 1.45, 918, RSRW-1, RED(C)
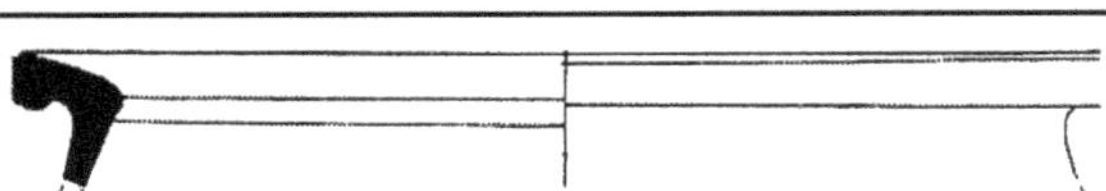 T.149	**T.149**:- Rim sherd of a pot with external projecting extremely everted rim with upper edge of lip sharp and lower edge rounded. (I.D = 19cm). SJN-B, TT2, SW, 4, 1.72, 923, RSRW-2, RED(C)

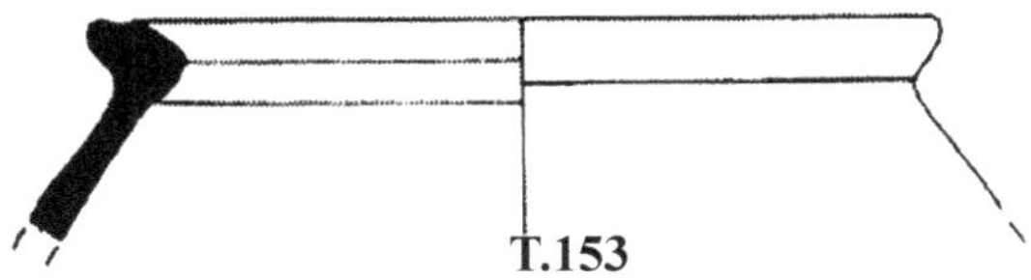 T.153	**T.153**:- Rim sherd of a pot with bilateral projecting rim with shallow grooves on upper side of the lip. External projection is tapering. (I.D = 14cm) SJN-B, TT2, SE, 2, 0.85, 1510, RSRW-1, RED(C)
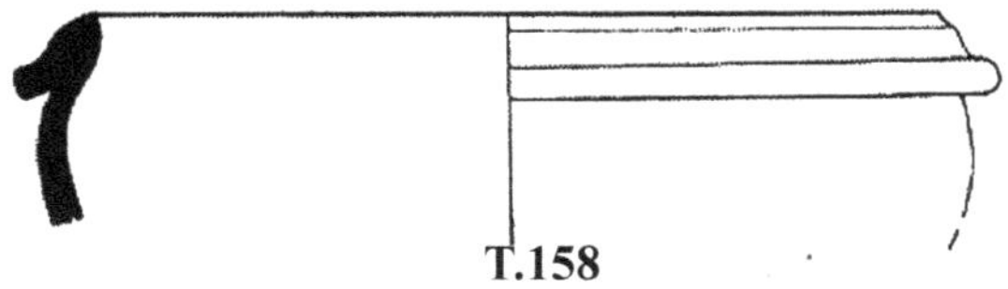 T.158	**T.158**:- Rim sherd of a pot with inverted tapering rim with a ledge on the exterior just below. Red lip (I.D = 14cm). SJN-B, TT1, NW, 3, 1.15, 320, RSRW-1, RED(C)
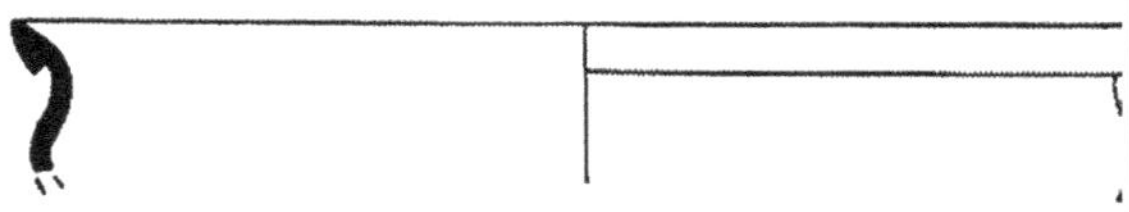 T.162	**T.162**:- Rim sherd of a carinated pot with external projecting triangular rim with an undercut. (I.D = 26cm) .SJN-B, TT1, SW, 2, 0.92, 112, RSRW-1, RED(C)
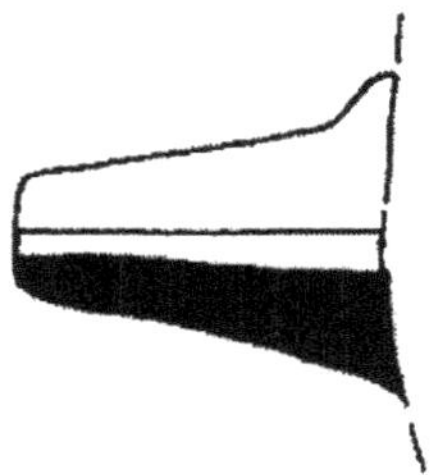 T.165	**T.165**:- Straight-sided, conical spout (L = 4.5cm) SJN-B, TT2, SE, 5, 2.41, 1541, RSRW-1, RED(C)
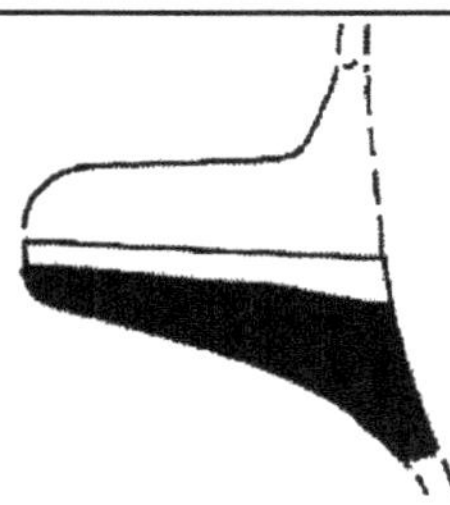 T.165.1	**T.165.1**:- Straight-sided, conical spout (L = 4.5cm) SJN-B, TT2, SW, 4, 1.58, 921, RSRW-1, RED(C)
 T.166	**T.166**:- Straight-sided spout with bulbous tip (L = 4cm) SJN-B, SW, 2, 0.86, 111, RSRW-1, RED(C)
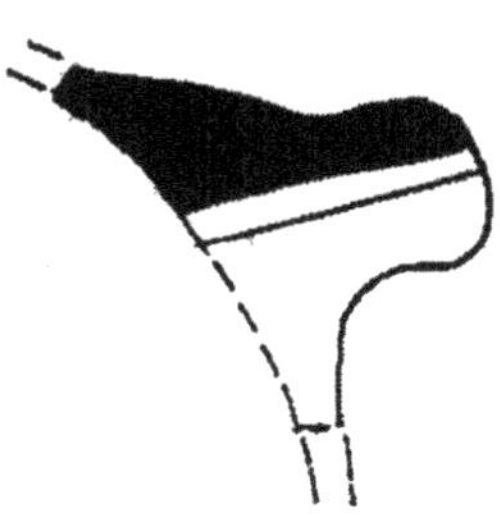 T.166.1	**T.166.1**:- Straight-sided spout with bulbous tip (L = 4cm) SJN-B, TT1, SE, 3, 0.96, 713, RSRW-1, RED(C)

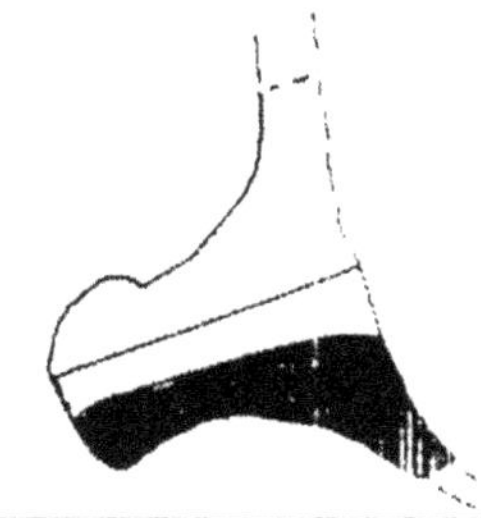 **T.284**	**T.166.3**:- Straight-sided spout with bulbous tip (L = 4cm) SJN-B, TT2, SE, 3, 1.36, 1521, RSRW-1, RED(C)
 **T.219**	**T.219**:- Rim sherd of a pot with bilateral projecting rim with both projections tapering (I.D = 28cm). SJN-B, TT1, NE, 2, 0.76, 505, RSRW-1, RED(C)
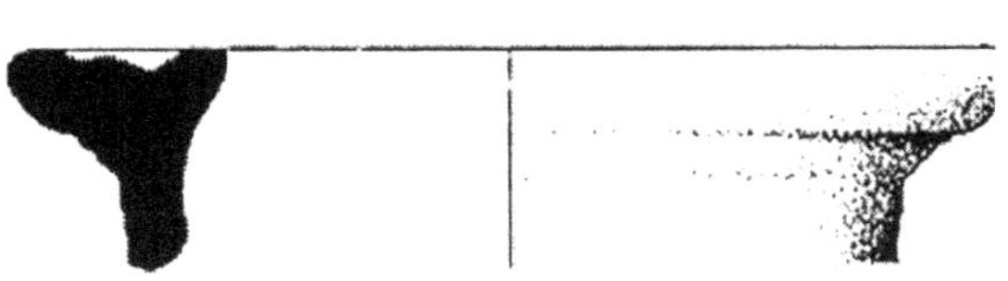 **T.414**	**T.414**:- Rim sherd of a small pot having external projecting short triangular rim having two prominent ridges and two broad grooves on the top. (I.D = 8cm) SJN-B, TT4, SW, 2, 3.42, 51139, RSRW-2, RED(C)
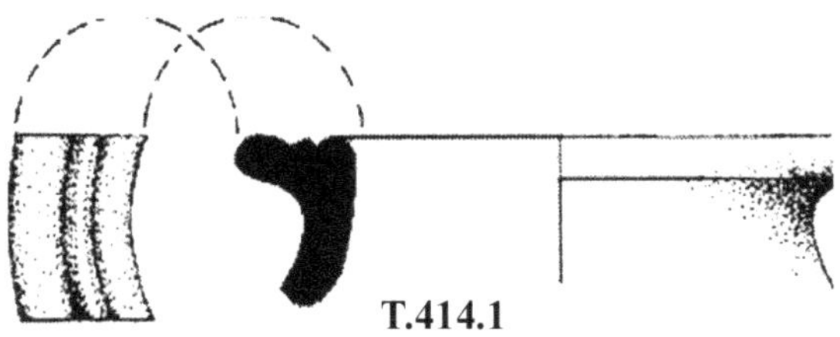 **T.414.1**	**T.414.1**:- Rim sherd of a pot having external projecting rim similar to above except that the neck is slightly higher. (I.D = 7cm) SJN-B, TT4, NE, 2, 0.72, 51519, RSRW-1, RED(C)
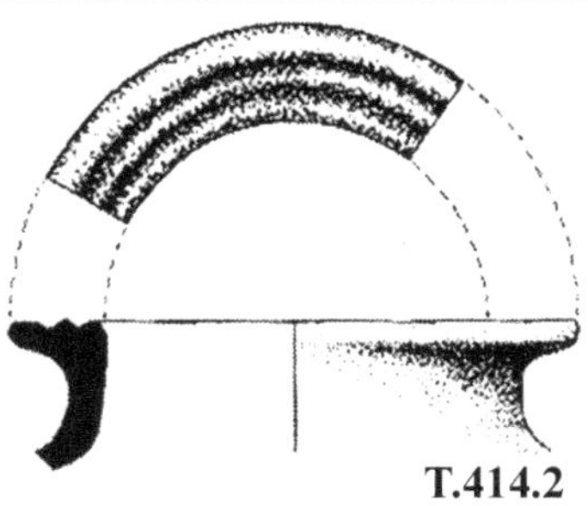 **T.414.2**	**T.414.2**:- Rim sherd similar to T.414 and T.414.1. This pot has a prominent slightly inverted mouth and the grooves make the rim appear almost like a ledge. (I.D = 12cm) SJN-B, TT4, NE, 2, 0.72, 51519, RSRW-1, RED(C)
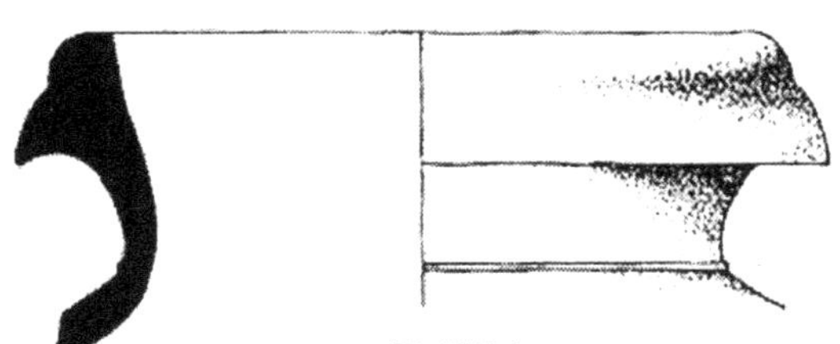 **T.415.1**	**T.415.1**:- Rim sherd of a high necked pot having a slightly inverted externally projecting triangular rim with the lower lip sharply beaked and a groove on the side (external). T.415 has 2 grooves. The shoulder and neck junction has a narrow ridge. (I.D = 10cm). SJN-B, TT4, NE, 5, 1.96, 51549, RSRW-1, RED(C)
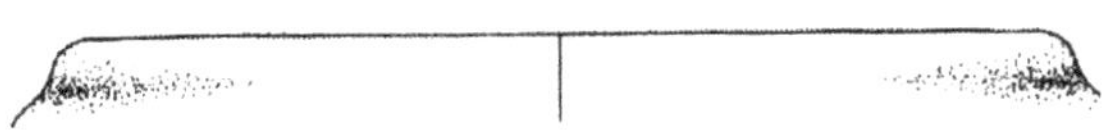 **T.416**	**T.416**:- Rim sherd of a pot having inverted tapering rim which thicken slightly on the exterior before the lip. (I.D = 18cm) SJN-B, TT4, SW, 2, 2.91, 51137, RSRW-1, RED(C)

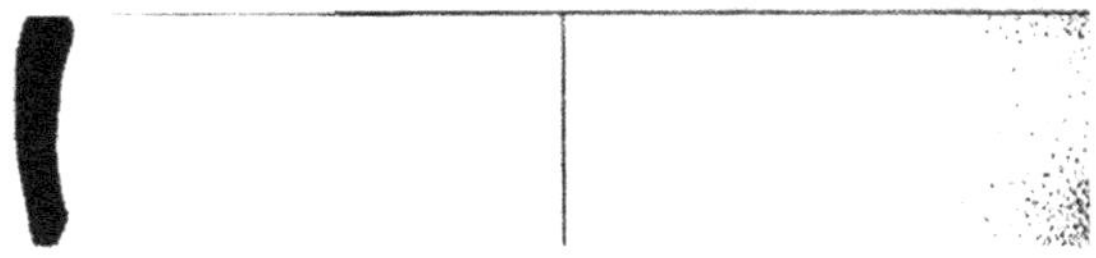 **T.417**	**T.417**:- Rim sherd of a bowl with simple square rim and low ridge running along the centre of the lip. The rim is slightly inverted. (I.D = 15cm) SJN-B, TT4, SW, 2, 3.67, 51140, RSRW-1, RED(C)
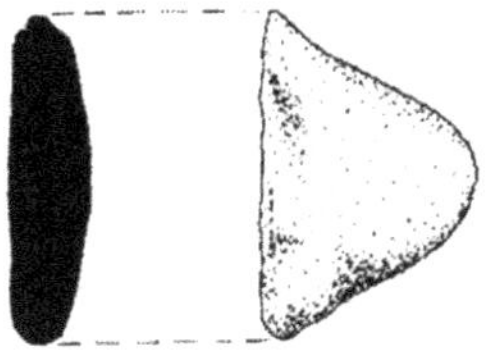 **T.419**	**T.419**:- Lug handle, triangular in shape with external angle rounded. It is not evenly thick and a shallow incised line can be seen on one side extending from the external angle almost to the luted edge. SJN-B, TT4, SW, 2, 2.68, 51136, RSRW-1, RED(C)
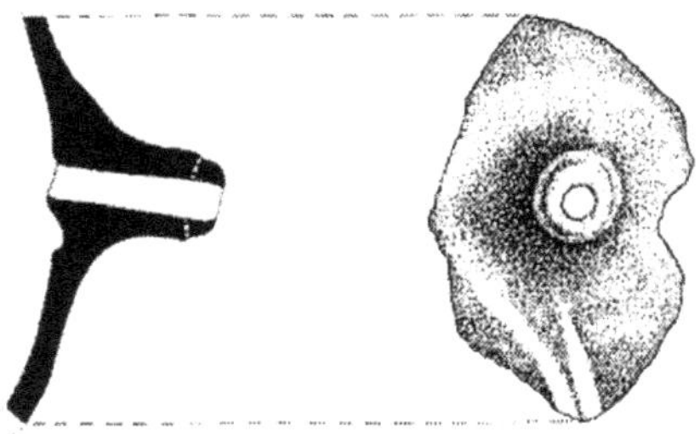 **T.420**	**T.420 D.196:-** Body sherd of a pot having a dark red slip on the exterior and having a spout with an appliquéd dot under it. Traces of white bands can be seen on the body above and below. SJN-B, TT4, SW, 2, 3.74, 51141, RSRW-1, RED(C)
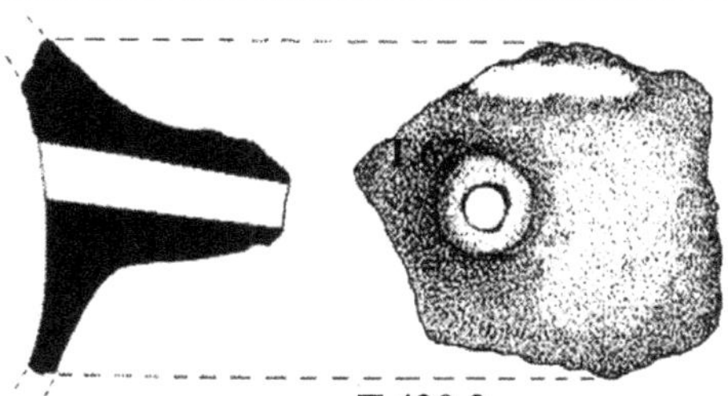 **T.420.2**	**T.420.2**:- Spout (conical) with white bands on vessel body. (L = 4.5cm) SJN-B, TT4, SW, 2, 3.67, 51140, RSRW-1, RED(C)
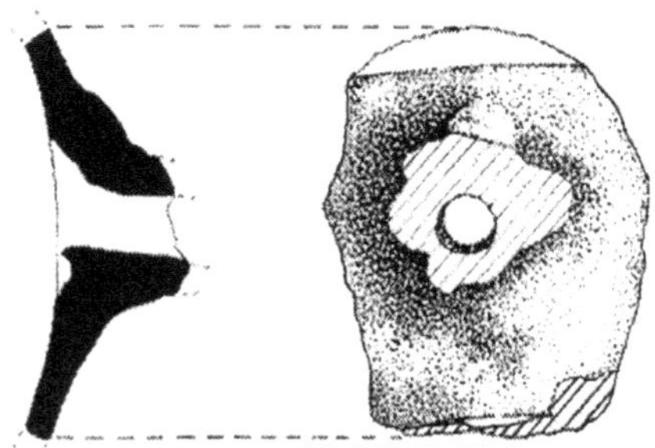 **T.420 D.196**	**T.420 D.196**:- Body sherd of a pot having a dark red slip on the exterior and having a spout with an appliquéd dot under it. Traces of white bands can be seen on the body above and below. SJN-B, TT4, SW, 2, 3.74, 51141, RSRW-1, RED(C)
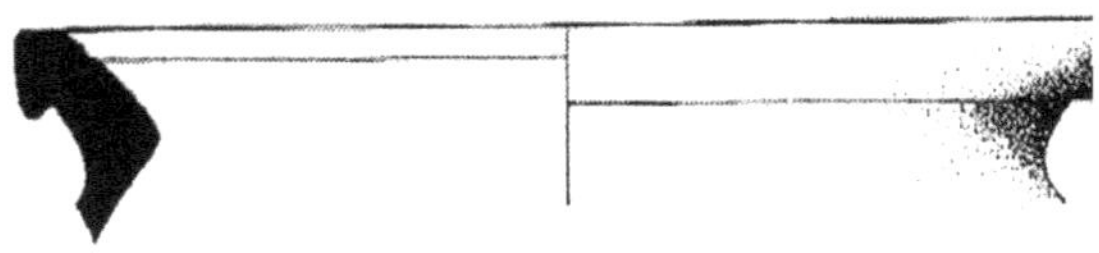 **T.435**	**T.435**:- Rim sherd of a pot having external projecting extremely everted rim and a constricted neck. The rim has a wide flange with a broad prominent groove on top, about 1/3rd of the way from the lip. The rim is quadrangular, narrowing just before the rounded lip turns downwards in a bead. (I.D = 24cm). SJN-B, TT4, NE, 5, 2.11, 51552, RSRW-1, RED(C)

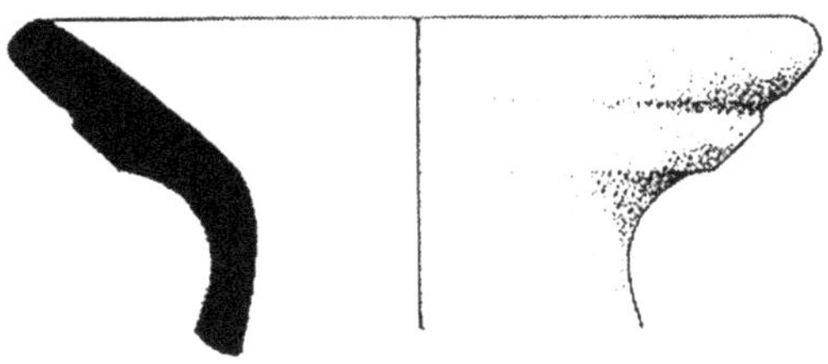 T.436	**T.436**:- Rim sherd of a pitcher/vase having a high straight sided neck and featureless rim. A broad flat ridge seem to go round the neck midway between the rim and shoulder (I.D =10cm) SJN-B, TT4, NE, 6, 2.98, 51571, RW-1, RED(C)
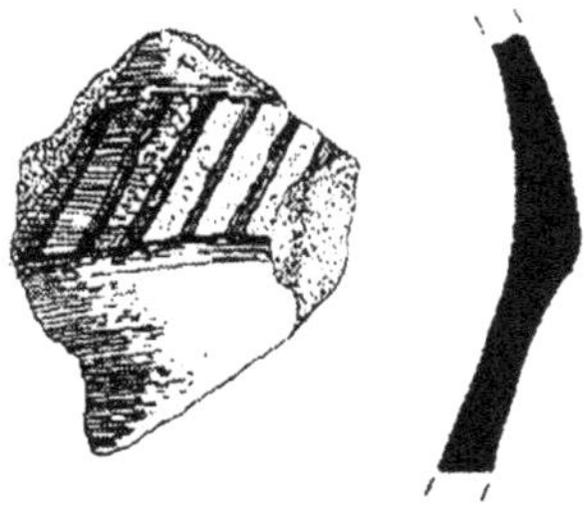 D.64.1	**D.64.1**:- Impressed and appliquéd bands with diagonal design. SJN-B, TT2, SE, 3, 1.52, 1522, RSRW-1, RED(C)
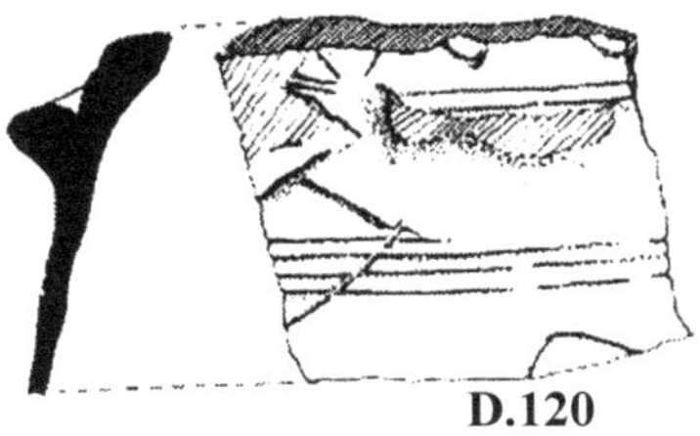 D.120	**D.120**:- Two horizontal parallel which bands within which are seen two rows of dots. Painted.
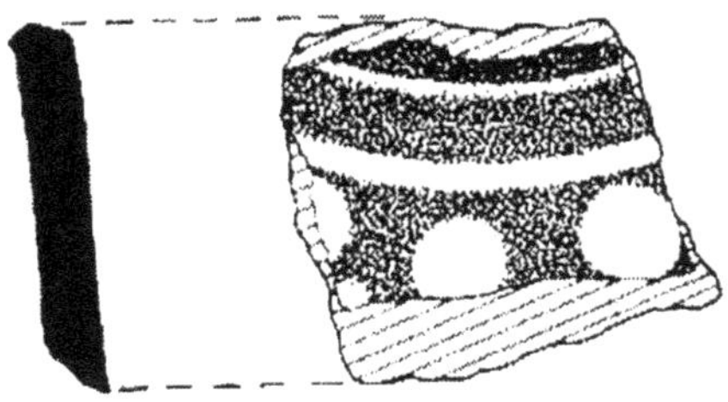 D.195	**D.195**:- Shoulder fragment of a pot having low ridge near the neck just below which is a painted white band and three large white dots. SJN-B, TT4, SW, 2, 3.11, 51138, RSRW-1, RED(C)
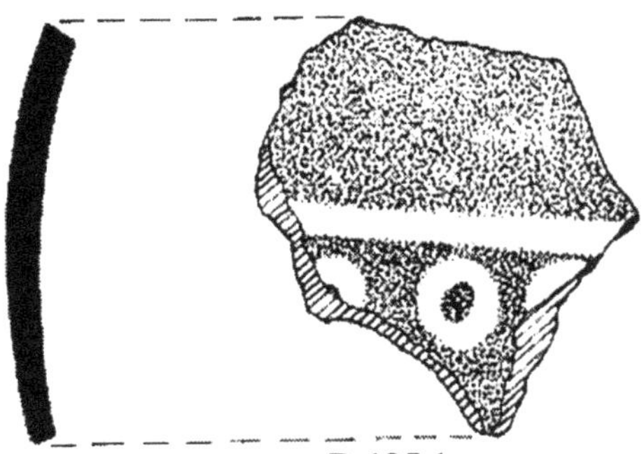 D.195.1	**D.195.1**:- Body sherd of a pot with similar painting except that the dots are not filled, they are white circles. SJN-B, TT4, SW, 2, 3.74, 51141, RSRW-1, RED(C)
 T.183	**T.183**:- Rim sherd of a pot with external projecting everted square rim with a groove on the inner side of the lip (I.D =20cm). SJN-B, TT1, SW, 1, 0.57, 103, RSRW-1, RED(C)

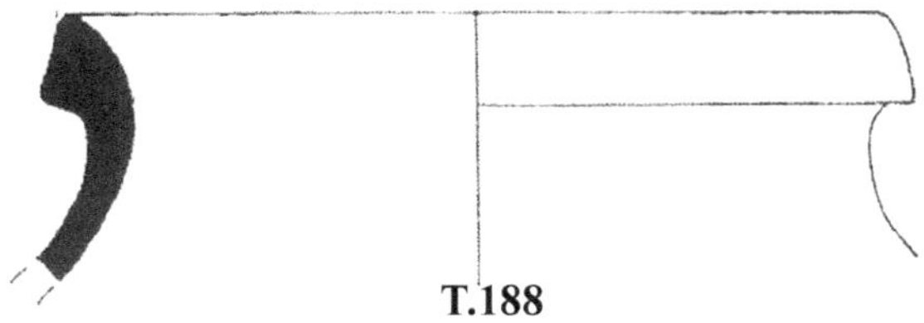 **T.188**	**T.188**:- Rim sherd of a pot with external projecting triangular rim (I.D = 13cm) SJN-B, TT1, SE, 2, 0.96, 713, RSRW-1, RED(C)
 **T.193**	**T.193**:- Rim sherd of a pot with external projecting inverted triangular rim (I.D = 18cm). SJN-B, TT1, SW, 2, 0.81, 109, RSRW-1, RED(C)
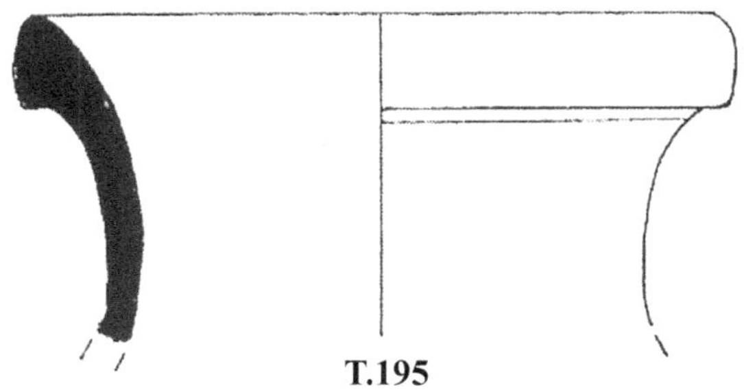 **T.195**	**T.195**:- Rim sherd of a high necked pot with external projecting triangular rim (I.D = 14cm). SJN-B, TT1, NW, 3, 1.07, 318, RSRW-2, RED(C)
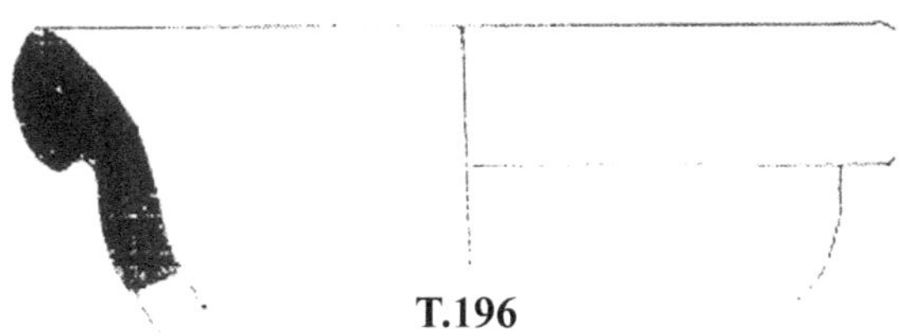 **T.196**	**T.196**:- Rim sherd of a pot with external projecting triangular rim with undercut (I.D =12cm). SJN-B, TT2, SE, 4, 1.78, 1527, RSRW-1, RED(C)
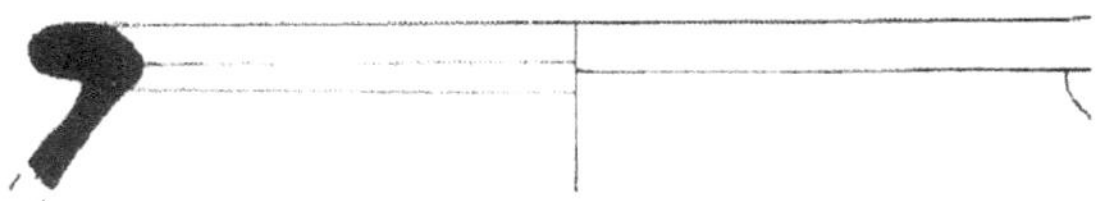 **T.209**	**T.209**:-Rim sherd of a pot with external projecting extremely everted rounded rim (I.D = 19cm). SJN-B, TT2, SW, 2, 0.98, 909, RSRW-1, RED(C)
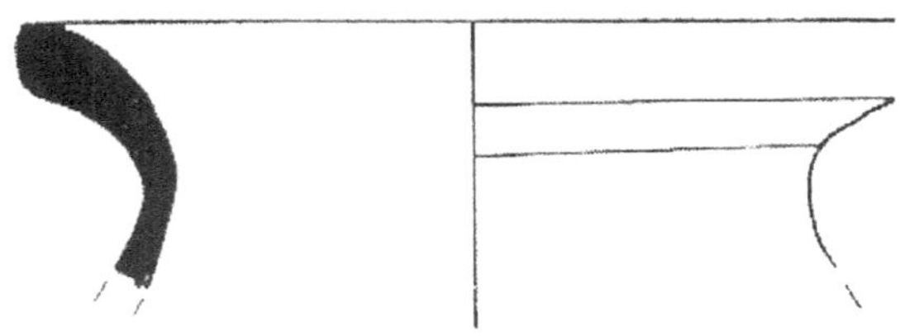 **T.214**	**T.214**:- Rim sherd of a pot with external projecting everted rounded rim (I.D = 12cm). SJN-B, TT2, SE, 2, 1.05, 1514, RSRW-1, RED(C)
 **Bs.82.1**	**Bs.82.1**:- Nipple-base of a large storage jar, similar to Bs 82. This base was found along with other pieces at the bottom of the trench TT4. It has yet to be reassembled. No rim is found. It is larger than Bs 82 and does not have bun-like raised feature inside the well/moat. The base appears bulbous. SJN-B, TT4, NE, 6, 3.60-3.79, 51583-87, BSRW-2, RED(C)

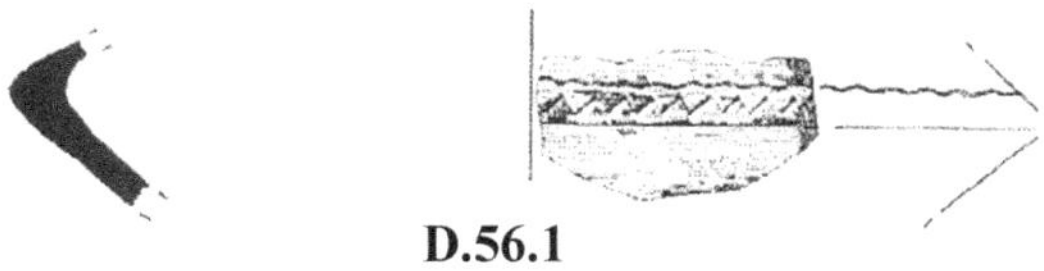 **D.56.1**	**D.56.1**:- Bands of triangular punctate marks with grooves SJN-B, TT2, SW, 4, 1.54, 920, RSRW-2, RED(C)
 **D.197**	**D.197**:- Body sherd (part of shoulder). The decoration consists of an incised design. A band of scallops/garland is seen under which are two broad grooves and thumbnail impressions vertically placed The slip is light red in colour. SJN-B, TT4, SW, 2, 3.11, 51138, RSRW-2, RED(E)
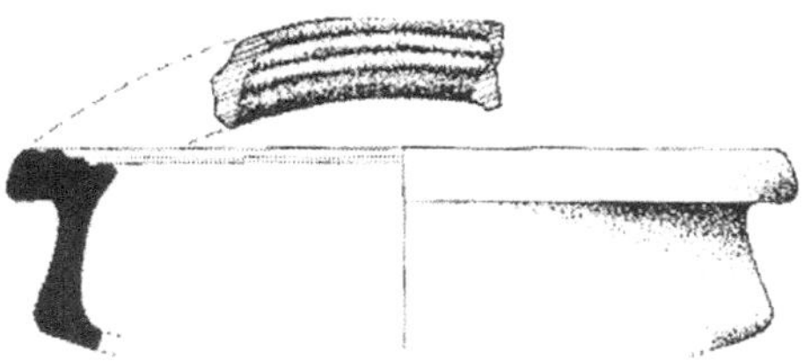 **T.413 D.194**	**T.413 D.194**:- Rim sherd of a carinated pot or handi having an external projecting rectangular rim having two high ridges and two shallow grooves on the top. The lower edge of the lip is rounded and slightly beaked. The shoulder has two low ridges just before the sharp carination. The carination has short oblique lines incised along the edge. (I.D = 12cm) SJN-B, TT4, SW, 1, 1.67, 51131, RSRW-2, RED(C)
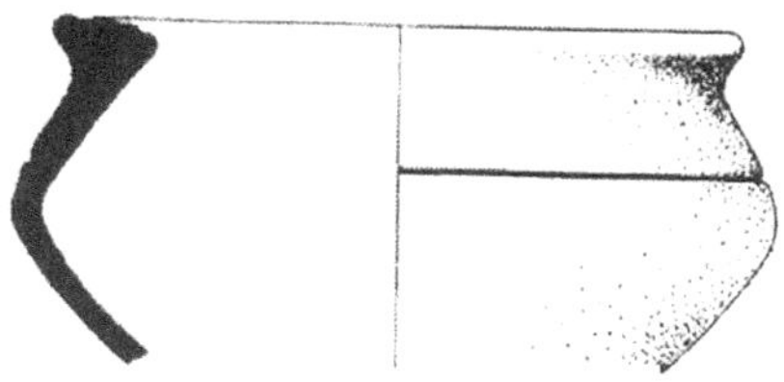 **T.414**	**T.414**:- Rim sherd of a small pot having external projecting short triangular rim having two prominent ridges and two broad grooves on the top. (I.D = 8cm) SJN-B, TT4, SW, 2, 3.42, 51139, RSRW-2, RED(C)
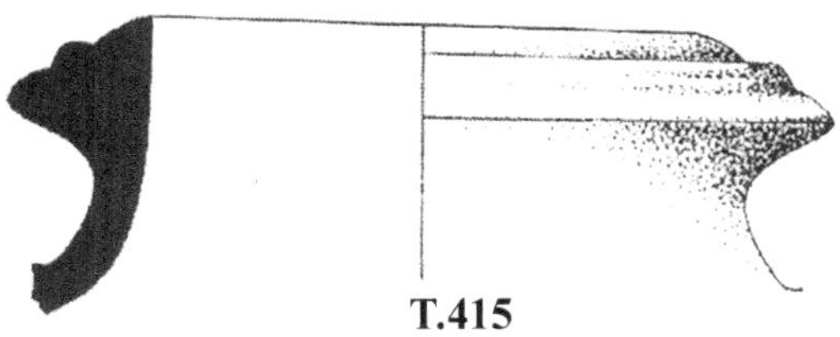 **T.415**	**T.415**:- Rim sherd of a pot having external projecting triangular rim with very sharp and high ridges on the side. The mouth of the rim is sharp and tapering. (I.D = 14cm) SJN-B, TT4, SW, 2, 3.11, 51138, RSRW-2, RED(C)
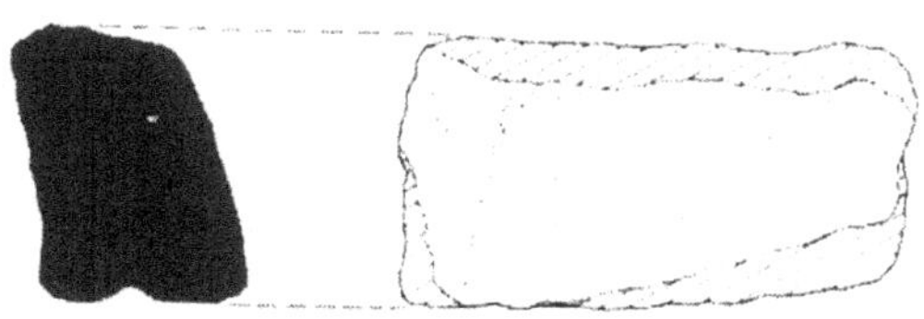 **T.418**	**T.418**:- Rim sherd of straight sided basin having uneven thickness. The rim is featureless and square with the external edge pinched a little higher at the mouth. (I.D = approximately 43cm) SJN-B, TT4, SW, 2, 2.25, 51134, RSRW-2, RED(C)
BLACK SLIPPED RED WARE	
 **T.114**	**T.114**:- Rim sherd of a plate with vertical squared rim, thick base; the plate is slipped on the inner surface and unslipped on the outer surface. The inner surface is black with a red band evident on the rim. The external surface of the base if rusticated. (I.D = 32cm) SJN-B, TT2, SE, 4, 2.05, 1533, BSRW-2, RED(C)

 T.115	**T.115**:- Rim sherd of a plate with an everted rounded rim. The rim is thicker than the base. This slip is applied to the inner surface. The base is rough on the exterior. (I.D = 28cm). SJN-B, TT1, NE, 3, 1.24, 517, BSRW-1, RED(C)
 T.119	**T.119**:- Rim sherd of a bowl with inverted square rim with a small external projection and a pronounced ledge on the exterior. The slip extends from the rim to the under side of the ledge. (I.D = 14cm) SJN-B, TT1, NW, 3, 1.79, 330, BSRW-2, RED(C)
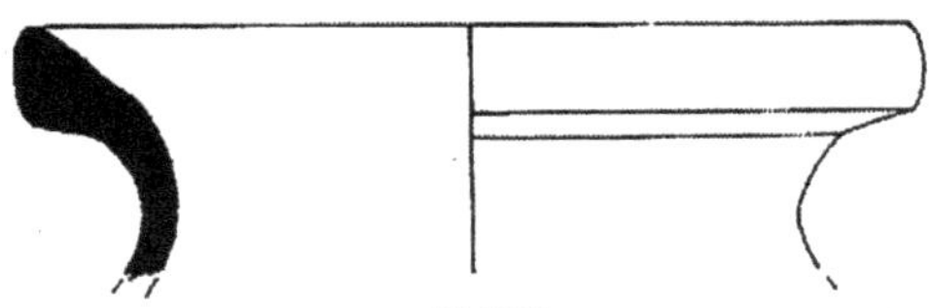 T.123	**T.123**:- Rim sherd of a high necked pot with external projecting triangular rim Slip is only on the exterior surface. (I.D =12cm) SJN-B, TT2, SW, 3, 1.42, 917, BSRW-1, RED(C)
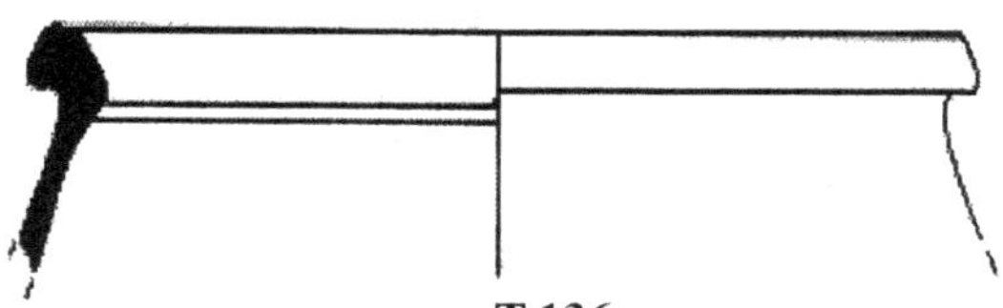 T.136	**T.136**:- Rim sherd of a wide-mouthed pot with external projecting extremely everted squared rim with a groove along the tip and along the shoulder. Grey slip (I.D = 20cm). SJN-B, TT2, SW, 4, 1.58 - 1.64, 921+ 922, BSRW-1, RED(C)
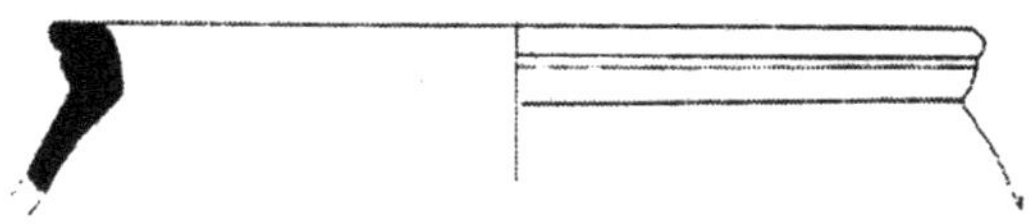 T.140	**T.140**:- Rim sherd of a wide-mouthed pot with constricted neck and an external projecting everted square rim and grooved on the underside of the lip. Grey slip. (I.D = 20cm). SJN-B, TT2, SE, 3, 1.31, 1519, BSRW-1, RED(C)
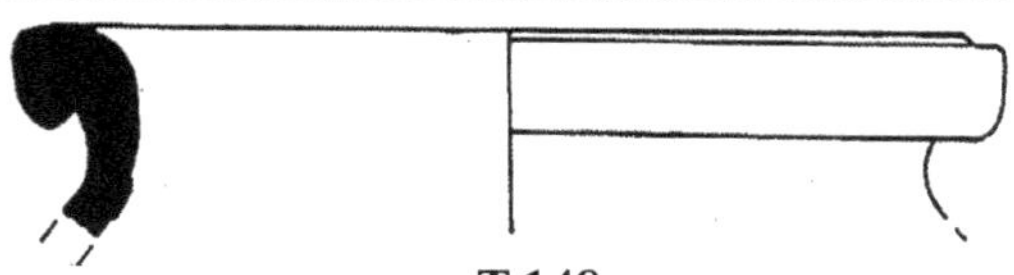 T.148	**T.148**:- Rim sherd of a pot with external projecting everted rounded rim. Grey slip (I.D = 18cm). SJN-B, TT2, SW, 4, 1.45, 918, BSRW-1, RED(C)
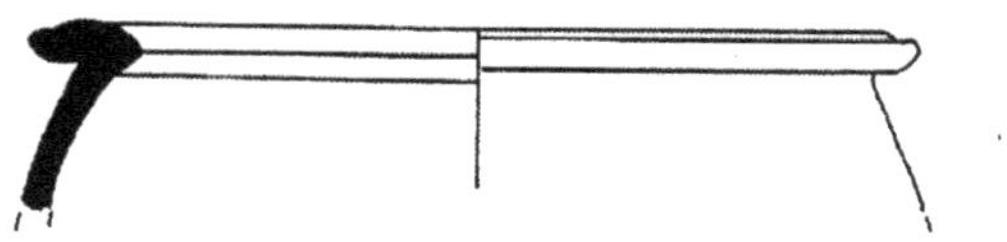 T.152	**T.152**:- Rim sherd of a pot with bilateral projecting rim with greater external projection. (I.D = 18cm). External projection is rounded. SJN-B, TT1 SW BSRW-1 RED(C)
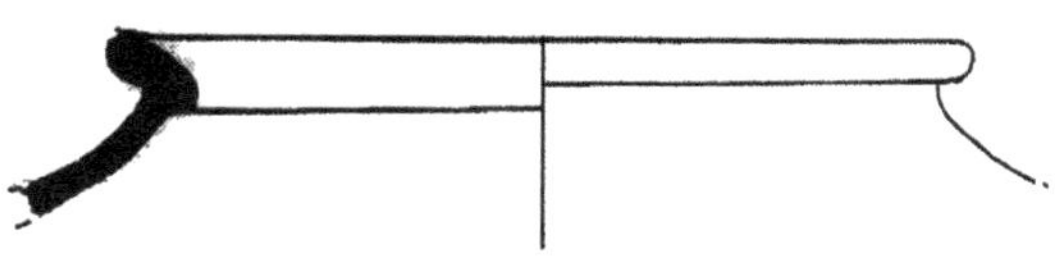 T.154	**T.154**:- Rim sherd of a pot with bilateral projecting rim. (I.D = 18cm) external projection is rounded. SJN-B, TT1, NW, 2, 1.02, 316, BSRW-2, RED(C)

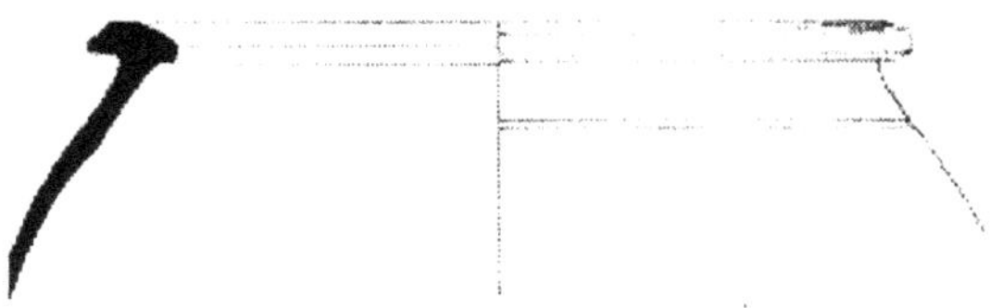 **T.155**	**T.155**:- Rim sherd of a pot with bilateral projecting rim, the external projection of which is beaked. (I.D = 24cm). SJN-B, TT2, SE, 3, 1.31, 1519, BSRW-2, RED(C)
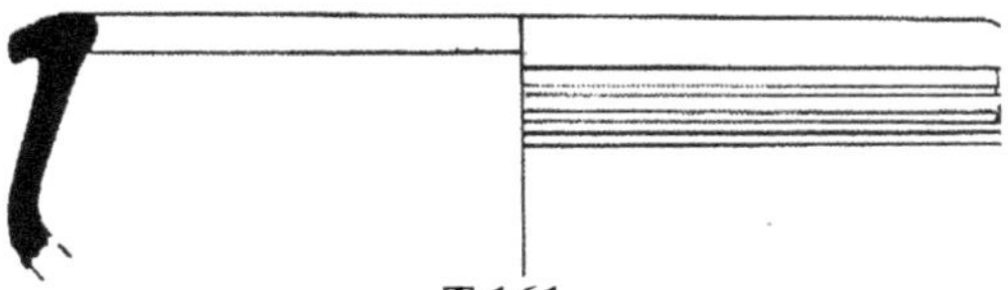 **T.161**	**T.161**:- Rim sherd of a carinated pot with external projecting extremely everted squared rim. Grey slip (I.D = 20cm). SJN-B, TT1, NE, 3, 1.15, 514, BSRW-1, RED(C)
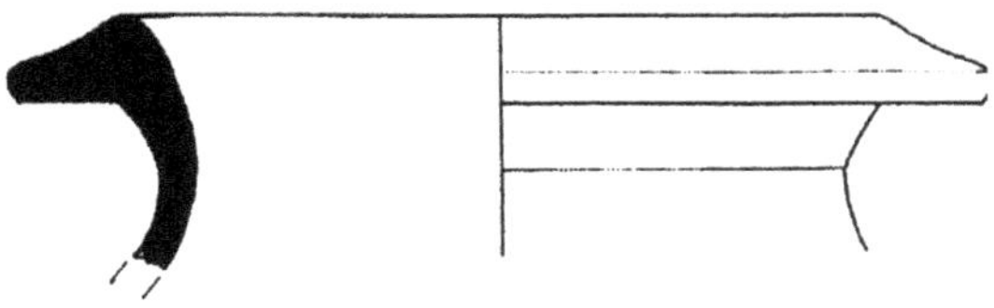 **T.202**	**T.202**:- Rim sherd of a pot with external projecting beaked rim with a wide lip. There is a ridge at the orifice (I.D = 16cm). SJN-B, TT2, SW, 3, 1.36, 916, BSRW-2, RED(C)
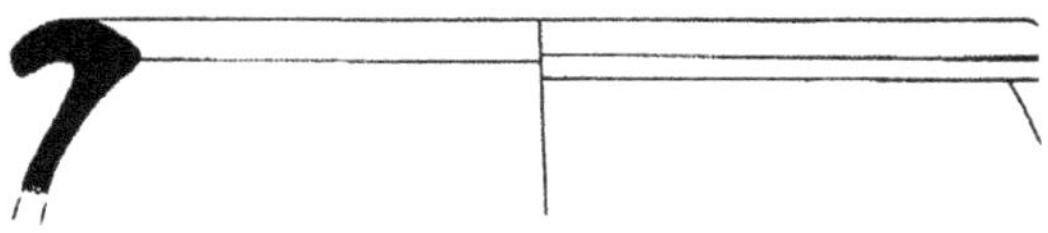 **T.222**	**T.222**:-Rim sherd of a pot with bilateral projecting rim with the external projection rounded and a groove on the upper side (I.D = 22cm). SJN-B, TT1, NW, 2, 0.82, 310, BSRW-2, RED(C)
 **T.225.1**	**T.225.1**:- Rim sherd very similar to T.225 (I.D = 22cm)SJN-B, TT1, SW, 2, 0.8, 108, BSRW-2, RED(C)
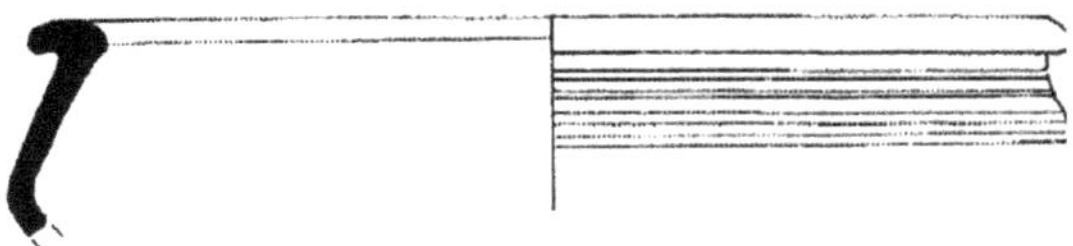 **T.230**	**T.230**:- Rim sherd of a carinated pot with external projecting rounded rim and corrugation on the shoulder (I.D = 21cm). SJN-B, TT1, NE, 2, 0.57, 503, BSRW-2, RED(C)
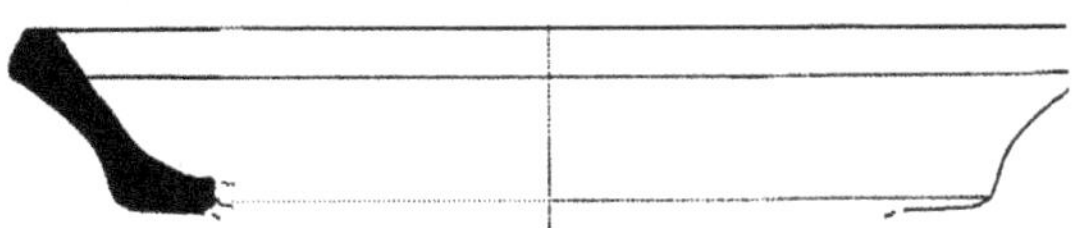 **T.239**	**T.239**:- Rim sherd of a plate with everted squared rim and flat base. There is a carination at the base. (I.D = 34cm) SJN-B, TT2, SE, 2, 0.85, 1510, BSRW-2, RED(C)
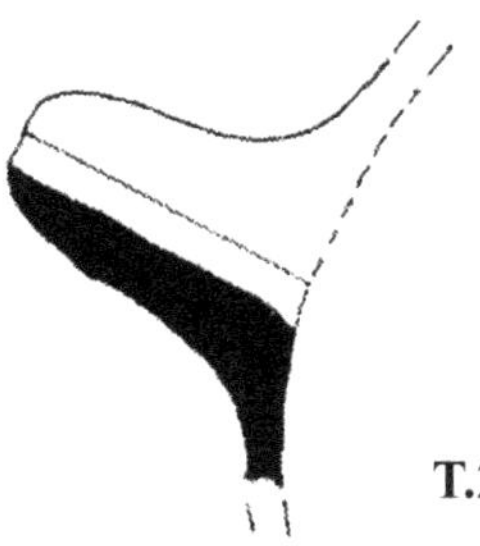 **T.255.2**	**T.255.2**:- Straight –sided spout SJN-B, TT2, SE, 3, 1.24, 1517, BSRW-1, RED(C)

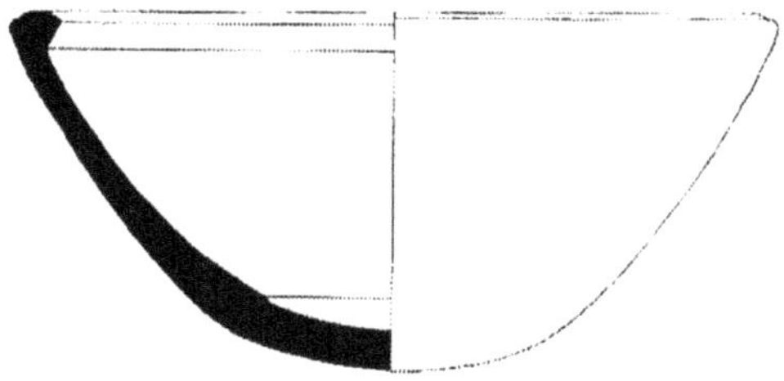 **T.421**	**T.421**:- Rim to base of a straight sided bowl or dish having a contiguous rounded base and an internal projecting inverted rim. The rim is short and triangular. Two grooves can be seen on the upper part of the rim and the mouth. The inside of the vessel at the base has a ring like offset. (I.D = 20.5cm) SJN-B, TT4, SW, 2, 2.91, 51137, BSRW-1, RED(C)
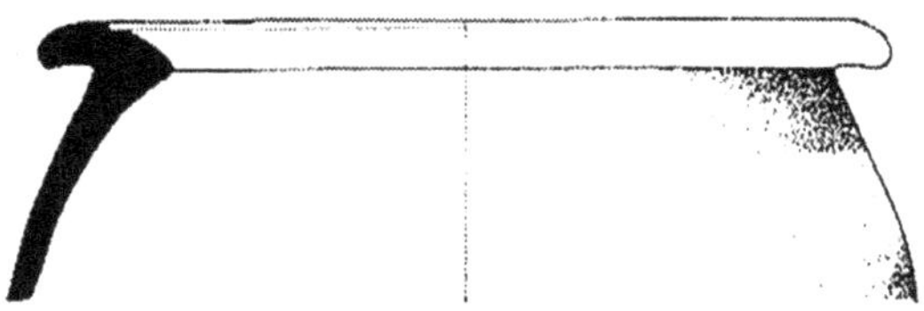 **T.439 D.202**	**T.439 D.202:-** Rim sherd of a pot having a constricted neck. The rim is external projecting extremely everted square, with a broad white band painted on the shoulder at the junction of the neck. A similar band is painted in a shallow groove along the lip on top of the rim. The slip is chocolate brown-black. (I.D = 14cm). SJN-B, TT4, SW, 2, 4.45m, 51144, BSRW-2, RED(C)
 **D.55.1**	**D.55.1**:- Bands of triangular punctate marks SJN-B, TT2, SW, 4, 1.45, BSRW-1, RED(C)

RED WARE

 Bs.39	**Bs.39:-** Contiguous rounded base SJN-B, TT1, SW, 1, 0.62, 104, RW-1, RED(C)
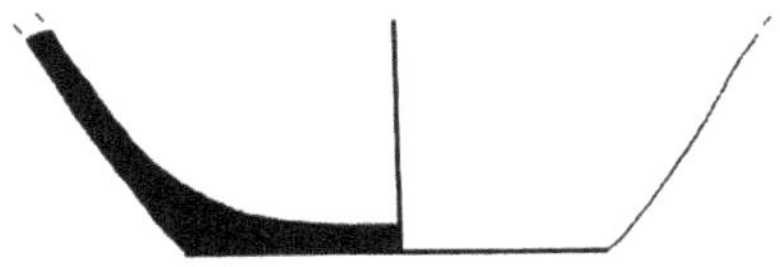 Bs.40	**Bs.40:-** Contiguous flat base SJN-B, TT2, SE, 4, 1.6, 1524, RW-1, RED(C)
 Bs.40.1	**Bs.40.1:-** Contiguous flat SJN-B, TT1, NW, 3, 1.07, 318, RW-1, RED(C) (Bs.41)
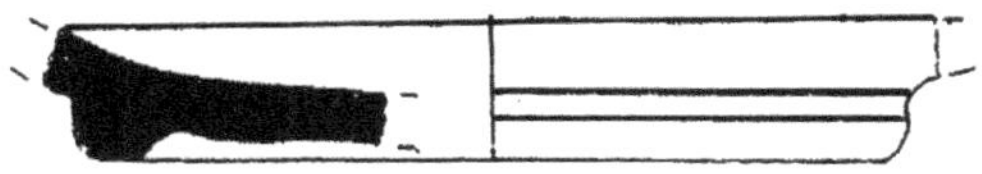 Bs.41	**Bs.41:-** Discoid base SJN-B, TT1, SW, 2, 0.8, 108, RW-1, RED(C)
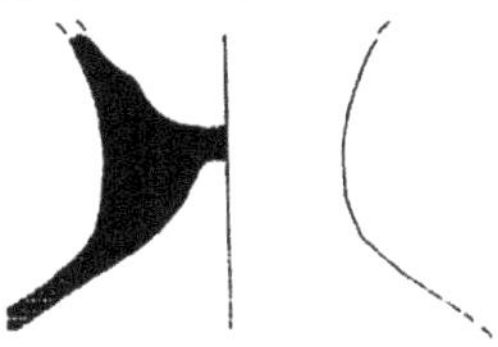 Bs.42	**Bs.42:-** Hour-glass shaped stand base. SJN-B, TT1, NW, 3, 1.18, 321, RW-1, RED(C)
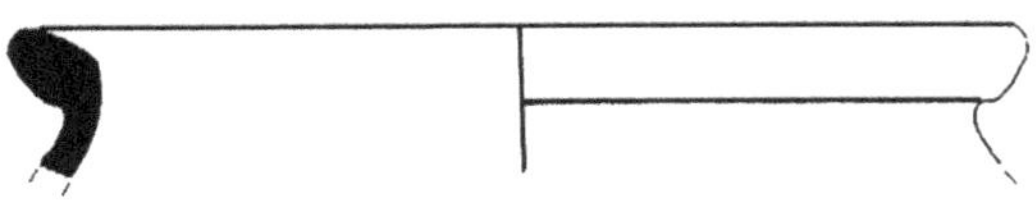 T.213	**T.213:-** Rim sherd of a pot with external projecting extremely everted rounded with sharply under cut rim (I.D = 26cm).SJN-B, TT1, NW, 2, 0.9, 313, RW-1, RED(C)
 T.225	**T.225:-** Rim sherd of a pot with bilateral projecting rim with square external projection and a groove on the upper side (I.D = 22cm)
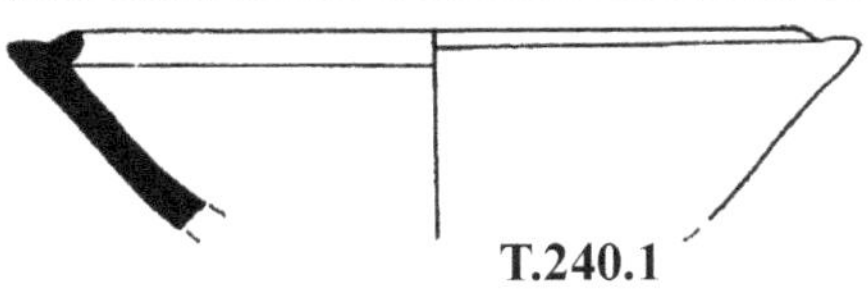 T.240.1	**T.240.1:-** Rim sherd of a carinated bowl similar to T.240, except that it has an inverted rounded rim (I.D = 14cm). SJN-B, TT1, NW, 3, 1.79, 330, RW-1, RED(C)
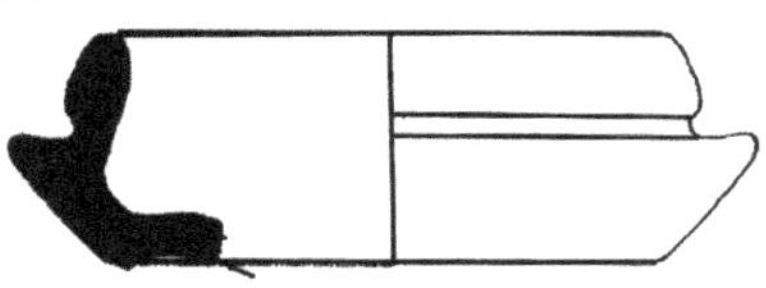 T.242	**T.242**:- Rim sherd of a small bowl with an inverted tapering rim just under which the body thickness abruptly to form a ledge (I.D = 7cm) .SJN-B, TT1, SW, 2, 0.86, 111, RW-1, RED(C)

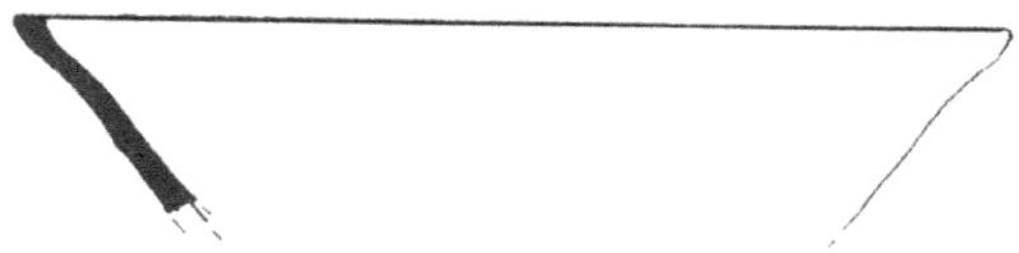 **T.243**	**T.243:-** Rim sherd of a stright sided bowl with an everted tapering rim (I.D =14cm). SJN-B, TT2, SE, 5, 2.56, 1543, RSRW-1, RED(C)
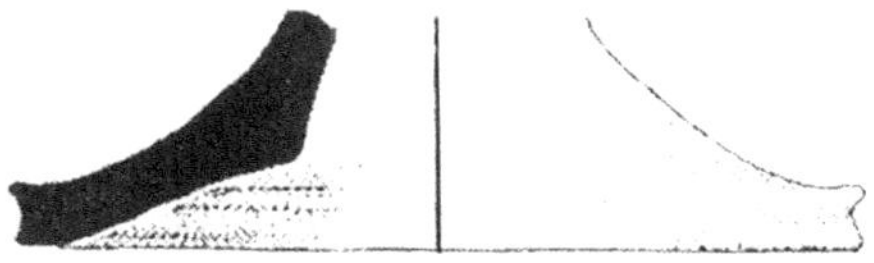 **Bs.48**	**Bs.48:-**Part of pedestal stand with everted squared rim with a wide groove along the outer edge. A sharp ridge is evident on the inner side before it starts going up vertically. Two shallow grooves are seen on the underside
 **Bs.49**	**Bs.49:-**Part of pedestal bifed stand.
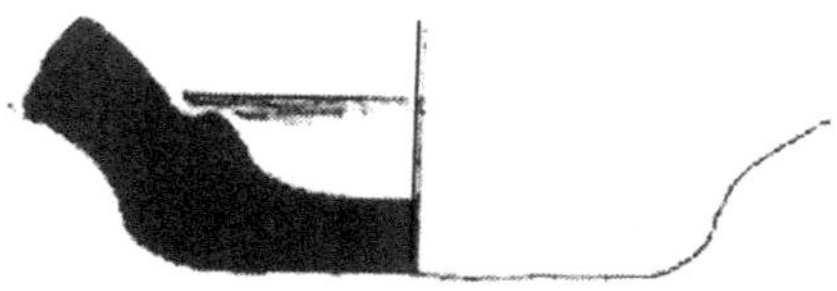 **Bs.50**	**Bs.50:-**Contiguous flat base with a sharp ridge on the inner side and a deep depression
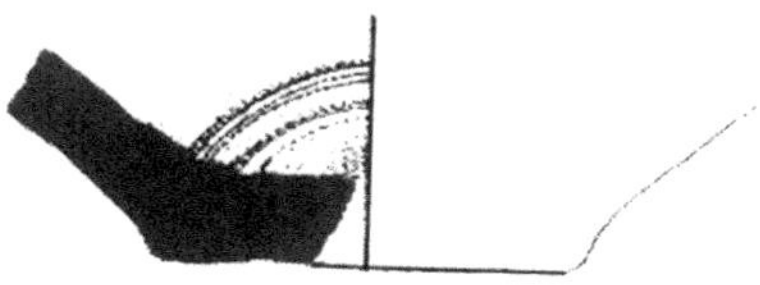 **Bs.51**	**Bs.51:-** Contiguous flat base with deep sharp grooves forming a ring on the inside and a deep depression in the centre.
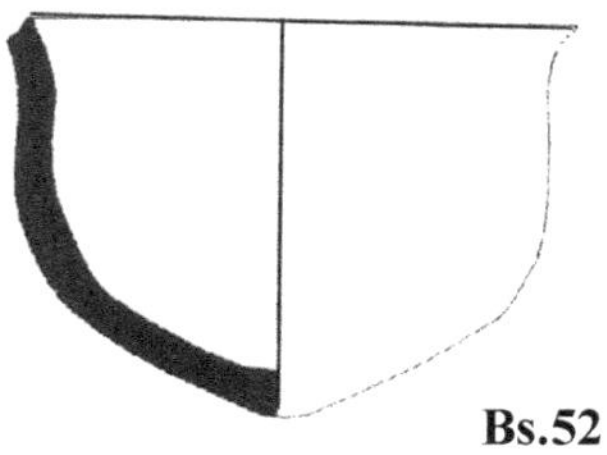 **Bs.52**	**Bs.52:-**Contiguous, rounded, ending in a point.
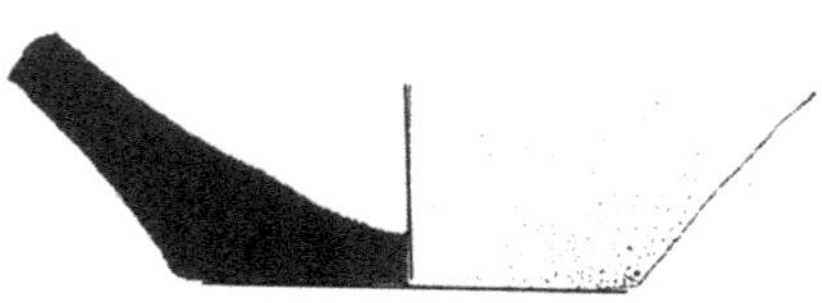 **Bs.79**	Bs.79:- Contiguous flat base of a straight sided bowl having a coarse red fabric. All evidence of glaze has worn off with the exception of small white patch which is still adhering at the centre of the body. SJN-B, TT4, SW, 2, 3.94, 51142, GRW, RED(B)
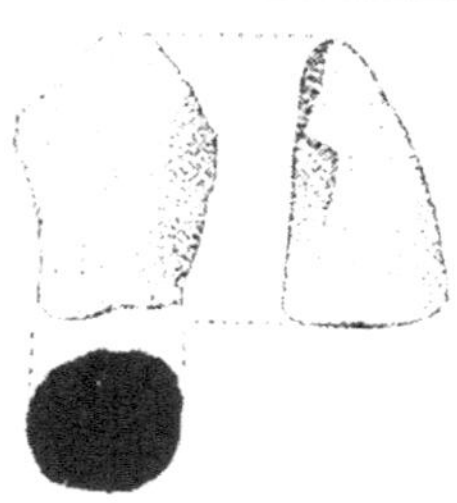 **T.325**	**T.325:-** Handle round in section. Flat at the end where it was luted to the body

 **T.325.1**	**T.325.1:-** Handle similar to above. The end where is luted to the body has a groove around it.
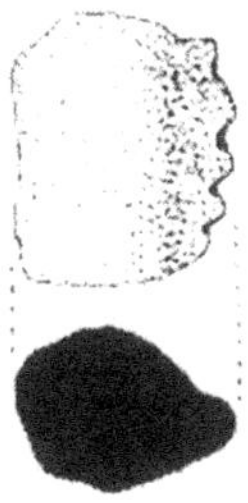 **T.373.1**	**T.326**:- Handle round in section with vertical ridge running along the external side. The ridge has a series of horizontal cut marks along it.
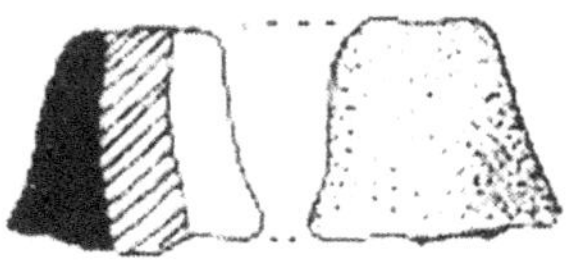 **T.327**	**T.327:-**Short cylindrical spout.
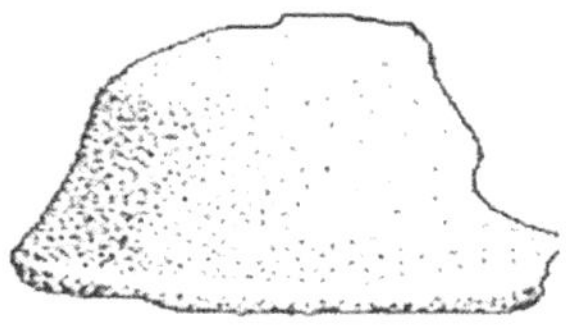 **T.328**	**T.328:-** Solid conical finial.
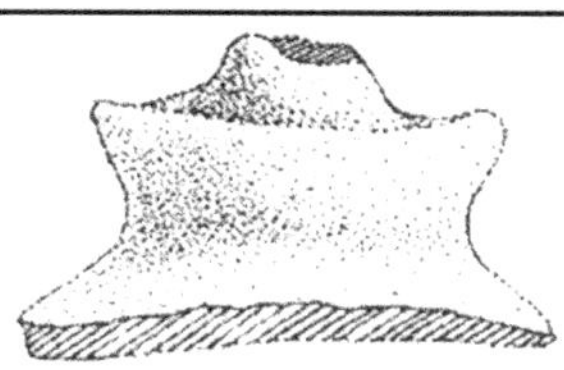 **T.329**	**T.329:-** Finial having narrow waist, flaring upwards to form a flat platform topped by a rounded knob.
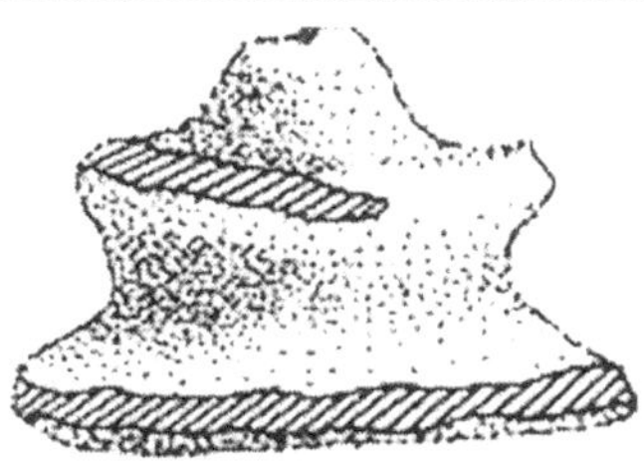 **T.393.2**	**T.329.1**:- Finial, similar to T.329
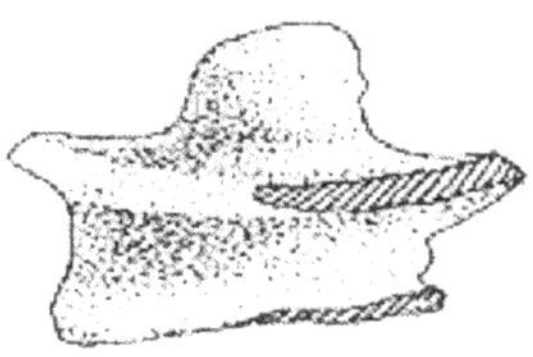 **T.329.2**	**T.329.2:-** Finial, similar to T.329

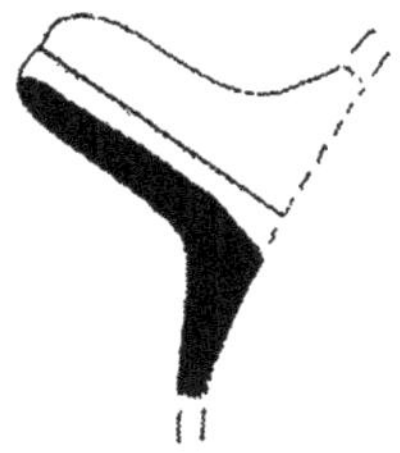 **T.255**	**T.255**:- Straight –sided spout SJN-B, TT2, SE, 2, 0.85, 1510, RW-1, RED(C)
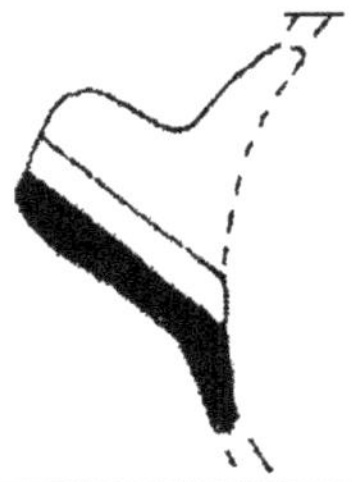 **T.258.3**	**T.258.3:-** Short spoutSJN-B, RW-1, RED(C)
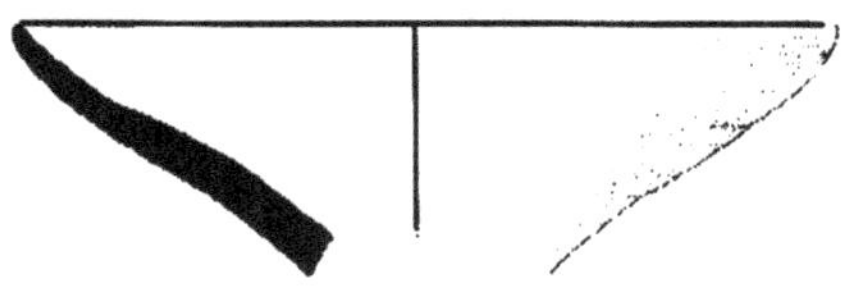 **T.304**	**T.304:-**Rim sherd of a straight side bowl with everted tapering rim (I.D = 22cm).
 **T.305**	**T.305**:- Rimsherd of a straight sided bowl with internal projecting squared rim thickening slightly at the tip. (I.D = 20cm).
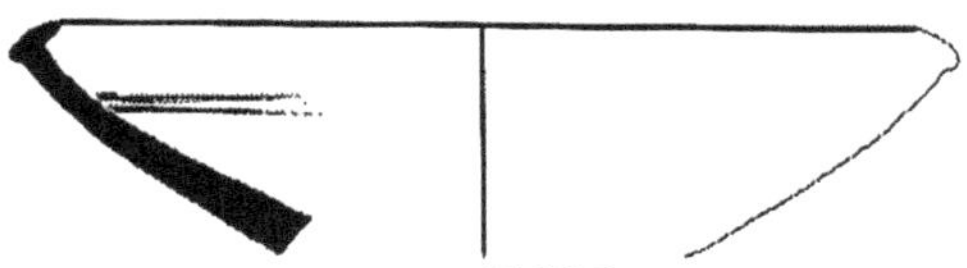 **T.306**	**T.306:-** Rim sherd of a straight- sided bowl with internal projecting tapering rim with two sets of horizontal parallel grooves on the body. (I.D = 17cm).
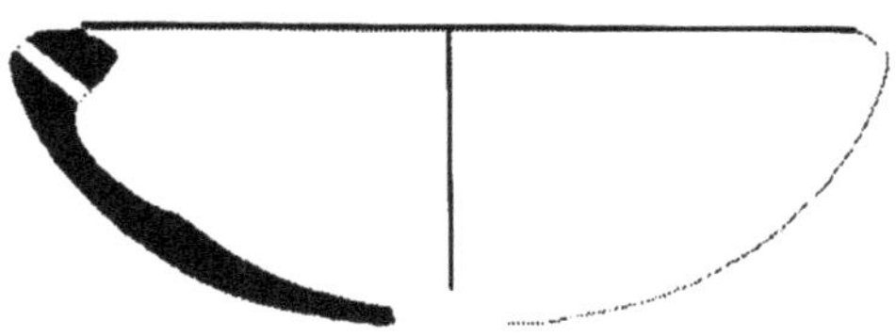 **T.307**	**T.307:-** Rimsherd of a spouted bowl (6 piaces refit) with inverted rounded rim with a carination just under the rim. A short spout is seen on the external side of the rim. The spout is luted on (I.D = 16cm).
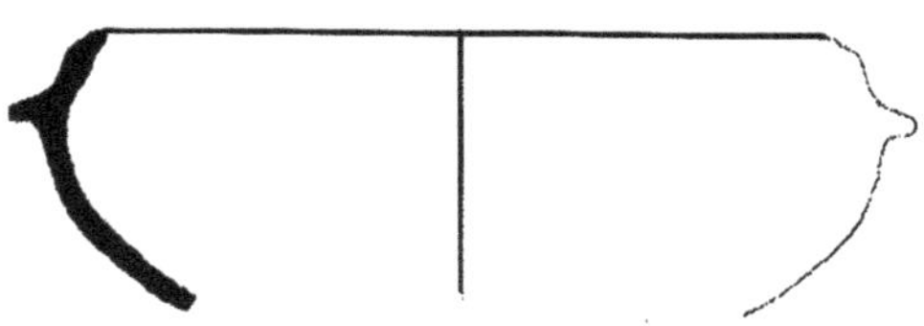 **T.308**	**T.308:-**Rim sherd of a (convex sided bowl) cooling pot with inverted squared rim and with a ledge on the external side (I.D = 16cm).
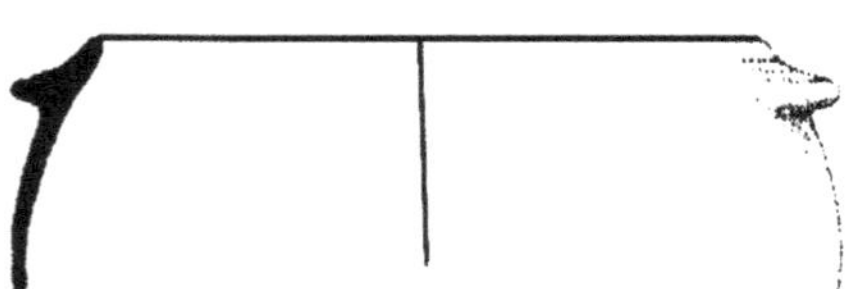 **T.309**	**T.309**:- Rim sherd of a pot with inverted rounded rim with a ledge having three grooves on the exterior. (I.D =19cm).

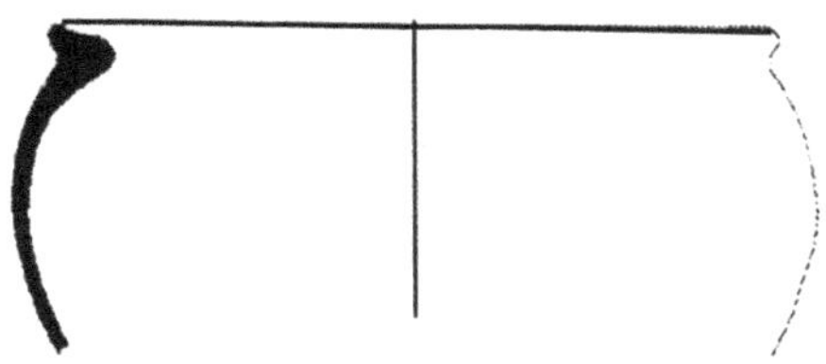 T.310	**T.310**:- Rim sherd of a pot with a constricted neck. The rim is external projecting extremely everted tapering rim, sloping downwards with a groove on the top. Some shallow horizontal grooves are evident on the body (I.D = 12.5cm).
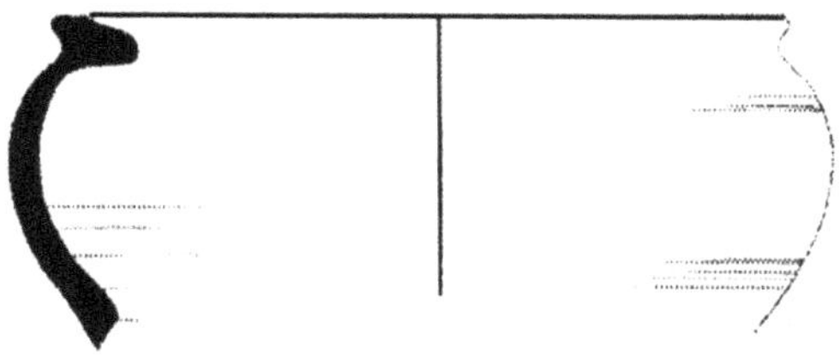 T.311	**T.311**:- Rim sherd of a pot with bilateral projecting squared rim on the external side and rounded on the inner side. (I.D =11cm). Two horizontal grooves can be seen on body.
 T.314	**T.314:-**Rimsherd of a small carinated pot with external projecting taper in rim. The carination is blunt (I.D = 4cm).
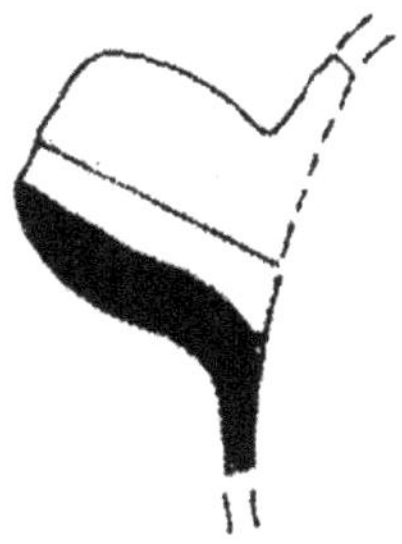 T.255	**T.255**:- Straight –sided spout SJN-B, TT2, SE, 2, 0.85, 1510, RW-1, RED(C)
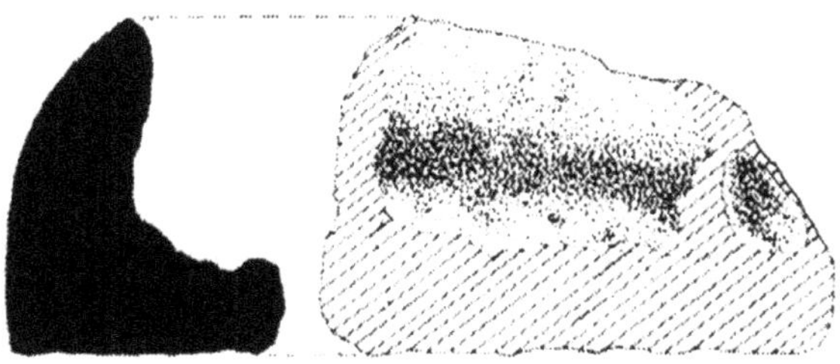 T.388	**T.424:-** Rim to base of an extremely coarse and thick hand made dough plate having a flat base and a thick short inverted rim which is tapering and which is interrupted by a deep thumb impression. SJN-B, TT4, NE, 5, 1.86, 51547, RW-2, RED(F)
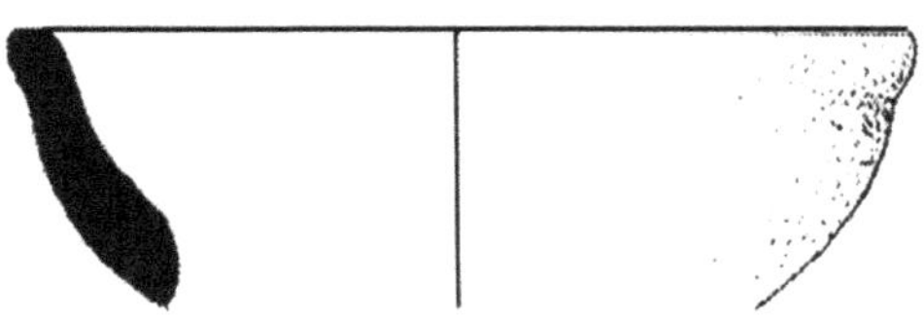 T.425	**T.425:-** Simple square rim of a small shallow lamp or bowl. The edges are slightly rounded. (I.D = 9cm) SJN-D, OX1, NE, 2, 1.55, 31714, RW-3, RED(C)

MISCELLANEOUS

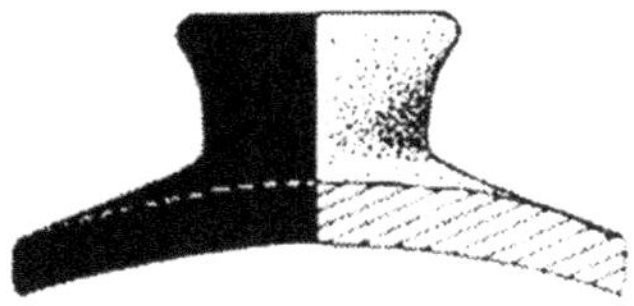 **T.430**	**T.430**:- This piece is the finial of the lid having sloping sides and a flat topped stud handle which is narrower at its base. It is quite sturdy .SJN-D, O1,SE, 2, 1.29, 29306, MISC-1, GREY(A)
 **Bs.83**	**Bs.83**:-. This is a contiguous flat base of a straight sided thickset vessel. SJN-B, TT4, NE, 5, 2.16, 51553, MISC-2, GREY(B)
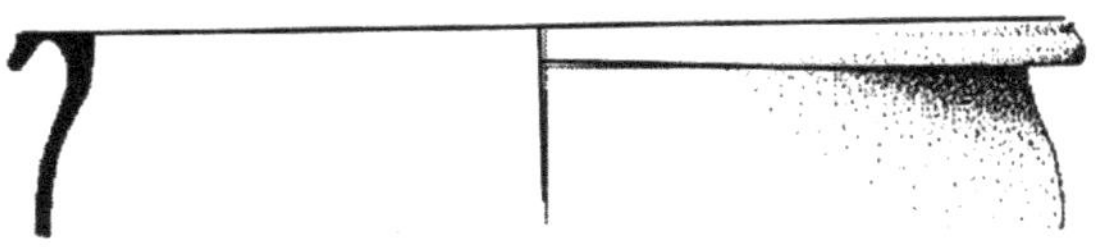 **T.431**	**T.431**:- This is a rim sherd of a cooking vessel having a short vertical neck from which the rim projects externally at a sharp angle. It again curves down and overhangs the neck.The lip is tapering and sharp. (I.D = 20cm) SJN-B, TT4, SW, 2, 0.93, 51123, MISC-3, RED(F)
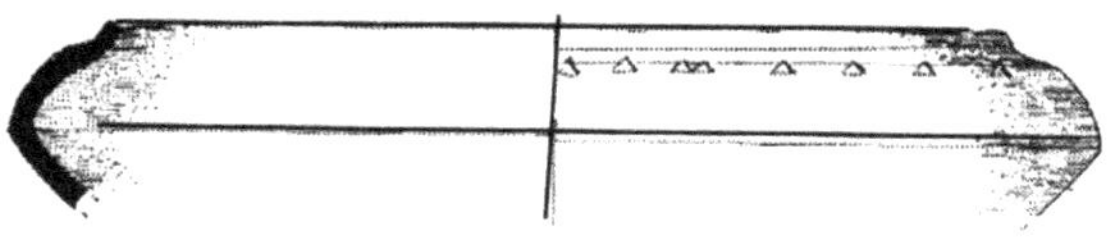 **D.45**	**D.45**:- Punctate marks with raised clay on one sideSJN-B, TT2, SE, 4, 2.31, 1538, MISC-3, RED (F)
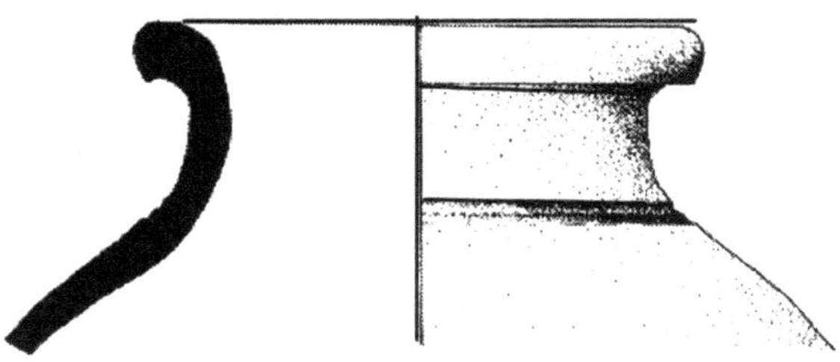 **T.432**	**T.432**:- This is an extremely fine well levigated micacious fabric with well levigated clay. Rim to shoulder of a globular pot having extremely everted rounded beaked rim and two broad grooves on the shoulder near the neck. I.D appears to be 12 cm. SJN-B, TT4, NW, 3, 0.95-1.06m, 51328/30/32, MISC-4, RED(G)
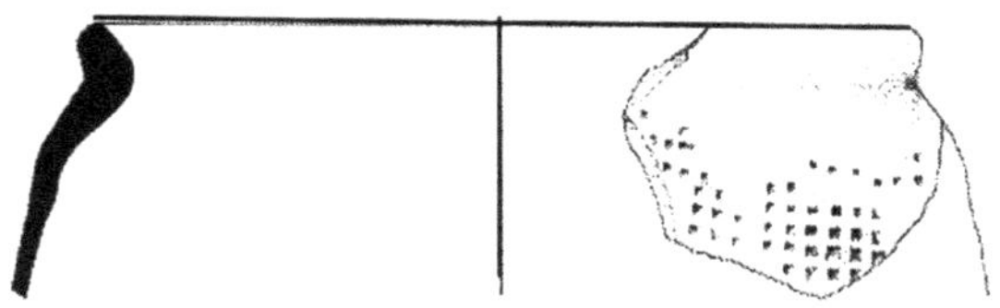 **T.433 D.200**	**T.433 D.200:-** This fabric is almost the same as the previous sample except that it is fired to pink colour and appears to have a creamy pink wash on the exterior. The fabric has large amounts of mica. The vessel appears to be moulded. This is a rim sherd or a convex sided bowl with a slight carination having an everted triangular rim with rounded edges. On the body a moulded mesh design of small squares can be seen. The surfaces are covered with fine mica. (I.D = 15cm) SJN-B, TT4, NE, 6, 3.56, 51582, MISC-5, RED(G)

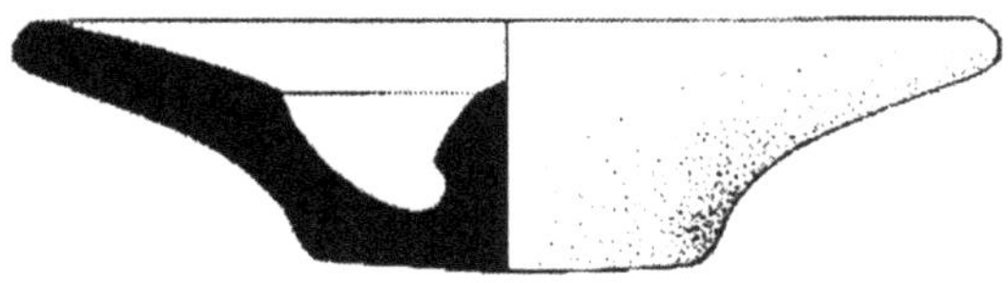 **T.434**	**T.434**:- This sample is an almost complete lid which has a contiguous flat base and flaring sides which projects outwardly to a rounded rim. A shallow groove runs along the lip on the upper side of the rim. A deep circular well is formed in the centre of the lid in which is seen a raised, dome shaped stud. The external diameter is 17 cm and the diameter of the well and the base if 7.5 cm SJN-B, TT4, NE, 6, 2.83, 51568, MISC-6, RED(H)
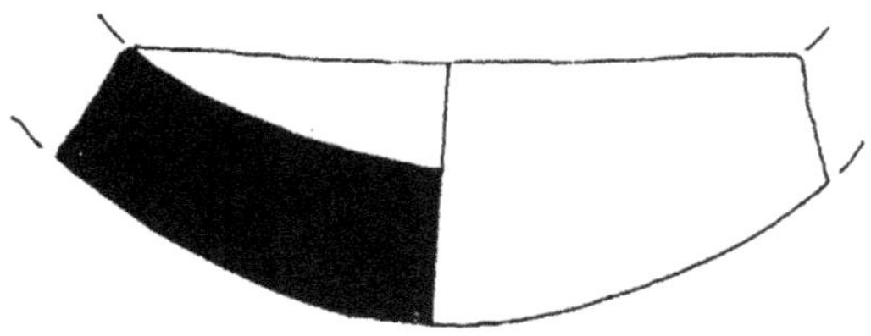 **Bs.39.2**	**Bs.39.2**:- Contiguous rounded base SJN-B, TT1, SW, 2, 0.7, 105, MISC-7, RED(I)
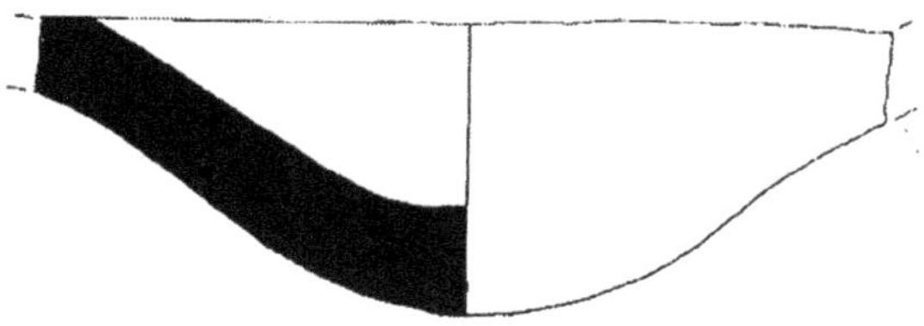 **Bs.39.1**	**Bs.39.1**:- Contiguous rounded base SJN-B, TT2, SE, 3, 1.36, 1520, MISC-7, RED(I)
 **T.251**	**T.251**:- Simple semi-circular lug handle. SJN-B, TT2, SW, 2, 0.94, 908, MISC-7, RED(I)
 **BS.84**	**Bs.84**:- Part of an object (?) which resembles a ring stand. The underside is flattish. The inner rim is thick and rounded. It tapers down to the external edge or rim which is somewhat square. Two grooves run along the upperside of the interior rim. Two perforations obliquely placed go through the ring stand. The purpose of this is unknown. Two such fragments have been found. The external diameter of the ring is 20cm and the internal ring is 10cmSJN-B, TT4, SE, 2, 1.11, 51722, MISC-7, RED(I)

CHINESE AND FAR EASTERN WARES

CHANGSHA UNDERGLAZE PAINTED WARE

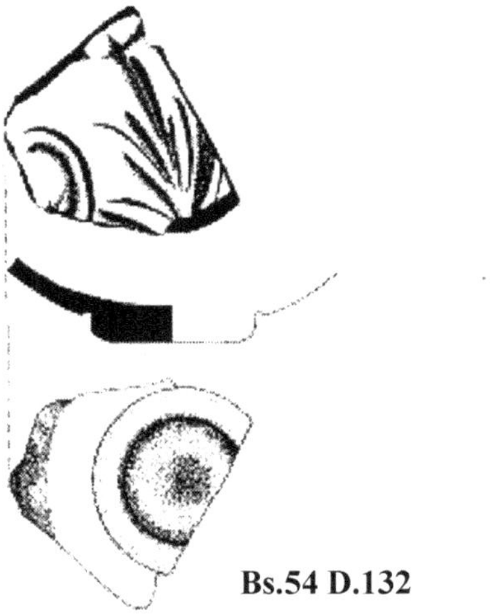 Bs.54 D.132	**Bs.54 D.132:-** Non contiguous ring base of a bowl having a square section and a carination on the side. A broad angular shallow groove is seen inside the ring. The exterior of the vessel is partially glazed. The interior has a light brown glaze with a painted design consisting of spikey fern-like leaves painted in green and brown. The painting is executed with feathery strokes. (D = 6cm) SJN-B, TT4, NW, 2, 0.66, 51317, CHANG, CH-1
CELADON	
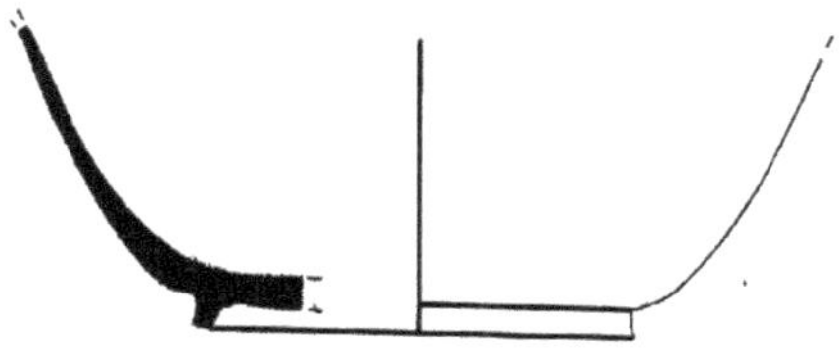 T.3	**T.3**:- Rim to base of a convex sided bowl, light green in colour (I.D = 17cm). Ring base. Simple everted tapering rim with spaced scallops. Three spur marks can be seen on the inner surface of the base. SJN-B, TT1, NW, 1, 0.68, 305, CEL-1, C(A)
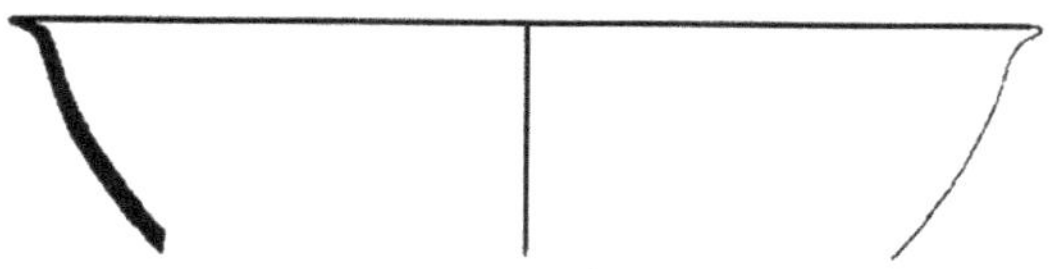 T.438	**T.438**:- Rim sherd of a bowl having an everted tapering rim. The glaze is greyish green (I.D = 17cm). SJN-B, TT4, SW, 2, 3.94, 51142, CEL-1, C(A)
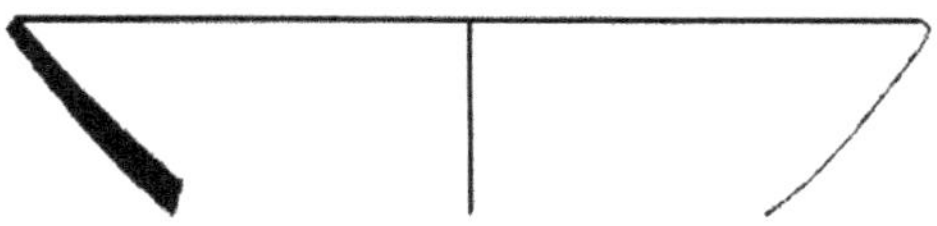 T.339	**T.339**:- Rim of a bowl having a simple tapering rim and an olive-green glaze. (I.D = 13 cm) SJN-B, TT4, SW, 2, 0.93, 51123, CEL-1, C(A)
 Bs.5	**Bs.5**:- Ring base of convex sided bowl with spur marks on the interior and on the ring foot. Olive green glaze. (D = 10cm) SJN-B, SURFACE, NIL, CEL-1, C(A)
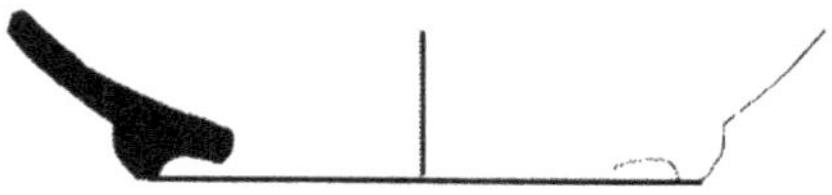 Bs.57	**Bs.57** Ring base of a dish having vertical sides and square sections. The interior of the vessel having a circular shallow groove from which another vertical shallow groove rises towards the rim (like a rib). The exterior has a shallow gadroon effect. (D = 9cm) SJN-B, TT4, SE, 2, 0.37, 51702, CEL-1, C(A)

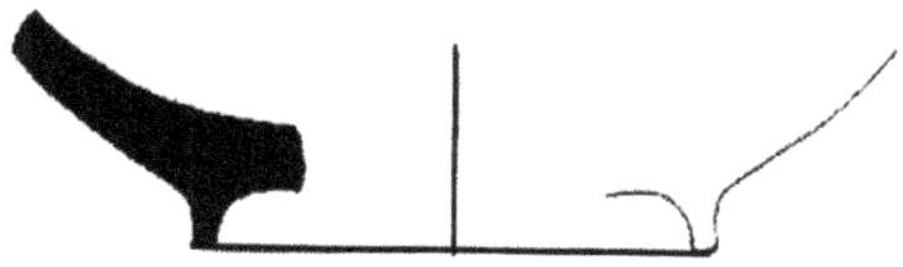 **Bs.58**	**Bs.58**:- Non contiguous ring base slightly tapering. Pale green glaze on all surfaces. SJN-B, SURFACE, CEL-1, C(A)
 **Bs.59**	**Bs.59**:- Base of a bowl, shallow-footed non contiguous splayed ring with channel rim. SJN-B, TT4, NW, 1, 0.17, 51302, CEL-1, C(A)
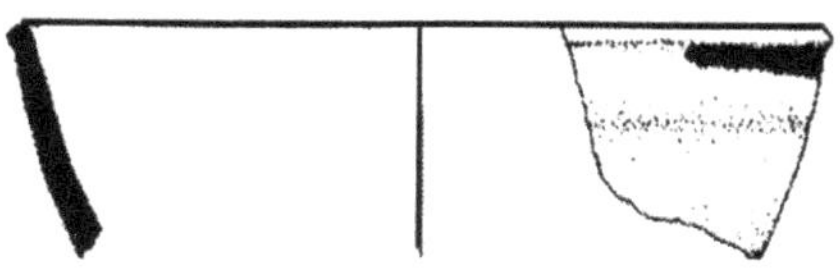 **T.340**	**T.340 D.137:-** Rim sherd of a bowl having an everted tapering rim. The internal glaze is a creamy-brown with broad ochre-brown band extending along the rim. The external glaze is a light brownish-green with a similar ochre band extending around the rim. The glaze is not very evenly applied . (I.D = 11cm) SJN-B, TT4, NE, 2, 1.09, 51531, STONE-1, S(A)
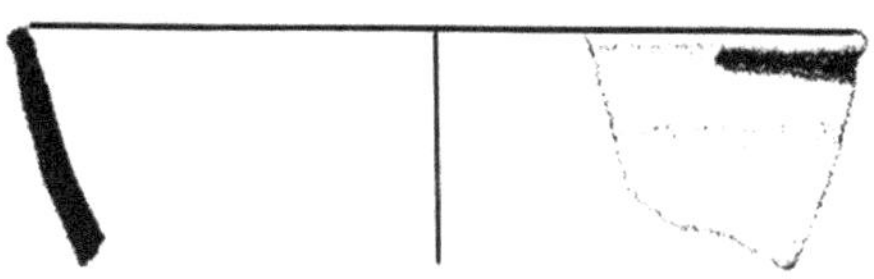 **T.340.1**	**T.340.1**:- Rim sherd of a bowl similar to T.340. The internal and external glaze is a brownish-olive-green. A dark brown discolouration (perhaps due to firing) can be seen on the exterior just under the lip and on the interior, lower part of the body. (I.D = 8cm) SJN-B, TT4, NE, 2, 1.05, 51529, STONE-1, S(A)
 **T.340 D.137**	**T.340 D.137**:- Rim sherd of a bowl having an everted tapering rim. The internal glaze is a creamy-brown with broad ochre-brown band extending along the rim. The external glaze is a light brownish-green with a similar ochre band extending around the rim. The glaze is not very evenly applied . (I.D = 11cm) SJN-B, TT4, NE, 2, 1.09, 51531, STONE-1, S(A)
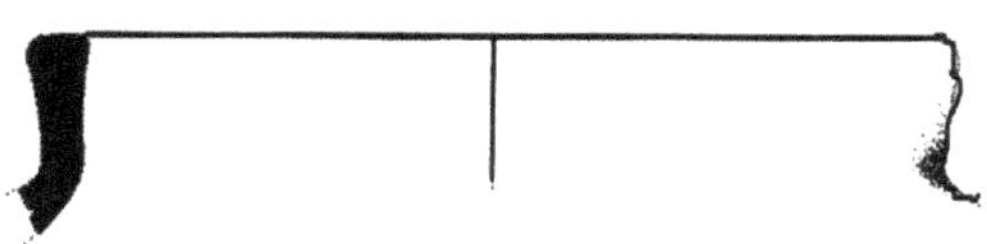 **T.342**	**T.342**:- Rim sherd of a large thick, storage jar or vase having a straight square rim. The curve of the shoulder is evident. The lip is glazed black and one spur mark can be seen on top of the rim. The glaze of the vessel itself is green with black inclusions. (I.D = 23cm) SJN-B, SURFACE, STONE-2, S(B)

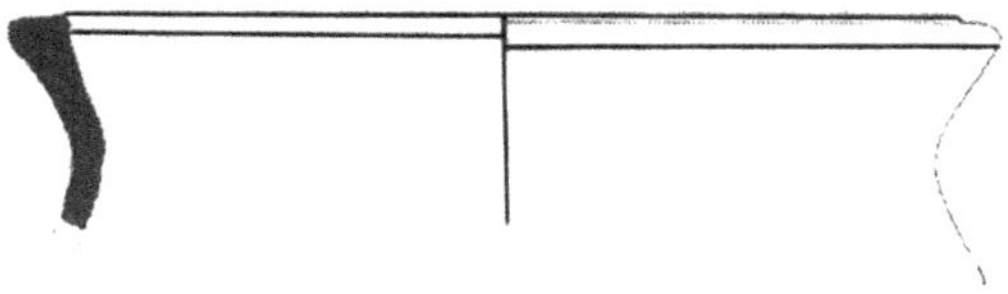 **T.8**	**T.8**:- Rim sherd of a jar with external projecting square rim, which broadens slightly at the top with rounded edges. It has a groove on the inner side. The jar appears to have a long neck which is inwardly sloping and which flares towards the shoulders. Creamish green in colour. (I.D = 10cm). SJN-B, TT1, NW, 3, 2.1, 335, STONE-3, S (A)
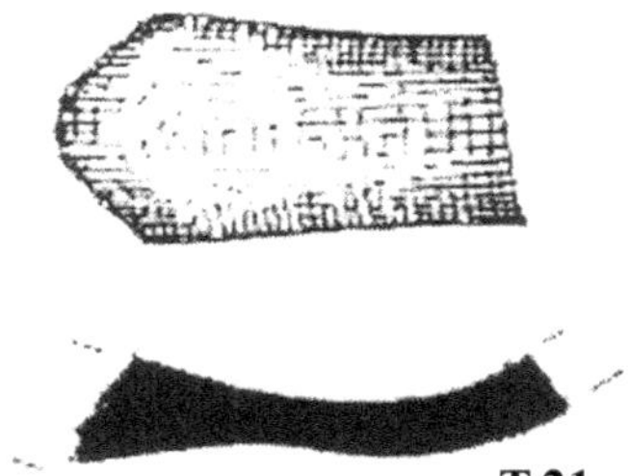 **T.21**	**T.21**:-Faux handle; flat. Green glaze highly eroded. (L = 2.8cm) SJN-B TT2, SE, 1, 0.49, 1502, STONE-3, S(C)
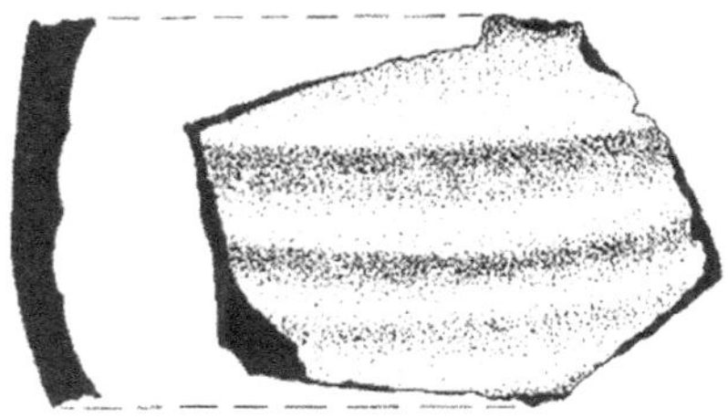 **T.343**	**T.343**:- Body sherd of a thick storage jar or a vase having a thin glaze unevenly applied. On the external surface the glaze stops above the base. An oval cloud of light green with a mark at the centre could be a luted handle. SJN-B, TT4, NE, 4, 1.66, 51542, STONE-3, S(B)
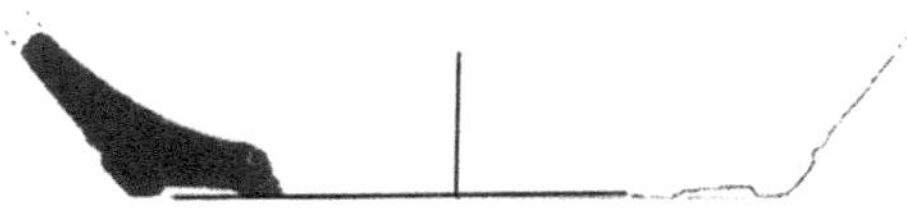 **Bs.60**	**Bs.60**:- Base of a bowl having a low ring (non-contiguous) and olive-green glaze covers the entire vessel. The ring foot (low) is square in section with a groove run above it at the junction with the body. Three white spur marks can be seen on the interior of the vessel. (D = 9cm)SJN-B SURFACE, STONE-3, S(C)
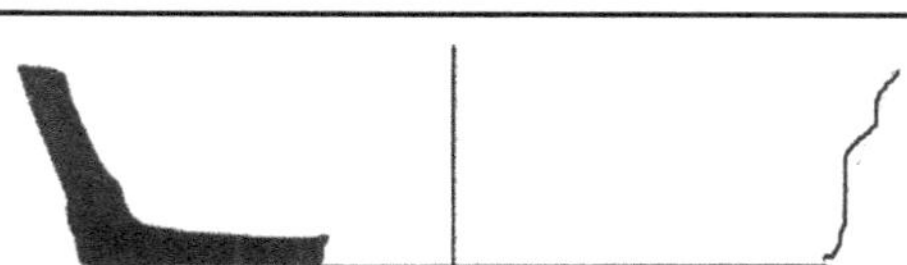 **Bs.45**	**Bs.45**:- Contiguous flat base of a storage jar/basin. No Glaze.STONE-3
PORCELAIN	
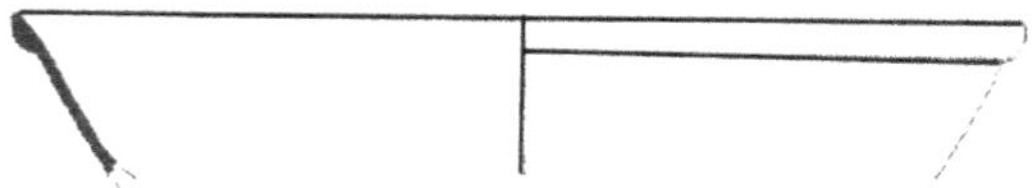 **T.6**	**T.6**:- Rim of a dish, simple everted, rounded rim with thickening on the exterior as a result of which there is a groove formed on the exterior. Glossy creamish green glaze. Striation marks of the fast wheel can be seen. (I.D = 20cm) SJN-B, TT1, SE, 1, 0.32, 703, PORC-1, P(A)
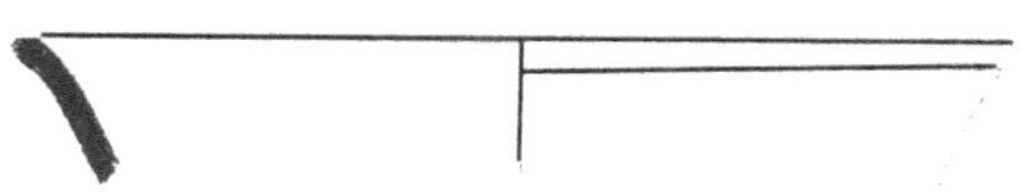 **T.41**	**T.41**:- Rim sherd of a dish with external projecting extremely everted rounded rim. Creamish fabric, creamy white glaze on exterior and interior surface. (I.D = 26cm). SJN-B, TT2, SE, 3, 1.12, 1515, PORC-2, P (B)

 **T.265**	**T.265**:- Rim sherd of a dish having simple everted rounded rim. White in colour (I.D = 16cm). SJN-B, SURFACE, PORC-1, P(A)
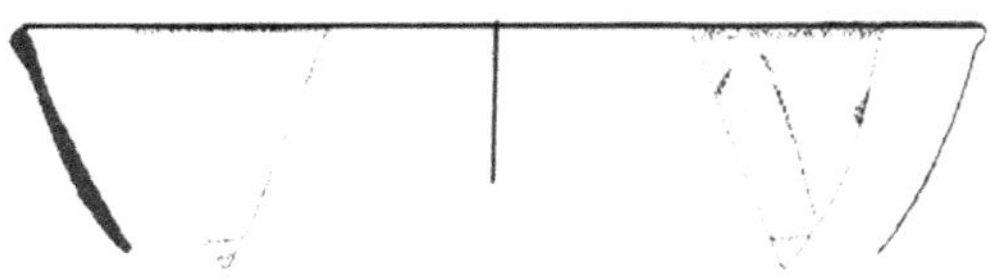 **D.83 T.266**	**T.266.D.83**:-Rim sherd of a convex sided bowl having external projecting rounded rim. Thin in section. Dark green along the rim and with a green oblique line (part of a design) on a greyish white background on the external surface (I.D = 9cm). SJN-K, B3, NE, 1, 0.16, 7701, PORC-4, P(A)
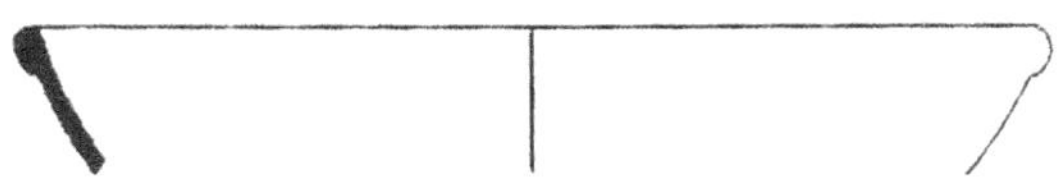 **T.333**	**T.333**:-Rim sherd of a bowl with a folded over/rolled rim similar to the glass bowls. A greyish white glaze on both surfaces. (I.D = 23cm) SJN-B, TT4, SW, 2, 0.93-1.46, 51123/30, PORC-1, P(A)
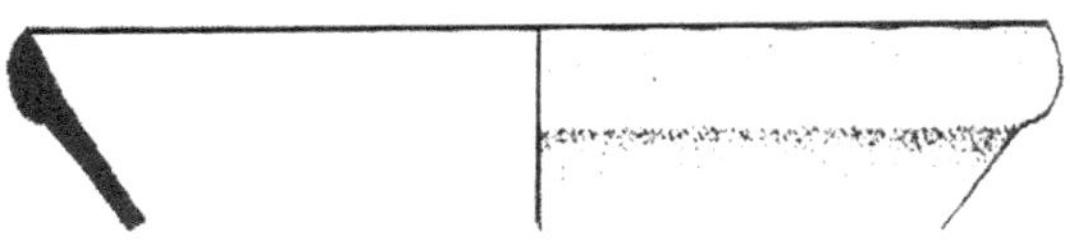 **T.333.2**	**T.333.2**:- Rim sherd of a bowl similar to the previous types. This rim is inverted and the fold or rolled of the rim is sharp. The hollow is only visible in one section. The outer surface of the rim has a damage glaze (I.D = 26cm) SJN-B, TT4, NE, 2, 0.95, 51526, PORC-1, P(A)
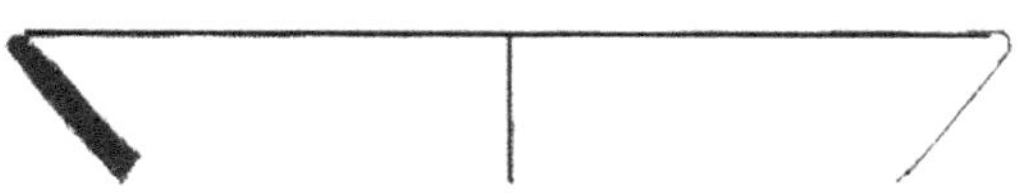 **T.334**	**T.334**:- Rim sherds of a dish / bowl having simple tapering rim and white glaze I.D 13 cm) SJN-B, TT4, NW, 2, 0.36m, 51306, PORC-1, P(A)
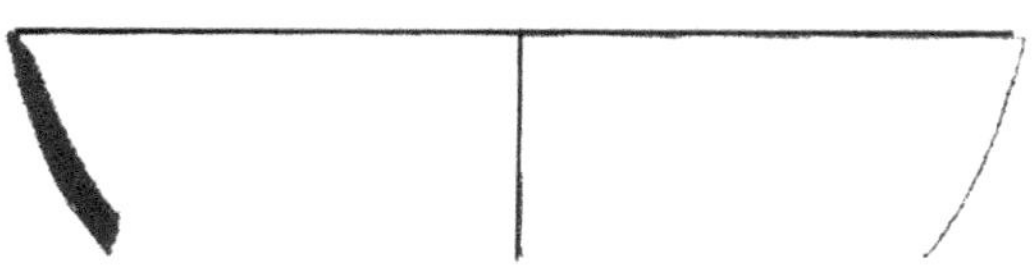 **T.335**	**T.335**:-Rim sherd of a bowl having an everted tapering rim and a white glaze. (I.D = 10cm) SJN-B, TT4, NE, 2, 0.49, 51511, PORC-1, P(A)
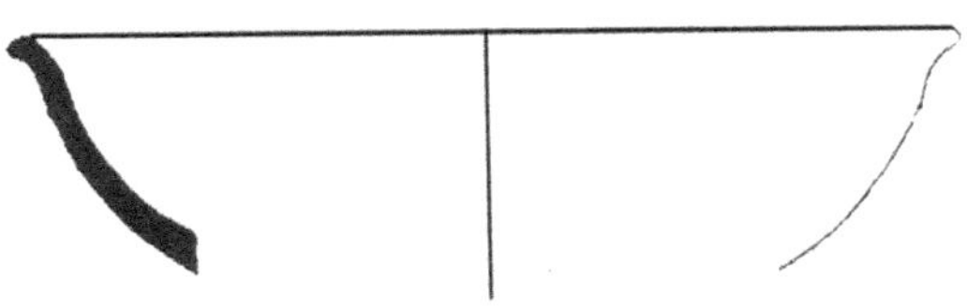 **T.332**	**T.332**:-External projecting everted rounded rim of a bowl or a dish having vertical ridge on the inside like a rib. The glaze is creamy-white. The rim fragment is very small. Hence it is very difficult to measure the I.D (14 cm approximately). SJN-B, TT4, NE, 3, 1.06, 51532, PORC-2, P(B)

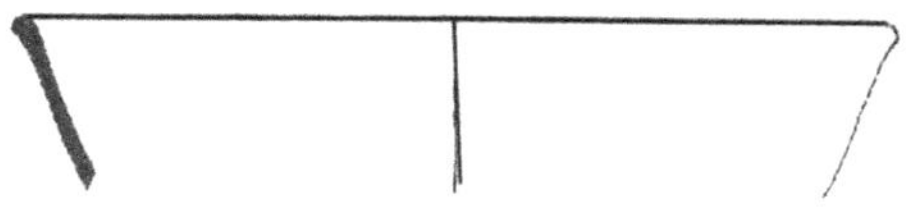 T.337	**T.337**:- rim sherd of bowl having a rounded rim. This singular piece has a white fabric and a shining dark emerald green glaze. This sample is from undisturbed context in the trench. (I.D = 14cm) SJN-B, TT4, NE, 3, 1.43, 51538, PORC-3, P(A)
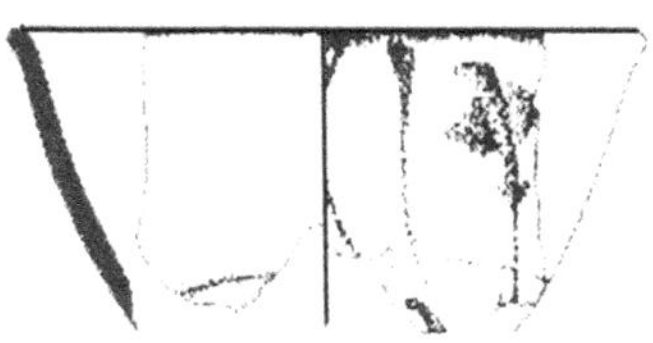 T.338	**T.338 D.133**:- Rim sherd of a convex sided bowl having a painted floral design in blue and a green band running along the lip. The decoration is on the exterior. The rim is everted and rounded. (I.D = 13cm) SJN-B, TT4, NW, 1, 0.17, 51302, PORC-4, P(A)
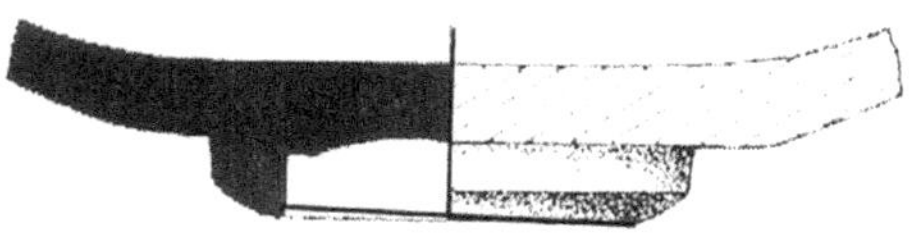 Bs.55	**Bs.55**:- Base of a shallow disk or bowl having a milky white fabric is a white glaze. The exterior is partially glazed. The base has a non contiguous ring foot and on the interior of the vessel can be seen a circular broad unglazed band, encrusted with black impurities. This may be due to stacking. This base was found inside the Bhandar/central well of the Dokhma. SJN-D, OX1, NE, 2, 1.75, 31721, PORC-1, P(A)
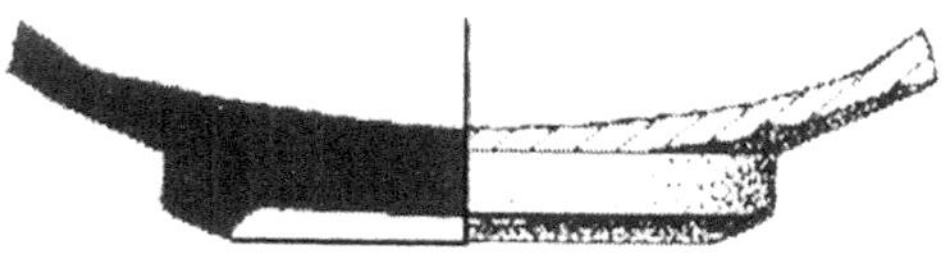 Bs.56	**Bs.56**:-Non contiguous ring base of a shallow dish or plate with a shallow groove running along the inside. The foot is square in section. The base is unglazed. SJN-B, TT4, NW, 2, 0.36m, 51306, PORC-1, P(A)
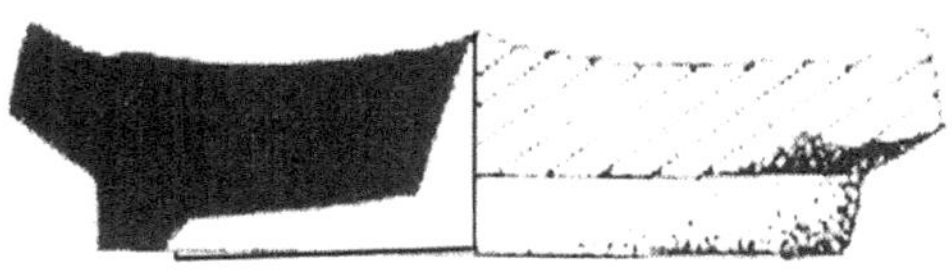 Bs.57	**Bs.57**:- Non contiguous ring base of a dish having vertical sides and square sections. The interior of the vessel having a circular shallow groove from which another vertical shallow groove rises towards the rim (like a rib). The exterior has a shallow gadroon effect. (D = 9cm) SJN-B, TT4, SE, 2, 0.37, 51702, CEL-1, C(A)
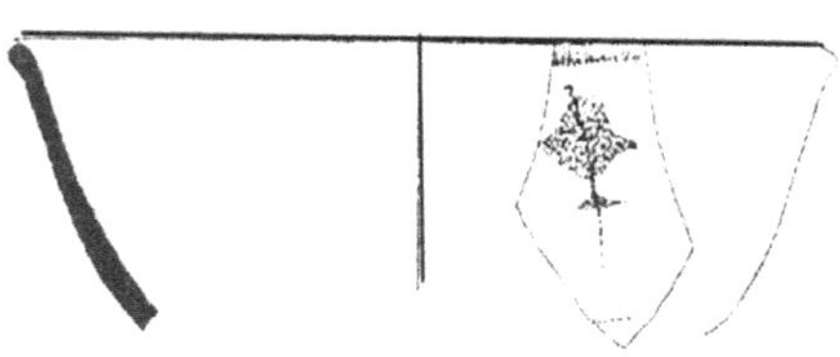 D.83.1	**D.83.1**:- Rim sherd a bowl with green band along the rim and a dark green design just below the rim on the external surface. The design consists of a vertical line over which a bivalve (oyster) shape has been painted and a horizontal slash cuts the vertical line just below.

 D.86	**D.86**:- White background with dark green loop and some blueish traces. SJN-K, A5, SE, 1, 0.3, 5501, FRIT, FRIT
 D.87	**D.87**:- Very small fragment of a deep purple and a bright green design. Probably Leaf/floral motif. SJN-K, A4, NE, 1, 0.32, 4501, PORC-E, P(B)
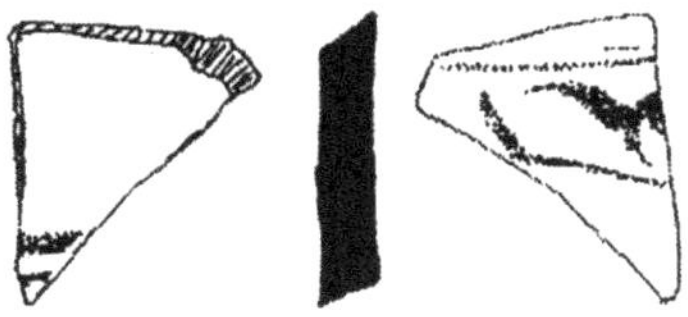 D.89	**D.89**:- Pale blue background with a blue line and two leaves. Probably part of a plant/floral design. SJN-K, A4,SE, 1, 0.27, 4701, PORC-4, P(A)
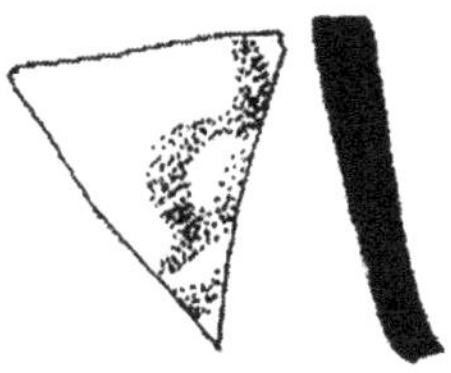 D.91	**D.91**:-Very small fragment. Pale blue background with a blue loop.
 D.93	**D.93:-** Pale blue background. Blue splotch of colour with darker blue radiating lines probably part of a floral/ other design since they are radiating from a small circle. SJN-K, B3, NE, 2, 0.44, 7708, PORC-4, P(A)
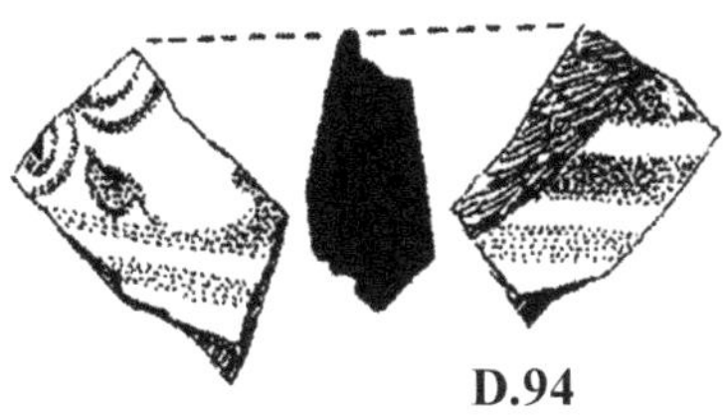 D.94	**D.94**:- Pale blue background. Crackled effect. Same blue lines and a dark blue thick curved band are evident. SJN-K, A4,SE, 1, 0.27, 4701, PORC-4
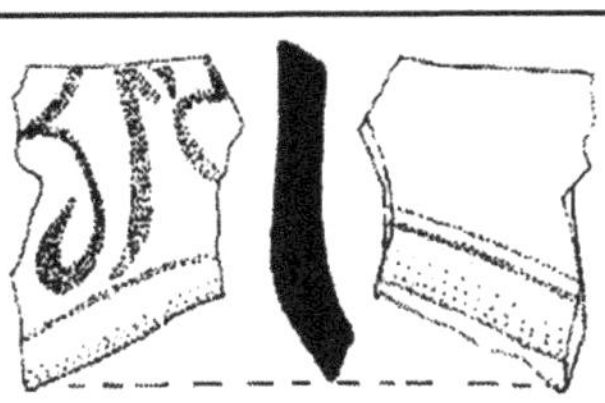 D.95	**D.95**:- Pale blue background. Fragment of a probably calligraphic design in blue with a blue horizontal band underneath. Two curved and one vertical band can be see. On inner surface two horizontal light blue bands can be seen. SJN-K, A4, NE, 1, 0.32, 4501, PORC-4, S(A)
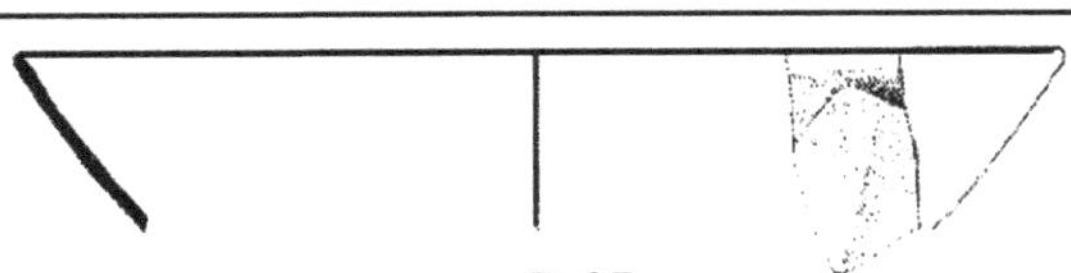 D.97	**D.97**:- Rim sherd of a convex sided dish. White with a moulded lotus petal on the outer surface. The petal has a ridge in the centre.

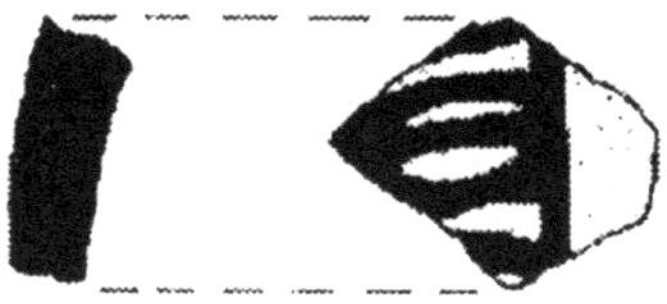 **D.134**	**D.134**:- Small body sherd with blue painting on a white (greyish) background. SJN-B, TT4, SW, 2, 0.97, 51124, PORC-4, P(A)
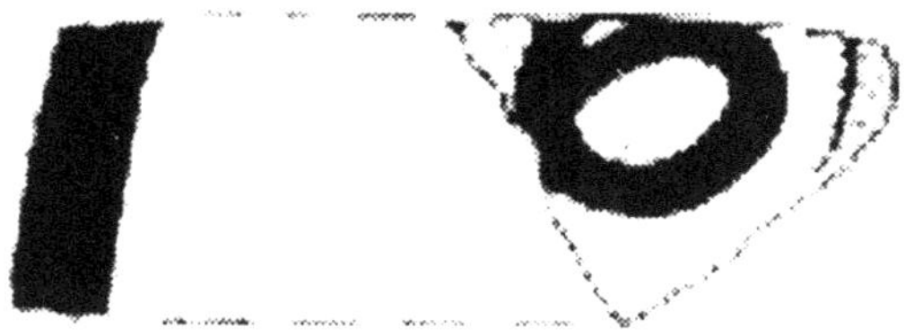 **D.135**	**D.135**:- small body sherd having a white background and a dark green loop/circle on the exterior. SJN-B, TT4, SW, 1, 0.13, 51101, PORC-4, P(A)
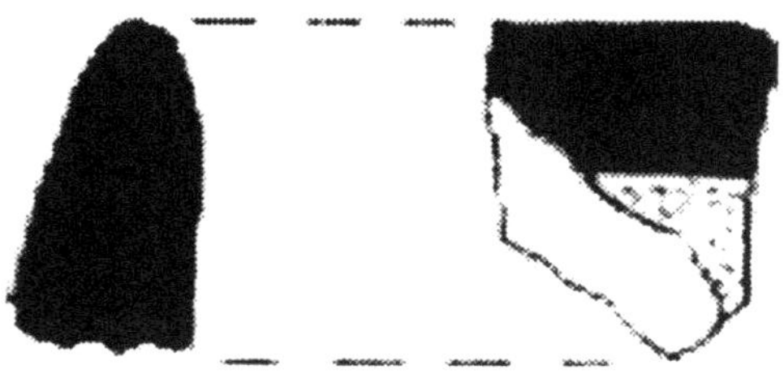 **D.136**	**D.136**:- Small rim, simple rounded with white glaze on inside and a broad-blue band running along the exterior on the lip. SJN-B, TT4, SW, 1, 0.13, 51101, PORC-4, P(A)

CHAPTER III - DATA ANALYSIS

"Whereas some of the more loquacious lot –
I think a Sufi pipkin – waxing hot –
"All this of Pot and Potter – tell me then,
Who is the potter, pray, and who the pot?"

- *Rubaiyat (Omar Khayyam)*

This chapter aims to present the quantification data of the Sanjan pottery and provides an interpretation of the material on this basis. The Sanjan assemblage has come from differing contexts. As explained earlier, the disturbed nature of the site and the many limitations of the excavations have proved a setback in the study of the material. Hence the selection of material for statistical analysis has to be done so that biased or disturbed sampling can be avoided as much as possible. This has severely limited the scope for a thorough analysis. The following points should be kept in mind while considering the statistical analysis of this material:

Counts and percentages have been calculated. However it is not possible to include the entire data base in this study. Hence analysis and conclusions have been presented in text where it is not possible to present tables and graphs.

Trench TT4 NE, which was dug to maximum depth, has been used as index for determining the ceramic sequence and quantification for the site. Data pertaining to this quadrant are presented here, with cross-reference and presentation of similar data from other relatively undisturbed context where necessary. However, in the tables and graphs representing the overall distribution and quantum of material in the entire trench TT4, caution should be exercised in the interpretation of data where layer is mentioned since the well deposit in the South West quadrant was allotted to layer 2 during the registration of the pottery and related finds. While the well belonged to layer 2, the deposit within should not be considered as a stratigraphic layer, comparable to layer 2 in the other quadrants. To avoid this drawback, two sets of tables have been presented whenever overall quantification of TT4 has needed to be done – inclusive of SW and exclusive of SW. The individual quadrant profiles are presented in separate tables.

All wares and shapes have not occurred in the NE quadrant and this limitation has been taken into account. Cross-referencing has been done with other quadrants of TT4 and, if necessary, trenches TT1 and TT2 in order to establish the sequence for the Sanjan Bandar as best possible. However, pottery from a relatively small area has been used as sampling and so caution should be exercised in regarding this as representative of the entire site. TT1 and TT2 have been used for cross referencing when required, and have not been used for detailed analysis due to their disturbed stratigraphy and problematic layer identification. The material has been included in establishing the overall ceramic profile of Sanjan Bandar (Tables 1, 2, 2.1, 2.2, 2.3 and 3).

Pottery from the Koli Khadi, section scraping on the embankment and the excavation of the well in TT3 has not been included, except in Table 1. Representative sherds have been used for illustration and morphological study, but a bulk of this material was not available to the author for final quantification and so little statistical analysis has been attempted. The data presented is based on available material only and may be erroneous. Pottery from the clearance work and excavation of the Sanjan Dokhma occurred in a secondary context, either mixed in the mud-mortar of the structure or from the surface and occasionally from the central well or bhandar. This pottery has also not been subjected to statistical analysis.

It must also be noted that the trenches used for this study, particularly TT4, were located on the 'Bandar' or port area, almost immediately on the river bank. If the waterfront was a highly populated or commercial area with intense activity and movement, as is suggested by the structural evidence, it may have influenced the occurrence and distribution pattern of the assemblage. Hence certain ware classes and vessel types may be more prominently represented and others less so.

Only rim sherds have been taken into account in the analysis since it was found that the initial collection of the pottery was somewhat biased towards particular ware classes. For example, the glazed classes and the Chinese and Far Eastern wares (referred to hereafter as 'Chinese' for reasons of brevity) were collected regardless of their diagnostic or non-diagnostic nature whereas the unglazed classes were not retained unless they were diagnostic. Hence counts of body sherds could present an unbalanced picture of the assemblage. Similarly, bases are better represented in some classes, perhaps due to their sturdy nature. Rims were collected for most classes with few exceptions and were therefore considered the most representative sherd type. However, there are some classes of pottery which have, by default, been excluded from the tables because rim shapes did not occur. For example, torpedo jar rims were few as compared to the body sherds which were found in abundance. An attempt is made to explain and clarify this in the text accompanying the tables and graphs. Some ware classes such as Changsha Underglaze Painted Ware were represented only by base or body sherds and so may have been left out of the sampling. Similarly, some miscellaneous classes also have not been represented in the counts due to a lack of rim sherds. It should not be taken to mean that ware classes did not occur earlier or later in the stratigraphy simply because they are not reflected here. This problem is discussed later in the section on depth-wise occurrence of ware classes where, to avoid misrepresentation and to present a more honest picture, lists of ware classes as they occur at varying depths (regardless of whether they are rims or not) is given. This is done without counts or percentages being calculated so that ware classes may at least be recorded as present at a certain level or depth while considering the occurrence sequence and a tentative chronology (see Lists A, B and C).

Percentages have been calculated on basis of layer total as well as overall total for quadrant and trench. However, it needs to be emphasized that layer demarcation was hampered by the disturbed nature of the site and methodological limitations. Hence the pottery occurring within a particular layer does not necessarily form a marked ceramic group. Variation in the occurrence of pottery wares at specific depths, in fact, suggests ceramic levels. For this reason, depths and the thickness of the layers as demarcated by the excavators are both mentioned. The author has attempted to take the depths into account along with the occurrence of ware types, in order to overcome the errors in layer identification during excavation.

It is pertinent to note again at this point that the depth of a lot which is recorded in the excavation methodology refers to the material that occurs between the previous lot number and the present one so that the pottery comes from the deposit in-between the two and not from the mentioned level alone. In this analysis, the occurrence of wares in the index trench TT4 NE has been discussed as occurring by depth, with TT4 NW and TT4 SE as cross-reference. Also, it should be kept in mind that all quadrants were not dug to the same depth.

The indigenous classes of pottery form a bulk of the assemblage. As is well known, Indian utilitarian wares have a long tradition rarely exhibiting stylistic or profile changes. An attempt is made to classify the ware shapes of some of the more prominent slipped ware classes according to the levels in which they occur, so that by association with chronologically dateable wares, they may be assigned a tentative chronology. It may also be possible to note the shapes which continue for the longest periods and which have a shorter life. The shapes which occurred in trench TT4 were used as they came from definite context. Hence some shapes which did not occur in this trench have been left out of the study.

EVES analysis could not be done in the course of this study but it is hoped that such an analysis will be possible in future when the methodology for the same is available. This analysis would have taken into account the brokenness of various pottery classes in the calculations thereby avoiding a statistical bias. As Kennet points out, the differing levels of brokenness can affect the ratio of pottery types (Kennet 2004: 68).

It may be pertinent to note here that, until recently, the study of pottery in India has mainly been descriptive and almost never includes a quantitative or statistical analysis of the material. As such, this is a new approach and it is impossible to make any inter-site comparisons with other excavated sites in India. Inter-site comparisons with sites outside India is hampered by the fact that the methodology of pottery collection and conventions of quantification used elsewhere are vastly different from those in India. The analysis presented in this study aims at a basic presentation of the Sanjan data, with all its limitations in mind.

Ceramic profile of Sanjan – total sherd count, diagnostics, glazed and unglazed wares:
(Tables 1, 2, 2.1, 2.2, 2.3 and 3)

Table 1 presents the total number of potsherds collected and registered during all three excavation seasons and from all the different locations where work was carried out. However, this is not the definitive figure since a large amount of pottery from section-scraping, ringwell context, wells and other disturbed context had been discarded without counting in the field. Only the diagnostic sherds have been retained and accounted for. Surface collection is also not included in this total. Of the 1,23,728 sherds that were registered, only rims, bases and distinctive body sherds were retained as diagnostics. Since this analysis concentrates on the material from the Sanjan Bandar area, it may be pertinent to note that the total number of registered sherds from TT4, TT1 and TT2 is 79,039. Of these, 8534 (10.80%) are diagnostic sherds (see Table 2).

Tables 2.1, 2.2 and 2.3 present the counts and percentage of diagnostic sherds as they appear in each trench. TT4 has a higher percentage of body sherds as compared to the other two trenches, whereas in TT1 the rim sherds constitute a higher percentage. But TT2 which was a smaller trench than either TT4 or TT1, and which was, in actuality, an extension of TT1 has an extraordinarily high quantity of base sherds. The rim sherds in TT1 are almost equal in proportion to the base sherds in TT2.

Ceramic profile of Sanjan –

total sherd count, diagnostics, glazed and unglazed wares

TT4	TT1	TT2	TT3/EMB/SJN-K	DOKHMA	Total
22,656	36,116	20,267	42,934	1755	1,23,728

Table 1. Total collection of pottery at Sanjan (by count)

	TT4	TT1	TT2	Total
RIM	1644	2336	301	**4281**
BASE	158	133	497	**788**
BODY	2084	1360	21	**3465**
total	**3886**	**3829**	**819**	**8534**

Table 2 Total number of diagnostic sherds in TT4, TT1 and TT2 (by count)

TT4	count	%
RIM	1644	42.31
BASE	158	4.07
BODY	2084	53.62
Total	**3886**	**100.0 0**

Table 2.1

TT1	count	%
RIM	2336	61.01
BASE	133	3.47
BODY	1360	35.52
total	**3829**	**100.0 0**

Table 2.2

TT2	count	%
RIM	301	36.75
BASE	497	60.68
BODY	21	2.57
total	**819**	**100.00**

Table 2.3

Total number of diagnostic sherds in TT4, TT1 and TT2 (by count and percentage)

Table 3. Glazed and unglazed wares (by count and percentage)

TRENCH	TT4	%	TT1	%	TT2	%	Grand Total	%
GLAZED	153	9.31	243	10.40	19	6.31	415	9.69
UNGLAZED	1491	90.69	2093	89.60	282	93.69	3866	90.31
TRENCH TOTAL	**1644**	**100.0 0**	**2336**	**100.0 0**	**301**	**100.00**	**4281**	**100.00**

This distribution pattern may be activity specific to this area, since TT2 SE was almost entirely taken up by a deep pit which started from close to the surface and continued through to the bottom of the trench. The number of base sherds retrieved from TT4 and TT1 are very low by comparison. Rim sherds from all locations and context were collected with no element of subjectivity in selection and are the diagnostic type used for all analysis that follows. Pits, structural activity and other disturbances may also have affected the distribution of these sherds.

Table 3 presents the proportion of glazed and unglazed wares as they occur in TT4, with TT1 and TT2 for cross-reference. By glazed wares are meant all wares, regardless of origin, which have glaze as surface treatment. Other means of decoration or ornament may or may not be present. Unglazed wares are all those classes of pottery, regardless of origin, which have no glaze although they may or may not have other surface treatment such as slip or wash. Miscellaneous wares of unknown provenance are also included in the unglazed wares. Of the 4281 rims collected from the three trenches, 9.69% are glazed and 90.31% are unglazed. It must be clearly understood that the 9.69% of glazed pottery consists of Chinese as well as West Asian wares having glaze and not West Asian glazed wares alone. These wares are present at Sanjan as maritime imports. They would, therefore, be more expensive. Besides, the supply of these wares would have been far more limited as compared to the indigenous and local wares which would have greater affordability and availability.

Occurrence of Chinese, Indigenous, West Asian and Unknown wares:
(Tables 4, 4.1, 4.2, 4.3, 4.4, 4.5, 4.6, 4.7, 4.8 and 4.9)

The pottery classes have been divided into four broad groups on the basis of their origin, regardless of whether they are glazed or unglazed. The proportions of these as they occur in trench TT4 and in its individual quadrants are seen in Tables 4 and 4.1. They present the occurrence of rim sherds of these groups in TT4 inclusive of the material from the well in TT4 SW and tables 4.2 and 4.3 present their occurrence excluding TT4 SW. The tables also present the layer-wise quantification of these groups so that distribution patterns as well as comparisons can be noted. The limitations of the layer identification have been discussed earlier in this chapter as well as in Chapter 1. It is reiterated that these layers do not represent specific ceramic strata, horizon or groups. It is important to present the depth of the layers as marked by the excavators since almost all the tables from here on refer to the layers.

TT4 - NE : Layer 1 = 0 to 0.29m
Layer 2 = 0.30 to 1.19m
Layer 3 = 1.20 to 1.55m
Layer 4 = 1.56 to 1.78m
Layer 5 = 1.79 to 2.70m
Layer 6 = 2.71 to 3.86m

TT4 - NW : Layer 1 = 0 to 0.17m
Layer 2 = 0.18 to 0.89m
Layer 3 = 0.90 to 1.16m

TT4 - SE : Layer 1 = 0 to 0.27m
Layer 2 = 0.28 to 1.18m

Tables 4 and 4.1 present the occurrence of these groups in TT4, inclusive of the material from the SW quadrant. However the SW quadrant which consisted of the well and the deposit of which was entirely attributed to layer 2, constitutes 34.43% of the entire collection of TT4. This well deposit cannot be equated with stratigraphic layer 2 in the other quadrants. If TT4 is considered to have three relatively clear quadrants and the material from TT4 SW is disregarded as suspect, then the proportions change. Tables 4.2 and 4.3 present the data exclusive of TT4 SW.

These tables make it clear that Chinese wares appear much after the West Asian wares which are present from layer 6 onwards. The contact with West Asia clearly predates contact with the East. West Asian ceramics make up 8.53% of the entire assemblage from TT4. They decrease to nearly half in layer 5 and even less in layer 4. Chinese wares make an appearance in layer 4. It must be noted that body sherds of STONE-3 are found in the uppermost part of layer 5 (see List A) but are not reflected in the analysis. Layer 3 shows an increase in the Chinese component to 2.75%. There is a sudden and sharp increase of West Asian pottery in layer 2 (14.08%). Chinese pottery also shows a marginal increase in this layer, though not as dramatic as the West Asian wares. Chinese wares increase to 3.70% in layer 1. Indigenous wares make up a bulk of the assemblage throughout the sequence. It is also noticeable that while West Asian wares show a decrease in layers 5, 4 and 3, the Indigenous wares show a corresponding increase. A relatively high percentage of Unknown wares, all unglazed, occur in Layer 6. They show a decreasing trend. It should be noted that many of the other Unknown wares are not represented in this analysis due to a paucity of rims (see Lists A, B and C).

Tables 4.4 and 4.5 present the data for the index quadrant TT4 NE. TT4 NE was dug to the lowest level (layer 6, 3.86m) (see Tables 4.4 and 4.5). Indigenous pottery, not surprisingly makes up a bulk of the assemblage (89.35%). West Asian wares, which occur from layer 6 continuously to layer 1, constitute 8.35% of the quadrant collection. Chinese wares make up 1.15% of it. It is the changing

Occurrence of Chinese, Indigenous, West Asian and Unknown wares:

Tr. TT4	Layer 6	Layer5	Layer 4	Layer 3	Layer 2	Layer 1	Tr. Total
(count) Chinese			1	2	13	2	**18**
Indigenous	66	208	77	175	874	24	**1424**
West Asian	5	8	1	5	168	4	**191**
Unknown	3	3		1	4		**11**
TOTAL	**74**	**219**	**79**	**183**	**1059**	**30**	**1644**

Table 4. Occurrence of ware groups (by count) in TT4 (inclusive of SW)

Tr. TT4	Layer 6	Layer 5	Layer 4	Layer 3	Layer 2	Layer 1	% of Tr. Total
% Chinese			1.27	1.09	1.23	6.67	**1.09**
Indigenous	89.19	94.98	97.46	95.63	82.53	80.00	**86.62**
West Asian	6.76	3.65	1.27	2.73	15.86	13.33	**11.62**
Unknown	4.05	1.37		0.55	**0.38**		**0.67**
TOTAL	**100.00**	**100.00**	**100.00**	**100.00**	**100.00**	**100.00**	**100.00**

Table 4.1. Occurrence of ware groups (by percentage) in TT4 (inclusive of SW)

Tr. TT4	Layer 6	Layer5	Layer 4	Layer 3	Layer 2	Layer 1	Quad. Total
(count) Chinese			1	2	7	1	**11**
Indigenous	66	208	77	174	418	23	**966**
West Asian	5	8	1	5	70	3	**92**
Unknown	3	3		1	2		**9**
TOTAL	**74**	**219**	**79**	**182**	**497**	**27**	**1078**

Table 4.2 Occurrence of ware groups (by count) in TT4 (exclusive of SW)

Tr. TT4	Layer 6	Layer 5	Layer 4	Layer 3	Layer 2	Layer 1	% of Tr. Total
% Chinese			1.27	1.10	1.41	3.70	**1.02**
Indigenous	89.19	94.98	97.47	95.60	84.10	85.19	**89.61**
West Asian	6.76	3.65	1.27	2.75	14.08	11.11	**8.53**
Unknown	4.05	1.37		0.55	0.40		**0.83**
TOTAL	**100.00**	**100.00**	**100.00**	**100.00**	**100.00**	**100.00**	**100.00**

Table 4.3 Occurrence of ware groups (by percentage) in TT4 (exclusive of SW)

Tr. TT4 NE	Layer 6	Layer5	Layer 4	Layer 3	Layer 2	Layer 1	Quad Total
(count) Chinese			1	2	4		**7**
Indigenous	66	208	77	72	122	1	**546**
West Asian	5	8	1	1	35	1	**51**
Unknown	3	3			1		**7**
TOTAL	**74**	**219**	**79**	**75**	**162**	**2**	**611**

Table 4.4 Occurence of ware groups by count in TT4 NE.

Tr. TT4 NE	Layer 6	Layer 5	Layer 4	Layer 3	Layer 2	Layer 1	Quad Total
% Chinese			1.27	2.67	2.47		**1.15**
Indigenous	89.19	94.98	97.46	96.00	75.31	50.00	**89.35**
West Asian	6.76	3.65	1.27	1.33	21.00	50.00	**8.53**
Unknown	4.05	1.37			0.62		**1.15**
TOTAL	**100.00**	**100.00**	**100.00**	**100.00**	**100.00**	**100.00**	**100.00**

Table 4.5 Occurrence of ware groups (by percentage) in TT4 NE.

Tr. TT4 NW	**Layer 3**	**Layer 2**	**Layer 1**	**Quad. Total**
(count) Chinese		3	1	**4**
Indigenous	102	196	14	**312**
West Asian	4	15	2	**21**
Unknown	1			**1**
TOTAL	**107**	**214**	**17**	**338**

Tr. TT4 NW	**Layer 3**	**Layer 2**	**Layer 1**	**% of Quad. Total**
% Chinese		1.40	5.88	**1.18**
Indigenous	95.33	91.59	82.36	**92.31**
West Asian	3.74	7.01	11.76	**6.21**
Unknown	0.93			**0.30**
TOTAL	**100.00**	**100.00**	**100.00**	**100.00**

Table 4.6 and 4.7 Occurrence of ware groups in TT4 NW (by count and percentage)

Tr. TT4 SE	Layer 2	Layer 1	Quad. Total
(counts)			
Chinese			
Indigenous	100	8	**108**
West Asian	20		**20**
Unknown	1		**1**
TOTAL	**121**	**8**	**129**

Tr. TT4 SE	Layer 2	Layer 1	% of Quad. Total
%			
Chinese			
Indigenous	82.64	100.00	83.72
West Asian	16.53		15.50
Unknown	0.83		0.78
TOTAL	**100.00**	**100.00**	**100.00**

Table 4.8 and 4.9 Occurrence of ware groups inTT4 SE (by count and percentage)

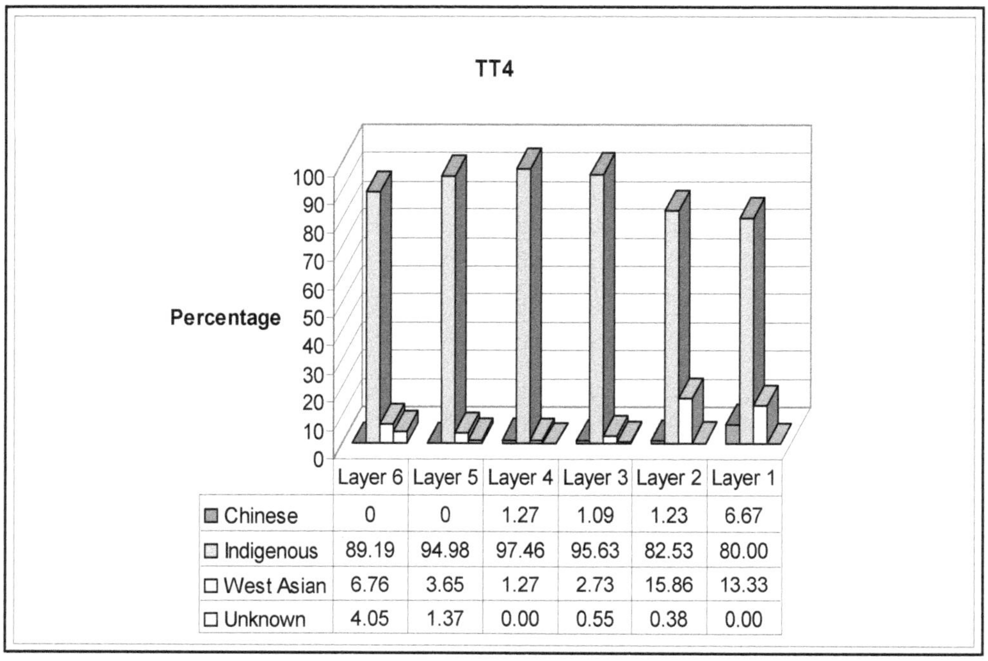

	Layer 6	Layer 5	Layer 4	Layer 3	Layer 2	Layer 1
Chinese	0	0	1.27	1.09	1.23	6.67
Indigenous	89.19	94.98	97.46	95.63	82.53	80.00
West Asian	6.76	3.65	1.27	2.73	15.86	13.33
Unknown	4.05	1.37	0.00	0.55	0.38	0.00

Fig. A Occurrence of ware groups in TT4 (inclusive of TT4 SW)

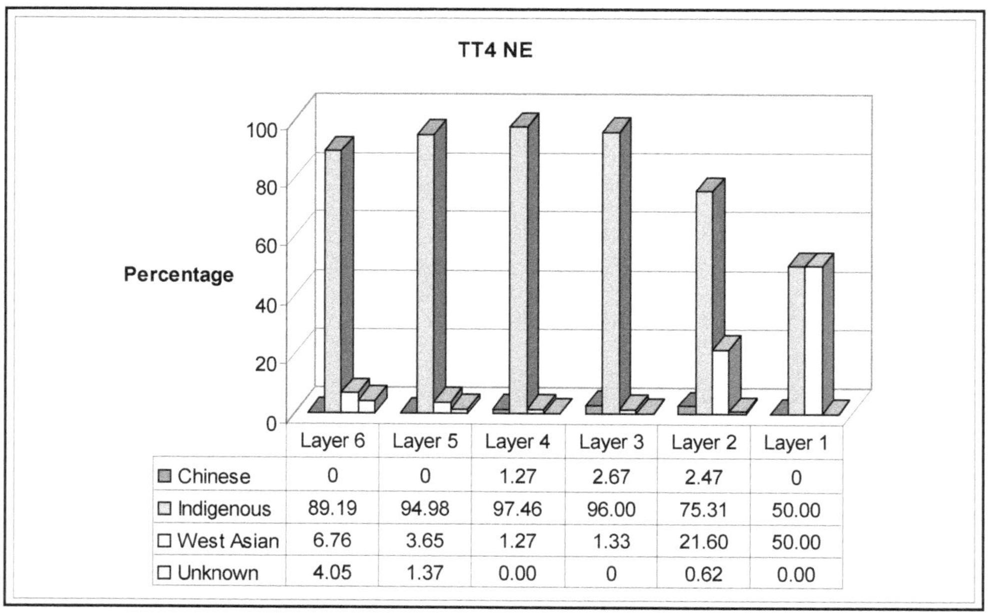

	Layer 6	Layer 5	Layer 4	Layer 3	Layer 2	Layer 1
Chinese	0	0	1.27	2.67	2.47	0
Indigenous	89.19	94.98	97.46	96.00	75.31	50.00
West Asian	6.76	3.65	1.27	1.33	21.60	50.00
Unknown	4.05	1.37	0.00	0	0.62	0.00

Fig.B Occurrence of ware groups in TT4 NE

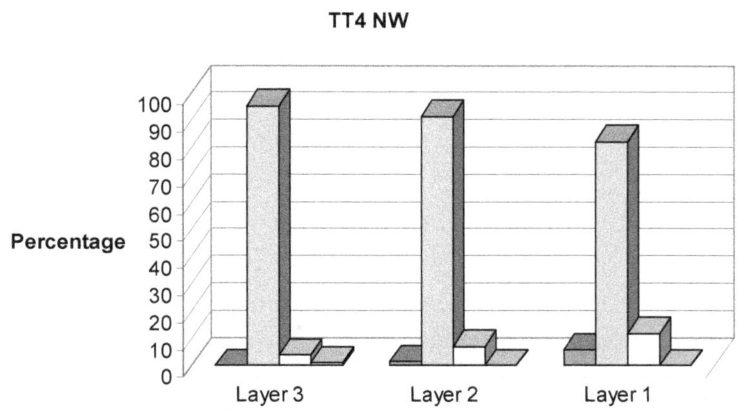

	Layer 3	Layer 2	Layer 1
Chinese	0.00	1.40	5.88
Indigenous	95.33	91.59	82.36
West Asian	3.74	7.01	11.76
Unknown	0.93	0.00	0.00

Fig.C Occurrence of ware groups in TT4 NW

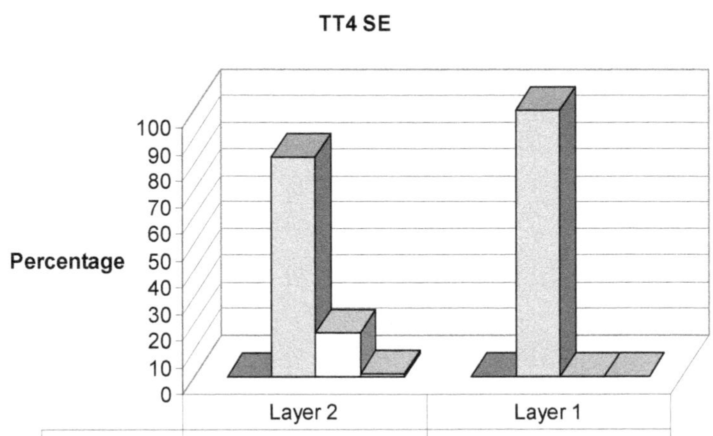

	Layer 2	Layer 1
Chinese	0.00	0.00
Indigenous	82.64	100.00
West Asian	16.53	0.00
Unknown	0.83	0.00

Fig.D Occurrence of ware groups in TT4 SE

proportions between layers and wares that appear to have a pattern. This is seen in TT4 NW and TT4 SE as well (see Tables 4.6, 4.7, 4.8 and 4.9). Indigenous wares increase from 89.19% in layer 6 to 96.00% in layer 3 while West Asian wares appear to show a decrease from 6.76% in layer 6 to 1.33% in layer 3. In layer 2, we see a dramatic increase in West Asian wares – 20.60% while the Indigenous wares decrease to 75.31%. Chinese wares which appear in layer 4 with 1.27%, increase in layers 3 and 2 to almost twice the percentage (2.67% and 2.47%). Chinese wares in layer 4 appear equal the West Asian wares as represented by rim sherds. In layer 3 the Chinese wares appear to be better represented than the

TT1 and TT2 :

(Tables 5, 5.1, 5.2, 5.3, 5.4, 5.5, 6, 6.1 and 6.2)

Trenches TT1 and TT2 were heavily disturbed by structural activity, pits, etc. The stratigraphy has been problematic, as mentioned earlier. Tables 5, 5.1, 5.2, 5.3, 5.4, 5.5, 6, 6.1 and 6.2 present the proportions of these ware groups as they occur in these trenches in order to provide a more complete profile for Sanjan Bandar. They are not taken up for detailed analysis. TT1 SW and TT1 NW were disturbed by ring-wells and other structural activity and TT2 SE was almost entirely taken up by a deep pit which started from just below the surface and

TTI and TT2 - Occurrence of Ware Groups

Tr. TT1	Layer 4	Layer 3	Layer 2	Layer 1	Quad. Total
(count) Chinese		2	1	2	**5**
Indigenous	28	548	884	608	**2068**
West Asian	3	72	95	84	**254**
Unknown		1	2	6	**9**
TOTAL	**31**	**623**	**982**	**700**	**2336**

Table 5 Occurrence of ware groups in TT1 (by count)

Tr. TT1	Layer 4	Layer 3	Layer 2	Layer 1	Quad. Total
% Chinese		0.32	0.10	0.29	**0.21**
Indigenous	90.32	87.96	90.02	86.86	**88.53**
West Asian	9.68	11.56	9.67	12.00	**10.87**
Unknown		0.16	0.20	0.86	**0.39**
TOTAL	**100.00**	**100.00**	**100.00**	**100.00**	**100.00**

Table 5.1 Occurrence of ware groups in TT1 (by percentage)

Tr. TT1 NE	Layer 4	Layer 3	Layer 2	Layer 1	Quad. Total
(count) Chinese			1		**1**
Indigenous	11	159	179	119	**468**
West Asian	2	21	24	14	**61**
Unknown					**1**
TOTAL	**13**	**180**	**205**	**133**	**531**

Table 5.2 Occurrence of ware groups in TT1 NE (by count)

Tr. TT4	Layer 4	Layer 3	Layer 2	Layer 1	Quad. Total
% Chinese			0.49		**0.19**
Indigenous	84.62	88.33	87.32	89.47	**88.14**
West Asian	15.38	11.67	11.71	10.53	**11.49**
Unknown			0.49		**0.19**
TOTAL	**100.00**	**100.00**	**100.00**	**100.00**	**100.00**

Table 5.3 Occurrence of ware groups in TT1 NE (by percentage)

Tr. TT1 SW	Layer 3	Layer 2	Layer 1	Quad. Total
(count) Chinese			2	**2**
Indigenous	108	236	127	**471**
West Asian	27	21	15	**63**
Unknown	1			**1**
TOTAL	**136**	**257**	**444**	**537**

Table 5.4 Occurrence of ware groups in TT1 SE (by count)

Tr. TT1 SE	Layer 3	Layer 2	Layer 1	Quad. Total
% Chinese			1.39	**0.37**
Indigenous	79.41	91.83	88.19	**87.71**
West Asian	19.85	8.17	10.42	**11.73**
Unknown	0.74			**0.19**
TOTAL	**100.00**	**100.00**	**100.00**	**100.00**

Table 5.5 Occurrence of ware groups in TT1 SE (by percentage)

Tr. TT2	Layer 6	Layer5	Layer 4	Layer 3	Layer 2	Layer 1	Quad Total
(count) Chinese							
Indigenous	1	12	92	98	63	5	**271**
West Asian	1	10	8	8	1		**28**
Unknown				1	1		**2**
TOTAL	**2**	**22**	**101**	**107**	**64**	**5**	**301**

Table 6 Occurrence of ware groups in TT2 (by count)

Tr. TT2	Layer 6	Layer 5	Layer 4	Layer 3	Layer 2	Layer 1	Quad Total
(count) Chinese							
Indigenous	50.00	54.55	91.09	91.59	98.44	100.00	**90.03**
West Asian	50.00	45.45	7.92	7.48	1.56		**9.30**
Unknown			0.99	0.93			**0.66**
TOTAL	**100.00**	**100.00**	**100.00**	**100.00**	**100.00**	**100.00**	**100.00**

Table 6.1 Occurrence of ware groups in TT2 (by percentage)

Tr. TT2 SW	Layer 4	Layer 3	Layer 2	Layer 1	Quad. Total
(count) Chinese					
Indigenous	77	29	11	1	**118**
West Asian	1	5	1		7
Unknown	1	1			2
TOTAL	**79**	**35**	**12**	**1**	**127**

Table 6.2 Occurrence of ware groups in TT2 SW (by count)

Tr. TT2 SW	Layer 4	Layer 3	Layer 2	Layer 1	Quad. Total
% Chinese					
Indigenous	97.47	82.85	91.67	100.00	**92.92**
West Asian	1.27	14.28	8.33		**5.51**
Unknown	1.27	2.57			**1.57**
TOTAL	**100.00**	**100.00**	**100.00**	**100.00**	**100.00**

Table 6.3 Occurrence of ware groups in TT2 SW (by percentage)

went down to the base of the quadrant. These quadrants have not been presented individually in these tables although the overall trench profile includes the data from them.

These quadrants of trenches TT1 and TT2 are used for providing a cross-reference to TT4 and for providing comparable data from a different location on the same site. It is in no way implied that the layers in TT1 and TT2 correspond to those in TT4.

Ceramic composition of Layers :
(Tables 7, 7.1, 7.2, 7.3, 7.4)

Tables 7, 7.1, 7.2, 7.3 and 7.4 present a closer look at the individual ware classes that make up the four groups as represented by rim sherds and their occurrence in different layers. They present the rim sherd count as well as the percentage of each ware class as it occurs in each layer. The layer total of rim sherds for the entire trench / quadrant is given in the last row at the end of the table. The vertical columns to the extreme right present the trench / quadrant rim total of all layers and percentage of a ware class in the overall assemblage. The percentage of a ware within a layer is calculated on the basis of the layer total. For example, in Table 7, BSGW = 27 rim sherds in layer 6, make up 36.49% of the layer total which is 74 rims. Columns to the extreme right show that the total count of BSGW in the trench is 684 rims, making up 41.61% of the total trenc collection which is 1644 rim sherds. Indigenous wares which number 66 rims in layer 6 constitute 89.19% of the layer total of 74 rim sherds. Similarly, CEL-1 = 3 rims in the overall trench assemblag of 1644 rims, forming 0.18% of the collection. A total of 18 Chinese rim sherds in TT4 constitute 1.09% of the 1644 rim sherds found. Ware classes not represented in these tables due to a lack of rim sherds but which are present in the collection have been recorded in Lists A, B and C and should be viewed along with the statistical tables so that a more complete understanding of the layer composition is possible.

Table 7 presents the overall ceramic composition of TT4. Table 7.1 presents the same profile with the exclusion of TT4 SW so that the material from the well deposit which was wrongly attributed to layer 2 may be disregarded and a more credible ceramic profile for the trench may be seen. It is clear from Table 7.1 that in layers 6 and 5 Indigenous wares and West Asian wares constitute a major part of the assemblage. Layer 6 has a significant number of Unknown wares. Amongst the Indigenous wares, it is noticed that wares with black slip are more prominent. BSGW shows an increase from 36.49% in layer 6 to 56.04% in layer 3 decreasing to 38.23% in layer 2. BSRW on the other hand shows a decreasing trend from layer 6 to layer 2. The most prominent red slipped ware is RSRW-1, showing a continuous increase from 16.22% in layer 6 to 21.52% in layer 4. This trend changes and in layers 3 and 2 the percentages drop to 12.64% and 9.96%. BSGW is the most prominent ware class of the ceramic assemblage, accounting for 42.76% of the 1078 rim sherds in this trench. It occurs continuously throughout the sequence.

The West Asian wares show an interesting pattern of occurrence. They appear from layer 6 onwards continuously throughout the sequence, constituting 8.53% of the total trench assemblage. The ware classes in layer 6 are mostly unglazed, with the exception of one WGW sherd which appears to be an intrusion. (see List A). There appears to be a decrease in West Asian wares from 6.76% in layer 6 to3.65% in layer 5, although glazed wares such as TGW-1, TGW-3 and TGW-5 are commonly seen in this layer along with unglazed classes. This decreasing trend continues through layer 4 (1.27%) and again increases to 2.75% in layer 3. (It should be noted that while these figures create the impression that very few West Asian sherds were encountered in these layers, this is definitely not so. As mentioned earlier, body sherds and base sherds have not been used for this analysis and are not reflected in these tables. These layers did contain sherds not accounted for here. For example TGW body and base sherds appear in layer 6 but are not reflected

TT4	Layer 6	%	Layer 5	%	Layer 4	%	Layer 3	%	Layer 2	%	Layer 1	%	Grand Total	% of trench
(Chinese)														
CEL-1									3	0.28			3	0.18
PORC-1									6	0.57			6	0.36
PORC-2					1	1.27	1	0.55					2	0.12
PORC-3							1	0.55					1	0.06
PORC-4											1	3.33	1	0.06
STONE-1									4	0.38	1	3.33	5	0.30
TOTAL					**1**	**1.27**	**2**	**1.09**	**13**	**1.23**	**2**	**6.67**	**18**	**1.09**
(Indigenous)														
BSGW	27	36.49	101	46.12	35	44.30	103	56.28	412	38.90	6	20.00	684	41.61
BSRW-1	18	24.32	38	17.35	7	8.86	23	12.57	144	13.60			230	13.99
BSRW-2	1	1.35	8	3.65	7	8.86	2	1.09	36	3.40			54	3.28
GW-1									4	0.38	2	6.67	6	0.36
GW-2									1	0.09			1	0.06
MRW-1							1	0.55					1	0.06
MRW-2									4	0.38			4	0.24
RSGW	2	2.70	1	0.46	4	5.06	8	4.37	36	3.40	4	13.33	55	3.35
RSPW-1					1	1.27							1	0.06
RSRW-1	12	16.22	47	21.46	17	21.52	23	12.57	105	9.92			204	12.41
RSRW-2	2	2.70	1	0.46	2	2.53	6	3.28	33	3.12	1	3.33	45	2.74
RW-1	3	4.05	7	3.20	3	3.80	8	4.37	71	6.70	8	26.67	100	6.08
RW-2			2	0.91	1	1.27			2	0.19	1	3.33	6	0.36
RW-3									1	0.09	1	3.33	2	0.12
SBW-1	1	1.35							12	1.13	1	3.33	14	0.85
SBW-2			3	1.37			1	0.55	6	0.57			10	0.61
SBW-3									2	0.19			2	0.12
WSRW									5	0.47			5	0.30
TOTAL	**66**	**89.19**	**208**	**94.98**	**77**	**97.47**	**175**	**95.63**	**874**	**82.53**	**24**	**80.00**	**1424**	**86.62**
(West Asian)														
BiGP									6	0.57	1	3.33	7	0.43
BW	2	2.70	2	0.91			1	0.55	1	0.09			6	0.36
BW(SPOT)									1	0.09			1	0.06
BW(TJ)			1	0.46									1	0.06
CPW									2	0.19			2	0.12
CSGW									1	0.09			1	0.06
ESW(R)									10	0.94			10	0.61
ESW(W)									25	2.36			25	1.52
FSGW									3	0.28			3	0.18
GBW									1	0.09			1	0.06
GRW									2	0.19			2	0.12
HsG(B)									17	1.61	2	6.67	19	1.16
HsG(M)									2	0.19			2	0.12
HsG(P)									12	1.13			12	0.73
HsG(W)									1	0.09			1	0.06

Table 7 - Ceramic composition of Layers in TT4 (inclusive of SW)

(continued on the next page)

continued from the last page ...

TT4	Layer 6	%	Layer 5	%	Layer 4	%	Layer 3	%	Layer 2	%	Layer 1	%	Grand Total	% of trench
LISV-1									2	0.19			2	0.12
LISV-4									1	0.09			1	0.06
LPW									1	0.09			1	0.06
LPW-1							1	0.55	9	0.85			10	0.61
SWGW									20	1.89			20	1.22
TGW-1			1	0.46					13	1.23	1	3.33	15	0.91
TGW-2							1	0.55	2	0.19			3	0.18
TGW-3			1	0.46			1	0.55		0.00			2	0.12
TGW-4							1	0.55	2	0.19			3	0.18
TGW-5			1	0.46					1	0.09			2	0.12
UGPW			1	0.46									1	0.06
UPW									2	0.19			2	0.12
WGP									1	0.09			1	0.06
WGPW									2	0.19			2	0.12
WGW	1	1.35							26	2.46			27	1.64
WSPW	2	2.70											2	0.12
WSPW-1									2	0.19			2	0.12
WSPW-2			1	0.46	1	1.27							2	0.12
TOTAL	**5**	**6.76**	**8**	**3.65**	**1**	**1.27**	**5**	**2.73**	**168**	**15.86**	**4**	**13.33**	**191**	**11.62**
(Unknown)														
MISC-10									1	0.09			1	0.06
MISC-3			1	0.46					2	0.19			3	0.18
MISC-4			1	0.46			1	0.55					2	0.12
MISC-5	1	1.35											1	0.06
MISC-6	2	2.70	1	0.46									3	0.18
MISC-8									1	0.09			1	0.06
TOTAL	**3**	**4.05**	**3**	**1.37**			**1**	**0.55**	**4**	**0.38**			**11**	**0.67**
L.TOTAL	**74**	**100.00**	**219**	**100.00**	**79**	**100.00**	**183**	**100.00**	**1059**	**100.00**	**30**	**100.00**	**1644**	**100.00**

Table 7 - Ceramic composition of Layers in TT4 (inclusive of SW)

TT4	Layer 6	%	Layer 5	%	Layer 4	%	Layer 3	%	Layer 2	%	Layer 1	%	Grand Total	% of trench
(Chinese)														
CEL-1														
PORC-1									4	0.80			4	0.37
PORC-2					1	1.27	1	0.55					2	0.19
PORC-3							1	0.55					1	0.09
PORC-4											1	3.70	1	0.09
STONE-1									3	0.60			3	0.28
TOTAL					**1**	**1.27**	**2**	**1.10**	**7**	**1.41**	**1**	**3.70**	**11**	**1.02**
(Indigenous)														
BSGW	27	36.49	101	46.12	35	44.30	102	56.04	190	38.23	6	22.22	461	42.76
BSRW-1	18	24.32	38	17.35	7	8.86	23	12.64	55	11.07			141	13.08
BSRW-2	1	1.35	8	3.65	7	8.86	2	1.10	5	1.01			23	2.13
GW-1									4	0.80	2	7.41	6	0.56
GW-2									1	0.20			1	0.09
MRW-1							1	0.55					1	0.09
MRW-2													0	
RSGW	2	2.70	1	0.46	4	5.06	8	4.40	27	5.43	4	14.81	46	4.27
RSPW-1					1	1.27							1	0.09
RSRW-1	12	16.22	47	21.46	17	21.52	23	12.64	49	9.86		0.00	148	13.73
RSRW-2	2	2.70	1	0.46	2	2.53	6	3.30	9	1.81	0	0.00	20	1.86
RW-1	3	4.05	7	3.20	3	3.80	8	4.40	63	12.68	8	29.63	92	8.53
RW-2			2	0.91	1	1.27			1	0.20	1	3.70	5	0.46
RW-3									1	0.20	1	3.70	2	0.19
SBW-1	1	1.35							8	1.61	1	3.70	10	0.93
SBW-2			3	1.37			1	0.55	4	0.80			8	0.74
SBW-3									1	0.20			1	0.09
WSRW														
TOTAL	**66**	**89.19**	**208**	**94.98**	**77**	**97.47**		**95.60**	**418**	**84.10**	**23**	**85.19**	**966**	**89.61**
(West Asian)														
BiGP									2	0.40	1	3.70	3	0.28
BW	2	2.70	2	0.91			1	0.55	1	0.20			6	0.56
BW(SPOT)									1	0.20			1	0.09
BW(TJ)			1	0.46									1	0.09
CPW														
CSGW														
ESW(R)														
ESW(W)									1	0.20			1	0.09
FSGW									3	0.60			3	0.28
GBW														
GRW														
HsG(B)									10	2.01	2	7.41	12	1.11
HsG(M)									2	0.40			2	0.19
HsG(P)									10	2.01			10	0.93
HsG(W)									1	0.20			1	0.09

Table 7.1 - Ceramic composition of Layers in TT4 (exclusive of SW) *(continued on the next page)*

continued from the previous page ...

TT4	Layer 6	%	Layer 5	%	Layer 4	%	Layer 3	%	Layer 2	%	Layer 1	%	Grand Total	% of trench
LISV-1														
LISV-4									1	0.20			1	0.09
LPW									1	0.20			1	0.09
LPW-1							1	0.55	1	0.20			2	0.19
SWGW									5	1.01			5	0.46
TGW-1			1	0.46					11	2.21			12	1.11
TGW-2							1	0.55	2	0.40			3	0.28
TGW-3			1	0.46			1	0.55					2	0.19
TGW-4							1	0.55	1	0.20			2	0.19
TGW-5			1	0.46					1	0.20			2	0.19
UGPW			1	0.46					1	0.20			2	0.19
UPW														
WGP									3	0.60			3	0.28
WGPW														0.00
WGW	1	1.35							12	2.41			13	1.21
WSPW	2	2.70											2	0.19
WSPW-1														
WSPW-2			1	0.46	1	1.27							2	0.19
TOTAL	**5**	**6.76**	**8**	**3.65**	**1**	**1.27**	**5**	**2.75**	**70**	**14.08**	**3**	**11.11**	**92**	**8.53**
(Unknown)														
MISC-10									1	0.20			1	0.09
MISC-3			1	0.46									1	0.09
MISC-4			1	0.46			1	0.55					2	0.19
MISC-5	1	1.35											1	0.09
MISC-6	2	2.70	1	0.46									3	0.28
MISC-8									1	0.20			1	0.09
TOTAL	**3**	**4.05**	**3**	**1.37**			**1**	**0.55**	**2**	**0.40**			**9**	**0.83**
L.TOTAL	**74**	100.00	**219**	100.00	**79**	100.00	**182**	100.00	**497**	100.00	**27**	100.00	**1078**	100.00

Table 7.1 - Ceramic composition of Layers in TT4 (exclusive of SW)

TT4 NE	Layer 6	%	Layer 5	%	Layer 4	%	Layer 3	%	Layer 2	%	Layer 1	%	Grand Total	% of trench
(Chinese)														
PORC-1									2	1.23			2	0.33
PORC-2					1	1.27	1	1.33					2	0.33
PORC-3							1	1.33					1	0.16
STONE-1									2	1.23			2	0.33
TOTAL					**1**	**1.27**	**2**	**2.67**	**4**	**2.47**			**7**	**1.15**
(Indigenous)														
BSGW	27	36.49	101	46.12	35	44.30	36	48.00	54	33.33	1	50.00	254	41.57
BSRW-1	18	24.32	38	17.35	7	8.86	8	10.67	19	11.73			90	14.73
BSRW-2	1	1.35	8	3.65	7	8.86	1	1.33	1	0.62			18	2.95
GW-1									1	0.62			1	0.16
MRW-1							1	1.33					1	0.16
RSGW	2	2.70	1	0.46	4	5.06	7	9.33	9	5.56			23	3.76
RSPW-1		0.00			1	1.27							1	0.16
RSRW-1	12	16.22	47	21.46	17	21.52	15	20.00	17	10.49			108	17.68
RSRW-2	2	2.70	1	0.46	2	2.53			6	3.70			11	1.80
RW-1	3	4.05	7	3.20	3	3.80	4	5.33	12	7.41			29	4.75
RW-2			2	0.91	1	1.27							3	0.49
SBW-1	1	1.35							3	1.85			4	0.65
SBW-2			3	1.37									3	0.49
TOTAL	**66**	**89.19**	**208**	**94.98**	**77**	**97.47**	**72**	**96.00**	**122**	**75.31**	**1**	**50.00**	**546**	**89.36**
(West Asian)														
BiGP									2	1.23	1	50.00	3	0.49
BW	2	2.70	2	0.91			1	1.33	1	0.62			6	0.98
BW(TJ)			1	0.46									1	0.16
FSGW									1	0.62			1	0.16
HsG(B)									6	3.70			6	0.98
HsG(M)									1	0.62			1	0.16
HsG(P)									6	3.70			6	0.98
SWGW									3	1.85			3	0.49
TGW-1			1	0.46					5	3.09			6	0.98
TGW-2									1	0.62			1	0.16
TGW-3			1	0.46									1	0.16
TGW-4									1	0.62			1	0.16
TGW-5			1	0.46									1	0.16
UGPW			1	0.46									1	0.16
WGP									1	0.62			1	0.16
WGW	1	1.35							7	4.32			8	1.31
WSPW	2	2.70											2	0.33
WSPW-2			1	0.46	1	1.27							2	0.33
TOTAL	**5**	**6.76**	**8**	**3.65**	**1**	**1.27**	**1**	**1.33**	**35**	**21.60**	**1**	**50.00**	**51**	**8.35**
(Unknown)														
MISC-10									1	0.62			1	0.16
MISC-2			1	0.46									1	0.16
MISC-3			1	0.46									1	0.16
MISC-4	1	1.35											1	0.16
MISC-5	2	2.70	1	0.46									3	0.49
TOTAL	**3**	**4.05**	**3**	**1.37**					**1**	**0.62**			**7**	**1.15**
L. TOTAL	**74**	**100.00**	**219**	**100.00**	**79**	**100.00**	**75**	**100.00**	**162**	**100.00**	**2**	**100.00**	**611**	**100.00**

Table 7.2 - Ceramic composition of Layers in TT4 NE

TT4 NW	Layer 3	%	Layer 2	%	Layer 1	%	Grand Total	% of trench
(Chinese)								
PORC-1			2	0.93			2	0.59
PORC-4					1	5.88	1	0.30
STONE-1			1	0.47			1	0.30
TOTAL			**3**	**1.40**	**1**	**5.88**	**4**	**1.18**
(Indigenous)								
BSGW	66	61.68	93	43.46	2	11.76	161	47.63
BSRW-1	15	14.02	30	14.02			45	13.31
BSRW-2	1	0.93					1	0.30
GW-1			1	0.47	2	11.76	3	0.89
GW-2			1	0.47			1	0.30
RSGW	1	0.93	14	6.54	3	17.65	18	5.33
RSRW-1	8	7.48	15	7.01			23	6.80
RSRW-2	6	5.61	2	0.93			8	2.37
RW-1	4	3.74	30	14.02	4	23.53	38	11.24
RW-2			1	0.47	1	5.88	2	0.59
RW-3			1	0.47	1	5.88	2	0.59
SBW-1			5	2.34	1	5.88	6	1.78
SBW-2	1	0.93	3	1.40			4	1.18
TOTAL	**102**	**95.33**	**196**	**91.59**	**14**	**82.35**	**312**	**92.31**
(West Asian)								
HsG(B)			4	1.87	2	11.76	6	1.78
HsG(P)			2	0.93			2	0.59
HsG(W)			1	0.47			1	0.30
LISV-4			1	0.47			1	0.30
LPW-1	1	0.93	1	0.47			2	0.59
TGW-1			1	0.47			1	0.30
TGW-2	1	0.93	1	0.47			2	0.59
TGW-3	1	0.93					1	0.30
TGW-4	1	0.93					1	0.30
TGW-5			1	0.47			1	0.30
UPW			1	0.47			1	0.30
WGPW			1	0.47			1	0.30
WGW			1	0.47			1	0.30
TOTAL	**4**	**3.74**	**15**	**7.01**	**2**	**11.76**	**21**	**6.21**
(Unknown)								
MISC-4	1	0.93					1	0.30
TOTAL	**1**	**0.93**					**1**	**0.30**
L.TOTAL	**107**	**100.00**	**214**	**100.00**	**17**	**100.00**	**338**	**100.00**

Table 7.3 - Ceramic composition of Layers in TT4 NW

TT4 SE	Layer 2	%	Layer 1	%	Grand Total	% of trench
(Indigenous)						
BSGW	43	35.54	3	37.50	46	35.66
BSRW-1	6	4.96			6	4.65
BSRW-2	4	3.31			4	3.10
GW-1	2	1.65			2	1.55
RSGW	4	3.31	1	12.50	5	3.88
RSRW-1	17	14.05			17	13.18
RSRW-2	1	0.83			1	0.78
RW-1	21	17.36	4	50.00	25	19.38
SBW-2	1	0.83			1	0.78
SBW-3	1	0.83			1	0.78
TOTAL	**100**	**82.64**	**8**	**100.00**	**108**	**83.72**
(West Asian)						
BW(SPOT)	1	0.83			1	0.78
ESW(W)	1	0.83			1	0.78
FSGW	2	1.65			2	1.55
HsG(M)	1	0.83			1	0.78
HsG(P)	2	1.65			2	1.55
LPW	1	0.83			1	0.78
SWGW	2	1.65			2	1.55
TGW-1	5	4.13			5	3.88
WGP	1	0.83			1	0.78
WGW	4	3.31			4	3.10
TOTAL	**20**	**16.53**			**20**	**15.50**
(Unknown)						
MISC-8	1	0.83			1	0.78
TOTAL	**1**	**0.83**			**1**	**0.78**
L.TOTAL	**121**	**100.00**	**8**	**100.00**	**129**	**100.00**

Table 7.4 - Ceramic composition of Layers in TT4 SE

here. The decreased numbers could, to some extent,be the result of distribution or occurrence patterns and sampling methodology (see List A). A sudden leap to 14.08% is seen in layer 2. The increase is not only in the quantity of pottery but also in the ware classes. A wide range of glazed as well as unglazed classes are seen in this layer. Open as well as closed vessel forms are present. The quantity of pottery in layer 2 in this trench is significantly high, making up for 46.10% of the trench total.

The ceramic composition of layers in the three individual quadrants of TT4 is given in tables 7.2, 7.3 and 7.4. Many wares have not been represented due to a lack of rim sherds in a particular layer. Hence it is necessary to refer to Lists A, B and C so that ware classes not represented or misrepresented in the analysis based on rim occurrence alone may be put in perspective.

TT4 NE provides data for the entire sequence from layer 6 to layer 1 since relatively less disturbance was seen in this quadrant (see Table 7.2 and List A). PORC-2 is seen as the earliest Chinese class, occurring in layer 4. However, it must be noted that body sherds of STONE-3 occur earlier from the upper depths of layer 5 and continue into layer 4 (see List A). Chinese wares continue through layers 3 and 2 showing an increased percentage. In layer 3 one can note that the Chinese and West Asian wares appear equal with 1.27% each. Apparently the introduction of West Asian wares to the site predates the introduction of Chinese wares since the former are seen from layer 6 onwards. Indigenous wares constitute 89.36% of the Quadrant assemblage. Of these, BSGW forms the most prominent class, increasing from 36.49% in layer 6 to 48% in layer 3. A sudden decrease is seen in layer 2 – 33.33%. Only two rim sherds were retained in layer 1, one of which was BSGW. It is no surprise then that it constitutes 50% of the quadrant total for this layer. However, a number of West Asian ware classes were present but are not seen in these tables as they were body and base sherds (see Table A). BSRW-1 shows a steady decrease through the sequence. RSRW-1 shows an increase from layers 6 to 3, decreasing to 10.49% in layer 2. The decrease in indigenous pottery in layer 2 is off-set by a corresponding increase in West Asian wares and, to a more limited extent, Chinese wares.

West Asian wares appear along with Indigenous wares from layer 6 onwards, constituting 8.35% of the quadrant assemblage. With the exception of one small WGW rim sherd, and some body sherds of TGW, all the other wares are unglazed in layer 6 (see List A). The WGW sherd is an intrusion. While WSPW is a sturdy utilitarian ware consisting of mostly storage and transportation forms, BW has a finer fabric and smaller vessel forms are seen. The presence of West Asian wares at this early level is significant, standing at 6.76% of the Quadrant total for this layer. A notable decrease is seen in layer 5, with the percentages of this group dropping to 3.65%. In reality, body sherds of WSPW-2, BW(TJ), and BW(SV) are present in large numbers but do not reflect in these tables. TGW appears more consistently in Layer 5 along with torpedo jars and BW. While West Asian wares appear to decrease steadily in layers 5, 4 and 3, Indigenous wares show an increase. Layer 2 shows a sudden and sharp increase in the quantity as well as variety of West Asian wares. The percentage rises to 21.60% of the layer total of the quadrant. This increase is reflected in other quadrants as well, though not as dramatically. A large number of glazed wares appear while the unglazed wares continue. Utilitarian forms as well as tablewares are represented. WGW is the most prominent with 4.32%, followed by HsG(P) and HsG(B) at 3.70% each. TGW-1 constitutes 3.09% of the layer total. Wares with pink fabric are prominently seen along with buff fabrics. SWGW makes up for 1.85% of the layer assemblage. BiGP continues from layer 2 into layer 1.

It must be noted that an early structural feature consisting of a double row of bricks was encountered at a depth of 2.02m in this quadrant and this reduced the excavation area. Hence the size of the excavated area after depth 2.02m is far smaller than above it. The data for quadrants TT4 NW and TT4 SE is presented in Tables 7.3 and 7.4 so as to provide a more comprehensive picture. Many ware classes not represented in TT4 NE may be seen in these quadrants.

TT4 NW was dug to layer 3. This is not to imply that the entire layer 3 was excavated in this quadrant. It implies that the excavation of this quadrant stopped at a depth of 1.16m in layer 3. The brick platform, designated Stucture 1 by the excavators, in this quadrant made it impossible to excavate fully. Most pottery collected from this quadrant came from the excavation work around the platform and the exposure of the platform itself. The quadrant's proximity to the well in TT4 SW should also be kept in mind. The data of this quadrant is presented in Table 7.3 and List B. Chinese wares are absent in layer 3 but occur in layer 2 (1.40%), increasing to 5.88% in layer 1. Indigenous wares form a bulk of the quadrant assemblage at 92.31%, although layer-wise a decreasing trend can be seen. As in the NE quadrant, black slipped wares predominate. BSGW makes up for 47.63% of the quadrant assemblage but shows a decrease from layer 3 to layer 1. BSRW remains somewhat consistent through layers 3 and 2, as does RSRW-1. It is interesting to note that RSGW shows an increase from 0.93% in layer 3 to

17.65% in layer 1. A similar increase is seen in TT4 SE (see Table 7.4), but not in TT4 NE. RW-1 remains the most prominent unslipped ware class, increasing from 3.74% in layer 3 to 23.53% in layer 1. The West Asian wares form 6.21% of the quadrant assemblage, increasing from 3.74% in layer 3 to 11.76% in layer 1. LPW-1 and TGW-2, TGW-3 and TGW-5 are present in layer 3. Layer 2 has a wider range of ware classes, both glazed as well as unglazed, constituting 7.01% of the quadrant assemblage. This is almost double the quantity of layer 3. HsG(B) is the only West Asian glazed ware in Layer 1.

TT4 SE was excavated to a depth of 1.18m, until a brick floor was encountered. The data from this quadrant is presented in Table 7.4 and List C. Most of the material in this quadrant comes from layer 2. Several unrecorded ware classes in layer 1, particularly Chinese and Unknown classes, are recorded in List C. As noticed earlier, among the Indigenous classes, the black slipped wares are significantly more in quantity than the red slipped ones. Among the slipped wares BSGW and RSGW-1 have the highest percentages in layer 2, at 35.54% and 14.05% respectively. RW-1 is the most prominent unslipped class (17.36%), increasing to 50% of the layer 1 collection in this quadrant. West Asian wares are only seen in layer 2, making up for 15.50% of the quadrant collection. WGW and TGW-1 are quantitatively higher than the other wares. Both glazed as well as unglazed wares. MISC-8 is the only Unknown ware seen in the collection. It occurs in layer 2.

Occurrence of ware classes by depth :
(Tables 8, 8.1 and 8.2 and Lists A, B and C)

All the above tables reflect the occurrence of rim sherds only and create the misimpression that a particular class did not occur or that it occurred at a specific depth. In order to present a far more truthful picture and to record the presence of ware classes at exact depths, it is necessary to list the occurrence all ware classes, regardless of diagnostic type, separately (see List A, B and C). Hence the focus of this section of the analysis takes into account ware classes not represented by rims in order to build a ceramic sequence. It is important to consider the occurrence of ware classes at varying depths in order to establish a pattern, if any, and to view these patterns in context of structures, features and any other known data. The following tables (Tables 8, 8.1 and 8.2) present the occurrence of ware classes as represented by rim sherds at different depths in TT4 NE as well as TT4 NW and TT4 SE so that the classes which do not occur in NE may be seen in the context of the other two quadrants. Lists A, B and C present the occurrence of ware classes, regardless of diagnostic type, depth-wise for each of the three quadrants and should be considered along with the tables in order to complete the picture.

While there is no doubt that the manner in which a vessel breaks may influence the numbers and types of sherds that occur, or that activities specific to an area may influence the occurrence of ware classes, it is significant to note the trend of ware classes as they appear, continue and stop through the sequence. Hence, the ware classes are recorded in Lists A, B and C without taking counts or percentage into account. Occurrence patterns of certain classes of pottery have been given special attention and have been discussed in more detail as they may be important in determining a ceramic sequence and suggestive of a tentative chronology.

Tables 8, 8.1 and 8.2 have been taken into consideration, along with Lists A, B and C. TT4 NE is the only quadrant which was excavated to a depth of 3.79m. The ceramic profile of this quadrant is considered in the following analysis, along with data from TT4 NW and TT4 SE. Observations made on the three quadrants are used to build the ceramic profile for the whole trench TT4.

Depth range of specific ware classes and patterns of occurrence :
(Tables 9, 9.1 and 9.2)

An analysis of tables 8, 8.1 and 8.2 and list A, B and C clearly shows the ceramic sequence and makes it possible to establish patterns of occurrence. Table 9, based on this analysis presents the depth ranges for some of the important West Asian ware classes in TT4, in an attempt to understand the sequence, and to ascertain if any specific ware classes can be seen to form a bracket. Table 9.1 presents the depth range for some of the prominent Indigenous classes. The depth range for some of the Chinese classes is given separately in Table 9.1, only so that they may be seen in the context of the other ware classes. No counts or percentages are taken into account in any of these tables. Some small sherds may have occurred as intrusions or due to disturbance, particularly close to the surface. In such cases the discrepancy has been noted and the next depth at which it has been recorded has been considered. Asterix marks indicate a single small body sherd which may be out of context. TT4 NE is the index quadrant. The depth range records the earliest depth at which a ware class makes an appearance and the last depth at which it was recorded, so that the extent of its continuity can be noted. Similar data from adjoining quadrants TT4 NW and TT4 SE are provided in the centre columns for cross reference. The column to the right takes all three quadrants into account and presents the lifespan of a ware class in the overall profile of the trench. The maximum depth recorded in each quadrant is mentioned as all quadrants were not excavated to the same depth.

It is clear from Lists A, B and C and from Tables 9, 9.1 and 9.2 that no glazed pottery or red slipped pottery occurs between 3.79m and 3.46m. While this may be, in some part, due to the pit noted by the excavators at the base of the quadrant (Gupta et al 2005: 59), the deposit above 3.56m does not show signs of disturbance as the pottery found in it shows a uniformity of occurrence and a gradual progression. Black slipped wares and WSPW-2 form the assemblage at the earliest level. The introduction of glazed wares to the sequence is seen in the appearance of TGW-1, TGW-4 and TGW-5 at 3.46m and almost simultaneously at 3.43m, the first red slipped sherds of RSRW-1 make their appearance. This is significant for two reasons – one, because TGW has a longer history than any of the other West Asian glazed wares and two, because none of these ware classes are seen in the lower deposit even as intrusions, leading one to conclude that the pit noted by the excavators did not start above this depth. Unglazed buff ware classes BW(TJ) and BW, also with a known long history, are introduced soon after at 3.38m, followed closely by BW(SV) at 3.32m. It is important to note that the unglazed West Asian ceramics are almost all sturdy, middle to large sized container vessels such as hole-mouthed jars, torpedo jars and storage vessels. BW is finer textured and has both utilitarian forms such as flasks and jars as well as smaller forms. The TGW seen in the early part of the sequence is also undecorated. TGW-3 and BW(SV) appear simultaneously. The glazed sherds do not appear as a regular occurrence. The unglazed ware classes are continuously seen. Another unglazed West Asian ware class is added to the repertoire at 3.14m – LISV-3. No major change is seen in the composition of the succeeding deposits for a long time. As mentioned earlier, the single small WGW sherd found at 2.78m is clearly an intrusion. (Not a single other sherd occurs anywhere in the deposit till 0.88m, i.e. almost 2m above.) The quantity of pottery shows a steady increase. Among the Indigenous wares, RSRW-2 and RW-1 are first seen at 2.98m. MISC-5 and MISC-6 are the two unknown classes that are seen till depth 2.75m. BW(SPOT) is introduced at 2.83m and is short-lived, disappearing at 2.47m. WSPW-1 and TGW-2 are seen for the first time at 2.35m. TGW-5 is not seen after 2.16m. The corpus of the collection does not significantly change a lot. RSGW is introduced at 2.38m and continues for a very long time after. The occurrence of TGW classes shows an increase after 2.35m. In the earlier deposit, they were present but not frequent. After 2.16m, their appearance is continuous and sustained.

One must note here that an early structural level was found at a depth of 2.02m. As mentioned earlier, this double row of bricks was resting on the deposit marked 2.02m. The thickness of the bricks has not been mentioned by the excavators (Gupta et al 2005: 59). If one considers that the thickness of the brick feature is approximately 10cm., then the deposit marked at approximately 1.90m is directly above and resting on the structure. This is important when one notes that Chinese wares are first noted at a depth of 1.86m. The ceramic assemblage associated with this structure does not show any major change from the previous deposit, other than the introduction of Chinese wares. It is important to note that the quantity of pottery and frequency of TGW as well as the unglazed classes in the deposit just before this feature show a significant increase. This is not clearly reflected in the earlier tables quantifying the pottery from quadrant TT4 NE and trench TT4 as these were based on rim sherds alone. The deposit immediately above the structure also does not show any change. The appearance of one small UGPW sherd at 1.96m is the only new factor. It is the first glazed sherd with a pink fabric. Unfortunately it is too badly eroded to identify with certainty. No glazed wares with pink fabric are seen after this for a long time, indicating that perhaps this small sherd may have been intrusive.

The first Chinese sherd to be found is STONE-3 at a depth of 1.86m and again at 1.80m, followed by PORC-2 at 1.73m. Chinese wares are seen quite regularly after this. PORC-3, a unique single sherd is found at 1.43m. The West Asian wares show no change, with the exception of ESW(W) at 1.66m. It continues through the deposit till about 0.60m, although two sherds of the ware are found in TT4 NW at 0.51m. LISV-1 is introduced at 1.55m.

The same trend continues till 1.24m when the first white glazed ware with buff fabric is found. This sherd belongs to CPW, an important and chronologically significant class which belongs to the Samarra Horizon group. In TT4 SE, this ware occurs at 1.16m along with WGW. From this point on, one sees the appearance of the Samarra Horizon classes. Another white glazed ware class, WGPW, though not having a buff fabric, is seen as early as 1.13m in TT4 NW. LPW-1 appears just after at 1.02m in TT4 NW. It does not occur in TT4 NE. It is significant to note that HsG(P) appears simultaneously with LPW-1 at 1.02m (The single small sherds of HsG(P) and HsG(B) found at 1.18m and 1.16m in TT4 SE may not signal the actual introduction of the ware as they may be intrusive) HsG(P) appears between 1.02m – 0.95m and HsG(B) between 0.99m and 0.95m. Their occurrence is continuous after this. HsG(M) and HsG(W) appear at 0.99m and 0.92m respectively. The last of the white glazed buff

	TT4,NE		CHINESE								INDIGENOUS											
LAYER	DEPTH	PORC-1	PORC-2	PORC-3	STONE-1		BSGW	BSRW-1	BSRW-2	GW - 1	MRW - 1	RSGW	RSPW-1	RSRW-1	RSRW-2	RW-1	RW-2	SBW-1	SBW-2		BiGP	BW
1	0.15m						1															
	0.19m																					
	0.22m																					
	0.25m																					
	0.29m																				1	
	L. TOTAL						**1**														**1**	
	%						50.00														50.00	
2	0.31m																					
	0.34m						2											1				
	0.39m						3					1										
	0.42m						3															
	0.46m																				1	
	0.49m	1												1		1		1				
	0.53m						2								1	1		1				
	0.55m						2									1						
	0.58m													1								
	0.60m						2	2														
	0.63m														2						1	
	0.65m						3					3				1						
	0.69m						2	4							1							
	0.72m						3	2				1		5	1							
	0.78m						1															
	0.80m						5	2						3								
	0.85m						2															
	0.88m						1	1				1		2		2						
	0.93m						5	2						3								
	0.95m	1							1					1		3						
	0.99m						3															
	1,86m																					
	1.03m						4	2		1		1										1
	1.05M				1		6	2				1										
	1.06m		1																			
	1.07m						1	1						1		1						
	1.09m				1		3	1				1			1	1						
	1.12m						1									1						
	1.19m																					
	L. TOTAL	**2**	**1**		**2**		**54**	**19**	**1**	**1**		**9**		**17**	**6**	**12**		**3**			**2**	**1**
	%	1.23	0.61		1.23		33.13	11.66	0.61	0.61		5.52		10.43	3.68	7.36		1.84			1.23	0.61

Table 8 - Occurance of ware classes (by depth) in TT4 NE

						WEST ASIAN													UNKNOWN					
	BW (TJ)	FSGW	HsG(B)	HsG(W)	HsG(P)	SWGW	TGW-1	TGW-2	TGW-3	TGW-4	TGW-5	UGPW	WGPW	WGW	WSPW	WSPW-2		MISC-10	MISC-3	MISC-4	MISC-5	MISC-6		TOTAL
																								1
																								1
																								2
																								100.00
					1																			1
			1																					4
			1											1										6
														1										4
																								1
																								4
																								5
					1								1	2										7
																								1
					1	1	1																	7
			1		1																			5
																								7
					1	1								1										10
						1	1																	14
		1																						2
			2																					12
																								2
							1							2										10
																								10
			1		1																			8
				1				1																5
							1																	10
																								10
																								1
										1														5
																		1						9
																								2
							1																	1
		1	**6**	**1**	**6**	**3**	**5**	**1**		**1**			**1**	**7**				**1**						**163**
		0.61	3.68	0.61	3.68	1.84	3.07	0.61		0.61			0.61	4.29				0.61						100.00

(continued on next double-spread pages)

	TT4,NE		CHINESE								INDIGENOUS											
LAYER	DEPTH	PORC-1	PORC-2	PORC-3	STONE-1		BSGW	BSRW-1	BSRW-2	GW - 1	MRW - 1	RSGW	RSPW-1	RSRW-1	RSRW-2	RW-1	RW-2	SBW-1	SBW-2		BiGP	BW
3	1.24cm																					
	1.24m											2										
	1.28m													2								
	1.34m						9		1			3		1								
	1.43m			1			6	4			1	1		8								
	1.50m						11					1		3		2						1
	1.55m						10	4						1		2						
	L. TOTAL			**1**			**36**	**8**	**1**		**1**	**7**		**15**		**4**						**1**
	%			1.35			48.65	10.81	1.35		1.35	9.46		20.27		5.41						1.35
4	1.60m						5	1					1	3	1							
	1.66m						3	1				1		4		1						
	1.70m						1					1		5	1	1						
	1.73m		1				9	2	3					2			1					
	1.77m						11	2				1		1								
	1.78m											1										
	L. TOTAL		**1**				**29**	**6**	**3**			**4**	**1**	**15**	**2**	**2**	**1**					
	%		1.54				44.62	9.23	4.62			6.15	1.54	23.08	3.08	3.08	1.54					
5	1.80m						6	1	4					2		1						
	1.86m						7	3	1					3	1	1	1					1
	1.91m						4					1		4		1						
	1.96m						14	3	2					5								
	1.99m																					
	2.02m						5	1														1
	2.05m						9	1	1					1		3	1		1			
	2.11m						4	1						4								
	2.16m						7							5								
	2.21m						6	2								1						
	2.27m						2	2	1													
	2.32m						5							1								
	2.35m						5	2						5								
	2.40m						1	9						2								
	2.43m						1	2	1					2								
	2.47m						10	4						4					1			
	2.53m						6	5														
	2.58m						6		2					1								
	2.62m						2							5								
	2.63m																		1			
	2.65m						4							1								
	2.70m						3	3						4		1						
	L. TOTAL						**107**	**39**	**12**			**1**		**49**	**1**	**8**	**2**		**3**			**2**
	%						45.92	16.74	5.15			0.43		21.03	0.43	3.43	0.86		1.29			0.86

Table 8 - Occurance of ware classes (by depth) in TT4 NE

						WEST ASIAN													UNKNOWN					
	BW (TJ)	FSGW	HsG(B)	HsG(W)	HsG(P)	SWGW	TGW-1	TGW-2	TGW-3	TGW-4	TGW-5	UGPW	WGPW	WGW	WSPW	WSPW-2		MISC-10	MISC-3	MISC-4	MISC-5	MISC-6		TOTAL
																								2
																								2
																								14
																								21
																								18
																								17
																								74
																								100.00
																								11
																								10
																								9
																1								19
																								15
																								1
																1								**65**
																1.54								100.00
																								14
																								18
																								10
												1												25
									1															1
																						1		8
							1																	18
																								9
											1					1								14
																								9
																								5
																								6
																								12
																								12
																			1					7
	1																							20
																								11
																								9
																								7
																								1
																								5
																				1				12
	1						**1**		**1**		**1**	**1**				**1**			**1**	**1**		**1**		**233**
	0.43						0.43		0.43		0.43	0.43				0.43			0.43	0.43		0.43		100.00

(continued on next double-spread page)

	TT4,NE		CHINESE								INDIGENOUS											
LAYER	DEPTH	PORC-1	PORC-2	PORC-3	STONE-1		BSGW	BSRW-1	BSRW-2	GW - 1	MRW - 1	RSGW	RSPW-1	RSRW-1	RSRW-2	RW-1	RW-2	SBW-1	SBW-2		BiGP	BW
6	2.75m						1	3														1
	2.78m						1					1		1								
	2.83m						5	4				1		3				1				
	2.88m													1	1	1						
	2.93m													2								
	2.98m													2	1	2						
	3.06m						2	2	1					1								
	3.09m						1															
	3.14m						2	1														
	3.18m						5	1						1								
	3.28m																					
	3.32m						6															1
	3.38m						1															
	3.43m													1								
	3.46m							7														
	3.51m						1															
	3.56m						2															
	3.60-3.79m																					
	3.60m																					
	L. TOTAL						**27**	**18**	**1**			**2**		**12**	**2**	**3**		**1**				**2**
	%						36.49	24.32	1.35			2.70		16.22	2.70	4.05		1.35				2.70
	Grand Total	**2**	**2**	**1**	**2**		**254**	**90**	**18**	**1**	**1**	**23**	**1**	**108**	**11**	**29**	**3**	**4**	**3**		**3**	**6**
	%	0.33%	0.33%	0.16%	0.33%		41.57%	14.73%	2.95%	0.16%	0.16%	3.76%	0.16%	17.68%	1.80%	4.75%	0.49%	0.65%	0.49%		0.49%	0.98%

Table 8 - Occurance of ware classes (by depth) in TT4 NE

						WEST ASIAN													UNKNOWN					
	BW (TJ)	FSGW	HsG(B)	HsG(W)	HsG(P)	SWGW	TGW-1	TGW-2	TGW-3	TGW-4	TGW-5	UGPW	WGPW	WGW	WSPW	WSPW-2		MISC-10	MISC-3	MISC-4	MISC-5	MISC-6		TOTAL
														1	2									6
																								14
																								3
																								2
																								5
																						2		8
																								1
																								3
																								7
																								7
																								1
																								1
																								7
																								1
																					1			3
														1	**2**						**1**	**2**		**74**
														1.35	2.70						1.35	2.70		100.0
	1	**1**	**6**	**1**	**6**	**3**	**6**	**1**	**1**	**1**	**1**	**1**	**1**	**8**	**2**	**2**		**1**	**1**	**1**	**1**	**3**		**611**
	0.16%	0.16%	0.98%	0.16%	0.98%	0.49%	0.98%	0.16%	0.16%	0.16%	0.16%	0.16%	0.16%	1.31%	0.33%	0.33%		0.16%	0.16%	0.16%	0.16%	0.49%		
	100.00%																							

Table 8 - Occurance of ware classes (by depth) in TT4 NE

Tr.Quad	TT4, NE		CHINESE								INDIGENOUS						
LAYER	DEPTH	PORC-1	PORC-4	STONE-1		BSGW	BSRW-1	BSRW-1	GW-1	GW-2	RSGW	RSRW-1	RSRW-2	RW-1	RW-2	RW-3	SBW-1
1	0.13m													1			
	0.17m		1								1			1			1
	L. TOTAL		**1**								**1**			**2**			**1**
	%		**14.29**								**14.29**			**28.57**			**14.29**
2	0.23m					2			2		2			2	1	1	
	0.28m					5					2			1			
	0.30m					7	3				1			4			1
	0.36m	1				10	3							4			
	0.39m					4	4				2			2			
	0.41m					9	2			1				1			
	0.44m	1				4					2						1
	0.48m					1											
	0.51m					4	3					4		3		1	
	0.54m																
	0.57m					1							1	3			
	0.59m					1					1	2					
	0.63m					1						2		1	1		
	0.65m					4						1					
	0.66m					2	1				1	1		1			1
	0.67m					2	3					1					
	0.72m			1		1					1			1			
	0.75m					3											
	0.78m					1	1										
	0.80m						1					1		1			
	0.82m					14	4				2	1		2			
	0.84m					10			1					1			1
	0.86m					1	3										
	0.89M					5	2					2		1			1
	L. TOTAL	**2**		**1**		**92**	**30**		**3**	**1**	**14**	**15**	**1**	**28**	**2**	**2**	**5**
	%	**0.95**		**0.47**		**43.60**	**14.22**		**1.42**	**0.47**	**6.64**	**7.11**	**0.47**	**13.27**	**0.95**	**0.95**	**2.37**
3	0.92m					3					2		1	4			
	0.95-1.06m																
	0.95m					8	3					1	2				
	0.99m					10	1					2		1			
	1.02m					14	7				1			1			
	1.04m					11	2	1				3	3	1			
	1.06m					4							1	1			
	1.13m					19	2					2					
	TOTAL					**69**	**15**	**1**			**3**	**8**	**7**	**8**			
	%					**57.5**	**12.5**	**0.83**			**2.5**	**6.67**	**5.83**	**6.67**			
	Grand Total	**2**	**1**	**1**		**161**	**45**	**1**	**3**	**1**	**18**	**23**	**8**	**38**	**2**	**2**	**6**
	%	**0.59%**	**0.30%**	**0.30%**		**47.63%**	**13.31%**	**0.30%**	**0.89%**	**0.30%**	**5.33%**	**6.80%**	**2.37%**	**11.24%**	**0.59%**	**0.59%**	**1.78%**

Table 8.1 - Occurance of ware classes (by depth) in TT4 NW - *(View this table on a double-spread page)*

			WEST ASIAN														UNKNOWN		
	SBW-2		HsG(B)	HsG(P)	HsG(W)	LISV-4	LPW-1	TGW-1	TGW-2	TGW-3	TGW-4	TGW-5	UPW	WGPW	WGW		MISC-4		1R Total
			1																2
			1																5
			2																**7**
			28.57																**100.00**
	1																		11
																			8
																			16
																			18
																			12
																			13
			1																9
				1				1											3
			1			1	1							1	1				20
																			5
																			4
			1																6
													1						6
																			7
																			6
																			4
																			3
									1										3
																			3
			1	1								1							26
																			13
																			4
																			11
	1		**4**	**2**		**1**	**1**	**1**	**1**			**1**	**1**	**1**	**1**				**211**
	0.47		**1.90**	**0.95**		**0.47**	**0.47**	**0.47**	**0.47**			**0.47**	**0.47**	**0.47**	**0.47**				**100.00**
	3				1														14
																	1		1
																			14
										1									15
							1												24
											1								22
																			6
									1										24
	3				**1**		**1**		**1**	**1**	**1**						**1**		**120**
	2.5				**0.83**		**0.83**		**0.83**	**0.83**	**0.83**						**0.83**		**100.00**
	4		**6**	**2**	**1**	**1**	**2**	**1**	**2**	**1**	**1**	**1**	**1**	**1**	**1**		**1**		**338**
	1.18%		**1.78%**	**0.59%**	**0.30%**	**0.30%**	**0.59%**	**0.30%**	**0.59%**	**0.30%**	**0.30%**	**0.30%**	**0.30%**	**0.30%**	**0.30%**		**0.30%**		**100.00%**

Table 8.1 - Occurance of ware classes (by depth) in TT4 NW - *(View this table on a double-spread page)*

	Tr. Quad			**INDIGENOUS**									
	TT4, SE												
LAYER	**DEPTH**	**BSGW**	**BSRW-1**	**BSRW-2**	**GW-1**	**RSGW**	**RSGW-1**	**RSGW-2**	**RW-1**	**SBW-2**	**SBW-3**		
1	0.27m												
	L. TOTAL												
	%												
2	0.37m												
	0.42m	3				1			4				
	0.49m	1	2			2			5				
	0.53m						2						
	0.57m	1											
	0.62m	1							2				
	0.66m	2		2							1		
	0.73m	12					2		2	1			
	0.79m	3					1						
	0.7m	2		2				1	1				
	0.83m												
	0.88m	1				1	2		2				
	0.90m	1											
	0.92m	2					3						
	0.94m	2											
	0.98m	3							1				
	1.02m	5	1		2		2		1				
	1.04m	2	2				1		3				
	1.08m												
	1.11m	3											
	1.16m	2	1			1	3		4				
	1.18m						1						
	L. TOTAL	**46**	**6**	**4**	**2**	**5**	**17**	**1**	**25**	**1**	**1**		
	%	**35.66**	**4.65**	**3.10**	**1.55**	**3.88**	**13.18**	**0.78**	**19.38**	**0.78**	**0.78**		
	Grand Total	**46**	**6**	**4**	**2**	**5**	**17**	**1**	**25**	**1**	**1**		
		35.66%	**4.65%**	**3.10%**	**1.55%**	**3.88%**	**13.18%**	**0.78%**	**19.38%**	**0.78%**	**0.78%**		

Table 8.2 - Occurance of ware classes (by depth) in TT4 SE - *(View this table on a double-spread page)*

			Tr. Quad								UNKNOWN		
BW(SPOT)	**ESW(W)**	**FSGW**	**HsG(M)**	**HsG(P)**	**LPW**	**SWGW**	**TGW-1**	**WGP**	**WGW**		**MISC-8**		**1R Total**
													8
								1					11
													2
													1
1									1				5
					1								6
													17
						2	4		2				12
													6
				1									7
													1
													5
													2
													4
	1		1										13
				1									9
													3
		2					1		1				15
											1		2
1	**1**	**2**	**1**	**2**	**1**	**2**	**5**	**1**	**4**		**1**		**129**
0.78	**0.78**	**1.55**	**0.78**	**1.55**	**0.78**	**1.55**	**3.88**	**0.78**	**3.10**		**0.78**		**100.00**
1	**1**	**2**	**1**	**2**	**1**	**2**	**5**	**1**	**4**		**1**		**129**
0.78%	**0.78%**	**1.55%**	**0.78%**	**1.55%**	**0.78%**	**1.55%**	**3.88%**	**0.78%**	**3.10%**		**0.78%**		**100.00%**

Table 8.2 - Occurance of ware glasses (by depth) in TT4 SE - *(View this table on a double-spread page)*

TT4, NE **Recorded Ware Classes (by depth)**

LAYER	DEPTH	CHINESE	INDIGENOUS	WEST ASIAN	UNKNOWN
1	0.15m		BSGW	HsG(B)	
	0.19m	STONE-5		HsG(B)	
	0.22m	STONE-4			
	0.25m	STONE-4, STONE-3			
	0.29m	STONE-6		BiGP	
2	0.31m	STONE-5		HsG(B), HsG(P), UPW	
	0.34m	PORC-1, STONE-6	SBW-1, BSGW, GRW	HsG(B), HsG(P), TGW-1	
	0.39m	STONE-6	BSGW, RSGW,BSRW-1	TGW-4, WGW, MGP, HsG(B) LISV-4	
	0.42m		BSGW	WGW, LISV-2, LISV-4, LISV-5	
	0.46m	STONE-6		MGB,BiGP, HsG(B), HsG(P), HsG(W), LISV-4,WSPW-2	
	0.49m	STONE-6, PORC-1	RSRW-1, SBW-1, RW-1	WGW,WGPW, BW, LISV-2	
	0.53m		GRW, SBW-1, BSGW, RSRW-2,RW-1	LISV-4	MISC-9
	0.55m	STONE-6	BSGW, RW-1	WGW, WGPW, HsG(P), UPW	
	0.58m	STONE-2, STONE-6	RSRW-1	WGW, SWGW	
	0.60m		BSGW, BSRW-1, RSRW-1	TGW-1, SWGW, CPW, HsG(P), UGPW, ESW(W), BW(TJ)	
	0.63m		RSRW-1	TGW-4, WGW, HsG(B), HsG(B), BiGP	
	0.65m		BSGW, RSRW-1, RSGW, RW-1	MGP	
	0.69m		BSGW, BSRW-1, RSRW-2,	WGW, SWGW, HsG(P), BW(TJ), LISV-4	
	0.72m		BSGW, RSGW, RSRW-1, RSRW-2, BSRW-1,	SWGW	
	0.78m		BSGW, RSGW		
	0.80m	STONE-2	BSGW, RSRW-1 BSRW-1	TGW-1, SWGW, CPW, HsG(B), HsG(M), GBW, WSPW-2, BW(TJ), LISV-1, LISV-3, LISV-5	
	0.85m	STONE-6	BSGW		
	0.88m		BSGW, RSGW, RSRW-1, BSRW-1, RW-1,	WGW, TGW-1, LISV-1, LISV-5	
	0.93m		BSGW, RSRW-1, BSRW-1	TGW-1, BW(TJ), LISV-4	
	0.95m	PORC-1	BSGW, BSRW-2, RSRW-1, RW-1	BiGP, TGW-2, HsG(P), HsG(B), HsG(M), BW(TJ), BW(SV), LISV-3, WSPW-2, BW	MISC-7
	0.99m		BSGW	TGW-2, TGW-4, HsG(M), BW	MISC-1, MISC-4
	1.03m		BSGW, BSRW-1, RSGW, GW-1	TGW-1, TGW-3, TGW-4, BW, LISV-5	
	1.05m	STONE-1, STONE-2	BSGW, BSRW-1, RSGW	TGW-1, TGW-4, BiGP, LISV-3	
	1.07m		BSGW, BSRW-1, RSRW-1, RW-1	TGW-1, TGW-4, WSPW-2	MISC-7, MISC-9
	1.09m	STONE-1, STONE-2	BSGW, BSRW-1, RSGW, RSRW-2, RW-1	TGW-1, TGW-4, BW(TJ), BW(SV), BW,	MISC-10
	1.12m		BSGW, RW-1	WSPW-2	MISC-7
	1.19m		RW-1	TGW-1, BW(TJ)	
3	1.24m		RSGW, RSRW-1,	CPW, TGW-1, BW(TJ), LISV-2, LISV-4, WSPW-2	
	1.28m		RSRW-1	LISV-2, LISV-5	MISC-6
	1.34m		BSGW, RSGW, RSRW-1, BSRW-2,	TGW-1, TGW-4, BW(TJ), ESW(W)	
	1.43m	PORC-3	BSGW, RSGW, BSRW-1,MRW-1	TGW-1, TGW-4, BW(TJ), ESW(W), BW	MISC-9
	1.50m		BSGW, RSGW, RSRW-1, RW-1	TGW-1, TGW-2, TGW-4, BW, WSPW-2	MISC-9
	1.55m	STONE-3	BSGW, RSRW-1, BSRW-1, RW-1	TGW-1, TGW-4, BW(TJ), BW, WSPW-1, UPW, LISV-1	MISC-9
4	1.60m	STONE-3	BSGW, RSRW-1, RSRW-2, BSRW-1,RSPW-1	TGW-3, TGW-4, WSPW-2	
	1.66m	STONE-3	BSGW, RSGW, RSRW-1, RW-1	TGW-1, TGW-4, ESW(W), WSPW-2	MISC-4, MISC-6
	1.70m		BSGW, RSGW, RSRW-1, RSRW-2,		
			BSRW-1, RW-1	TGW-4, WSPW-2, LISV-2	MISC-6, MISC-9
	1.73m	PORC-2	BSGW, RSRW-1, BSRW-1, BSRW-2, RW-2, GW-2,	TGW-2, TGW-4, BW(SV), WSPW-2	MISC-7, MISC-9
	1.77m		BSGW, RSRW-1, BSRW-1, RSGW	BW(TJ), WSPW-2,	MISC-4, MISC-6, MISC-9

List A - Occurrence of ware classes in TT4 NE (all diagnostics without count)

continued on the next page

continued from the last page ...

LAYER	DEPTH	CHINESE	INDIGENOUS	WEST ASIAN	UNKNOWN
5	1.80m	STONE-3	BSGW, BSRW-2, RSRW-1, RW-1	TGW-4, BW(TJ), BW(SV), BW, WSPW-2,	MISC-7, MISC-9
	1.86m	STONE-3	BSGW, BSRW-1, BSRW-2, RSRW-1,	TGW-1, TGW-2, TGW-4, BW(TJ),	
			RW-1, RW-2	BW(SV), WSPW-2	
	1.91m		BSGW, RSGW, RSRW-1, RW-1	TGW-3, TGW-4 BW(TJ), BW(SV), BW, WSPW-2	
	1.96m		BSGW, RSRW-1, BSRW-1, RW-1	TGW-3, TGW-4, UGPW, BW(TJ),	
				BW(SV), BW, WSPW-1, WSPW-2	
	1.99m			TGW-3	
	2.02m		BSGW, BSRW-1, RW-1	TGW-1, TGW-4, BW(TJ), BW(SV), BW, WSPW-2	MISC-6
	2.05m		BSGW, BSRW-1, BSRW-2 RSRW-1, RW-1, RW-2, SBW-2, SBW-4	TGW-1, TGW-4, BW, WSPW-1, WSPW-2	
	2.11m		BSGW, BSRW-1, RSRW-1	TGW-1, TGW-3, GBW, BW(TJ), BW, WSPW-2, WSPW-3,	MISC-7
	2.16m		BSGW, RSRW-1	TGW-4, TGW-5, BW(TJ), BW(SV), BW, WSPW-2, WSPW-3	MISC-2
	2.21m		BSGW, BSRW-1, RW-1	WSPW-2	
	2.27m		BSGW, BSRW-1, RSRW-1	BW(TJ), BW(SV), WSPW-2, WSPW-3,	
	2.32m		BSGW,RSRW-1,	BW(TJ), BW(SV), WSPW-2	MISC-3
	2.35m		BSGW, RSRW-1, BSRW-1	TGW-2,TGW-3, BW(TJ), BW(SV), BW, WSPW-1,WSPW-2	
	2.40m		BSGW, RSRW-1, BSRW-1	BW(SV), BW	
	2.43m		BSGW, RSRW-1, BSRW-1,BSRW-2	BW(TJ), BW(SV), BW,WSPW-2,	MISC-3
	2.47m		BSGW, RSRW-1, BSRW-1	BW(SPOT), BW(SV), WSPW-2,	MISC-3, MISC-5
	2.53m		BSGW, BSRW-1	TGW-3, BW(TJ), BW(SPOT), BW(SV), WSPW-2	MISC-5
	2.58m		BSGW, RSRW-1, BSRW-1	TGW-3, WSPW-2	
	2.62m		BSGW, RSRW-1, RW-2, GW-1	BW(SPOT), BW(SV), WSPW-2,	MISC-9
	2.63m		SBW-2		
	2.65m		BSGW, RSRW-1, RSGW	BW(SV), BW, LISV-3	MISC-4
	2.70m		BSGW, RSRW-1, BSRW-1, RW-1	BW(SV), BW, WSPW-2	MISC-4
6	2.75m		BSGW, BSRW-1	BW(SV), BW,	MISC-4, MISC-7
	2.78m		BSGW, RSGW, RSRW-1,	WGW,WSPW-2	
	2.83m		BSGW, RSGW, RSRW-1, BSRW-1, SBW-1	BW(SV), BW(SPOT), WSPW-2, WSPW-1, LISV-3	MISC-5, MISC-6
	2.88m		RSRW-1, RSRW-2, RW-1	BW(SV)	
	2.93m		BSGW, RSRW-1,	WSPW-2, BW(SV)	
	2.98m		RSRW-1, RSRW-2, RW-1		
	3.06m		BSGW, BSRW-1, BSRW-2, RSRW-1, GW-1,	BW(SV) WSPW-2	MISC-6
	3.09m		SBW-2	BW(SV), WSPW-2	
	3.14m		BSGW, BSRW-1	BW(TJ), BW(SV), UPW, LISV-3	MISC-6
	3.18m		BSGW, RSRW-1, BSRW-1	BW(SV), WSPW-2	
	3.32m		BSGW BW, WSPW-2	TGW-3, BW(TJ), BW(SV),	
	3.38m		BSGW,	BW(TJ), BW	
	3.43m		RSRW-1	WSPW-2	MISC-6
	3.46m		BSRW-1	TGW-1, TGW-4, TGW-5, WSPW-2	
	3.51m		BSGW	WSPW-2	
	3.56m		BSGW,	WSPW-2	MISC-5
	3.60-3.79m		BSRW-2		

List A - Occurrence of ware classes in TT4 NE (all diagnostics without count)

Tr.Quad LAYER	TT4, NW DEPTH	CHINESE	INDIGENOUS	WEST ASIAN	UNKNOWN
1	0.13m	PORC-1, PORC-2	RW-1	HsG(B)	
	0.17m	PORC-1,PORC-4, CEL-1	RSGW, SBW-1, RW-1	HsG(B), LISV-1, LISV-3	
2	0.23m		BSGW, RSGW, GW-1, RW-1, RW-2, RW-3	LISV-2	
	0.28m		BSGW, RSGW, RW-1		
	0.30m		BSGW, RSGW, BSRW-1, RW-1, SBW-1	HsG(B)	
	0.36m	PORC-1	BSGW, BSRW-1, RW-1	LISV-5	
	0.39m	PORC-1	BSGW, RSGW, BSRW-1, RW-1	HsG(P), LISV-5	
	0.41m		BSGW, BSRW-1, RW-1, GW-2		
	0.44m	PORC-1, CEL-1,	BSGW, RSGW, SBW-1,	WGW, HsG(B), BiGP, BW(TJ), WSPW-3, LISV-5	
	0.48m		BSGW,	TGW-1, HsG(P), LISV-1, LISV-5	
	0.51m		BSGW, RSRW-1, BSRW-1,	TGW-1, TGW-3, TGW-6, LPW-1,	
			RW-1, RW-3,	CPW, WGW, HsG(W), HsG(B), HsG(M), BiGP, WGPW, WSPW-3, ESW(W), UPW, LISV-2, LISV-3, LISV-4	
	0.54m			TGW-1, LISV-5	MISC-7
	0.57m		BSGW, RSRW-2, RW-1,	TGW-1, HsG(P)	
	0.59m		BSGW, RSGW, RSRW-1	TGW-1, ESW(W), WSPW-1,	MISC-9
	0.63m		BSGW, RSRW-1, RW-1, RW-2	TGW-1, SWGW, HsG(B), HsG(P), BW(TJ), LISV-2, LISV-3, LISV-4	
	0.65m		BSGW, RSRW-1, SBW-2, SBW-3	WGW,UPW	
	0.66m	CHANG	BSGW, RSGW, RSRW-1, BSRW-1, RW-1, SBW-1,	TGW-1, HsG(P), BW(TJ), BW	
	0.67m		BSGW, RSRW-1, BSRW-1	TGW-1, TGW-4, CPW, BW, BW(TJ),	
				WSPW-1, LISV-3, LISV-5	
	0.72m	STONE-1	BSGW, RSGW, RW-1	TGW-1, TGW-2, BW, HsG(P), LISV-1, LISV-2	
	0.75m	STONE-2	BSGW	TGW-1, TGW-4, LISV-1, LISV-2, LISV-5	
	0.78m	STONE-2	BSGW, BSRW-1,	TGW-2, WGW, BW, LISV-3	
	0.80m	STONE-2	BSRW-1, RSRW-1, RW-1	TGW-1, TGW-4, BW(SV)	MISC-4
	0.82m	PORC-1	BSGW, RSGW, RSRW-1, BSRW-1, RW-1	TGW-1, TGW-4, TGW-5, WGW, BW(TJ), HsG(P), HsG(M), HsG(B), WSPW-1, WSPW-2WGPW, LISV-1	
	0.84m	PORC-1	BSGW, BSRW-1, RW-1,	TGW-1, TGW-4, BW(TJ), BW,	
			GW-1, SBW-1	ESW(W), LISV-5	
	0.86m	STONE-4	BSGW, BSRW-1	TGW-4, BW(TJ), BW	
	0.89M	PORC-1, STONE-1	BSGW, RSRW-1, BSRW-1, RW-1, SBW-1	TGW-1, BW(TJ), ESW(W), LISV-1, LISV-3, LISV-4, LISV-5	
3	0.92m		BSGW, RSGW, RSRW-2, RW-1, SBW-2, SBW-4	TGW-1, TGW-2, BW(TJ), BW, ESW(W), HsG(W)	MISC-4
	0.95m		BSGW, RSRW-1, RSRW-2, BSRW-1	TGW-4	MISC-4, MISC-7
0.99m	STONE-2	BSGW, RSRW-1, BSRW-1, RW-1	TGW-1, TGW-3, BW, HsG(P), HsG(B)		
	1.02m		BSGW, BSRW-1, RSGW, RW-1 HsG(P), WSPW-2	TGW-1, TGW-3, TGW-4, LPW-1,	
	1.04m	STONE-3	BSGW, RSRW-1, BSRW-1, BSRW-2, RW-1	TGW-1, TGW-3, TGW-4, BW, LISV-1	MISC-4
	1.06m		BSGW, RSRW-2, RW-1,	TGW-1	
	1.13m		BSGW, RSRW-1, BSRW-1, RW-1	TGW-2, TGW-3, BW(TJ), WGPW	

List B - Occurrence of ware classes in TT4 NW (all diagnostics without count)

Tr.Quad LAYER	TT4, SE DEPTH	CHINESE	INDIGENOUS	WEST ASIAN	UNKNOWN
1	0.27m	PORC-5, STONE-4			MISC-6
2	0.37m	CEL-1			MISC-6, MISC-10
	0.42m	PORC-1	BSGW, RSGW, SBW-1, SBW-3, RW-1		
	0.49m		BSGW, BSRW-1, RSGW, RW-1	WGW, BiGP	
	0.53m		RSRW-1		MISC-11
	0.57m		BSGW, RSGW		
	0.62m		BSGW, RW-1	WGW, ESW(P), BW(SPOT)	MISC-10
	0.66m		BSGW, BSRW-2, SBW-3	WGW, LPW-1, BiGP, WSPW-1	
	0.73m		BSGW, BSRW-2, RW-1, RSRW-1, SBW-2, RSGW	TGW-1, TGW-2, WGW, LISV-1	
	0.79m		BSGW, RSRW-1	HsG(B), HsG(P), WGPW, WSPW-2, WSPW-3, BW(TJ), SWGW, WGW, TGW-1, TGW-4	
	0.83m			TGW-1, WGW, BW(TJ)	
	0.88m		BSGW, RSGW, BSRW-1, RSRW-1, RW-1, RW-2	HsG(P), LISV-3	
	0.90m		BSGW	WGPW	
	0.92m		BSGW, RSRW-1	TGW-1	
	0.94m		BSGW	TGW-1, TGW-4, BW(TJ), UPW	
	0.98m		BSGW, RW-1	HsG(P), LPW-1, TGW-1	
	1.02m	PORC-1	BSGW, RSRW-1, BSRW-1, GW-1, RW-1	TGW-1, TGW-4, HsG(B), HsG(M), BW(TJ), LISV-2	
	1.04m	PORC-1	BSGW, RSRW-1, BSRW-1, RW-1	TGW-1, HsG(B), HsG(P), LISV-1	
	1.08m			TGW-1, HsG(P)	
	1.11m		BSGW		MISC-7
	1.16m		BSGW, RSGW, RSRW-1, BSRW-1, RW-1	CPW, TGW-1, TGW-4, WGW, HsG(P), HsG(B), ESW(W), WSPW-2	
	1.18m		RSRW-1	TGW-2, HsG(B), BW(TJ)	MISC-8

List C - Occurrence of ware classes in TT4 NW (all diagnostics without count)

Depth range of specific ware classes and patterns of occurrence :

	TT4 NE	TT4 NW	TT4 SE	TT4
Ware Class	*(Maximum depth = 3.79m)* **Depth range**	*(Maximum depth = 1.16m)* **Depth range**	*(Maximum depth = 1.18)* **Depth range**	*(Maximum depth = 3.79m)* **Overall Depth Range**
WSPW-1	2.35m - 1.55m	0.82m - 0.59m	0.66m	2.35m - 0.59m
WSPW-2	3.56m - 0.80m	1.02m - 0.82m	1.66m - 0.79m	3.56m - 0.79m
WSPW-3	2.27m - 0.79m	0.79m	0.79m	2.27m - 0.79m
BW	3.38m - 0.95m	1.04m - 0.66m		3.38m - 0.66m
BW(TJ)	3.38m - 0.60m	1.13m - 0.44m	1.18m - 0.79m	3.38m - 0.44m
BW(SV)	3.32m - 0.95m	0.80m		3.32m - 0.80m
BW(SPOT)	2.83m - 2.47m		0.62m*	2.83m - 2.47m
ESW(W)	1.66m - 0.60m	0.92m - 0.51m*	1.16m - 1.02m	1.66m - 0.60m
LISV-1	1.55m - 0.80m	0.89m - 0.48m	1.04m - 0.73m	1.55m - 0.48m
LISV-2	1.70m - 0.42m	0.75m - 0.51m	1.02m	1.70m - 0.51m
LISV-3	3.14m - 0.80m	0.89m - 0.51m	0.88m	3.14m - 0.51m
LISV-4	1.24m - 0.39m	0.63m - 0.51m		1.24m - 0.39m
LISV-5	1.28m - 0.42m	0.89m - 0.36m		1.28m - 0.36m
TGW-1	3.46m - 0.60m	1.06m - 0.48m	1.16m - 0.73m	3.46m - 0.48m
TGW-2	2.35m - 0.95m	1.13m - 0.72m	1.18m - 0.73m	2.35m - 0.72m
TGW-3	3.32m - 1.03m	1.13m - 0.51m		3.32m - 0.51m
TGW-4	3.46m - 0.34m*	1.04m - 0.67m	1.16m - 0.79m	3.46m – 0.67m
TGW-5	3.46m - 2.16m	0.82m*		3.46m - 2.16m
TGW-6		0.51m		0.51m
CPW	1.24m - 0.60m	0.67m - 0.51m	1.16m	1.24m - 0.51m
LPW-1		1.02m - 0.51m	0.98m	1.02m - 0.51m
WGW	0.88m - 0.39m	0.82m - 0.44m	1.16m - 0.62m	1.16m - 0.39m
SWGW	0.79m - 0.58m	0.63m	0.79m	0.79m - 0.58m
BiGP	1.05m - 0.29m*	0.51m - 0.44m	0.66m - 0.49m	1.05m - 0.44m
WGPW	0.55m - 0.49m	1.13m - 0.51m	0.90m - 0.79m	1.13m - 0.49m
HsG(B)	0.95m - 0.15m	0.99m - 0.13m	1.18m* - 0.79m	0.99m - 0.13m
HsG(P)	0.95m - 0.31m	1.02m - 0.39m	1.16m* - 0.79m	1.02m - 0.31m
HsG(M)	0.99m - 0.80m	0.82m - 0.51m	1.02m	1.02m - 0.51m
HsG(W)	0.46m	0.92m - 0.51m		0.92m - 0.46m

Table 9 Depth range of important West Asian ware classes in TT4
(Asterix marks indicate a single small body sherd which may be out of context)

	TT4 NE	TT4 NW	TT4 SE	TT4
Ware Class	*(Maximum depth = 3.79m)* **Depth range**	*(Maximum depth = 1.16m)* **Depth range**	*(Maximum depth = 1.18m)* **Depth range**	*(Maximum depth = 3.79m)* **Overall Depth Range**
BSGW	3.56m - 0.15m	1.13m - 0.23m	1.16m - 0.42m	3.56m - 0.15m
BSRW-1	3.46m - 0.39m	1.13m - 0.30m	1.16m - 0.49m	3.46m - 0.30m
BSRW-2	3.60/3.79m - 0.95m	1.04m	0.70m - 0.66m	3.60/3.79m - 0.66m
RSGW	2.83m - 0.39m	1.02m - 0.17m*	1.16m - 0.42m	2.38m - 0.39m
RSRW-1	3.43m - 0.49m	1.13m - 0.51m	1.18m - 0.53m	3.43m - 0.49m
RSRW-2	2.98m - 0.53m	1.06m - 0.57m	0.70m	2.98m - 0.53m
RW-1	2.98m - 0.49m	1.06m - 0.13m	1.16m - 0.42m	2.98m - 0.13m
RW-2	2.62m - 1.73m	0.63m - 0.23m	0.88m	2.62m - 0.23m
RW-3		0.51m - 0.23m		0.51m - 0.23m
GW-1	3.06m - 1.03m	0.84m - 0.23m*	1.02m	3.06m - 1.02m
GW-2	1.73m*	0.41m*		

Table 9.1 Depth range of certain Indigenous ware classes in TT4

	TT4 NE	TT4 NW	TT4 SE	TT4
Ware Class	*(Maximum depth = 3.79m)* **Depth range**	*(Maximum depth = 1.16m)* **Depth range**	*(Maximum depth = 1.18m)* **Depth range**	*(Maximum depth = 3.79m)* **Overall Depth Range**
STONE-1	1.09m - 1.05m	0.89m - 0.72m		1.09m - 0.72m
STONE-2	1.09m - 0.80	0.99m - 0.75m		1.09m - 0.75m
STONE-3	1.86m - 1.55m	1.04m		1.86m - 1.04m
STONE-4	0.25m - 0.22m	0.86m*	0.27m	0.27m - 0.22m
STONE-5	0.31m - 0.19m			0.31m - 0.19m
STONE-6	0.85m - 0.25m			0.85m - 0.25m
CHANG		0.66m*		0.66m*
PORC-1	0.95m - 0.34m	0.89m - 0.13m*	1.04m - 0.42m	1.04m - 0.34m
PORC-2	1.73m - 1.06m	0.13m*		1.73m - 1.06m
PORC-3	1.43m*			1.43m
PORC-4		0.17m*		0.17m*
PORC-5			0.27m*	0.27m*

Table 9.2 Depth range of certain Chinese ware classes in TT4
(Asterix marks indicate a single small body sherd which may be out of context)

wares to appear is SWGW at 0.79m. From the time the white glazed classes and the sgraffiato wares appear, the quantity of pottery shows a sudden and sharp increase in all the quadrants. The unglazed wares are still seen but in less quantity than in the lower deposits. They appear to be tapering off. The Indigenous wares continue without too much variation, with black slipped wares apparently more dominant than the red. Interestingly, a number of Chinese wares make an appearance at the same depths as the Samarra Horizon classes and the sgraffiato wares. STONE-3 does not appear after 1.04m, but STONE-1 and STONE-2 appear at 1.09m and PORC-1 at 0.95m. STONE-6 appears at 0.85m.

By 0.80m the unglazed ware classes like WSPW-2, WSPW-3 and BW(SV) totally stop. BW(TJ) seems to continue for a while but in reduced numbers. TGW-2 also stops at 0.72m although TGW-1, TGW-3 and TGW-4 continue. It must be noted that the brick platform in TT4 NW, designated Structure 1by the excavators, was located at the base of layer 2 (Gupta et al, 2005: 58). Since layer 2 in this quadrant is marked up to 0.89m, it is presumed that the base of the platform rests on it. Depth recorded from surface to the top of the platform is 0.51m. The floor in TT4 SE which was encountered at a depth of 1.18m may be considered slightly earlier than the platform. It is unclear if thetwo are directly associated. It appears that the platform may be of a slightly later date. Both features clearly belong to the same depths as the Samarra Horizon ware classes and the sgraffiato classes.

Almost all the white glazed wares cease at a about 0.58m – 0.51m, with the exception of WGW which continues a little longer till 0.39m.ESW(W) alsostops at 0.51m. Interestingly, HsG(M) and HsG(W) also cease by about 0.46m, while HsG(P) stops at 0.31m. TGW classes totally cease after 0.48m. BiGP also ends by 0.44m, the same as BW(TJ) Most LISV classes also stop 0.51m with the exception of LISV-5 which continues till 0.36m and two small body sherds close to the surface. It must be noted that the only sherd of CHANG occurs at 0.66m in TT4 NW. Among the Indigenous classes BSRW-2 ends at 0.66m and RSRW-2 at 0.53m. RSRW-1 stops at 0.49m while RSGW is not seen after 0.39m. BSRW-1 is last found at 0.30m. The only wares to continue into the uppermost part of the deposit are BSGW, RW-1, RW-2 and RW-3. STONE-5 appears at 0.31m and continues till 0.19m. PORC-1stops at 0.31m. The only wares to occur in the uppermost deposits are STONE-4, STONE-5, STONE-6 and HsG(B), along with the indigenous classes. Most miscellaneous wares are short lived with the exception of MISC-4 and MISC-6. Almost no miscellaneous wares occur after 0.53m or so.

It is possible to identify three distinct patterns of ceramic occurrence on the basis of the analysis given above.

1. From 3.79m to 1.25m – Introduction of Indigenous wares, West Asian unglazed wares and TGW, most of the miscellaneous wares. In the later part of this deposit, the introduction of Chinese wares.

2. From 1.24m – 0.39m – Introduction of the Samarra Horizon wares, sgraffiato wares, other glazed West Asian wares with buff and pink fabrics and continuation of ware classes from earlier deposit. Continuation of Chinese wares and introduction of some new classes. Continuation of Indigenous wares. Gradual decrease and ending of unglazed West Asian wares, TGW, and other glazed classes with pink and buff fabric. End of the Samarra Horizon wares and some sgraffiato wares.

3. From 0.38m to Surface: End of HsG(P) in the early part of this deposit. Continuation of HsG(B) to the surface. No other West Asian wares. Very few Indigenous wares (mostly black slip). Continuation of Chinese wares (different classes). Overall decrease in pottery.

The association of these three patterns of occurrence can be associated with the structures and features in the trench. Three levels can be clearly recognized.

Level I : This level extends from the lowest depths (3.79m) up to the introduction of white glazed ware classes or the Samarra Horizon classes and sgraffiato wares. The earliest part of this level is marked by mostly indigenous black slipped wares and an unglazed West Asian ware, WSPW-2. This short spell is immediately followed by the introduction of glazed ware TGW-1, TGW-4 and TGW-5. Other unglazed ware classes, mostly buff fabric, red slipped indigenous wares and other classes are introduced and the quantity of pottery continues to increase steadily. No major changes are seen in this level. TGW which was infrequent in the initial part of this level increases tremendously in the later part and remains the only glazed ware through this level. The West Asian wares are mostly unglazed utilitarian ware classes for storage and transportation. Most Indigenous wares are introduced during this level. Several Miscellaneous classes also appear but do not seem to have continuity for any considerable time.

Level I (a) : The later part of this level is marked by two significant factors – the presence of the early structural feature in TT4 NE at 2.02m and the introduction of Chinese wares (STONE-3 and PORC-2) in the deposit above it. No other major changes are seen in the assemblage. The quantity of TGW shows an increase. But no other glazed wares are seen. This later part of the Level, marked by the arrival of Chinese wares and the structure which this deposit is associated may be considered as Level I(a), starting at approximately 2.02m and extending up to 1.25m.

Level II : This level begins at approximately 1.24m with the introduction of the white glazed ware classes or the Samarra Horizon classes, followed soon after by the sgraffiato ware classes and other glazed wares with pink fabric. The Indigenous ware classes continue from Level I into this level, as do most of the other West Asian wares. The unglazed classes and TGW from Level I taper off and stop in the later part of this level. Structure 1, the brick platform in TT4 NW, the cobble and brick floor in TT4 SE and probably the well associated with it, belong to this level. STONE-3 stops at 1.04m but other Chinese wares are introduced. These include STONE-1,STONE-2,STONE-4,STONE-6,PORC-1 and CHANG. All West Asian glazed wares, with the exception of two sgraffiato classes, stop by about 0.39m. Most Indigenous ware classes also stop at about the same time.

Level III : This level begins at approximately 0.38m and extends to the surface. It is marked by a total decrease in pottery. Very few of the Indigenous wares from the earlier levels are seen. BSGW, BSRW-1, RW-1 and RW-2 are seen but no red slipped wares are present. No West Asian ware classes occur, other than HsG(B) which shows a continued presence. HsG(P) ends by 0.31m. STONE-4 and STONE-6 from Level II end by 0.22m. STONE-5, PORC-4 and PORC-5 are seen as single occurrences only in these uppermost deposits close to the surface.

A brief note on the chronology of ware types :

West Asian ware classes which are considered chronologically specific are briefly reviewed in this section to establish, if possible, a tentative chronology for the Levels mentioned above. Chinese wares cannot be used in this study as their identification is still to be conclusively done. The ware classes discussed here may be considered the ceramic markers for each Level. (see Chapter II for detailed descriptions and discussions on ware classes and suggested chronology at other sites). While it is true that not all scholars agree upon the dating of wares, this discussion only aims to present the chronological framework within which a ware has been bracketed. It is in no way a comprehensive discussion. The wares are taken up for discussion in order of their appearance in the Levels mentioned above. The depth at which they first appear is mentioned.

Level I : WSPW-2 : (depth: 3.56m) This ware class occurs in the earliest deposit in TT4 and continues through Level I and Level II. It is a utilitarian ware which mainly consists of container vessels like hole-mouthed jars and amphorae. At 'Ana it is dated to the late Sasanian – Umayyad period, continuing into the Abbasid period (Northedge et al 1988: 77, 82).

TGW : (depth: 3.46m) This is one of the most prominent wares found at Sanjan. It starts very early in the sequence and continues through Level I and most of Level II. This ware has a very long tradition, stretching from Partho-Sasanian times to the Early Islamic period. Northedge and Kennet discuss monochrome glazing from Partho-Sasanian times and specifically identify the type that occur in Islamic strata earlier than the Samarra Horizon or on sites abandoned before its introduction, as havinga deep turquoise to sea green color. They are of the opinion that the ware is Pre-Samarra Horizon or post-Sasanian, extending through the Samarra Horizon up to the Ottoman period.(Northedge and Kennet1994: 21 - 23). This observation of color is significant considering that the TGW sub-classes that occur in the early levels of TT4 are TGW-1, TGW-4 and TGW-5. It is also significant that the barbotine decoration on some of these is noted by the authors. At 'Ana, Qal'a Island (Iraq) it is associated with the Samarra Horizon and dated to c. 836 – 892 AD (Northedge et al 1988: 83, 91). Whitehouse suggests that it belongs to Period I, c. 803-4 at Siraf. (Whitehouse 1971: 4). At Bhanbore in Pakistan it is dated to the Sasanian period (Pakistan Archaeology 1; 1964). TGW has been reported from Manda, Shanga and Kilwa (see Chapter II) and has been dated to between the 8th and the 10th or 11th centuries.

BW : (depth: 3.38m) Buff ware has a long history, going back to pre-Islamic times. Costa and Wilkinson are of the opinion that the ware can be dated to between 800 and 1100 AD (Costa and Wilkinson1987:187). Kevran reports similar finds to those at Sanjan from Suhar. She suggests that they go back to Sasanian times (3rd century to early 7th century). According to her, sherds with barbotine designs and other decorations could be Sasanian, while others could belong to the Early Islamic period. (Kevran 2004: fig11no.19. 278– 79). At Manda, Chittick reports this ware along with gudulia and eggshell wares as belonging to Period I. He doubts the early date (late

7th and early 8th century) proposed by Rosen-Ayalon at Susa but agrees that some of the motifs go back to the Umayyad period or even earlier (Chittick 1984:94). Horton calls these wares Fine Creamwares at Shanga, suggesting that they were made at Siraf. He dates them to the 8th century by corroborating the dates with Siraf and Susa (Horton 1996: 297-98).

BW(TJ): (depth: 3.38m) These torpedo jars also have a very long history going back to the Sasanian period. Adams makes a distinction between the Sasanian and the Islamic types, noting that the Sasanian ones have more elongated and conical bodies while the Islamic ones are more globular (Adams 1970:100). Kevran reports them from Suhar and it is worth noting at this point that one of the jars reported by her in layers 12 – 14 Level 5 appears to be an exact match to Bs.77.1 found at Sanjan (TT4 NE, depth = 2.43m). Kevran dates it to about 670 – 890 AD (Kevran 2004: fig 24 no.2. 307). While it is possible that the illustrated vessel and the Sanjan sherd may not necessarily be as similar as it seems, the early dates for torpedo jars is interesting.

ESW(W) : (depth: 1.66m) This is one of the ware classes considered as Pre-Samarra Horizon. It extends into the period of the Samarra Horizon. Northedge places this class with CPW, BW (with barbotine design), TGW, SWGW, and early Monochromes to the last fifteen years of the 9th century (Northedge 1996: 240 – 44). At 'Ana it has been placed in Group 4, dated to the early Abbasid period, 836 – 892 AD (Northedge et al 1988: 90 – 91). Kevran finds it at Suhar Level V, layer 10 and dates it to 860 -890 AD. At Susa, she dates it to between 700 and 900 AD. Costa and Wilkinson suggest that bases with raised centers were current in the 9th century at Nishapur but by the 10th century they had developed the flat base form (Costa and Wilkinson 1987: 187).

All the wares mentioned above function as markers of Level I and I (a). Level I belongs to a period prior to the introduction of Samarra Horizon wares. Given the thickness of the deposit that makes up this Level and the stratigraphy which shows continuous occupation with no breaks, it may be possible to assign this Level to the 8th century AD and perhaps even to the late 7th century AD. Almost all the West Asian ceramics that occur in this level have an antiquity that goes back to the late Sasanian period. The bulk of these wares represent container, storage and transport vessels, suggestive of the movement of goods and commercial contact with West Asia. Level I may be said to date between the late 7th or early 8th century to the late 8th century AD.

The arrival of Chinese wares marks Level I(a). These wares cannot be used as chronological indicators in this study as their identification is yet to be done. However, considering that the West Asian wares provide evidence for a late 7th – early 8th century date for the start of Level I, it may be reasonable to suggest a late 8th century AD to early or middle 9th century AD date for Level I(a).

Level II : CPW : (depth: 1.24m) According to most scholars, CPW is one of the earliest wares belonging to the Samarra Horizon. While the debate on exact dates for the Samarra Horizon wares and the order of their appearance still continues, it must be noted that many scholars had grouped most white glazed wares together, whether plain, splashed or painted. This has created some confusion regarding the order of their introduction. According to Kevran, CPW is dateable to the mid 8th century at Susa (Kevran 1977: 89). She suggests that the blue on white painted versions at Susa may appear around 750 AD and continue till 880 AD, and that other colors, such as brown, start around 830 AD (Kevran 2004:315). Tampoe places them at the end of the 8th century at Siraf (Tampoe 1989: 88). Mason and Keall assign it to Period 4 at Siraf, dateable from 850 – 950 AD (Mason and Keall 1991: 57). Northedge and Kennet are of the opinion that it went out of production soon after the founding of Samarra in 836 AD (Northedge and Kennet 1994: 25).

WGW : (depth: 1.16m) White glaze has been one of the important markers of the Samarra Horizon. It has been frequently debated whether the plain white glazed wares predate the painted and splashed ones or if it the other way around. A large number of scholars have placed both the plain as well as the decorated varieties in the same class and provided a date for the group. At Siraf the ware is placed in Phase A, c. 850 AD. At Manda it is dated to the mid 9th century (Chittick 1984: 77) and at Kilwa to Period Ia (9th century–c.100-1000 AD (Horton 1996: 277 -78). Carswell and Pritchett report it from Mantai in Phase 5, dateable to the 9th–10th century (Carswell and Pritchett 1974: 303). It is interesting to note that WGPW, a white glazed ware with pink fabric appears almost simultaneously with WGW.

BiGP : (depth: 1.05m) This is one of the ware classes with a pink fabric. The color splashes on the glaze have been the distinguishing characteristic of this ware, leading to scholars referring to it as Splashed Ware. Most scholars believe it should be included in the Samarra Horizon group of wares and it is seen as a precursor to the sgraffiato wares. Northedge places these wares in the 9th century and the ones with sgraffiato in the 10th century. (Northedge 1994: 33 – 34)At Sanjan this ware occurs just slightly before the sgraffiato wares. Northedge and Kennet suggest it was introduced before 861 AD at Samarra (Northedge and Kennet 1994 : 33). At Tell Abu Sarifa,this ware is found in Level V (800–950 AD) (Adams 1970:108,119). Watson is of the opinion that the splashed wares belong to the 9th century and the splashed wares with sgraffiato belong to the 9th – 10th century (Watson 2004: 47 – 48).

LPW-1 : (depth: 1.02m) This ware is not as commonly found as TGW or the sgraffiato classes at sites outside the Persian Gulf and the Middle East region. It is one of the important ceramic markers for the Samarra Horizon. Mason lists it as one of the elite ceramics in the group 1 and suggests that Basra was the sole producer of LPW between 800 and 975 AD (Mason 1997: 103 – 107). Northedge and Kennet date monochrome lustre to between 885 and 895 AD at Samarra (Northedge and Kennet 1994: 32 – 33). Kennet suggests three stages for the Samarra Horizon, placing LPW in stage 3. While Kennet, Northedge and Whitehouse all seem to agree that LPW and early sgraffiato are introduced after 885–895 AD, Kevran places it slightly earlier at 861 in Susa (Kennet 2004: 31–32).

HsG(P) and HsG(M) (depth: 1.02m), HsG(B) (depth: 0.99m) and HsG(W) (depth: 0.92m) : the first two sgraffiato classes appear simultaneously with LPW in TT4, and the latter follow almost immediately. Northedge and Kennet are of the opinion that sgraffiato forms part of the late Samarra Horizon assemblage, dateable to the early 10th century (Northedge and Kennet 1994: 32 – 33). Grube also believes that the transition from Splashed wares to sgraffiato takes place in the early tenth century. He suggests that they went out of production in the Gulf by 1150 AD (Grube 1994: 119, 121). Mason and Keall place it between 1050 and 1150 AD (Mason and Keall 1991: 62 - 63). Kevran reports them from Levels V and VI at Suhar, dating them to 860 – 1100 AD (Kevran 2004: 315). Whitehouse refers to hatched sgraffiato as Style III or late sgraffiato (Whitehouse 1979: 49 – 61). At 'Ana it is placed in Group 6 along with LPW and Splashed ware or BiGP, dateable to the late 10th and early 11th century (Northedge et al 1988: 94). Adams dates it to 950 – 1100 AD in Level VI at Tell Abu Sarifa (Adams 1970: 8 – 10).

SWGW : (depth: 0.79m) The grouping together of all white glazed wares, painted or splashed and sometimes even plain, has meant that many scholars have attributed a single date for all the different types. For example Kennet places both splashed as well as painted types in one class and assigns them a date prior to plain white glazed ware, i.e.after 803–4 and before 835-6 AD (Kennet 2004: 31–33). However, Mason and Keall assign it to Period 5 at Siraf, dated to c. 950 – 1050 AD, later than WGW (Mason and Keall 1991: 57). Kevran dates it to c. 830 AD at Suhar and places CPW earlier at c.750 AD, based on the evidence from Susa (Kevran 2004: 315). At Sanjan it appears after the introduction of sgraffiato wares.

The Samarra Horizon wares and sgraffiato wares are the markers of Level II. Most scholars agree that the white glazed Samarra Horizon wares belong to the early or middle 9th century AD and extend into the early 10th century AD, by which time graffiato wares become current. Hatched sgraffiato ware continues into the middle 12th century. Level II is very clearly marked by the Samarra Horizon wares and graffiato. The introduction and continuation of these wares in Level II, and their disappearance in the upper part of this Level, clearly indicate an early or middle 9th century AD to middle or late 12th century AD date for this level

Level III : No chronologically relevant ware classes occur in this level, with the exception of HsG(B) which continues from the previous level and which has been discussed above.

The deposit above Level II indicates a brief period of continued occupation which is marked by the absence of almost all the West Asian ware classes seen in the earlier levels. No new West Asian or Indigenous wares appear. Some Chinese wares and some Indigenous wares with black slip are the only components of this Level. No specific ceramic markers can be identified. Level III belongs to a period when the Samarra Horizon had definitely ended and the hatched sgraffiato wares had almost all but stopped, with only one type, HsG(B) still in circulation. It may be considered the terminal occupation of Sanjan Bandar. The dates suggested for Level III are middle to late 12th century AD,extending perhaps into the early 13th century AD.

The suggested periodisation of these levels has been set out in Table 10 along with the ceramic composition of each Level.

Suggested chronology for Indigenous vessel forms : (Tables 11, 11.1, 11.2 and 11.3)

Indigenous wares are notoriously difficult to date given their long, and frequently unchanging, history. They constitute a bulk of the assemblage. The shapes of three important indigenous wares at Sanjan, BSGW, BSRW-1 and RSRW-1 are listed in Tables 11, 11.1 and 11.2 in order to note the morphological and stylistic changes which may be chronologically relevant. As noted in the

Suggested Chronology :

Level	Period	Ceramic Assemblage
Level III (0 – 0.38m)	Middle/Late 12th century AD To Early 13th century AD	PORC-1, PORC-4, PORC-5, STONE-4, STONE-5, STONE-6, CEL-1, HsG(B) BSGW, BSRW-1, RSGW, SBW-1, RW-1, RW-2, RW-3, GRW
Level II (0.39m – 1.24m)	Early/Middle 9th century AD To Middle/Late 12th century AD	STONE-2, STONE-6, PORC-1, BSGW, BSRW-1, RSGW, RSRW-1, RSRW-2, SBW-1, SBW-2, RW-1, RW-2, RW-3, GW-1, GW-2, TGW-1, TGW-2, TGW-3, TGW-4, TGW-5, CPW, WGW, LPW-1, SWGW, BiGP, WGPW, HsG(P),HsG(B),HsG(M), HsG(W), MGP, MGB, UGPW, ESW(W), ESW(P), BW, BW(TJ),BW(SV), WSPW-1, WSPW-2,WSPW-3 MISC-1, MISC-4, MISC-6, MISC-7, MISC-8, MISC-9, MISC-10,MISC-11
Level I (a) (1.25 m - 2.02m)	Middle/Late 8th century AD To Early/Middle 9th century AD	STONE-3, PORC-2, PORC-3, BSGW, BSRW-1, BSRW-2, RSGW, RSRW-1, RSPW-1, RW-1, RW-2, GW-2, MRW-1TGW-1, TGW-2, TGW-3, TGW-4, BW, BW(TJ),BW(SV), WSPW-1, WSPW-2, ESW(W), LISV-1, LISV-2, LISV-5MISC-4, MISC-6, MISC-7, MISC-9
Level I (2.03m - 3.79m)	Late 7th/Early 8th century AD To Middle/Late 8th century AD	BSGW, BSRW-1, BSRW-2, RSGW, RSRW-1, RSRW-2, RW-1, RW-2, GW-1, SBW-2, SBW-4, TGW-1,TGW-4,TGW-3,TGW-5, BW, BW(TJ), BW(SV), BW(SPOT), WSPW-1, WSPW-2, LISV-3MISC-2, MISC-3, MISC-4, MISC-5, MISC-6,MISC-7, MISC-9

Table 10 Periodisation and ceramic composition of the Levels in TT4

Suggested chronology for Indigenous vessel forms : (Tables 11, 11.1, 11.2 and 11.3)

BSGW

Level	New Shapes	Continuing Shapes
Level III (0 – 0.38m)	T.221, T.224, T.237, D.50, D.63	T.82,T.91,T.136,T.157, T.184, T.220, D.58
Level II (0.39m – 1.24m)	T.80, T.84, T.90, T.96.1, T.98, T.100, T.102, T.106, T.107, T.113, T.119, T.121, T.125, T.127, T.127.1 T.129, T.132, T.138, T.139, T.155, T.156, T.157, T.158, T.160 T.161, T.162, T.174, T.179, T.192.1, T.193, T.202, T.207, T.208, T.211, T.212, T.222, T.225, T.236, T.309, T.312, T.410 Bs.37 D.43, D43.2, D.53.3, D.55.1, D.55.4, D.62, D.56,	T.81, T.82, T.83, T.85, T.87, T.88, T.89, T.91, T.92, T.92.1, T.92.2, T.93, T.95, T.95.1, T.96, T.97, T.104, T.110, T.112, T.135, T.136, T.140.1, T.149, T.180, T.188, T.220, T.223, D.43.1, D.43.2, D.43.3
Level I (a) (1.25m - 2.02)	T.86, T.96, T.122, T.135, T.136, T.140.1, T.178, T.188, T.191, T.220, T.223, T.230, T.308, T.311, T.407 D.43.3, D.71, D.77.1	T.87, T.88, T.89, T.92.1, T.92.2, T.95, T.95.1, T.110, T.147, T.180 T.406 D.43.1, D.43.2
Level I (2.03 - 3.79m)	T.81, T.82, T.83, T.85, T.87, T.88, T.89, T.91, T.92, T.92.1, T.92.2, T.92.3, T.94, T.95, T.95.1, T.97, T.99, T.104, T.110, T.112, T.144, T.146, T.147, T.148, T.149, T.160, T.180, T.168, T.183, T.184, T.193, T.195, T.198, T.406 D.43.1, D.43.2, D.43.4, D.58	

Table 11 Occurrence of BSGW vessel shapes in TT4

BSRW-1

Level	New Shapes	Continuing Shapes
Level III (0 – 0.38m)	T.312, T.309 Bs.253.2	T.132, T.179, T.221
Level II (0.39m – 1.24m)	T.88, T.100, T.122, T.132, T.135, T.136,T.140,T.147,T.152,T.174, T.178, T.220, T.221, T.237, T.310, T.315 D.55.1, D.55.3, D.74	T.147, T.155, T.179, T.230
Level I (a) (1.25m – 2.02)	T.125, T.131, T.144, T.179, T.196 T.200, T.222, T.236.1, T.242 D.57.1, D.76	T.195
Level I (2.03 – 3.79m)	T.95, T.95.1, T.147, T.155, T.177, T.180, T.195, T.196, T.207, T.212, T.230, T.231, T.240, T.245 D.55.4	

Table 11.1 Occurrence of BSRW-1 vessel shapes in TT4

RSRW-1

Level	New Shapes	Continuing Shapes
Level III (0 – 0.38m)		
Level II (0.39m – 1.24m)	T.127, T.186, T.231, T.250, T.252.1, T.258, T.309, T.312, T.412, T.414.1, T.414.2 Bs.41, Bs.82	T.144, T.202
Level I(a)(1.25m – 2.02) T.119, T.113, T.135, T.139,	T.95.1, T.96.1, T.105, T.109, T.196, T.229, T.415.1 T.143, T.144, T.162, T.166.3 T.169, T.174, T.176, T.184, T.197, T.200, T.202, T.204, T.209, T.211, T.214, T.236, T.238, T.255, T.258.2, T.258.3 T.318, T.321 Bs.39.1, D.47, D.55.2, D.64	T.122, T.123, T.128, T.129, T.13,
Level I(2.03 – 3.79m)	T.122, T.123, T.128, T.129, T.130, T.161, T.167, T.195, T.196, T.201, T.229,T.259, T.415.1, T.435 D.71	

Table 11.2 Occurrence of RSRW-1 vessel shapes in TT4

SHAPES	OTHER WARES
T.95.1	BSGW
T.113	BSGW
T.119	BSGW, BSRW-1
T.122	BSRW-1, BSRW-2, RW-1
T.123	RSRW-2
T.122	RW-1, BSRW-1
T.126	RW-1
T.127	RSRW-2
T.129	WSRW
T.135	BSGW, BSRW-1
T.143	RW-1
T.144	BSGW, BSRW-1, RW-3
T.161	RW-1, BSGW
T.162	RW-1,RW-2, BSGW
T.174	RSGW, BSRW-1, BSGW, RSRW-2
T.176	BSRW-1
T.184	BSGW, RSGW
T.186	RW-1
T.195	BSRW-1, BSRW-2, RW-1
T.196	BSRW-1, BSRW-2,BSGW, RSRW-2, RW-1
T.197	RSGW, RSRW-2, RW-1
T.200	RW-1
T.202	BSGW
T.209	BSGW, BSRW-1
T.211	BSGW, BSRW-1
T.220	BSGW, BSRW-1
T.229	RSGW
T.231	RSGW, RW-1, BSGW, BSRW-1, BSRW-2
T.236	BSGW
T.309	BSGW, BSRW-1
T.312	BSGW, BSRW-1, RSRW-2
T.321	BSGW
T.412	RW-1
T.415.1	BSGW
Bs.41	GRW
D.71	RW-1, BSGW

Table.11.3 RSRW-1 shapes which occur in other wares

previous chapter, many shapes are frequently common to more than one class.

Shapes common to both the black slipped wares are :

T.88, T.95, T.95.1, T.144, T.147, T.100, T.122, T.135, T.136, T.140.1, T.158, T.178, T.220, T.230, T.236, T.125, T.132, T.135, T.136, T.155, T.180,T.207, T.212, T.222, T.309 and T.312.

A very large number of the RSRW-1 shapes are common to the other slipped and unslipped classes. These are listed in Table.11.3. A number of the shapes are present in both the red as well as the black slipped wares.

The common shapes in different ware classes imply a common tradition and probably a common source. The difference in the grey and the red fabric is mostly due to human intervention. The petrographic analysis places the fabrics in the same Petrofabric group (Chapter II, and Appendix iv). The provenance for these wares may then be the same, having the same clay source. The shapes can be assigned dates on the basis of the Levels in which they first occur. Associated non-indigenous wares help in providing a date for specific shapes. Some shapes like T.95.2, T.95, etc. appear to have a very long tradition, starting from the earliest levels and continuing through the sequence. Others seem to have a very brief life.

Sample	BS	Age of sample based on the value of T1/2 = 5570 ± 30 YBP	Calibrated age YR.BP	Calibrated age range YR.BP
SB 0401 TT4 NE 2.0m	2539	1190 ± 90	1120	1260 - 970
SB 0402 TT4 NE 0.55m - 0.50m	2546	850 ±	740	920 - 670

Radiocarbon dates for trench TT4

Radiocarbon dates :

The Radiocarbon dates presented above have been made available to the author by the world Zarathushti Cultural Foundation

The Radiocarbon dates provided by the Birbal Sahni Institute of Palaeobotany, Lucknow for the Sanjan samples are presented in Table 10. Many scholars have reservations as to the reliability of Carbon dating and its accuracy. However, these dates are presented here in order to place the chronology of the ceramic profile and the identified Levels in perspective. Sample no. SB 0401 which came from TT4, NE depth 2.0m provides a date c. 830 AD and sample no. SB 0404 from TT4 NE depth 0.55–0.50m provides the datec.1210 AD when calculated. These dates do not appear altogether unrealistic in the context of the Levels that have been defined for the site. An early 9th century date for the later part of Level I is not far from the dates suggested by the ceramic assemblage – a late 8th and early 9th century date. This sample is collected from the deposit directly associated with the early structure and provides a date for it. The deposit below this may then be assigned to an earlier period spanning from the early 8th century (perhaps even a little earlier) to late 8th century. The date provided by sample no. SB 0404 appears to be slightly later than the date suggested by the ceramic assemblage. This sample comes from depth 0.55m – 0.50m, the depths at which the Samarra Horizon wares are ending and a little before the sgraffiato wares taper off. The Samarra Horizon wares are not reported to continue as long as the early 13th century at any site. Accepting this date would also imply that the deposit overlying this depth belongs to approximately the middleof the 13th century. The ceramic sequence supports a late 12th to early 13th century date for Level III, not for the upper deposit of Level II.

The Radiocarbon dates have been presented here as a reference and to provide corroborative evidence, if any, for the ceramic composition and sequence of the Level.

The occupation at Sanjan Bandar :

The occupation at Sanjan Bandar shows evidence of being a continuous one. The site clearly provides evidence of an extensive and constant contact with West Asia and some contact with China and the Far East. The contact with West Asia clearly predates that with China. The large storage vessels, amphorae and jars are indicative of transportation and trade of goods from the 8th century AD, if not earlier. The presence of ware classes dating back to the late Sasanian and early Islamic periods throughout the deposit of Level I and I(a) are indicative of this. The introduction of Chinese wares is seen to occur at a time when the contacts with West Asia have already been well established and appear to be quite intense. The date for the introduction of the Chinese wares at the site is tentatively placde at the end of the 8th century AD or beginning of the 9th century AD, given the thickness of the deposit that precedes it and the composition of the ceramic assemblage which consists of many wares considered to belong to the pre-Samarra Horizon. The absence of any glazed wares other than TGW in Levels I and I (a) lead one to conclude that they predate the introduction of Samarra Horizon. Level II sees the introduction of a large number of Samarra Horizon ware classes and sgraffiato wares. The assemblage of this deposit shows a dramatic increase in the variety of ware classes – especially glazed wares – and also in quantity. The presence of elite and expensive wares such as CPW, WGW, LPW-1, etc indicates a demand for luxury wares. It is also indicative of the consumer base which could absorb such expensive goods. In the previous levels the utility wares dominated the West Asian component of the assemblage. In this level, the utility wares are present but the luxury wares are equally prominent, if not more so. The later part of this level sees a gradual discontinuation of the wares from Levels I, I(a) and most of the wares from the earlier part of Level II. The presence of the Samarra Horizon ware classes and the sgraffiato wares, and their gradual disappearance, securely places this level to the period spanning the early to mid 9th century AD to the middle or late 12th century AD. Level III is marked by the absence of almost all the West Asian classes of the previous levels, a decrease in Indigenous pottery and some continued presence of Chinese wares. No new ware classes, either Indigenous or West Asian are seen. There is also no evidence of sustained trade or contact with West Asia any more. Level III is not represented by a very thick deposit and may be considered the terminal period of occupation. It clearly belongs to the period when the Samarra Horizon wares and sgraffiato ware classes had all but stopped, as had TGW. It has been dated to the middle or late 12th century AD, perhaps extending into the early 13th century.

* * *

CHAPTER IV

EVALUATION AND INTERPRETATION OF THE EARLY MEDIEVAL SETTLEMENT AT SANJAN

"They change their climate, not their soul, who rush across the seas."

- Horace

Early Medieval maritime trade – A brief review -

The Indian Ocean, extending from the East African coast in the west to the Pacific Ocean in the East, covers a huge expanse and forms a complex entity between these two extremes. It is composed of units, each of which has its own complexities. The many seas of the Indian Ocean have geographical, climatic and cultural variations of their own. The lands surrounding this immense body of water have unique cultural, social and historical characteristics. None of these units is static. The dynamics of this complex system have been mastered and exploited by merchants and mariners from antiquity to present day. Climate-driven trade is complex and high-risk in nature. Nowhere is this seen more clearly than in the Indian Ocean where the monsoon winds have defined the nature of maritime activity. The rhythms of the ocean and the monsoon winds, which the Romans, Greeks and Persians of the pre-Islamic period had learnt to use to great advantage, became life-lines for civilisations along the Indian Ocean littoral in the Islamic period. The first few centuries after the advent of Islam can be seen as a time when mastery over the ocean was perfected. Mariners and merchants who sailed from one extreme of the Ocean to the other, in Arab dhows and Chinese junks understood that it was the monsoon winds that governed the patterns of trans-oceanic trade. That these voyages were full of peril is attested to in the numerous accounts left by mariners, geographers, travellers, traders and story-tellers. Interaction and commerce between the many regions of the Indian Ocean was already established by the early centuries of the Christian era, as the evidence of Roman trade shows. But it was with the rise of Arab power and the fall of the Persian Empire that the entire Indian Ocean opened up to unprecedented commercial traffic.

The basic components of trade, production – transportation / distribution – consumption, required a system which was designed to take into account the requirements of long-distance interregional and intercultural exchange. Dependence on the vagaries of climate made the enterprise high-risk. Goods produced at one place had to be transported to the port, sometimes over long distances from the hinterland. They would have to be stored until such time as the ship was ready to sail. Packing and loading had to be done in a manner to ensure they would survive the long voyage. The ship itself would need to be stocked and prepared for this journey. The embarkation had to be timed just right to coincide with the monsoon winds. Perils of storms, ship-wrecks, pirates and landfall had to be anticipated. The delivery of the cargo to its destination was another complicated system which would have to take into account the local polity, geographical viability of the destination ports and contingencies which would often be outside the control of the traders and sailors. The port of disembarkation would not always be the final destination for the cargo. The goods would then have to be carried over land or by coasting vessels to the markets where they would find buyers. The return voyage would require the vessel to be laden with local goods which would have a demand in the markets back home. This seemingly simple activity was in reality a highly complex system which depended on something as variable as the climate. The further away the destination, the greater were the risks and the system more complicated, with increased ports-of-call and more human intervention. Two factors governed the whole exercise – demand and supply. The fluctuations in these two factors determined the economic viability of the enterprise.

Chaudhuri points out two types of voyages in the Indian Ocean – one, between "the commercial cities of a fairly homogenous area, such as the Red Sea and the Persian Gulf, united after the seventh century by the common bond of Islam" and two, the long trans-regional trip to India, South-east Asia and China (Chaudhuri 1985: 15). It was the second type of voyage that showed the mastery of the Early Medieval sailors over the elements. The cyclical nature of the monsoons defined the sailing schedule in the Indian Ocean and, by extension, determined the entire commercial activity and economy of the Medieval world. The ports of the Persian Gulf and the ports on the western Indian seaboard were linked by sea routes charted by the monsoon winds. Dhows on the China run from the great port cities of the Persian Gulf would call in at the ports on the western Indian sea-board before embarking on their eastward journeys, and again on their return. But Indian ports were not always a halfway-house for these ships. They were also destinations. The sailing season from Gujarat to Aden was September to May when the North-east Monsoon dominated, and the sailing season from Aden to Malabar was October to February (Ray 2003: 20). The entire process, from the production of the traded goods to their loading and departure from the home port, had to be fine-tuned with clock-work precision so that the vessels could catch the wind and embark on their long journeys at the right time. Any delay would upset the entire schedule. Similarly, the return voyage had to be planned with equal precision. The demand for goods from India was certainly one of the most important factors to drive this trade. Spices, teak, cotton, indigo, precious stones, aromatics and other goods from

India had a huge market abroad, just as the goods from West Asia, East Africa and China were in demand in India. The close interaction between both regions is attested to in numerous accounts left by travellers, sailors and merchants. Regular commercial traffic, and its resultant prosperity, is reflected in the development of ports all along the rim of the Indian Ocean. The presence of Indian pottery on sites in the Persian Gulf and records of Gujarat traders settled in the Middle East and on the East African coast confirm the close and intense cross-cultural nature of Indian Ocean trade, just as the presence of foreign pottery at Indian sites and records of foreign settlers on Indian soil attest to the movements of goods and people during this period.

Theories of ports and their relevance to the West coast of India :

The development of port sites on the west coast of India depended on certain factors which also determined the nature and quantum of the trading activity. The resources of the hinterland determine the locations of ports, just as much as geological suitability and geographical convenience. The west coast of India, with its accessibility to the resources of the hinterland, and its long coastline with a number of natural harbours, was an important participant in the trade of the Indian Ocean. Peninsular India lacked the vastness of the Gangetic Plains of North India, with their fertile alluvial deposits, and was not as conducive to extensive agriculture. As Ray points out, trading activities complemented agriculture in South and peninsular India, considering the availability of other resources (Ray 2003: 23 - 24). The long coastline of Western India, extending from Kutch and Saurashtra down to the Malabar, had been commercially active since proto-historic times if not earlier. The factors that determined port locations on this sea-board were primarily resource accessibility and geological suitability. Ray points out that the interface between riverine and maritime zones meant that market centres developed at or near the mouths of rivers, with access to the coast. This meant accessibility to the routes of the interior as well. Ray suggests that in ancient India there was no specific terminology for a port and that the term 'pattana' was used for both, a market and a port since the location of such a settlement at the mouth of a river usually meant that it had a dual function (Ray 2003: 20 – 23). This is particularly relevant in the context of Sanjan which was known as 'Sanyanapattana' or 'Samyanapattana' in epigraphic records (see Chapter I).

Pinto-Orton points out that many factors can determine the nature of a port. She discusses three models of ports and formulates a fourth model. Although she focuses mainly on the ports of the Early Historic period in Gujarat, the models are relevant to the medieval period as well (Pinto-Orton 2001: 19 – 28).

i) Port-of-Trade Model: This model was proposed by Leeds who suggested that Indian ports functioned as ports-of-Trade for 1500 years, beginning in the 7th century. A Port-of-Trade is defined as a neutral zone for traders, administered by professional organisation of the native authorities. Trading relationships were formalized by the government or through administrative channels. Travelling merchants came to exchange goods with resident buyers and sellers. Emphasis was placed on control of imports and their protection in distribution. Codes of law and assured security for traders facilitated commerce. The port-of trade was not a market for the local population andmay have been isolated from subsistence markets.

ii) Upstream-Downstream Model: Bronson suggested that a coastal port located on a river could control the immediate region as well as the upstream regions or hinterlands that supply goods to the port. Alternately the control centre could be located upstream from the port, at a nodal point connecting several branching marketing systems of the hinterland which then feed down-river to the port at the sea-coast.

iii) Hinterland-Maritime Space Model: Weigend and Sinclair suggest that ports are located in regions that produce desirable exports. Weigend compares a port to a knot where the "ocean and inland transport lines meet and intertwine". Local and long distance supplies and resources are carried on transport lines that intersect at ports. The 'Hinterland', 'Foreland' and 'Maritime Space' work together for the success of a port. By 'Hinterland' is meant the developed land space that is connected to the port by transport lines. The 'Foreland' comprises the ports that either receive or send cargo to the port in question. 'Maritime Space' is the water between ports and the shipping lanes that are organised and controlled by man for this body of water.

iv) A fourth model proposed by Pinto-Orton looks at the Gujarat sea-going trade and the transformation between subsistence markets and luxury trade. She suggests a coasting trade network, with merchants picking up agricultural produce at small locations and eventually arriving at a large port. This may be necessitated by unreliable transport systems on land. Two functions are fulfilled by small ports, one for the subsistence markets and one for the luxury markets. The definition of 'subsistence' and 'luxury' depends on the final destination of goods. Items of trade which may be necessary for subsistence in one place may be luxury goods in another. For example, rice, ghee, spices or herbs may be subsistence goods in India but would be luxuries elsewhere. She also suggests that port locations were determined more by the commercial potential of the hinterland than by the exploitation of natural harbours. This may explain why a port can be located at an unlikely spot without deep anchorage.

She discusses three levels of trade that may have operated in Gujarat. The first level is that of the rural market where subsistence items were passed along a chain of coastal markets, eventually reaching the large port. Second, when luxury goods shipped from a port were supported by demand from a foreign power. Third, when agricultural goods were transformed into exports for the international market. (Pinto-Orton 2001: 19 – 28)

In ancient times western India was divided into three provinces – Saurashtra, Anartta and Lata. Saurashtra referred to the whole Gujarat peninsula. Anartta extended from Palanpur to Vadnagar and covered North Gujarat. Lata was South Gujarat, from the river Mahi or the Narmada all the way to Daman or even further. The North Konkan region is an extension of the same coastline and has frequently been included in the Lata province at various points in history (Jain 1990: 11). This stretch of the western Indian coastline, from Khambat to Chaul, had access to the hinterland and its resources – an advantage exploited from the Early Historic period through the Medieval to present day. Not surprisingly, a number of prominent ports emerged as centres of trade with extensive contacts. Khambat (Kanbaya), Bharuch (Barygaza), Surat (Surbaya), Sanjan (Sindan), Thana (Tana, or Sristhana), Kalyan (Kalliana), Sopara, and Chaul (Chemulla or Saymur or Saimur) were some of the prominent ports which frequently find mention in history. Not all these ports functioned simultaneously and a pattern of dislocation of ports is discernable. Often, one port became prominent and commercially active for a while before falling into a decline, when some other port emerged as an important centre. This may be due to the fact that the sea-levels which were volatile and prone to fluctuations, affected the functioning of the ports as well as the settlement. A study of the palaeo-channels and palaeo-mudflats on the Gujarat coast indicate that several landlocked sites and centres such as Vallabhi and Modhera were once riverine ports, closer to the sea when the Gulf of Khambat was much wider (Thakker et al 1993, cited by Ray 2003: 78). Ray too makes a note of this dislocation of ports in western India - "A pattern that emerges in the historical period is the location of a series of ports along the coast, whose fortunes fluctuated over time. One example relates to the shifts on the west coast between the ports of Bharuch, Sopara, Kalyan, Chaul and Surat during the ancient and medieval periods." (Ray 2003: 78). A brief evaluation of the ports of western India in the light of the above models and the role of Sanjan in the trade network of the Indian Ocean can be effected by reviewing two other known Early Medieval ports which are presently under archaeological study – Chaul and Khambat.

Chaul :

Chaul (18° 33′ N, 73° 00′ E) (Gazetteer of the Bombay Presidency 1882 :Thana, Vol.XIV: 715) is known in history by various names – Semulla, Chemulla, Seymur, Saimur, Semur, etc. Located a short distance from the modern city of Bombay or Mumbai, Chaul is known to have functioned as a port from very early times. Its strategic position at the mouth of the Kundalika River gave it accessibility to the hinterland on the one hand, and to the commercial traffic of the Arabian Sea on the other, making it one of the most important and active ports on the west coast. Excellent berthing facilities and a deep anchorage were a natural advantage. One of the earliest references to the port of Chaul is in the Periplus of the Erythrean Sea (Huntingford 1980). Subsequently, it has found mention in numerous accounts of Muslim traders, sailors and geographers in the Early Medieval period. Al Masudi (915 AD) visited India and spent some time in Chaul. He mentions the towns of Simur, Subara and Tana in Lata country as belonging to the king Al-Balharay. He says the king protects the Muslims in his territories more than any other king in Al-Sind or Al-Hind. There are ordinary and Friday mosques in the territory. The ruler was a person called Janj. More interestingly, Simur is said to have 10,000 Bayasira Muslims. Besides, a number of people from Siraf, Basra, Uman, Baghdad and other places are said to live in Simur (Ahmad 1960: 107-8). Almost all the Muslim writers mention Simur and attest to the considerable Muslim population of the settlement. The foreigners settled at Simur and listed by Masudi in his description are all from the Persian Gulf, specifically from the ports which were the trading partners of Simur in the Indian Ocean network. The account by Buzurg Ibn Shahiyar al-Ram-Hurmuzi about voyages from Siraf to Saimur bears out this connection (Hourani 1951: 119-20). Ibn Haukal, Al Ishtakhri and others also mention this port city. The references almost always mention the sites of Sindan, Saymur and Kanbaya together.

Chaul has been excavated from 2001 to 2007, after extensive explorations in the area (Gogte 2003: 1 - 8; 2004 : 124 -32; Gogte et al 2006: 62 -79). This note is based on the published preliminary reports and on the author's own observations as part of the exploration and excavation team, and on the pottery collected at the site. The pottery has not yet been completely classified or analysed and these observations are preliminary in nature. The excavations were conducted at various locations since the site is under intense cultivation and the presence of coconut groves makes it difficult to find any space to excavate. Explorations suggest that the site is spread over 2 km along the Kundalika River and has a breadth of about 900m. The presence of temples, structural remains, sculptures, and inscriptions on Muslim gravestones provide surface evidence for the extent and the nature of the settlement. The excavations themselves produced evidence for continuous occupation at the site from the 2nd century BC to the 18th century AD. The presence of Roman amphorae sherds, Satavahana pottery, RPW sherds and glazed and unglazed wares from West Asia and China show a continuous occupation with extensive trade contacts. Fresh-water ring-wells found in one of the trenches (CHL-V) and brick-built structures in another (CHL-A) marked the Satavahana levels at the site. The medieval levels did not have any structural remains but yielded a very large number of glass beads or 'trade-wind' beads which, according to Gogte, were manufactured at Chaul and exported to East Africa, glazed ceramics and Chinese Blue and White Porcelain and stoneware vessel sherds were found. Glass was also found in considerable quantity. The stratigraphy in the trenches was hard to establish due to soil conditions and the disturbed nature of the site but an interesting pattern is discernable in the medieval occupation of the site.

A bulk of the glazed pottery assemblage appears to belong to the 10th / 11th century and after with very little material that can be said to belong irrefutably to the earlier period. Other than some TGW sherds, ESW(W), WSPW-2, and BW, no other wares seen in the Sanjan corpus occur at Chaul. HsG is, however, found in sufficient quantity to suggest trade links with the Persian Gulf in the 10th / 11th century. None of the Samarra Horizon wares or elite wares occur at Chaul. They were not found in the explorations either. Monochrome wares with buff and pink fabrics are found but a larger number of monochromes are of Khambat manufacture. The explorations as well as the excavations resulted in the collection of a very large number of Chinese sherds - Blue and White Porcelain of the Ming period, Stoneware (very few Dusun, mostly dark brown or black glazed stonewares, and some which could have a Southeast Asian provenance other than China) and CEL-2. The Ming porcelain and the celadon are of a fine quality, unlike the Sanjan PORC-4 sherds which are very much inferior. It appears that the trade between Chaul and the Persian Gulf was less intense during the first three centuries of the Islamic period and increased greatly from the 10th / 11th century onwards. The contact with the Far East, on the other hand, appears to have been very close and the volume of trade suggested by the presence of Chinese wares in the assemblage is high. However, no early Chinese wares such as Changsha or the cream porcelains of the T'ang period are seen in the collection either. The difference in the composition of the Sanjan and Chaul collections suggests either that the trading partners of Sanjan in the Persian Gulf were not the same as that of Chaul or that the demand for goods in the ports of Sanjan and Chaul varied perhaps because of socio-cultural reasons. The 14th – 17th century period appears to have been a period of intense commercial activity at Chaul. Sanjan, on the other hand, had lost importance by this time. The Portuguese presence at Chaul is well documented and the presence of large Portuguese structures such as forts and cathedrals attest to the fact that Chaul was still important as a port. The Portuguese presence at Sanjan or St. John, as they came to call it, is seen in the Limes found along the navigable channels in the mudflats from the sea to Sanjan Bandar, and the 'fort' which appears to be more of a watch-tower and control point on the river than a structure for defence. Sanjan was probably not a very important settlement for the Portuguese. Daman was the port that the Portuguese were to dominate and control until the 20th century.

The port at Chaul was connected to the hinterland by ancient trade routes that can be traced on the basis of the Buddhist caves. (Gogte et al 2006: 76 77). Various routes appear to converge at Chaul from the interior country, no doubt bringing and taking goods from the port. Chaul appears to follow the Hinterland-Maritime Space model of a port in many ways. The convergence of the trade routes over land is complemented by the presence of smaller ports like Palshet, Mandad, Bagmandala, etc which probably had coasting links with the larger port settlement of Chaul. The importance of Chaul in international trade was probably the reason why kings who controlled this stretch of the coast ensured that Chaul offered a hospitable climate for merchants, especially the Muslim traders. The interaction between the small er ports and the main port would have serviced the subsistence market as well as the luxury market, as Pinto-Orton suggests about the Gujarat coasts (Pinto-Orton 2001: 19 – 28). Local populations speak of coasting vessels from ports further south, like Mangalore, which would bring roof tiles and other goods to Chaul and Revdanda until as recently as the early or middle 20th century

The rise of the sea-level by 4 – 5m in the last 2500 years affected the functioning of the port at Chaul (Gogte et al 2006: 79). The development and emergence of Bombay as a port as well as a centre for ship-building eventually led to the decline of Chaul, until it finally lost importance.

Khambat :

One of the most frequently mentioned ports of West India in the accounts and chronicles of Muslim writers is Khambat. It is referred to as Kanbaya, Kambaya, etc. Although the antiquity of the site goes back to the Harappan period, Khambat, from all accounts, emerged and thrived as an important player in the maritime trade of the Indian Ocean in the Early Medieval period. R. N. Mehta's study of the topographical, archaeological and toponymical features of Khambat presents the case for the settlement very well.

He concludes a report on his explorations and study by stating that Khambat was developed on the bad-lands on the north of the Mahi river in the 6th or 7th century by mercantile communities from Nagara, the Early Historic port site which was rendered unviable by the silting up of the river. The mud-flats separated the port from the town. It is suggested that the markets were initially located on the western outskirts. The town plan showed a rectangular pattern that grew later to a pentagular plan when suburbs were included in it. Hindus and Jains made up the original population. The Parsis and Muslims who came later settled in the western part of the town. Its geographical proximity to the hinterland sites of Anhilwad Patan and Ahmedabad was an advantage. Mehta suggests that Khambat fell into a decline due to political factors and not because of silting, since the situation of the mud-flats and the bad-lands existed even at the time of its formation (Mehta 1975: 18 -27; also Mehta and Shah 1968). This is in variance with the opinion of other scholars who think that it was the silt formation which led to the decline. Janaki discusses the problem of the approach to the Khambat port and the presence of the mud-flats. The dangers and perils of the Gulf of Khambat were well known to sailors as far back as the time of the Periplus of the Erythrean Sea. Al-Masudi visited the port in 916 AD and described the town as well as the traded goods in the markets of Khambat. Ibn Haukal (968 – 976 AD) placed it six miles from the sea while Idrisi (1153 AD) referred to it as a coasting station. But in 1340 AD, Ibn Batuta observes the fury of the tides and notes that the vessels had to anchor at Gogha and Gandhar and it was small boats which would bring their cargos into the main harbour (Janaki 1980: 4-5). No matter the difficulty in navigating the perilious Gulf and the difficulty in entering the Khambat harbour, the port seems to have attracted large numbers of traders and trading communities who were undeterred by these issues. Large communities of Muslim traders from Persia and the Persian Gulf regions are known to have settled at the site as the tombs and cenotaphs, with names, genealogies and dates, attest (Lambourn 2003: 213 -40). Rich and varied resources and produce from the hinterland made their way to the port where they were laden on to ships carrying them to markets in the east and the west. Precious stones, cotton cloths, leather goods, ivory, incense and aromatics, precious woods, etc were exported from the port of Khambat to the ports of the Persian Gulf, the Arabian Peninsula, the East African coast and Southeast Asia. From the late 13th century to the 15th century, Khambat specialised in the export of an unusual item – Muslim grave-stones and grave memorials. The cemeteries of Samudera-Pasai attest to the demand for these memorials in Java and Sumatra and the Malacca straits (Lambourn 2003 : 221-89).

As discussed in Chapter II, recent explorations and surveys in Khambat by Bhan suggest that it was also a centre for the production of glazed earthenwares, a hitherto unknown fact about the settlement (Bhan 2006: 90 -96). The glazed ware industry in Khambat appears to have been located in a specific area of the town. The mound is under present occupation. Nevertheless, the surface is littered with pot sherds, triple spurs or kiln furniture, wasters, fused vessels which collapsed in the kiln, etc. From all appearances, the output of this industry was high, suggesting a large consumer base. This consumer base probably comprised of the local population as well as the export market. The large expatriate communities settled in Khambat probably had a demand for these products, which would have been cheaper and more easily available than the imports from the Persian Gulf. The local industry was catering to this component of the consumer base as is evident from the fact that the wares imitate the Gulf prototypes in glaze, shape and, sometimes, even design. The monochromes have a range of colours but the most popular colours are still the ones preferred in West Asia – turquoise, green and blue. The sherds with painted decoration use manganese black or brown, green and blue on a white background, rarely employing any other colours. The bowls and dishes all have the same hemispherical shape and heavy ring-foot base as the West Asian vessels. This attempt to replicate the ware types of the Persian Gulf as closely as possible is indicative of the client base the industry was catering to. That these wares were found at other coastal sites such as Bharuch and Goa by Carswell during his survey (Carswell 1979: 29, 30, 31-32) and by the excavators at Chaul indicates these goods formed part of the inventory for coasting trade. But it also appears that they made their way into the hinterland, as the Champaner material shows (personal observation of the pottery collection in the Department of Archaeology, M.S.University, Baroda; also, Kuldeep Bhan, personal communication). The sherds illustrated in the Nevasa report appear to be of Khambat origin (Sankalia et al 1960: 339, 343 Fig.164). It is difficult to date this industry with any precision but it evidently paralleled West Asian production, probably continuing until the 16th century or so.

Khambat, defined by geography very different from that of Chaul, functioned somewhat differently. It was favourably placed in relation to Malwa, Saurashtra and the plains of Gujarat, with routes that connected the settlement to the hinterland. The silting up of the navigable channels to Bharuch in the 7th century led to the emergence of Khambat as one of the more important ports of the Early Medieval period. Chaul, Debul, and Danda Rajapur were some of the ports further south on the coast with which Khambat had coasting trade contacts (Janaki 1980: 38). To put it simplistically, Khambat appears to be a port which fits into both, the Hinterland-Maritime Space model as well as Pinto-Orton's model of coasting trade. The appointment of administrators for the port, especially important Persian and Arab merchants appointed to high office, suggests a complex system which probably incorporates several aspects of the other models as well. One such 17th century merchant, Ali Akbar, was appointed by Shahjahan between 1646 and 1647 AD, as administrator to the ports of Khambat and Surat, because he was "a merchant and knowledgable about jewels and horses it is probable that he will be able to administer ports in a fitting manner" (Lamborne 2003: 213, 222). Ali Akbar is said to have had extensive trading contacts in the Persian Gulf and a close association with the Governor of Basra, a port to which his ships sailed. His expertise in horses is mentioned as a reason for his appointment. Lambourn suggests that the practise of appointing Persian or Arab merchants as administrators of ports belongs to an older pattern of administration. The import of horses from the Gulf was apparently one of the most important components of the trade with India. The merchants involved in this trade were considerably powerful having access to political centers of power . She mentions the case of another merchant at the end of the 13th century who was appointed vizier by the Pandya king, thanks to his close association with the a powerful horse trader who controlled Qais in the Persian Gulf. Yet another merchant is said to have administered the city of Goa for the Kadamba ruler, Jayakesin I (1050 -80 AD) and was appointed governor of Konkan. (Lamborne 2003: 235 - 36). In the same context, Lambourn also mentions the appointment of Mohammed Sugatipa as the Governor of Sanjan. It appears that the policy of linking commerce with polity was used frequently by kings controlling the ports on the west coast. This implies that the commercial traffic was not only overseen and controlled by the administration, but that it also defined the nature of the trade and the quantum of traffic. The port at Khambat served as a centre for both local as well as long-distance trade until it was eclipsed by Surat in the 17th century.

Sanjan

If the same models are applied to Sanjan, it appears that Sanjan was located at a point which gave it access to resources from further in-land. However, the Varoli River is not a large navigable stream, like the Kundalika River, that could have been used for transportation of goods from up-river regions to the port. Land routes would have necessarily been the transportation lines for the transference of goods. The location of the Sanjan Bandar / port, just before the stretch of rock out-crops on the river bed, at the head of the creek, suggests that it was the furthermost point of trade up-river from the sea. It could well be that the larger vessels of trans-oceanic trade may not have actually pulled into harbour at the Bandar itself but that they anchored at sea. Smaller crafts would have ferried the goods upstream to the port. The estuarine area may have had numerous small jetties and landing platforms, on smaller channels and streams. Two such small facilities probably existed from the 14th – 15th century onwards until the silting up of the Varoli and the shrinkage of the estuarine area affected them. Medhi Falia and Palgam, small settlements on either bank of the Varoli, are marked by two similar Portuguese Limes that overlook the stretch of mud-flats between them. Explorations at both sites revealed evidence for 14th century occupation or activity in the area. At the base of the settlement over-looking the mudflats at Palgam, it was possible to collect some blue-glazed monochrome ware of the Khambat type and one sherd with a paler fabric along with a sherd of Blue and White Porcelain. Medhi Falia is located across the creek. Explorations resulted in the collection of similar sherds a short distance from the Limes along the edge of the creek. Fishing boats can still navigate in parts of the mud-flats at high tide and a number of them could be seen stranded in the mud due to low tide conditions. It is difficult to confirm if such small landing facilities could have existed earlier. However, if the estuary was wider and the anchorage deeper in the past, it is likely that the creek provided a safe harbour to medium sized sailing vessels. The observations made by geo-archaeologists in the 2004 survey are of relevance in this context (Rajaguru and Deo 2004: 97; also, Chapter I in this study). If, as they suggest, the sea has receded by about 1km, then the configuration of the estuary and the profile of the coastline itself would have been different in the past. Although they do not mention when this recession could have happened, it is possible that the geography of the area may have been somewhat different at the time when Sanjan functioned as a port. Sanjan Bandar appears to have been the main centre. The control of the estuary and the entry and exit of ships may have been regulated from this settlement which had a vantage location. That the traffic at the port was regulated is suggested by the appointment of a Governor for the port. It may also be presumed that coasting vessels from Khambat to Chaul, or further south, and vessels sailing to Khambat from southern coasts may have stopped at Sanjan, including it in the coastal trade network. The nature of trade, then, would have included local or coastal markets as well as the international market – or, as Pinto-Orton terms it, subsistence markets as well as luxury markets. The Sanjan port would have incorporated some features of each of the models.

The archaeological evidence suggests that during the earlier period of the settlement, Sanjan was probably involved more with coastal trading and had some limited contact with the foreign markets. It is likely that the contact may not have been direct initially and that it gradually developed until the trade with ports across the ocean became a mainstay of the settlement. The traded goods in this initial stage appear to be items other than ceramics. Pottery of this period suggests the import of perishable goods - perhaps oils or wines, as the amphorae sherds and the storage vessel sherds indicate, and other items. The mention made in the Chinchani copper plate to the conquest of the neighbouring ports (Vetakula) by Mohammed Sugatipa for his king suggests that by the end of the 9th century Sanjan was a thriving port with a quantum of trade that merited the appointment of a Governor and that the harbours in the vicinity were under his control. The presence of Muslim and Persian Zoroastrian population as residents of the settlement implies that the foreign element in the trade at Sanjan was not a negligible one (see Chapter I). Agricultural, forest and mineral produce of the hinterland was probably exported to destinations across the Ocean, while luxury goods like fine glazed earthenwares, glass, wine, oils and, perhaps, horse were brought into Sanjan. It is not yet clear if the imports were meant to cater only to the local populations or if they were taken to in-land markets. In all probability, they did both – meeting the local demand as well as the demand of markets in-land. An important point to note is that no other site on the Indian west coast has yielded such a range or quantity of glazed earthenwares belonging to the Abbasid period. This may well be due to the fact that few Early Medieval sites have been excavated on the coast. Nevertheless, even at excavated sites like Chaul, none of the elite wares seem to occur, nor is the same range of imported wares seen. Surveys and explorations along the coast, such as Carswell's survey, report TGW or Chinese wares as surface finds but no report mentions the Samarra Horizon classes. The demand for these goods at Sanjan seems to be particular, catering to a specific clientele. Hinterland sites like Nasik or Nevasa also do not report any of the classes seen at Sanjan, other than monochrome or TGW and sometimes Chinese wares (Sankalia and Deo 1955: 86; Sankalia et al 1960: 339). If the imports at Sanjan did find their way to the hinterland markets, it is not possible to identify them until more Early Medieval sites are excavated. The port appears to have functioned until the middle or Late 12th century AD, or the early 13th century AD, at the latest.

Evaluation of the Historical data in light of the Archaeological evidence :

The authenticity of the historical sources can be evaluated against the archaeological data now available about Sanjan. Sources mentioned in Chapter I are taken up in this section and individually examined in the light of material evidence.

Literary Sources : The Kisse-i-Sanjan, as noted in Chapter I, narrates the story of the settlement. The narrative is ethno-centric, focusing on the Zoroastrian settlement at the site. However, it still provides some vital pieces of information. The question of whether a migration did take place, as the Kisse claims, is the first point to consider. That a Zoroastrian Persian community was settled at Sanjan is attested to by the presence of the Dokhma. This irrefutable archaeological evidence for the presence of the foreign community at the settlement establishes that either one or several migrations may have taken place and that the presence of the migrants was considerable enough to require the construction of a large mortuary structure. Human remains in such large quantities – 400 to 450 individuals - in the central well of the structure indicate a very large population (Walimbe and Mushrif, personal communication). Although the Dokhma is dated to a later period, the mention of the migrant community in the Kisse appears to be authenticated by its presence. The sherds found in the mud mortar of the construction indicate a date which coincides with Level II and Level III. The Kisse clearly states that it was only one group of Zoroastrians who chose to migrate. The fact that this group chose to spend a century in Kohistan and then fifteen years in Hormuz indicates that the reasons for the migration may have been other than religious persecution alone. The fact that other Zoroastrians did not relocate implies that the situation may have been difficult, but not impossible. The other important point that the Kisse makes is that the migration was planned and organised, with the clear intention of sailing to India, indicating that the land of Al Hind was not unknown to them. Their arrival and temporary stay at Diu cannot be verified. However, it is clear that some contingency drove the migrants to set sail once again, this time for the mainland. They are said to have made landfall at Sanjan more by accident than by design. This may not be true. If the migrating community was as organised as is implied by the presence of a wise priest-leader and if they owned sea-worthy crafts for trans-oceanic voyaging, it appears unlikely that their arrival at Sanjan was an accident. It is more likely that their time in Diu was spent in establishing contacts on the mainland and deciding upon the most suitable and hospitable place for relocation. The scepticism of the local ruler and his subsequent treaty with the migrants finds mention next. The Kisse describes Sanjan as a desert where the migrants built a city, naming it in memory of a town in Persia. This is refuted by epigraphic evidence. The epigraphic sources mention Sanjan as a settlement going back many centuries prior to the supposed migration (see Chapter I). Moreover, this would imply that the earliest level at Sanjan would have to be equated to the original settlement of the migrants. This seems unlikely since the Kisse itself says that Sanjan had a king, implying that a settlement did exist for him to rule over. The identity of the king cannot be traced until the date of this migration can be established with certainty. Unfortunately, the structure of the fire temple which the migrants are supposed to have built cannot be traced at Sanjan. The prosperity of the town and the migration of small groups to other sites along the coast are mentioned, and the excavations certainly reflect the economic prosperity of the Sanjan population. Well-built brick structures such as wells, platforms and floors, the presence of ring-wells, expensive glazed ceramics from West Asia and China, imported glass, and other luxury items are all evidence of a healthy economy. That agriculture was not the main occupation of the migrants is suggested by the relocation of small groups to other port sites along the coast, instead of the hinterland. Andre Wink is of the opinion that the mainstay of the migrant community was trade and that the migration was "not so much a flight as a readjustment of commercial patterns that had arisen long before Islam" (Wink 2002: 105). If the migrating community was in actuality a mercantile group, it would explain the increase in the commercial activities at Sanjan. The movement of small groups to Khambat, Bharuch, Randher and Navsari – all of which were ports – bears out in the epigraphic and other literary records for these sites. The mention of Randher near Surat is interesting because it is located very close to the site of Tena where Unvala found the remains of three Dokhmas, one of which, at least, has an antiquity that matches the Sanjan structure (Unvala 1951; also, Chapter I). The narrative now becomes a little obscure, mentioning seven hundred years of settlement in Sanjan, on the one hand, and five hundred years of settlement on the other, before Islam came to India, and Sultan Mahmud ruled at Champaner. Nevertheless, the suggestion that Sultan Mahmud was Mohammed Begada and that the settlement of Sanjan was sacked by him does not bear out in the archaeological record which clearly indicates that the terminal date for the settlement at Sanjan does not extend past the late 12th or early 13th century. The late 15th century date of Mohammed Begada is too distant in time from the termination date for the site, as seen in the archaeological record. The terminal period at Sanjan does not have any evidence for the 14th century or later. It is possible that homesteads and farmsteads continued to exist, perhaps with remnants of the Zoroastrian population, as can be seen by the continued use of the Dokhma till the mid 15th century and the settlement at Koli Khadi. The habitation on the Bandar has no such evidence. The fleeing population of Zoroastrians next moved to the caves of Bahrot in order to protect the Holy Fire. Caves which appear to belong to a period far earlier than the Medieval, are said to have housed the Holy Fire for twelve years. The presence of structural remains on the Bahrot hills and the caves themselves indicate the presence of an early settlement and perhaps the presence of at least one ancient trade route, linking Sanjan to the hinterland.

The chronology established by the ceramic evidence, Radio Carbon dates and structural remains for the settlement on Sanjan Bandar provide a framework to evaluate the events mentioned in the Kisse. The existence of a settlement prior to the migration is suggested by the mention of the king of Sanjan with who the migrants forged a treaty. It would appear, that the earliest level at Sanjan, Level I would belong to the period when Sanjan was a small, regional settlement, functioning as a port for coasting and perhaps even for trans-oceanic vessels. More importantly, the ceramics of this early level show an acquaintanceship with the Persian Gulf at an early date. The later part of this Level shows the development of this limited contact into a closer one, reflected as it is in the introduction and gradual increase of TGW as well as in the quantum of pottery and pottery types. Sanjan appears to be gaining prominence as a port. Level II shows a tremendous increase in the quantity as well as the variety of pottery. The introduction of tablewares and fine trade ceramics indicates not only an improvement in the economy at large, but also of a change in the demands of the market. The earlier Levels had a concentration of utility wares and the focus appears to be on the trade of items and goods other than pottery and luxury items. No tablewares or fine wares were seen. In this level, the trend appears to have changed. A market for fine and expensive goods, not only for utilitarian items, suggests that there was a clientele rich enough to afford luxury goods and, more importantly, had a demand for them. The overall change in the consumption pattern could be reflective of a change in the cultural component of the population. One may hypothesise that it was the migration that introduced the new cultural component to the settlement which changed both, the pattern of consumption, as well as the nature of the trade that the port was involved in. This is not improbable considering that the migrating community which may have comprised of merchants and traders would have had close contacts in the home country and access to its resources. The know-how of the new migrants would have added an impetus to the trade and proved an advantage in a growingly competitive market. The local king would have recognised the advantage of allowing such a group to settle and trade at the port in his control. While pottery cannot be seen as ethno-centric or said to provide irrefutable proof for the ethnicity of its users, a change in the consumer base and usage patterns may be reflected in an assemblage. While numerous other factors could be responsible for the changes seen in the assemblage, it is likely that the migration may have been one contributing factor. The increased volume of trade which follows in Level II is seen both, in the quantity of pottery, as well as in the presence of expensive and elite wares. The termination of the settlement around the end of the 12th 0r beginning of the 13th century belies the invasion of Sanjan by Mohammed Begada. If the settlement was indeed abandoned, the invader of Sanjan would have to be someone who predated Begada by at least a hundred and fifty to two hundred years. The identity of King Jadi Rana has invited much discussion and debate since no such king appears in Indian records. 'Jadi' is derived from Jadhav and 'Rana' is a title. It is possible that the king in question was a Jadav Mandlik or a feudatory of the Rashtrakutas who controlled the region. This study does not forward any suggestion about the identity of this king since it is in the realm of speculation with no archaeological evidence to support it.

Epistolary Evidence : The Rivayets or correspondence between the Zoroastrian settlers in India and those in Persia provide evidence, as discussed in Chapter I, that Sanjan was already abandoned no longer functioning at the time of their writing (1478). This confirms the fact that Mohammed Begada cannot be the Sultan who attacked Sanjan. The evidence from the excavations also support the fact that Sanjan does not have a deposit of datable to later than the mid 13th century.

Epigraphic Evidence : The Nagarjunakonda inscription of Abhira Vasushena (278 AD) and the Nahapana inscription (32 – 77 AD), cave no.10 in the Pandulena caves of Nasik both mention Sanjan as early as the first three centuries of the Christian era. Although the excavation trenches provide no evidence for such early deposits, the two sherds of RPW from the ring-well pits in TT1 indicate the presence of an older deposit or an earlier site in the vicinity of Sanjan. The ancient rock-cut caves of Bahrot which pre-date the Sanjan Settlement are also evidence of human activity in the area at a very early date.

The Buddhavarsa inscription (671 AD) mentions Western Chalukyan control over the region. The governance of this area by a brother of Pulakeshin II who had cordial relations with the Sasanians is relevant in the light of the migration of the Zoroastrian Persians. It strengthens the hypothesis that the migrants took an informed decision to settle at Sanjan since it was a known territory where they knew they would be welcome. It appears logical that they would seek refuge in friendly territory. The inscription confirms the existence of Sanjan prior to the migration.

The inscription of Amoghavarsa I (871 AD) establishes the Rashtrakuta presence at Sanjan. More importantly, the administrative jurisdiction of Sanjan is mentioned as covering a vast area, indicating the importance of the port and the settlement at this time. Level II, to which the date of this inscription belongs, is the most prosperous phase of the settlement. The inscription confirms the political a nd commercial importance of the settlement by this time.

The Chinchani grants, spaced over a hundred and twenty five years, provide corroborative evidence to the archaeological finds. The mention of Sanjanamandala implies that at the time of writing (926 AD), Sanjan was not a singular settlement but, as mentioned in the Amoghavarsa I inscription, it was an administrative unit and an important settlement with a Friday mosque, temples to the goddess Dasami and a cosmopolitan population. While there is no evidence for the mosque or the temple (numerous sculptures, including the goddess image found on the surface indicate that there may well have been such a structure), the most important reference is to the appointment of Mohammed Sugatipa or Madhumati, a 'Tajik' to the post of Governor by the Silahara king Krishna II sometime between 878 and 915 AD. As discussed earlier, the pattern of appointing governors, particularly of Arab or Persian origins, was an interesting practise. The fact that the Tajik had conquered the neighbouring ports or harbours and that they were now under the central authority of the Silahara king, reveals the close ties and interactions between the commercial, political and military elements. That Sanjan had become an important trading centre is borne out by the rich deposit of Levels I, I (a) and II. This is corroborated by the inscriptional evidence which indicates that Sanjan now merited a governor and that he oversaw the functioning of the port. The mention of the 'Hanjamanapaura' or the Zoroastrian Persians confirms that by this date the migrants had not only settled, but had established themselves as an important component of the population. The migration mentioned in the Kisse finds authentication in this record and establishes a date from which it is safe to infer that the migration took place certainly before the appointment of Sugatipa. It is difficult to establish whether the Tajik was a mercenary, a merchant or a horse-dealer. It does establish that the foreign settlers at Sanjan, Arab and Persian, must have been in large enough numbers to merit a Muslim Governor. Sanjan was probably at the height of prosperity at the time of Sugatipa's appointment. The continued prosperity of the port is corroborated in the next grant in the set which is datable to 1023 AD. The mention of the various communities includes the Tajiks and the 'Hamjamana', again attesting to the continued presence of the foreigners. The third grant which is dated to 1056 AD describes the geographical and political territoryof Sanjan. The area mentioned is very large, extending inland. The reference to Sanjan not as a 'mandala' or group of villages, but as 'Samyanapattana', a port and a market (see above) makes it clear that the status of Sanjan was increasing in importance along with the growth of its commercial activity. It can be confirmed by the archaeological data that, by this date in the 11th century, Sanjan was indeed thriving as a market and a port. Exploration and survey in the areas mentioned in the inscription are needed to confirm if there is any archaeological evidence to prove that Sanjan did cover such a vast territory.

Accounts of Geographers, Mariners and Travellers : The descriptions of Sanjan given by the many geographers, mariners and travellers refer to the settlement as Sindan. The confusion regarding many towns having the same name is cleared when one considers the location given for the town by the writers. Most writers like Al Biladuri, Al Ishtakri and Al Biruni give its location in terms of distance from other ports like Surbaya, Saymur, Debal Kanbaya and Broach. The descriptions clearly fit Sanjan and identify it as Sindan. While the teak, bamboo and other goods of trade mentioned by writers like Al Biladuri and Ibn Haukal do not survive in the archaeological record, the fact that habitation deposits corresponding to the dates of these writers have been found in the excavations is evidence that the site thrived between the 9th and the 12th centuries. A point of interest to note is the mention made by Al Biladuri of the subjugation of Sindan by Fazl, son of Mahan, a slave of the Samma house, in 820 ADand that hesent to the Khalifa Al Mamun (813 – 833 AD) the gift of an elephant. The account speaks of the construction of a Friday mosque and that the people of Sanjan reclaimed the town from Fazl's brother who had seized power and sent the Khalif the longest teak ever seen. This is not verifiable in the archaeological record. But the presence of Muslim traders in the town and the presence of a Friday Mosque are attested to by other writers and the epigraphic records. That a struggle for control over the port took place or that Muslim domination in Sanjan actually predated the appointment of Mohammed Sugatipa is not reflected in other records but does raise the question of political control over the ports and the relationships between local rulers, migrant trading communities and the influence they exerted over the ports. The writers mention the items traded not only in Sindan but also in other ports like Kanbaya, Samur, Surbaya, etc. None of them lists pottery as an item of import at Sindan.

An evaluation of the historical data in the light of the archaeological evidence helps in interpreting the settlement at Sanjan and understanding it.

Conclusions :

The analysis of the ceramic assemblage from the Early Medieval settlement of Sanjan which has been presented in this study and the foregoing discussion which evaluates the historical data in the light of archaeological evidence, leads to a fresh interpretation of the site. These conclusions are presented in brief :

1) The Early Medieval site at Sanjan which finds mention in literary, epigraphic and other historical sources was located on the north bank of the Varoli River and covered an area of about 1.5 x 1.5km. The most intense occupation was located on the Sanjan Bandar area which can be identified as the main port settlement. The total deposit with pottery occurrence was 3.79m in the excavated trench TT4, which was taken as an index in this study. The site was strategically located at the head of the creek, with access to the hinterland and its resources on the one hand and to the coast on the other.

2) The ceramic assemblage from the explorations and the excavations which was subject to study consisted of West Asian wares, Far Eastern wares and Indigenous wares. The morphological, typological and statistical analysis of these wares brought to light classes of pottery which were not known in the Indian context to date. The study of this assemblage and the occurrence pattern of ware classes made it possible to identify three ceramic levels and one sub-level in the occupation of the site.

3) Level I was the lowest level, representing the earliest phase of occupation. It extended from approximately 3.79m to 2.03m. The ceramic composition of this Level suggests a contact with West Asia. The goods imported into the port were items transported in amphorae and storage vessels. It is likely that many other goods were traded but did not survive in the archaeological record. The range of indigenous wares shows a predominance of black slipped vessels. No glazed ceramics occur, with the exception of TGW. The port appears to be involved in coasting trade much more actively than in trans-oceanic trade. No structural remains are found for this early occupation. The ware classes suggest a chronology which extends from the late 7th century AD to the middle to late 8th century AD, a period when the traditions of Sasanian wares were still current and glazed ceramics had not yet gained the importance which they enjoyed in the later centuries.

4) Level I (a) is actually the later part of Level I, extending from a depth of 2.02m up to 1.25m. The ceramic assemblage of Level I continues without too much change in this sub level. The most important change which is notice is the appearance of Chinese wares and their continuous presence from this point onwards. The ceramic composition of this level reveals a growth in the quantum as well as in the nature of trade. The introduction of a new component, Chinese wares, coupled with the continued presence of Turquoise Glazed Wares, Buff Ware, Torpedo jars or Persian Gulf amphorae and other West Asian classes with pre-Islamic history suggests that the settlement was not only expanding its sphere of trading activities but that trans-oceanic trade was gaining importance at the site. Direct and regular trade with the Persian Gulf is reflected in the increase in pottery. Indigenous pottery shows a similar increase both, in quantity and in variety. However, glazed tablewares do not form part of the collection. The suggested dates for this level are middle or late 8th century AD to early or middle 9th century AD.

5) Level II extends from 1.24m to 0.39m. This level is marked by the appearance of Samarra Horizon ware classes, sgraffiato classes and several glazed wares. There is a sharp increase in the variety as well as

quantity of pottery. Chinese wares continue to show an increase. The West Asian ware classes of the previous levels continue until the later part of this level, when they begin to taper off and eventually end. The Samarra Horizon classes also continue through the level, tapering off and ending in the later part of the level. Sgraffiato wares continue a little longer. This level represents the most active and prosperous period in the occupation. Trans-oceanic trade appears to be the mainstay of the port as the market opens up to Chinese as well as West Asian imports. The consumer base apparently consists of a population which can afford and has a demand for these luxury goods. The Samarra Horizon wares in this level help to date it with some accuracy. The overall quantum and range of pottery indicates intense and sustained trade which continued for a long period of time. The dates suggested for this level are early or middle 9th century AD to middle or late 12th century AD.

6) Level III: The deposit of this level extends from 0.38m to surface. A decline in the quantity of pottery and a marginal decrease in the West Asian wares is the only change seen. A simultaneous marginal increase is seen in the Chinese wares. The most noticeable decrease is seen in the indigenous wares. There is no evidence to suggest occupation after early to middle 13th century. The port appears to have stopped functioning after at this time.

7) A study of the historical data about Sanjan reveals that the Kisse-i-Sanjan or the story of Sanjan is a prime source which narrates the story of migration the Zoroastrian Persians to India and their settlement at Sanjan. By correlating the data from the text (which is not a history so much as an oral tradition), the dates and events mentioned in it, and the chronology suggested by the archaeological and ceramic record, it is possible to evaluate the authenticity of the narrative. That a community of Persian Zoroastrians did settle at Sanjan is attested to by the presence of the Tower of Silence which was excavated in 2004, as well as epigraphic records which speak of the 'Hanjamana' or anjuman (Persian for community). The migration is not reflected directly in the archaeological record but, by inference, it may be hypothesised that the site which was familiar to the migrants because of its contact with the Persian Gulf region may have been chosen for settlement due to its commercial potential. The arrival of the migrants and the pact that they forged with the local ruler cannot be ascertained but is not improbable, considering that the settlement of foreign trading communities in port settlements is a pattern seen at almost all sites in the Indian Ocean. That the migrant group was mercantile in nature is borne out by their migration to other coastal towns and ports, in preference to the hinterland. The settlement was not established by the migrant community as the Kisse, states since the epigraphic sources predating the Medieval period speak of the settlement at a date as early as 278 AD. An even earlier inscription mentions Nanangol or Nargol, which is located on the coast a short distance from Sanjan Bandar, at a date as early as 32 AD. While the excavated trenches had no such early levels the presence of the Red Polished ware sherds, even from disturbed contexts in the trenches implies the presence of earlier settlement at or near Sanjan. The Bahrot caves which are probably Buddhist, and predate the Medieval period, also attest to the fact that Sanjan existed prior to the migration. It is likely that the migrants may have established a colony but they certainly did not establish the town. The date of their arrival is a matter of debate amongst scholars. From the archaeological record it is possible to infer that the small port of Sanjan with coasting trade and some limited contact with the Persian Gulf during the early occupation levels and increasing commercial activity, found an impetus by the arrival of the Persian traders. This may be seen in the increase in the volume and nature of trade and in the change in the demands of the consumer base. A change in the cultural component could have resulted in the changed pattern of consumption. The migration could then be placed tentatively in the late 8th or early 9th century. This is a hypothesis and not a statement of fact, since pottery is not ethnocentric. It cannot be said with certainty that the migration of Persian Zoroastrians alone could have been responsible for the increase in commercial activity in the port. The long period of prosperity mentioned in the text is seen in the deposits of Level I, Level I(a) and Level II. The invasion of Sanjan by a Sultan Mahmud, supposedly Mohammed Begada of Champaner as the text implies, is not substantiated by the archaeological evidence since the terminal occupation at Sanjan does not extend into the 15th century. The ceramic evidence does not support such a late date. If the reason for the termination of the occupation is attributed to invasion or abandonment caused by it, the date of this event would have to be placed in the early 13th or, at the latest, middle 13th century AD. Silting of the Varoli, the formation of sandbars or other geological reasons could have also led to the termination of the occupation.

8) The date for the construction of the Dokhma appears to be late 12th century or so, on the evidence of the pottery mixed in the mud mortar. Its continued use until the mid 15th century could be attributed to the existence of farmsteads and Homesteads that may

have continued to function even after the termination of the main settlement on the Bandar. This pattern of settlement is seen at present as well.

9) The epigraphic records for Sanjan which fall within the Early Medieval period corroborate the dates reflected in the archaeological record. The prosperity of the port and its increased importance in maritime trade in the mid 9th to Early 13th centuries is attested to by the appointment of Governor for the port. Trade was thus regulated and organised. The importance of the foreign communities, especially the Muslims and Arabs, is attested to by the appointment of a Tajik governor. The growth of the settlement is also reflected in the increased administrative jurisdiction it controlled, as mentioned in these records.

10) The port of Sanjan incorporated various elements of the different models discussed earlier. It appears to have evolved from a small port involved coasting trade and limited foreign contact to a full-fledged utility with regulations in place to handle trans-oceanic trade. The import items catered certainly to the local population and perhaps to markets in the hinterland. As compared to Chaul and Khambat, Sanjan appears to have thrived during the Abbasid period as well as after. The ceramic assemblage from Chaul does not have the same components as Sanjan for the corresponding period, indicating that they had different trading partners in the Gulf and the market demands for the two settlements were different. The Samarra Horizon classes are missing in Chaul, while it is the sgraffiatos and later ware classes that abound. Sanjan does not continue to function as a port as long as Chaul. Khambat whose establishment is supposedly as early as Sanjan came into prominence after the 10th century and had a glazed ware industry which catered to local, inland and foreign markets. No ceramic evidence has yet been found to establish the pre -10th century history of Khambat.

11) The settlement at Sanjan establishes it as an important port on the West coast of India in the Early Medieval period (late 7th century AD to early or middle 13th century AD) with extensive contacts with West Asia and a more limited contact with China and the Far East. The ceramic evidence from the site establishes the role it played in the maritime trade of the Indian Ocean during this period.

* * *

BIBLIOGRAPHY

Adams, R. McC. 1970. Tell Abu Sarifa, A Sasanian-Islamic Ceramic Sequence from South Central Iraq. *Ars Orientalis* 8: 87 – 119

Ahmad, Maqbul. S. 1960. Al-Masudi, On the Kings of India. In *Al Masudi Commemorative Volume* (Ahmad, Maqbul. S and Rahman, S. eds) Indian Society for the History of Science and Institution of Islam Studies, Aligarh Muslim University: 97 – 116.

Allan, J. W. 1971. *Medieval Middle Eastern Pottery*. Ashmolean Museum. University of Oxford.

Anonymous. 1964. Excavations at Banbhore. *Pakistan Archaeology* 1: 49 -79.

Bhan, K. K. 2006. Towards an understanding of Medieval Glazed Pottery Manufacture from Lashkarshah, Khambat, Gujarat. *Man and Environment* XXXI, 2: 90 – 95.

Bhandarkar, S. R. Sanjan *Copper-plate Inscription of Amoghavarsha I Saka year 973 (871 A. D)*. Epigraphia Indica Vol. XVIII.

Bronson, B. 1977. Exchange at the Upstream and Downstream Ends : Notes Towards a Functional Model of the Coastal State in Southeast Asia. In *Economic Exchange and Social Interaction in South East Asia*, Michigan Papers on South and South East Asia (Hutterer R. ed.): 39-52

Carwell, J. 1977. China and Islam in the Maldive Islands. *Transactions of the Oriental Ceramic Society* 1975 – 76, 1976 – 77: 121 -198.

Carswell, J. 1978. China and Islam: A Survey of the coast of India and Ceylon. *Transactions of the Oriental Ceramic Society* 1977 – 78: 25 – 68.

Carswell, J and Prickett, M.1984. Mantai 1980 : A Preliminary Investigation. Ancient Ceylon, *Journal of the Archaeological Survey Department of Sri Lanka* 5: 3 -80.

Chaudhary, K. N. 1985. *Trade and Civilization in the Indian Ocean. An Economic History from the rise of Islam to 1750.* Cambridge.

Chittick, N. 1974. *Kilwa. An Islamic Trading City on the East African Coast.* Nairobi.

Costa, P. M. and Wilkinson T. J. 1987. The Hinterland of Sohar. *Journal of Oman Studies* 9: 184 – 209.

Cox, W. 1945. The Book of Pottery and Porcelain Vol. I. New York.

Dhabhar, B. N. 1999. *The Persian Rivayats of Hormazyar Framarz and others.* Bombay.

Dhalla, H and Nanji, R. 2006. New Perspectives on the Zoroastrian Migration to India – A Short Note. *Fezana Journal.* USA: 75 – 78.

Eduljee, H. 1991. *Qesseh-i-Sanjan.* K. R. Cama Oriental Institute, Mumbai.

Elliot, H. M. and Dawson, J. 1866 – 77. *History of India as told by its own Historians,* Vol. I (reprint 1952). London.

Gazetteer of the Bombay Presidency, Vol XVI (Thana) 1882 (reprint 2000). Mumbai. 301 – 05.

Gokhale, S. 2004. Coins found in the Excavations at Sanjan. *Journal of Indian Ocean Archaeology* 1: 107 – 12.

Glover, I. 2002. West Asian Sasanian-Islamic Ceramics in the Indian Ocean, South, Southeast and East Asia. *Man and Environment* Vol. XXVII 1: 165 -177.

Gogte, V. D. 2003. Discovery of the Ancient Port of Chaul. *Man and Environment* Vol. XXXIII 1: 1 – 8.

Gogte, V. D., Pradhan S., Dandekar A., Joshi S., Nanji R., Kadgaonkar S. and Marathe V. 2006. The Ancient Port at Chaul. *Journal of Indian Ocean Archaeology* 3: 62 – 80.

Grube, E. (ed.) 1994. *Cobalt and Lustre. The first centuries of Islamic Pottery.* The Nasser. D. Khalili Collection of Islamic Art. Vol. 9 London.

Gupta, S. P, Dalal K., Dandekar A., Nanji R., Mitra R. and Pandey R. 2002. A Preliminary Report on the Excavations at Sanjan (2002). *Puratattva* 32: 182 – 198.

Gupta, S. P, Dalal K., Dandekar A., Nanji R., Pandey R., and Mitra R. 2003. Early Medieval Indian Ocean Trade : Excavations at Sanjan, India. Circle of Inner Asian Art Newsletter 17: 26 – 34.

Gupta, S. P, Dalal K., Dandekar A., Nanji R., Aravazhi P., and Bomble S. 2004. On the footsteps of the Zoroastrian Parsis in India : Excavations at Sanjan on the West Coast – 2003. *Journal of Indian Ocean Archaeology* 1: 93 – 106.

Gupta, S. P, Dalal K., Nanji R., Dandekar A., Bomble S., Mushriff-Tripathi V., Kadgaonkar S., Chaudhuri G., Sharma P. and Abbas R. 2005. Preliminary Report of the third Season of Excavations at Sanjan – 2004. *Journal of Indian Ocean Archaeology* 2: 55 – 61.

Guy, J. 2001. Early Asian Ceramic Trade and the Belitung (Tang) Cargo. *Transactions of the Oriental Ceramic Society* Vol. 66: 13 – 67.

Guy, J. 2003. The Intan Shipwreck. A 10th century Cargo

in South East Asian Waters. In Song Ceramics: Art History, Archaeology and Technological Colloquim, *Art and archaeology* No. 22 (Pearson, S. ed.). London: 171 – 191.

Horton, M. 1996. Shanga. *The Archaeology of a Muslim Trading Community on the East Coast of Africa.* London.

Hourani, G. 1995. *Arab Sea-faring in the Indian Ocean in Ancient and Medieval Times.* (Carswell, J. ed.). New Jersey.

Huntingford, G. B. W. 1980. *The Periplus of the Erythrean Sea.* London.

Jain, V. K. 1990. *Trade and Traders in Western India (AD 1000 – 1300).* New Delhi.

Janaki, V. A. 1969. *Gujarat as the Arabs Knew It – A Study in Historical Geography.* Geography Research Paper Series No. 4. Baroda.

Janaki, V. A. 1980. *The Commerce of Cambay from the Earliest Period to the Nineteenth Century*. Baroda.

Joglekar, P. P. 2003. *Faunal Remains from Sanjan* (Unpublished Report submitted to the World Zarathushti Cultural Foundation).

Juma, A. 2004. *Unguja Ukuu on Zanzibar. An Archaeological Study of Early Urbanism.* Studies in Global Archaeology 3.

Karmerkar, M and Dhunjishah S. 2002. *From the Iranian Plateau to the Shores of Gujarat : The Story of Parsi Settlements and Absorption into India.* Mumbai.

Kennet, D. 1991. Excavations of the site of Al-Qusur. Failaka, Kuwait. P*roceedings of the Seminar for Arabian Studies* 21: 97 – 111.

Kennet, D. 2004. *Sasanian and Islamic Pottery from Ras al-Khaimah.* BAR International Series 1248. Oxford.

Kerr, R. 2002. Chinese Porcelain in the Gulf Region. *Journal of Oman Studies* 12: 125 – 31.

Kevran, M. 1996. Indian Ceramics in Southern Iran and Eastern Arabia: Repertory, Classification and Chronology. In *Tradition and Archaeology – Early Maritime Contacts in the Indian Ocean (Ray, H. P and Salles F. J eds.), Proceedings of the International Seminar Techno-Archaeological Perspectives of Sea-faring in the Indian Ocean (4th cent. B. C – 15th cent. A.D).* New Delhi: 37 – 58.

Kevran, M. 2004. Archaeological Research at Suhar 1980 – 1986. *Journal of Oman Studies* 13: 263 – 381.

Konow, S. 1982. Sanjan Plates of Buddhavarasa. *Epigraphia Indica* Vol. XIV (1917 -18) (Thomas, F.W. ed.): 144 -52.

Lambourn, E. 2003. Of Jewels and Horses: the Career and Patronage of an Iranian Merchant under Shahjahan. *Iranian Studies* Vol. 36 no.2: 13 – 240.

Lane, A. 1947. *Early Islamic Pottery. Mesopotamia, Egypt and Persia.* London.

Lane, A. 1957. *Islamic Pottery from the Ninth to the Fourteenth Centuries A. D.* London.

Leeds, A. 1961. The Ports of Trade in Pre-European India as an Ecological and Evolutionary Type. In *Symposium: Patterns of Land Utilization. American Ethnological Society* (Garfield V.E. ed.). University of Washington Press.

Mason, R. B and Keall, E. J. 1991. The Abbasid glazed Wares of Siraf and the Basra Connection: Petrographic Analysis. *Iran* Vol. XXIX: 51 – 66.

Mason, R. B. 1997. Medieval Iranian Lustre-Painted and Associated wares: Typology in Multidisciplinary Study. *Iran* Vol. XXXV: 103 – 107.

Medley, M. (1976) *The Chinese potter, a practical history of Chinese ceramics*. Oxford

Mehta, R. N and Patel A. J. 1967. *Excavations at Shamalaji.* Baroda.

Mehta, R. N and and Shah D. R. 1968. *Excavations at Nagara.* Baroda.

Mehta, R. N. 1975. Khambat (Cambay) : Topographical, Archaeological and Toponymical Perspectives. *Journal of M. S University* Vol. 24 (1): 18 – 28.

Menanat, D. 1898. *The Parsis in India Vol. I.* (Marzban M. M. trans. and ed. 1917). Bombay.

Minorsky, V. 1937. *Hidud al -'Alam: The Regions of the World – A Persian Geography 372 A.H. – 982 A.D.* London.

Mitra, R and Dalal K. 2005. A Report on the Glass Vessels from Sanjan – 2002. *Journal of Indian Ocean Archaeology* 2: 62 – 68.

Morgan, P. 1994. Abbasid Opaque White-glazed Wares. In *Cobalt and Lustre, The first Centuries of Islamic Pottery (Grube, E. ed) The Nasser D. Khalili Collection of Islamic Art Vol. 9*. London: 36 – 37.

Morgan, P. 1994. Sgraffiato Types and Distribution. In *Cobalt and Lustre, The first Centuries of Islamic Pottery (Grube, E. ed) The Nasser D. Khalili Collection of Islamic Art Vol. 9*. London: 119 – 23.

Mushriff, V. and Walimbe S. R. 2005. Human Skeletal Remains from Sanjan Excavations. *Journal of*

Indian Ocean Archaeology 2: 73 – 92.

Nanji, R. and Dandekar A.2005. Excavations at Sanjan Dakhma – 2004. *Journal of Indian Ocean Archaeology* 2: 69 -72.

Nanji, R. and Dhalla H. 2007. The Landing of the Zoroastrians in Sanjan : The Archaeological Evidence. In Parsis in India and the Diaspora (Hinnells, J. and Williams A. ed.). *Routledge South Asian Religion Studies;* London. 35 – 58.

Northedge, A. 1996. Friedrich Sarre's 'Die Keramik von Samarra' in Perspective. In *Continuity and Change in Northern Mesopotamia from the Hellenistic to the Early Islamic Period* (Bartl K. and Hauser S. ed.) BBVO 17: 229 – 58.

Northedge, A., Bamber A. and Roaf M. 1988. *Excavations at Ana Qala'a Island.* Iraq Archaeological Reports 1.

Northedge, A. and Kennet D. 1994. The Samarra Horizon. In *Cobalt and Lustre, The first Centuries of Islamic Pottery (Grube, E. ed.) The Nasser D. Khalili Collection of Islamic Art Vol. 9.* London: 21 – 35.

Orton-Pinto, N. 2001. *Ancient Sea-faring in Early Historic Gujarat.* (Unpublished Ph.D Thesis, University of Pennsylvania). UMI Microfilms.

Paymaster, R. B. 1915. *Kisse-i-Sanjan.* Mumbai.

Priestman, S. M. N. 2005. *Settlement and Ceramics in the Southern Iran: An Analysis of Sasanain and Islamic Periods in the Williamson Collection.* (Unpublished M. A. Thesis, University of Durham).

Rajaguru, S. N and Deo S. G. 2005. Coastal Geomorphology of Sanjan. *Journal of Indian Ocean Archaeology* 2: 93 – 98.

Ray, H. P. 2003. *The Archaeology of Seafaring in Ancient Southeast Asia.* London.

Rougeulle, A. 1996. Medieval Trade Networks in the Western Indian Ocean (8th - 14th cenury). Some Reflections from the Distribution Pattern of Chinese Imports in the Islamic World. In Tradition and Archaeology – Early Maritime Contacts in the Indian Ocean (Ray, H. P and Salles F. J eds.), *Proceedings of the International Seminar Techno-Archaeological Perspectives of Sea-faring in the Indian Ocean (4th cent. B. C – 15th cent. A.D).* New Delhi : 158 – 80.

Sachau, E. C. (ed.). 1983. *Alberuni's India Vol. I and II.* New Delhi.

Sankalia, H. D. 1983. Sanjan – A Miniature Bombay in c. A. D. 980. In *D. C. Sircar Festschrift, Sri Dinesachandrika Studies in Indology* (Mukherjee, B. N., Das B. R., Biswas S. S. and Singh S. P. eds.). Delhi: 209 – 13.

Sankalia, H. D., Deo S. B. and Ansari Z. D. 1960. *From History to Prehistory at Nevasa.* Poona.

Sankalia, H. D., Subbarao and Deo S. B. 1958. *Excavations at Maheshwar and Navdatoli.* Baroda.

Sankalia, H. D. and Dikshit M. G. 1952. *Excavations at Brahmapuri (Kolhapur) 1945 – 46.* Poona.

Sankalia, H. D. and Deo S. B. 1955. *Excavations at Nasik and Jorwe 1950 – 51.* Poona.

Sedov, A. V. 1996. Qana (Yemen) and the Indian Ocean: The Archaeological Evidence. In *Tradition and Archaeology – Early Maritime Contacts in the Indian Ocean (Ray, H. P and Salles F. J eds.), Proceedings of the International Seminar Techno-Archaeological Perspectives of Sea-faring in the Indian Ocean (4th cent. B. C – 15th cent. A.D).* New Delhi: 11 – 35.

Sinclair, R. 1967. Port-Hinterland-Foreland-Maritime Space: Relationships of the Port of Detroit. In *Feschrift Leopold G. Sheidel.* Vienna: 356-374.

Sircar, D. C. 1963. Veraval Inscription of Chalukya Vaghela Arjuna, 1264 A. D. *Epigraphia Indica* Vol. XXXIV (Sircar, D. C. ed.) (1961 – 62): 146.

Sircar, D. C. 1963. Nagarjunakonda Inscription of the time of Abhira Vasushena, Year 30. *Epigraphia Indica* Vol. XXXIV (Sircar, D. C. ed.) (1961 – 62): 197 – 204.

Sircar, D. C. 1962. Chinchani Copper-plates. Epigraphia Indica Vol. XXXII (1957 – 58) (Sircar, D. C. and Chhabra B. ed.).

Stillman, N. A. 1973. The Eleventh century Merchant House of Ibn 'Awkal (a Geniza Study). *Journal of the Economic and Social History of the Orient* 16: 15 – 88.

Tampoe, M. 1989. *Maritime Trade between China and the West. An Archaeological Study of the Ceramics from Siraf (Persian Gulf), 8th to 15th century A.D.* BAR International Series 555. Oxford.

Thakkar, P. S., Raval M. H. and Dasgupta A.R. 1993. *Study of Ancient Ports of Gujarat using Remote Sensing Data.* (Paper presented at the International Workshop on Application of Remote Sensing for Archaeology and Related Disciplines, Hydrabad.)

Unvala, J. M. 1951. *Report on the Dokhmas of Tena.* Bombay.

Watson, O. 2004. *Ceramics from Islamic Lands. The Al Sabah Collection.* London.

Weigend, G. 1958. Some Elements in the Study of Port Geography. *Geographical Review* 48: 185-200.

White house, D. and Williamson A. 1973. Sasanain Maritime Trade. *Iran* 11: 29 – 49.

Whitehouse, D. 1968. Excavations at Siraf. First Interim Report. *Iran* Vol. 6: 1 – 22.

Whitehouse, D. 1971. Excavations at Siraf. Fourth Interim Report. *Iran* Vol. 9: 1 – 17.

Whitehouse, D. 1973. Chinese Stoneware from Siraf: the Earliest Finds. In *South Asian Archaeology: Papers from the First International Conference of South Asian Archaeologists held in the University of Cambridge* (Hammond, N. ed.). London: 241 – 55.

Williamson, A. 1972. Persian Gulf Commerce in the Sasanian Period and the First Two centuries of Islam. *Bastan Shanasi va Honar-e-Iran* 9 – 10: 97 – 112.

Wink, A. 2002. *Al-Hind – The Making of the Indo-Islamic World* Vol. I. New York.

Zhang Jun-Yan, 1983. Relations between China and the Arabs in Early Times. *Journal of Oman Studies* 6: 91 – 109.

* * *

APPENDIX 1

TERMINOLOGY OF PERIODISATION

Period	Subdivision	Date (AD, AH / AD)
Parthian		150 - 226 AD
	Late Parthian	2nd cen. AD
Sasanian	Early	3rd - 4th cen. AD
	Middle	4th - 5th cen. AD
	Late	6th - 7th cen. AD
Early Islamic	Ummayad	11 AH / 632 AD - 132 AD/750 AD
	Early Abbasid	132 AD/750 AD-Early 3rd cen AH/9th cen AD
	Samarran	Early - Late 3rd cen AH / 9th cen.AD
	Post-Samarran	4th cen.AH/10th cen AD-Early 5th cen. AH/11th cen. AD
Middle Islamic	Late Abbasid	Early - Late 6th cen. AH / 12th cen. AD - 7th cen. AH 13th cen. AD
	II-Khanid	Late 7th cen AH / 13th cen AD - 8th cen AH / 14th cen.AD
Later Islamic	Ottoman and Mogul	Post 8th cen. AH / 14th cen. AD - 13th cen. AH / 19th cen. AD

APPENDIX 2

Persian	Chinese	Indian
226-641 Sasanid	581-618 Sui	89-305 Western Kshatrapas
	618-684 Early Tang	319-470 Guptas
661-750 Umayyad	684-756 High Tang	500-649 Valabhis
		470-900 Chalukyas
		580-734 Gurjaras
750-1258 Abbasid	756-827 Middle Tang	630-972 Rashtrakutas
	827-907 Late Tang	
	907-960 Five Dynasties	900-1350 Silharas
	960-1127 Northern Song	
1037-1231 Saljuq	1127-1279 Southern Song	
1227-1336 II Khanid	1271-1368 Yuan	
1385-1500 Timurid	1368-1644 Ming	

APPENDIX 3 - WARE CLASSIFICATION

No.	WARE	WARE CLASS	SUB-CLASS	FABRIC	PETRO. GROUP	COLOUR (surface)	COLOUR (fabric)
1.	Turquoise Glazed Ware	TGW	TGW-1	BUFF(A)	D, E		5Y 8/4, 5Y 8/2
			TGW-2	BUFF(B)	D		2.5Y 8/4
			TGW-3	BUFF(A)			5Y 8/4, 5Y 8/2
			TGW-4	BUFF(A)	E		5Y 8/4, 5Y 8/2
			TGW-5	BUFF(A)	E		5Y 8/4, 5Y 8/2
			TGW-6	BUFF(C)			
2	White Glazed Ware	WGW		BUFF(A)	D		5Y 8/4
3	Splashed White Glazed Ware	SWGW		BUFF(A)	D		5Y 8/4
4	Cobalt Painted Ware	CPW		BUFF(A)	E		5Y 8/4
5	Lustre Painted Ware	LPW	LPW-1	BUFF(A)	D		5Y 8/4, 5Y 8/3
			LPW-2	BUFF(A)			5Y 8/4, 5Y 8/3
			LPW-3	BUFF(A)			5Y 8/4, 5Y 8/3
6	Bichrome Glazed Pink Ware	BiGP		PINK(B)			7.5YR 7/4,
7	Hatched sGraffiato Ware (Polychrome)	HsG(P)		PINK(B)	A		7.5YR 7/4, 5YR 7/4
8	Hatched sGraffiato Ware (Bichrome)	HsG(B)		PINK(B)	A		7.5YR 7/4,
9	Hatched sGraffiato Ware (Monochrome)	HsG(M)		PINK(B)			7.5YR 7/4,
10	Hatched sGraffiato Ware (White)	HsG(W)		PINK(B)			7.5YR 7/4,
11	Hatched sGraffiato Ware (Yellow)	HsG(Y)		PINK(B)			7.5YR 7/4,
12	Champleve	CHAMP		PINK(B)			5YR 7/4
13	Monochrome Glazed Pink Ware	MGP		PINK(A) PINK(F)			
14	Monochrome Glazed Buff Ware	MGB		BUFF(A) BUFF(E) BUFF(F)	H		
15	Cuerda Seca Glazed Ware	CSGW		PINK(B)			10R 6/3
16	Unique Glazed Pink Ware	UGP		PINK(A) PINK(B) PINK(D) PINK(E)			
17	White Glazed Pink Ware	WGPW		PINK(A) PINK(B)			5YR 7/3
18	Incised Glazed Pink Ware	IGP		PINK(D)			5YR 6/3
19	Splashed Glazed Pink Ware	SGPW		PINK(B)			
20	Yemen?	YEMEN		PINK(B)			
21	Khambat Type Glazed Ware	KTGW		RED(C)			
22	Glazed Red Ware	GRW		RED(B)			
23	Frit	FRIT		FRIT	J		
24	Eggshell Ware(white)	ESW(W)		BUFF(G)	G		2.5 Y 8/2
25	Eggshell Ware(white)	ESW(P)		PINK(G)			5YR 8/2
26	Eggshell Ware(white)	ESW(R)		RED(A)			5YR 6/6
27	Buff Ware	BW		BUFF(A) BUFF(E) BUFF(F)			2.5Y 8/2

No.	WARE	WARE CLASS	SUB-CLASS	FABRIC	PETRO. GROUP	COLOUR (surface)	COLOUR (fabric)
28	Buff Ware (Torpedo Jars)	BW(TJ)		BUFF(A) BUFF(E)	C, F		
29	Buff Ware(Storage Vessels)	BW(SV)		BUFF(E)	C, H		
30	Buff Ware(Spot)	BW(SPOT)		BUFF(H)			
31	Unglazed Pink Ware	UPW		PINK(A) PINK(B) PINK(E) PINK(F)			
32	White Slipped Pink Ware	WSPW	WSPW-1	PINK(D) PINK(F)	C		
			WSPW-2	PINK(D) PINK(E)	C		
			WSPW-3	PINK(B)	E		
33	Slipped Brittle Ware	SBW	SBW-1	PG(A)		10R 4/8	2.5YR 6/6
			SBW-2	PG(A)	C		2.5YR 6/6
			SBW-3	PG(A)			2.5YR 6/6
			SBW-4	PG(B)			
			SBW-5	PG(B)			
34	Large Incised Storage vessels	LISV	LISV-1	MR(A)	I	5R 4/1	10R 5/6
			LISV-2	MR(B)		10R 6/6	10R 6/6
			LISV-3	MR(A)	I	10R 3/2	10R 4/4
			LISV-4	MR(C)		5YR 6/3,	2.5YR 6/6
			LISV-5	MR(D)	I	10R 5/8	10R 5/8
35	Mica Red Ware	MRW	MRW-1	RED(M)			
			MRW-2	RED(M)			
36	Red Polished ware	RPW		RED(N)			
37	Black slipped Red Ware	BSGW		GREY(A)	B	2.5 N 2/0	2.5Y N2/0, 2.5Y N3/0,
38	Red Slipped Grey Ware	RSGW		GREY(A) GREY(B) GREY(C)		2.5YR	2.5Y N2/0
39	Grey Ware	GW	GW-1	GREY(A)			
			GW-2	GREY(C)			
			GW-3	GREY(B)			
40	Red Slipped Red Ware	RSRW	RSRW-1	RED(C) RED(D)	B		10R 5/6
			RSRW-2	RED(C) RED(E)			10R 5/6
41	Black Slipped Red Ware	BSRW	BSRW-1	RED(C)	B	2.5 N 2/0	10R 5/6
			BSRW-2	RED(C)			
42	White Slipped Red Ware		WSRW	RED(C)			
43	Red Ware	RW	RW-1	RED(C)	B		5YR 5/6
			RW-2	RED(F)			
			RW-3	RED(C)			

No.	WARE	WARE CLASS	SUB-CLASS	FABRIC	PETRO. GROUP	COLOUR (surface)	COLOUR (fabric)
44	Miscellaneous	MISC	MISC-1	GREY(A)			
			MISC-2	GREY(B)			
			MISC-3	RED(F)			
			MISC-4	RED(G)			
			MISC-5	RED(G)			
			MISC-6	RED(H)			
			MISC-7	RED(I)			
			MISC-8	RED(J)			
			MISC-9	PINK(H)			
			MISC-10	RED(K0			
			MISC-11	RED(L)			
45	Changsha Underglaze Painted Ware	CHANGSHA		CH(A)			
46	Stoneware	STONE	STONE-1	S(A)			
			STONE-2	S(B)			
			STONE-3	S(C)			
			STONE-4	S(A)			
			STONE-5	S(D)			
			STONE-6	S(E)			
47	Celadon	CEL	CEL-1	C(A)			
			CEL-2	C(B)			
48	Porcelain	PORC	PORC-1	P(A)			
			PORC-2	P(B)			
			PORC-3	P(A)			
			PORC-4	P(A)			
			PORC-5	P(A)			

APPENDIX 4 : REPORT OF THE PETROGRAPHIC ANALYSIS OF CERAMICS FROM SANJAN

- Prof. K. Krishnan

Thin-sections of 45 representative ceramics from Sanjan were chosen to conduct a pilot study. The objectives of this study are:

(i) to characterize the ceramics from Sanjan into different petrographic groups based on its composition

(ii) to see how far these petrographic groups are comparable to the archaeological categories.

(iii) to understand the geological provenance of the raw-material,

(iv) to understand the texture of the clay paste,

(v) to understand the technique of manufacturing, and

(vi) to understand the microstructural features of different wares.

The archaeological classification of the ceramics brought to light varieties of wares - West Asian glazed and unglazed earthenwares, indigenous slipped and unslipped wares and Far Eastern wares. Representative samples were selected for thin-section analysis. The thin-sections were studied using a polarizing microscope [Leitz Laborlux (12 Pol D)]. The minerals were viewed in plane polarized light (PPL) and crossed polarized light (XPL) and identified with the help of their optical properties. Grain size measurements of the representative area of each fabric were done with the help of point counting (James Swift Automatic Point Counter with a stepping stage attachment to it). The point counting was done by setting the stepping stage movement to single jump.

Based on the mineralogy the thin-section samples were grouped into ten main groups, namely A, B, C, D, E, F, G, H, I and J. The criteria chosen for defining the group was the mineralogical composition of the grains.

The thin-sections falling in Fabric A had a fairly micaceous matrix. It is made of fine grained clay. The matrix is non-pleochroic in Plain Polarised Light (PPL) and fairly birefringent in Crossed Polarised Light (XPL). The grains were sub- angular to sub-rounded in shape. Their sizes varied from 80 μ to 120 μ. Its frequency fell between 5 to 10%. The grains show more or less a parallel orientation. Voids were very few. Most of them appeared to be areas of grain fallout. The mineralogy of the detritals included quartz, plagioclase feldspar (dominant), mica and sparse occurrence of patches of hematite. The non-plastic inclusions in these sections were well sorted (Fig 1). Their grain-size distribution character was nearly unimodal (Fig 1).

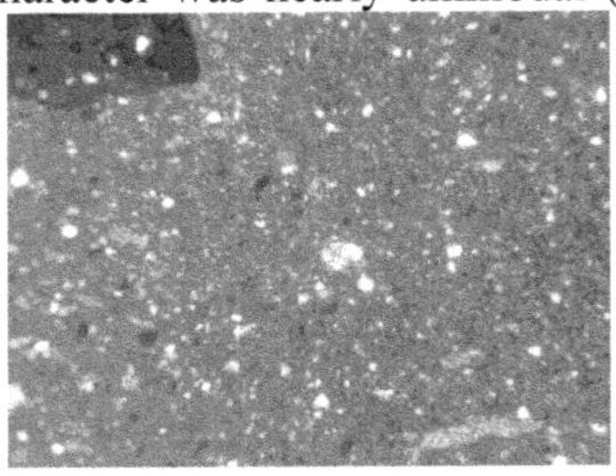

Figure 1 : Photomicrograph of specimen number 2 HsG (B) – Pink (B) falling in petrographic group Fabric A. Majority of the grains seen are quartz and plagioclase feldspar. These have sub-angular to sub-rounded shape. The longitudinal fragments show more or less a parallel orientation. Frequency of grains is 5% (field of view 1.4 sq. mm.).

The matrix of the samples falling under fabric B is coarse grained, less compact and extremely ferruginous. It is non-pleochroic and has a brownish color in PPL which turns to dark brown in XPL in XPL. The grain shape in this group varied from angular to sub-angular. The largest grains in this section were 200 μ, 300 μ, 360 μ, 400 μ, 600 μ and 1600 μ. The grains were well sorted. The frequency of non-plastic inclusions in various samples varied from 40 to 50%. The grain size distribution character in all samples was unimodal. The non-plastic inclusions in majority of the samples did not show any particular orientation except in two samples which showed more or less parallel orientation. The detritals in the thin-sections of the samples were constituted by quartz, plagioclase feldspar, microcline feldspar, augite, mica, bioclasts, cryptocrystalline silica, basalt fragments and argillaceous inclusions.

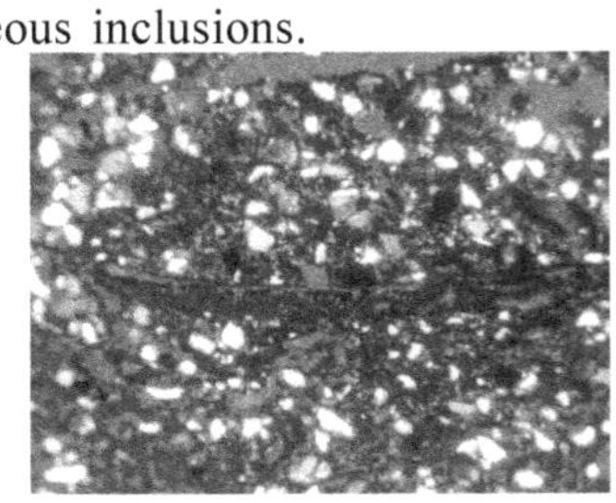

Figure 2 : Photomicrograph of specimen number 36 SGW-1 - Grey (A) falling in petrographic group Fabric B. The grains that are seen are quartz, plagioclase feldspar and bioclast (note the longitudinal one in the centre). The grains have angular to sub-angular shape. The grains do not show any orientation (field of view 1.4 sq. mm.)

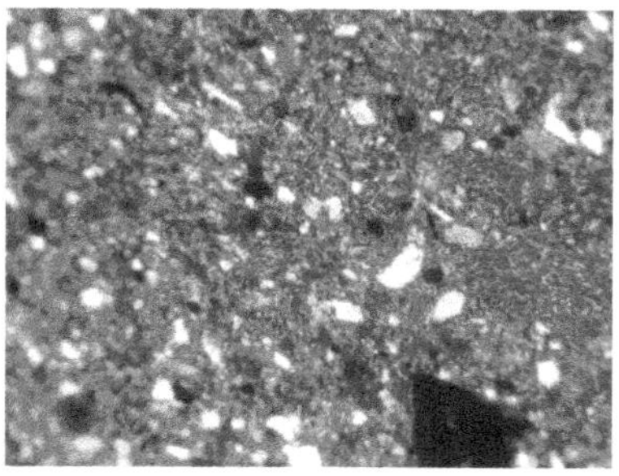

Figure 3 : Photomicrograph of specimen number 37 falling in petrographic group Fabric C. The grains have angular to sub-angular shape. They show more or less parallel orientation. Quartz and mica are visible in the photomicrograph (field of view 1.4 sq. mm.)

The matrix of the section falling under the group fabric C is medium grained. It is ferruginous and non-pleochroic in PPL and dark brown in XPL. The grains are mostly sub-angular. The largest grain sizes observed in the thin-sections were of 200 μ, 240 μ, 260 μ, 300 μ and 340 μ. They are well sorted and the frequency of grains varies from 20-50%. The grain-size distribution pattern is unimodal. The grains do not show any kind of orientation in most of the cases except in one case where it shows a parallel orientation. Voids are few. The dominant minerals are quartz, plagioclase feldspar, microcline feldspar, cryptocrystalline silica, argillaceous inclusions, augite, mica, very few bioclasts and few patches of hematite.

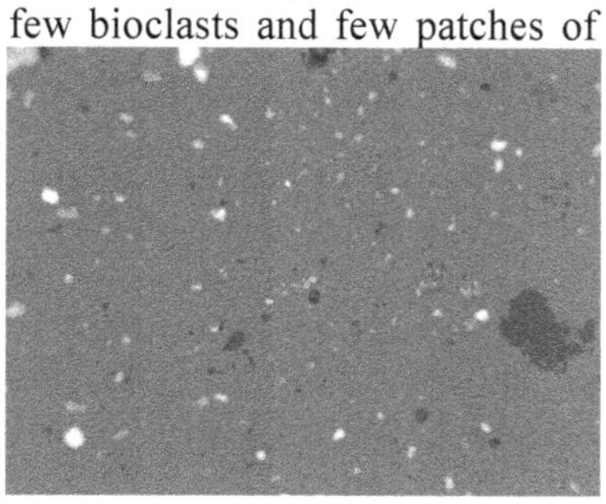

Figure 4 : Photomicrograph of specimen number 29 falling in petrographic group Fabric D. The grains have sub-angular to sub-rounded shape. Most of the grains seen in the photomicrograph are quartz (field of view 1.4 sq. mm.).

The thin -sections falling under the group fabric D had a fine grained smooth matrix. The samples were non-pleochroic in PPL and dark grey in XPL. The grains were sub-angular to sub-rounded in shape. The largest grains observed in the samples had sizes of 160 μ and 700 μ. The non-plastic inclusions were perfectly sorted to well sorted and had a frequency of 5 to 10%. The grain size distribution character was unimodal (Fig 5). Most of the grains showed more or less a parallel orientation. Voids were few, the majority being grain fallout. The detritals included quartz, mica, plagioclase feldspar and a few patches of hematite.

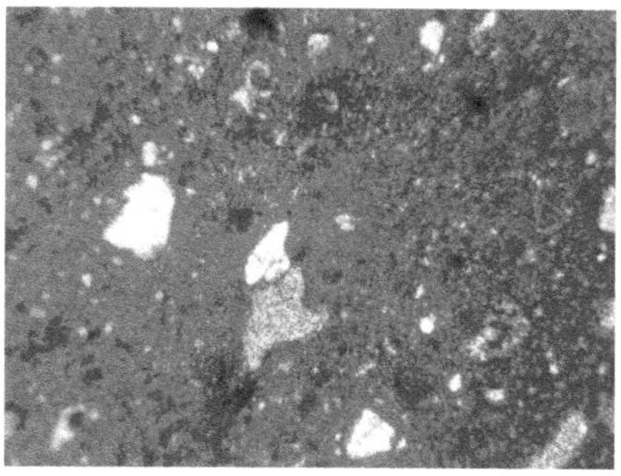

Figure 5 : Photomicrograph of specimen number 30 falling in petrographic group Fabric E. The grains have sub-angular to sub-rounded shape (field of view 1.4 sq. mm.).

The thin-sections falling under the category of fabric E had a fine grained grayish matrix. These samples are non-pleochroic in PPL and dark grey in XPL. The grains were angular to sub-angular to sub-rounded in shape. The highest grain size in each sample was 80 μ, 100 μ, 200 μ and 600 μ. The grains were perfectly sorted to well sorted and had a frequency of < 5-10%. The grain-size distribution pattern was mostly unimodal. Some of the sections had grains showing parallel orientation, with some having no definite orientation. Voids were few and most of them were grain fallouts. The non-plastic inclusions included quartz, plagioclase feldspar, microcline plagioclase, augite and mica. Patches of hematite was also observed.

The thin-section falling under the category of fabric F had a coarse grained ferrugenous matrix. This sample is non-pleochroic in PPL and dark red in XPL. The grains were angular to sub-angular in shape. The highest grain size in the sample was 300 μ. The grains were well sorted and had a frequency of 40%. The grain-size distribution pattern was unimodal. Grains did not show any definite orientation. Voids were many and most of them were grain fallouts. The non-plastic inclusions included quartz, plagioclase feldspar, augite, mica, crypto crystalline silica and bioclasts. Patches of hematite was also observed.

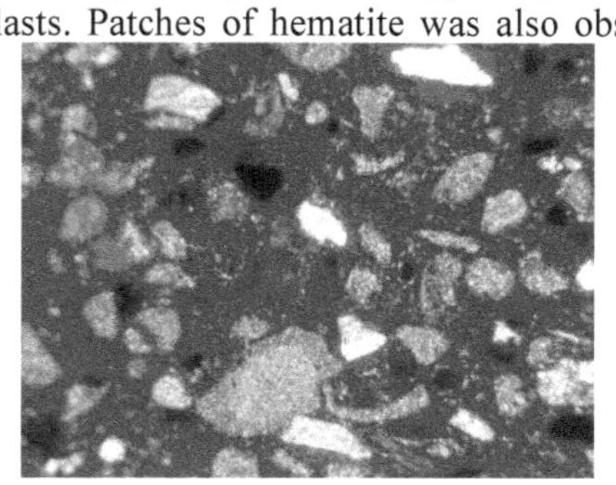

Figure 6 : Photomicrograph of specimen number 16 falling in petrographic group Fabric F. The grains have sub-angular to sub-rounded shape. The grains are well

sorted and do not show any definite orientation (field of view 1.4 sq. mm.).

The thin-sections falling under the category of fabric G had a fine grained smooth matrix. These samples are non-pleochroic in PPL and dark grey in XPL. The grains were angular to sub-angular in shape. The highest grain sizes in the samples were 120 μ and 160 μ. The grains were perfectly to well sorted and had a frequency of 5%. The grain-size distribution pattern was unimodal. Grains showed a parallel orientation. Voids were very few and most of them were grain fallouts. The non-plastic inclusions included quartz, plagioclase feldspar, and mica.

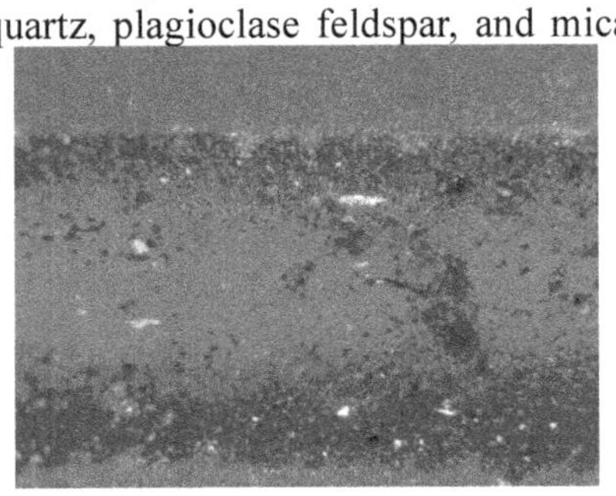

Figure 7: Photomicrograph of specimen number 9 falling in petrographic group Fabric G. The grains have angular to sub-angular shape. The grains are well sorted and show a parallel orientation (field of view 1.4 sq. mm.).

The thin-sections falling under the category of fabric H had a fine grained fairly micaceous matrix. These samples are non-pleochroic in PPL and grey in XPL. The grains were angular to sub-angular in shape. The highest grain sizes in the samples were 160 μ and 200 μ. The grains were well sorted and had a frequency of 20 to 30%. The grain-size distribution pattern was unimodal. Grains did not show any orientation. Voids were very few and most of them were grain fallouts. The non-plastic inclusions included quartz, plagioclase feldspar, cryptocrystalline silica, augite and mica. There are a few hematite patches also.

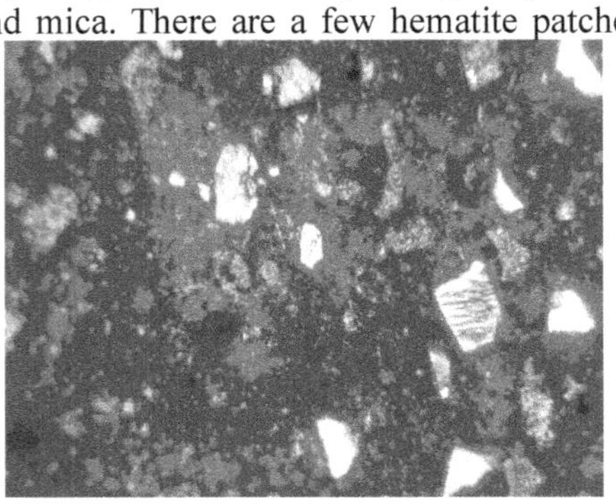

Figure 8 : Photomicrograph of specimen number 43 falling in petrographic group Fabric H. The grains have angular to sub-angular shape. The grains are well sorted and do not show any orientation (field of view 1.4 sq. mm.).

The thin-sections falling under the category of fabric I had a medium to fine grained matrix. One of them was a micaceous matrix and showed slight pleochroism in PPL. The other samples are non-pleochroic in PPL and red to brown in XPL. The matrix looked ferruginous. The grains were angular to sub-angular in shape and in one case large number of flaky grains were found. The highest grain sizes in the samples were 50 μ, 60 μ, 80 μ, 100 μ, 120 μ and 400 μ. The grains were perfectly to well sorted and had a frequency of < 5 to 30%. The grain-size distribution pattern was unimodal. Grains did not show any orientation. Voids were very few to many and most of them were grain fallouts. The non-plastic inclusions included quartz, plagioclase feldspar and mica. There are a few hematite patches also.

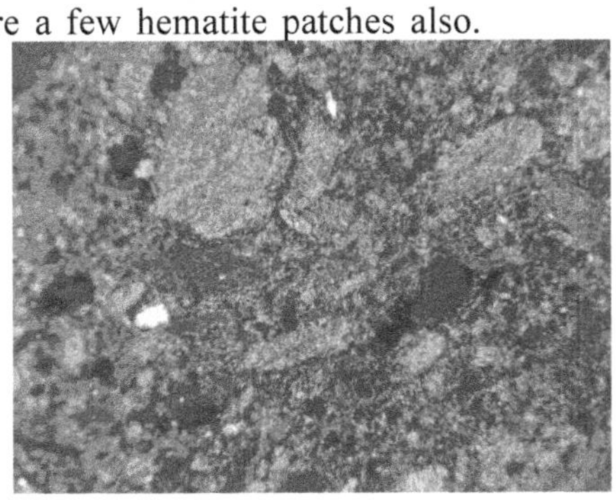

Figure 9 : Photomicrograph of specimen number 17 falling in petrographic group Fabric I. The grains have angular to sub-angular shape. These are well sorted and do not show any orientation. Flakes of mica and fragments of quartz are visible (field of view 1.4 sq. mm.).

The thin-section falling under the category of fabric J had a fine grained matrix. The sample is non-pleochroic in PPL and red in XPL. The matrix looked ferruginous. The grains were angular to sub-angular in shape. The highest grain size was 200 μ. The grains were perfectly to well sorted and had a frequency of 10 to 15%. The grain-size distribution pattern was unimodal. Grains did not show any orientation. Voids were many and most of them were grain fallouts. The non-plastic inclusions included quartz, plagioclase feldspar, cryptocrystalline silica and mica.

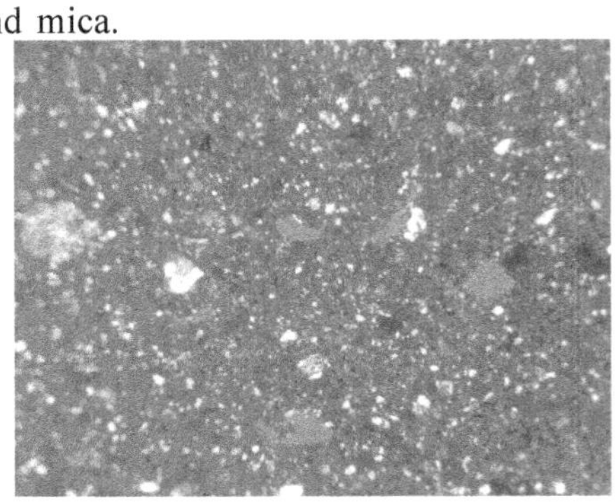

Figure 10 : Photomicrograph of specimen number 21 falling in petrographic group Fabric J. The grains have sub-angular to sub-rounded shape. Quartz and plagioclase

feldspars are visible in the photomicrograph (field of view 1.4 sq. mm.).

A few comments

The mineralogy of the non-plastic inclusions in each fabric is given below allows one to compositionally group them into different geological terrains :

(i) basaltic source- does contain basalt fragments and its components + bioclast

(ii) basaltic source Ia- has basaltic components and no bioclast

(iii) basaltic source Ib- has basaltic components and argillaceous inclusions

(iv) mica quartz- Very high mica (62.5) with quartz as the second (less than half)

(v) quartz- very high quartz (64.1), second feldspar (less than half)

(vi) high feldspar – feldspar high (47.7), quartz second (38.6), mica (4.5)

(vii) feldspar – feldspar (48.3), quartz (31), mica (3.4)

Fabric groups	Mineralogy	Texture
A	quartz, plagioclase feldspar (dominant), mica and sparse occurrence of patches of hematite	sub- angular to sub-rounded high feldspar IV
B	quartz, plagioclase feldspar, microcline feldspar, augite, mica, bioclasts, cryptocrystalline silica, basalt fragments and argillaceous inclusions	angular to sub-angular basaltic I
C	quartz, plagioclase feldspar, microcline feldspar, cryptocrystalline silica, argillaceous inclusions, augite, mica, very few bioclasts and few patches of hematite	Sub-angular to Basaltic Ib
D	quartz, mica, plagioclase feldspar and a few patches of hematite.	sub-angular to sub-rounded quartz-feldspar VI, VII
E	quartz, plagioclase feldspar, microcline plagioclase, augite and mica. Patches of hematite	angular to sub-angular basaltic Ia
F	quartz, plagioclase feldspar, augite, mica, crypto crystalline silica and bioclasts. Patches of hematite	angular to sub-angular basaltic Ia
G	quartz, plagioclase feldspar, and mica	angular to sub-angular feldspar V
H	quartz, plagioclase feldspar, cryptocrystalline silica, augite and mica. There are a few hematite	angular to sub-angular basaltic Ia
I	quartz, plagioclase feldspar and mica. There are a few hematite patches	angular to sub-angular mica*-quartz II
J	quartz, plagioclase feldspar, cryptocrystalline silica and mica.	angular to sub-angular quartz III

It appears that the raw material clay belong to four geological sources. These three sources may be said as

1) the clay close to a basaltic region,

2) a clay that has inclusions of basaltic minerals

3) rich in quartz

4) mica one

	RN/2	RN/9	RN/16	RN/17	RN/21	RN/26	RN/29	RN/30	RN/43	RN/37	RN/36
Quartz	38.6	31	13.9	30.6	64.1	40	53.6	16.7	14.9	40.2	76.3
Pl.feld	47.7	48.3	58.3	6.9	23.1	47.1	39.3	63.3	72.3	41.3	16.3
Mica	4.5	3.4	0.9	62.5	7.7	4.3	3.6	10		8.7	1.3
Hematite	9.1	11.4	19.4		2.6	1.4	3.6	10	6.4	5.4	1.3
cry.cry.sil			6.5		2.6	1.4				1.4	1.3
Bioclast			0.9								2.5
Augite						5.7			6.4	2.2	1.3
Argilla.										1.4	

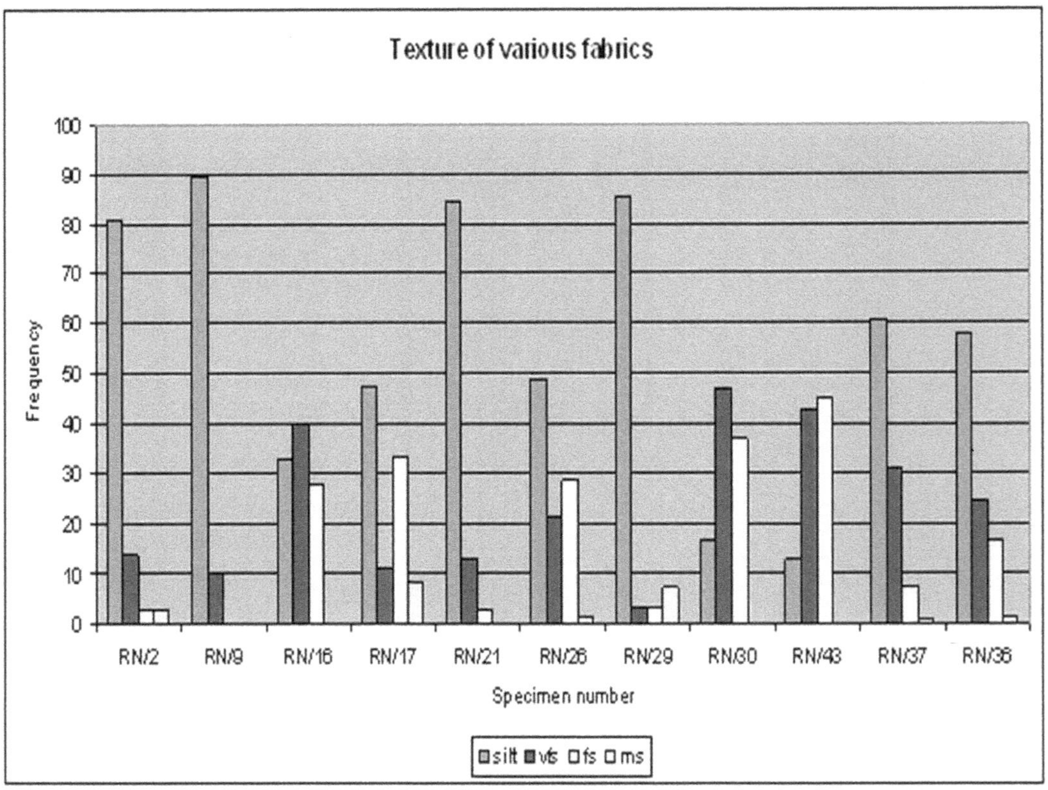

Textural group

	RN/2	RN/9	RN/16	RN/17	RN/21	RN/26	RN/29	RN/30	RN/43	RN/37	RN/36
silt	80.6	89.7	33	47.2	84.6	48.6	85.7	16.7	12.8	60.4	57.7
vfs	13.9	10.3	39.6	11.1	12.8	21.4	3.6	46.7	42.6	30.8	24.4
fs 2.8		27.4	33.3	2.6	28.6	3.6	36.7	44.7	7.7	16.7	
ms	2.8			8.3		1.4	7.1			1.1	1.3

(i) silt tempered (2, 9, 21, 29) (ii) sand tempered (16, 30, 43) (iii) silt-sand tempered (17, 26, 36)

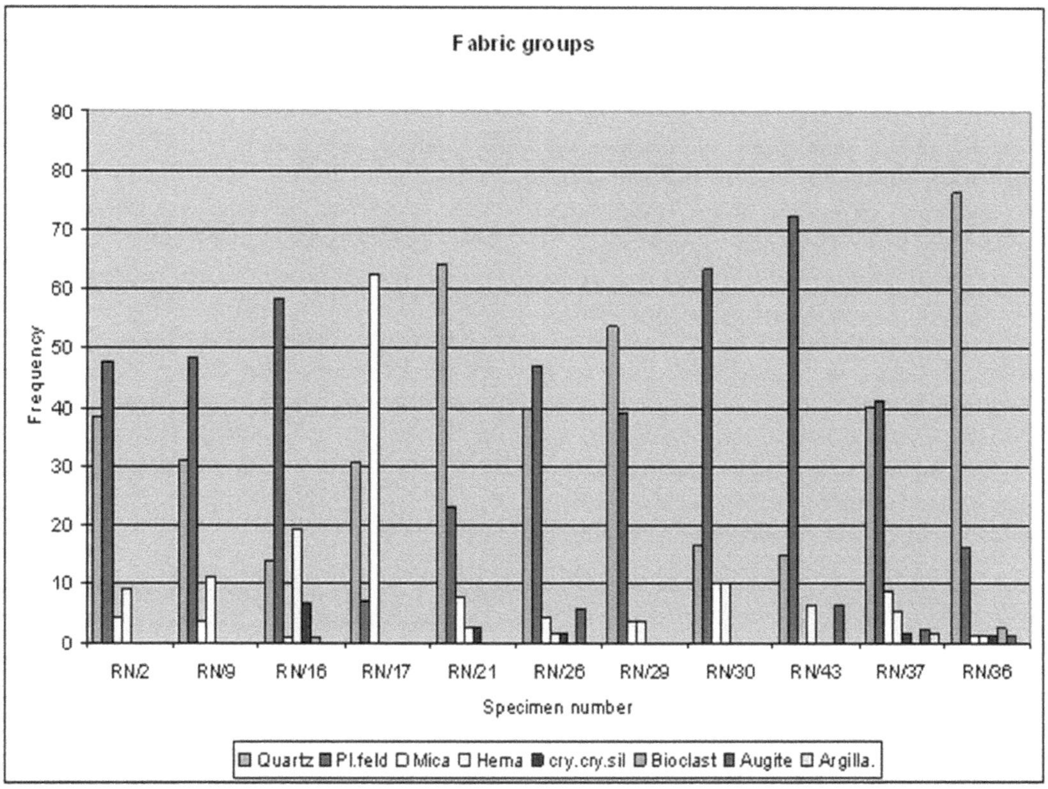

(Images at :_
http://www.wzcf.org/sanjan-excavations/pottery.html.)

* * *

APPENDIX 5 : FABRIC DESCRIPTION

Changsha :

Fabric : - CH(1) This fabric is greyish – buff stone ware thickly potted and several small black and white specks can be seen. In part the firing has caused the fabric to run pinkish and grey. A few air holes are evident. The surface is treated to a cream or white slip upon which the painting is executed. The potting is thick and glaze on exterior is partial. The translucent glaze has degraded to a creamy brown. There is fine crazing which be seen under the lens.

Porcelain :

Fabric : - P(A) This fabric is high fired white porcelain. It has very few inclusions. Sharp glass like fracture. Occasionally small black inclusions are seen. The clay is very well levigated and there are few air holes. The different porcelain classes that have this fabric are PORC – 1 (White glaze on white fabric), PORC – 3 (green glazed porcelain) and PORC – 4 (Blue and white porcelain).

Fabric : - P(B) This fabric is creamy white with small black inclusions and air-holes. It is not as fine as white porcelain. It resembles the 'CHANGSHA' fabric and appears to be an intermediate stage between porcelain and stone ware. The class of porcelain which has this fabric is PORC – 2. (cream white glaze on a cream white fabric).

Celadon :

Fabric: - C (A) This fabric is a porcelaineous stone ware having a fine grey well levigated and well fired body. The glaze is in different shades ranging from light olive-green to a pale-apple-green. Most of the shards at SJN – 2004 have a thin glaze which appears slightly translucent. The class that have this fabric is 'CEL'.

Fabric : - C(B) This fabric is extremely fine and compact, light grey in colour with a translucent green glaze. This appears to be a developed form of 'C(A)'. The class code of the ware is CEL-1. Apparently this class can have variation in the basic fabric.

Stone ware :

Fabric : - S (A) It is a semi-coarse grey ware, not always evenly fired. The fabric ranges from grey to pink. It can also be a creamy-grey colour. Sometimes fine inclusions and air-holes are visible. The class codes in S (A) are Stone – 1 (Bichrome glazed – usually in shades of brown and green). And in the same class, unglazed.

Fabric : - S (B) This fabric is extremely compact stone – like, and very high fired. A few black fine inclusions may be seen and almost no air-holes. The glaze is usually in shades of green. Occasional misfiring may make the fabric turn pink. The class code of the wares in this fabric are Stone – 2 (Dusun type), Stone – 3 (Yueh type).

Pink Ware :

Fabric : - Pink (A) The fabric is reddish-pink with very fine small white inclusions. The clay is well levigated and slightly sandy. A few air-holes can be seen. In some samples the fabric can be semi-coarse and fired to a greyish-pink instead of a dark reddish –pink. The classes in this fabric are MGP, BiGP, WGP and UPW. This fabric is however is not very common and only a few samples are seen.

Fabric : - PINK (B) This is a fine, soft, kaolinite like fabric with occasional fine black or white inclusions and few air-holes. The colour can range from a soft creamy pink to dark salmon-pink and sometimes grey due to firing. The fabric is usually well levigated, wheel thrown (fast) and the texture is often chalky. When glazed a slip can be seen underlying the glaze. Sometimes small quartzite like inclusions (crystalline). This fabric is more or less even. The classes within this fabric are MGP, WGP, Bi GP, STAR, H.sG (M), H.sG(P), H.sG(B), GPW(?), WSPW and UPW.

Fabric : - Pink (C) This is a semi-coars fabric having many large black inclusions and colour spots. The colour ranges from pink to grey. The surfaces often pitted due to the loss of impurities. They are not always well fired vessels. This fabric is not very common.

Fabric :- Pink (D) This fabric is reddish–pink, sandy, semicoarse and with a fine white inclusions and many air-holes. It is almost like a sandy red ware. But it is not so dark in colour. It can be thinly potted as in IGP (W) or thickly potted as in WSPW – 1. In the thickly potted vessel larger inclusions and grog can be seen and the core can range from a brownish-grey to a reddish – pink.

Fabric :- Pink (E) This fabric is mostly levigated and sturdy, despite air-holes. It is quite compact. Some fine black inclusions, occasionally grog and a few fine shiny specks of what could be mica or some crystalline inclusions can be seen. This fabric is very similar to Buff (E) torpedo jars and storage vessels. This fabric is not seen in the finer and smaller vessel shapes. The fabric can be fired from a pink-red to a grey. Mostly in large vessels in WSPW – 2 have bitumen coating on the inside and a white slip or wash to the exterior. The classes represented by this fabric are WSPW – 1 and WSPW – 2/

Fabric : - Pink (F) This fabric is very similar to Pink (B). It is soft, chalky kaolinite and has creamy- pink colour (almost like the soft chalky, white Buff ware). There are some white inclusions and perhaps grog. But the fabric is evenly and finely levigated despite this. The classes in this fabric is UPW and thickness of most shards implies that this fabric was used for manufacturing large storage vessels or vases.

Fabric : - Pink (G) this fabric is almost exlusively used in Eggshell Ware Class ESW (P). It is a pale reddish- pink with fine specks of mica on both surfaces and few sections. The mica has a reddish gold (copper) tone which may be a reflection of the surrounding baric, this has to be verified. The texture of this fabric is not as soft or smooth as pink (B) or even Buff (G). Sometimes a white or cream was is seen on the exterior.

Fabric : - Pink (H) This fabric is yet to be verified. It is a hard semicoarse fabric having a range from purplish pink to a greyish –red. The fracture is laminated and also sharp. There may be some inclusions of grog and few specks of mica. Some air holes are seen in section. Some small white implosions can be seen on the surface. The texture is brittle and hard. Sometimes a white thin wash can be seen on the vessel. The only samples in this fabric have been designated MISC – 9 since these wares are not been identified definitely.

Slipped Brittle Ware :

Fabric :- PG (A) This fabric is semi coarse, fired hard, and sand tempered. The texture is brittle and the grog, black inclusions and air-holes can be seen. The surface treatment on this fabric is usually slip. The colour of the fabric ranges from grey to pink. Often the slip appears to be applied with a brush or some applicant and the horizontal streaks can be seen. The fabric can be quite coarse and sometimes the inclusions make the surface (interior) pitted and rough. Occasionally a few specks of mica are seen but not commonly. The fabric can be quite sturdy and compact. The classes represented by this fabric are SBW– 1, SBW - 2, SBW- 3 (Rare).

Fabric : - PG (B) This fabric is more red in colour and coarser than PG(A). The grog is more evident and there is evidence of mica (red-gold) on some shards. Some inclusions are also seen. The surface treatment of this fabric is slightly different. Similar shards have been found at Chaul and the texture is leathery. Most of the shards appears to be thinly potted. It is not a common fabric at SJN. The classes represented are SBW – 3.

Buff Ware :

Fabric : - Buff (A) This fabric is a white earthen ware/ kaolinite having a creamy yellow colour. There are some air-holes and the fabric is semi – soft and chalky. There are some air – holes and occasional inclusions mostly white which could be either quartzite particles or even frit. The fabric can sometimes be fired to a light pink or creamy white and have a slightly harder texture. However it is usually a creamy yellow and quite soft. The clay appears to be well levigated. Usually these vessels are glazed and the glaze sometimes shows vitrification, air-bubbles or evidene of over firing. The classes represented in this fabric are numerous – TGW – 1, TGW – 3, TGW – 4, TGW – 5, LPW – 1, WGW, CPW, CSW, BW, BW(TJ).

Fabric : - Buff (B) It is a sandier fabric, slightly coarser in texture. There are many air holes and the fabric can be fired to a light pink. It is a variation of Buff (A). There are some black inclusions and occasionally white. Some botanical inclusions can be seen on some of the shards where the glaze is eroded. The colour ranges from creamy white to a pale pink or even a reddish- pink. The hardness of the fabric is not so much because of the clay but because of the firing. The classes represented in this ware are TGW – 1, TGW-2, TGW – 4, MGB.

Fabric :- Buff (C) This fabric is a very fine, soft, chalky, yellow with very well levigated clay. It can be fired to a very pale pink. There are almost no inclusions and air-holes. This is a rare fabric and may be a variation of previous two. The classes represented are TGW – 6 and MGB.

Fabric :- Buff (D) This fabric is a pale creamy yellow fired quite hard as compared to some of the other buff fabrics. It is compact and not chalky. There are a few air-holes and fine black inclusions but this fabric is more brittle and has been noticed in only one instance of WGW.

Fabric :- Buff (E) The fabric is creamy white and compact with some air holes and few small inclusions. It is fired hard and as a cement like texture. The core can be fired to a pinkish tone. Usually this fabric is thickly potted. Sometimes it is chalky and the surfaces are not treated to a glaze or a wash. Sometimes due to overfiring the fabric can be lose its density and have pumice stone texture. In the case of Buff (TJ) a white wash on the exterior may have been present but cannot be seen now. Bitumen however can be seen on the interior of these vessels. Other than BW(TJ) this fabric is seen in

BW and BW(SV).

Fabric : - Buff (F) This is a fine, soft, chalky fabric. The core is frequently fired to a very pale pink but the surfaces are white or creamy white. Sometimes it appears as if a white wash has been applied to a pink body but this is actually due to the firing. Very rarely small black inclusions can be seen or even air holes. The clay is very well levigated and approaches the fabric of Eggshell ware. In one instance a large white quartz like pebble can be seen in section but this is an accident. The classes in this fabric are BW.

Red wares :

Fabric : - Red (A) this fabric is extremely fine and well levigated, red in colour and very well fired. The fabric has some small flecks of mica. This particular fabric is restricted to ESW (R). It is very thinly potted with striation marks clearly seen. The thickness of the core does not exceeds 1mm. This fabric is almost like RPW.

Fabric :- Red (B) This is a very coarse fabric, extremely sandy and have large amounts of inclusions along with some whitish flecks (possibly vegetable temper) can be seen, sometimes giving the surface a pitted appearance. This fabric is encountered only in glaze Red Ware (GRW). They appear to be a thin wash prior to glazing but this is sometimes so uneven or thin that the vessel body shows through the glaze. Instead of appearing white or cream, it ends up looking pinkish-grey.

Fabric : - Red (C) This fabric is semi coarse, red to orange in colour, sometime fired to a grey and red. There is a lot of vegetable temper and small white specks can be seen. Mica is also evident. The texture is a little sandy and some lime inclusions can be seen. A few black inclusions are not uncommon. Sometimes this fabric can be quite dense. Firing is not always even. The classes represented are RSRW – 1, RSRW-2, BSRW – 1, BSRW – 2, RW – 1, RW – 3.

Fabric : - Red (D) This fabric is very similar to Red (C), since it is also sandy, bratty and has inclusions. It also fired similarly. However no vegetable temper can be seen and the most prominent characteristics of this fabric is the large talc inclusions in section and on the surface. One can easily scratch this with a thumbnail. The inside of the vessel usually have a rougher surface than the slipped exterior. It is not a common fabric. The class represented is RSRW – 1.

Fabric : - Red (E) This is a compact semi-fire clay usually well levigated with some mica specks and very few but large inclusions. One quartzite pebble in one of the samples is almost 1.5 mm. It fires from an orangy-red to grey. No black inclusions or white specks can be seen. The external surface is usually slipped. A light red or orange colour. The class representing this fabric is RSRW – 2.

Fabric : - Red (F) It is an extremely coarse badly levigated fabric having large angular inclusions (quartzite), small black pebbles white calcatric inclusions and perhaps talc. Some of these inclusions can measure upto 2 or 3 mm in size. There are many large air-holes. The texture is sandy but hard. The classes representing this fabric are RW – 2.

Fabric : - Red (G) It is an extremely fine, well levigated micaeous fabric. It is sandy and chalky. In some shards one or two large inclusions can be seen but it is relatively free of any such disturbances. The fabric is compact and has a laminated fracture. There is no slip but it is possible that the exterior may have been treated with a light wash. The colour ranges from dark pink-red to orange-red. The surfaces are covered with mica or sometimes just speckled with it. The mica is silver in colour. The fabric is so far seen only in MISC – 4 (fine red ware with mica) and MISC – 5 (pink ware with mica wash). Although MISC – 5 is called pink ware and has every appearance of being so, the fabric is almost exactly same as MISC – 4. It has therefore been taken as a firing anomaly.

Fabric : - Red (H) Very similar to Red (G), this fabric is also well levigated and has similar tones of colour. However there is less mica, there is also some white talc like inclusions and a gritty sandy texture. However, the most noticeable feature of this fabric is the presence of large pieces of grog seen on the surfaces. It is possible that the white inclusions may infact be crushed buff fabrics that has been used as grog/temper. The class represented is MISC – 6 (sandy red ware).

Fabric : - Red (I) This is fine sandy fabric, quite well levigated and soft. It is usually fired from pale-yellowish/ brownish-red to bluish-grey. Despite the air-holes and chalky texture, it appears quite sturdy. There is a noticeable amount of mica and vegetable temper. The fragments of some botanical remains can also be seen adhering to the surface. Some white inclusions may be seen and these appear to be quite soft, whitish-cream. The classes represented are MISC – 7 (fine red-grey ware).

Fabric : - Red (J) This is a variation of Red (H). It is more coarse and does not have as much noticeable mica. But there are large grog inclusions, quartzite particles and what appears to be lime inclusions. It fires to a red colour and sometimes has a dark red slip or wash. The inclusions are similar to those in pink (H). The class represented is MISC – 8.

Grey Wares :

Fabric :- Grey (A) This fabric is evenly fired, semi-coarse to coarse in texture. It has a large amount of inclusions and organic temper. There are many air-holes and some mica can also be seen. The colour can range from grey to a dark-blackish-grey. Sometimes (though not frequently) the grey core can have thin red horizons on either side. The compactness and levigation in this fabric is variable. The classes represented by this fabric areBSGW, RSGW, GW – 1.

Fabric :- Grey (B) It is an extremely coarse and sandy fabric with numerous small white inclusions of lime and organic temper which can still be seen in the clay. The firing is not even. The texture is hard. The colour ranges from blackish-grey to reddish-grey. There is not much mica in this fabric.. The classes in this fabric are BSGW, RSGW, GW – 3, and MISC – 2.

Fabric : - Grey (C) It is an extremely fine, evenly fired, well levigated grey fabric having steel grey colour and almost no inclusions. Very few air- holes can be seen and few specks of mica. It has a slight sand temper and the fabric is quite compact. The classes represented are GW – 2. (Fine Grey ware).

LISV :

Fabric : - MR (A) This fabric is hard and brittle. It is compact despite having some air-holes and pitting. There are may decrystallized lime inclusions and some grog. The colour ranges from purplish maroon to a dark brick red. No mica can be seen in this fabric. The surfaces are usually to be a thin white coating or a wash. There are many white inclusions in the fabric. The texture is semi coarse to coarse (though it is usually not as coarse as the red or grey wares). The density of this fabric varies but usually this fabric is quite hard and stony. The classes are LISV – 1 and LISV – 3.

Fabric : - MR (B) The colour of this fabric ranges from pinkish-red to a darker shade of the same. It is the variation of MR (A). Although there are few black and white inclusions, they are less than MR (A). The fabric is better levigated and the texture is rought and cement like. The surface appears exfoliated. Fine grog can be seen in section. Sometimes the colour of the fabric can range from an orangy-brick red. Some of these vessels may have a wash on the surface. The fabric is usually compact. Some air holes are seen. It may also have sand tempering. The class is LISV – 2.

Fabric : - MR (C) This fabric is a light orangy – red colour and it is softer then the other LISV classes. It has a chalky, soft inclusion and appears to be between MR (A) and Red (C). The class is represented by LISV – 4.

* * *

www.ingramcontent.com/pod-product-compliance
Ingram Content Group UK Ltd.
Pitfield, Milton Keynes, MK11 3LW, UK
UKHW061847041026
15591UKWH00015B/949